Contemporary Mathematicians

Gian-Carlo Rota
Editor

Birkhäuser
Boston · Basel

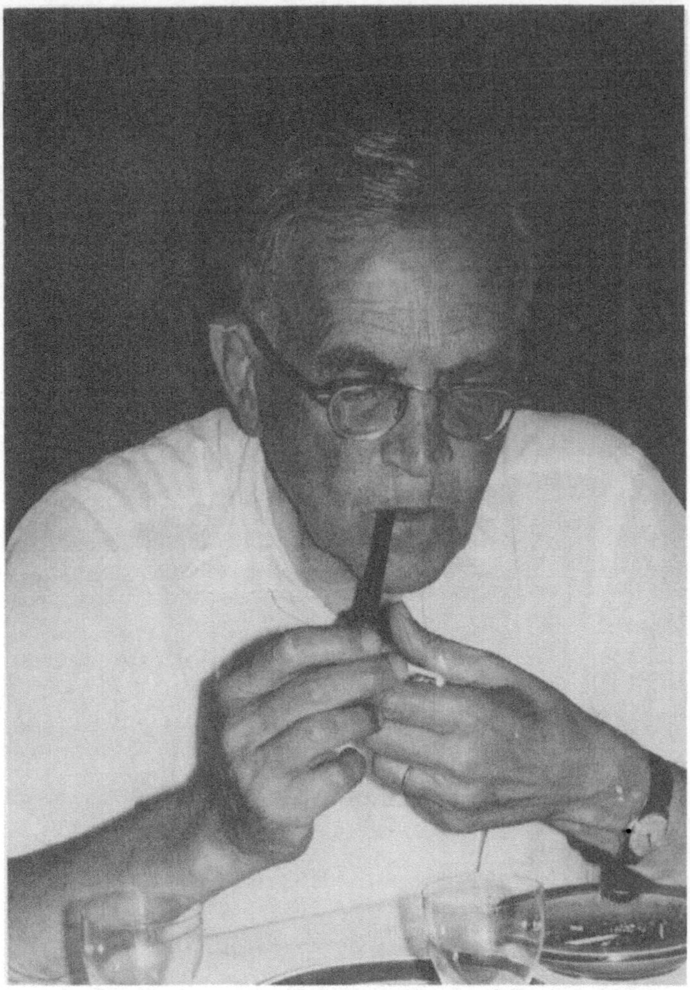

I. J. Schoenberg

I.J. Schoenberg
Selected Papers
Volume 1

Edited by
Carl de Boor

Birkhäuser
Boston · Basel
1988

I.J. Schoenberg
Carl de Boor (*Editor*)
University of Wisconsin-Madison
Madison, WI 53705
U.S.A.

Library of Congress Cataloging-in-Publication Data
Schoenberg, I. J.
[Selections. 1988]
Selected papers / I.J. Schoenberg ; edited by Carl de Boor.
p. cm.—(Contemporary mathematicians)
Includes bibliographies.

0-8176-3405-3 (v. 2)
1. Mathematics—Collected works. I. De Boor, Carl. II. Title.
III. Title: I.J. Schoenberg, selected papers. IV. Series.
QA3.S365A25 1988
510—dc19 88-2860

CIP-Titelaufnahme der Deutschen Bibliothek
Schoenberg, Isaac J.:
Selected papers / I. J. Schoenberg. Ed. by Carl de Boor. —
Boston ; Basel : Birkhäuser.
(Contemporary mathematicians)

NE: Schoenberg, Isaac J.: [Sammlung]
WG: 27 DBN 55.138120.5
6955/01 rr
Vol. 2 (1988)

WG: 27 DBN 88.021073.7 88.01.20
6955/02* rr

Softcover reprint of the hardcover 1st edition 1988

ISBN-13: 978-1-4612-8411-6 e-ISBN-13: 978-1-4612-3948-2
DOI: 10.1007/ 978-1-4612-3948-2

Foreword

These *selecta* contain 761 of the more than 2600 pages of I. J. Schoenberg's published articles. The selection made and the grouping in which the papers are presented here reflect most strongly Schoenberg's wishes.

The first volume of these *selecta* is drawn from Schoenberg's remarkable work on Number Theory, Positive Definite Functions and Metric Geometry, Real and Complex Analysis, and on the Landau Problem. Schoenberg's fundamental papers on Total Positivity and Variation Diminution, on Pólya Frequency functions and sequences, and on Splines, especially Cardinal Splines, make up the second volume.

In addition, various commentaries have been provided. Lettered references in these refer to items listed alphabetically at the end of each commentary. Numbered references refer to the list of Schoenberg's publications to be found in each volume. Those included in these *selecta* are starred.

It has been an honor to have been entrusted with the editorial work for these *selecta*. I am grateful to the writers of the various commentaries for their illuminating contributions and to Richard Askey for solid advice.

Carl de Boor

Table of Contents

Volume 1

C. de Boor: Foreword v

Table of Contents, Volume 2 ix

Publications of I. J. Schoenberg xi

I. J. Schoenberg: A brief account of my life and work 1

Number theory

[**1**] Über die asymptotische Verteilung reeller Zahlen mod 1 13

[**18**] On asymptotic distributions of arithmetical functions 43

[**22**] Regular simplices and quadratic forms 59

P. Erdős: Commentary 67

Linear inequalities

[**7**] On finite and infinite completely monotonic sequences 70

[**50**] (With T. S. Motzkin), The relaxation method for linear inequalities 75

J.-L. Goffin: The relaxation method for linear inequalities 87

Metric geometry

[**16**] Remarks to Maurice Fréchet's article 'Sur la définition axiomatique d'une classe d'espace distanciés vectoriellement applicable sur l'espace de Hilbert' 90

[**25**] Metric spaces and positive definite functions 100

[**26**] Metric spaces and completely monotone functions 115

[**28**] (With J. von Neumann), Fourier integrals and metric geometry 146

[**29**] Positive definite functions on spheres 172

[**49**] On a theorem of Kirzbraun and Valentine 186

S. Cambanis and D. Richards: Metric geometry 189

R. Askey: Commentary on a footnote 192

Real analysis

[**24**] On the Peano curve of Lebesgue 197

[**51**] An isoperimetric inequality for closed curves convex in even-dimensional Euclidean spaces 199

[**70**] On the Besicovitch-Perron solution of the Kakeya problem 221

Complex analysis

[**44**] (With T. S. Motzkin), On lineal entire functions of n complex variables . . . 229

[**53**] On the zeros of the generating functions of multiply positive sequences and functions . . . 239

[**55**] (With G. Pólya), Remarks on de la Vallée Poussin means and convex conformal maps of the circle . . . 265

[**60**] (With B. Epstein), On a conjecture concerning schlicht functions . . . 305

[**106**] Norm inequalities for a certain class of C^∞ functions . . . 308

S. Ruscheweyh: The Pólya-Schoenberg conjecture . . . 317

Extremum problems

[**100**] (With A. Cavaretta), Solution of Landau's problem concerning higher derivatives on the halfline . . . 320

[**105**] The perfect B-splines and a time-optimal control problem . . . 335

[**141**] The Landau problem I. The case of motions on sets . . . 350

[**149**] The Landau problem II. The case of motions on curves . . . 361

[**150**] The Landau problem III. Motions on special curves and time-optimal control problems . . . 373

History

[**157**] On polynomial interpolation at the points of a geometric progression . . . 391

Permissions . . . 405

Volume 2

Table of Contents, Volume 1 vii

Publications of I. J. Schoenberg ix

Cardinal splines

[31] Contributions to the problem of approximation of equidistant data by analytic functions. Part A: On the problem of smoothing or graduation. A first class of analytic approximation formulae 3

[32] Contributions to the problem of approximation of equidistant data by analytic functions. Part B: On the problem of osculatory interpolation. A second class of analytic approximation formulae 58

[110] Cardinal interpolation and spline functions IV. The exponential Euler splines . 88

[113] Cardinal interpolation and spline functions VII. The behavior of cardinal spline interpolants as their degree tends to infinity 111

[129] On the remainders and the convergence of cardinal spline interpolation for almost periodic functions 137

C. de Boor: Cardinal splines 164

Smoothing

[48] On smoothing operations and their generating functions 169

[76] Spline functions and the problem of graduation 201

Total positivity and variation diminution

[4] Über variationsvermindernde lineare Transformationen 207

[15] Zur Abzählung der reellen Wurzeln algebraischer Gleichungen 216

[45] (With M. Aissen and A. M. Whitney), On the generating functions of totally positive sequences I 235

[90] (With M. Marsden), On variation diminishing spline approximation methods 247

S. Karlin: Total positivity and variation diminishing transformations 269

Pólya frequency functions

[43] On Pólya frequency functions. I. The totally positive functions and their Laplace transforms 277

[39] On Pólya frequency functions. II. Variation diminishing integral operators of the convolution type 321

[47] (With Anne Whitney), On Pólya frequency functions. III. The positivity of translation determinants with an application to the interpolation problem by spline curves 332

[86] (With H. B. Curry), On Pólya frequency functions. IV. The fundamental spline functions and their limits 347

S. Karlin: Pólya frequency functions and sequences 384

The variational approach to splines

[72] Spline interpolation and best quadrature formulae 391
[75] On best approximations of linear operators 397
[80] On interpolation by spline functions and its minimal properties 407
C. de Boor: The variational approach to splines 428

Hermite-Birkhoff interpolation

[89] On Hermite-Birkhoff interpolation 432
G.G. Lorentz: Schoenberg's work on Hermite–Birkhoff interpolation 438

Permissions 441

Publications of I. J. Schoenberg

1* Über die asymptotische Verteilung reeller Zahlen mod 1, *Math. Z.* **28** (1928), 171–199.

2. Sur un théorème de Steiner relatif à la quadrature des courbes roulettes, *Ann. Sci. Univ. Jassy* **16** (1930), 6–13.

3. Über total monotone Folgen mit stetiger Belegungsfunktion, *Math. Z.* **30** (1929), 761–767.

4* Über variationsvermindernde lineare Transformationen, *Math. Z.* **32** (1930), 321–328.

5. (With G. A. Bliss), On separation, comparison and oscillation theorems for self-adjoint systems of linear second order differential equations, *Amer. J. Math.* **LIII** (1931), 781–800.

6. The minimizing properties of geodesic arcs with conjugate end points, *Ann. of Math. (2)* **32** (1931), 763–776.

7* On finite and infinite completely monotonic sequences, *Bull. Amer. Math. Soc.* (Feb. 1932), 72–76.

8. On finite-rowed systems of linear inequalities in infinitely many variables. I, *Trans. Amer. Math. Soc.* **34** (1932), 594–619.

9. On finite-rowed systems of linear inequalities in infinitely many variables. II, *Trans. Amer. Math. Soc.* **35** (1932), 452–478.

10. (With G. A. Bliss), On the derivation of necessary conditions for the problem of Bolza, *Bull. Amer. Math. Soc.* (Dec. 1932), 858–864.

11. Some applications of the calculus of variations to Riemannian geometry, *Ann. of Math.* **33** (1932), 485–495.

12. Convex domains and linear combinations of continuous functions, *Bull. Amer. Math. Soc.* (Apr. 1933), 273–280.

13. (With T. H. Hildebrandt), On linear functional operations and the moment problem for a finite interval in one or several dimensions, *Ann. of Math.* **34** (1933), 317–328.

14. A remark on the preceding note by Bochner, *Bull. Amer. Math. Soc.* (Apr. 1934), 277–278.

15* Zur Abzählung der reellen Wurzeln algebraischer Gleichungen, *Math. Z.* **38** (1934), 546–564.

16* Remarks to Maurice Fréchet's article 'Sur la définition axiomatique d'une classe d'espace distanciés vectoriellement applicable sur l'espace de Hilbert,' *Ann. of Math.* **36** (1935), 724–732.

17. On the zeros of the successive derivatives of integral functions, *Proc. Nat. Acad. Sci.* **21** (1935), 674–676.

18* On asymptotic distributions of arithmetical functions, *Trans. Amer. Math. Soc.* **39** (1936), 315–330.

19. Extensions of theorems of Descartes and Laguerre to the complex domain, *Duke Math. J.* **2** (1936), 84–94.

20. On certain two-point expansions of integral functions of exponential type, *Bull. Amer. Math. Soc.* (1936), 284–288.

21. On the zeros of successive derivatives of integral functions, *Trans. Amer. Math. Soc.* **40** (1936), 12–23.

22* Regular simplices and quadratic forms, *J. London Math. Soc.* **12** (1937), 48–55.

23. On certain metric spaces arising from Euclidean spaces by a change of metric and their imbedding in Hilbert space, *Ann. of Math.* **38** (1937), 787–793.
24* On the Peano curve of Lebesgue, *Bull. Amer. Math. Soc.* **44** (1938), 519.
25* Metric spaces and positive definite functions, *Trans. Amer. Math. Soc.* **44** (1938), 522–536.
26* Metric spaces and completely monotone functions, *Ann. of Math.* **39** (1938), 811–841.
27. On metric arcs of vanishing Menger curvature, *Ann. of Math.* **41** (1940), 715–726.
28* (With J. von Neumann), Fourier integrals and metric geometry, *Trans. Amer. Math. Soc.* **50** (1941), 226–251.
29* Positive definite functions on spheres, *Duke Math. J.* **9** (1942), 96–108.
30. On local convexity in Hilbert space, *Bull. Amer. Math. Soc.* **48** (1942), 432–436.
31* Contributions to the problem of approximation of equidistant data by analytic functions. Part A: On the problem of smoothing or graduation. A first class of analytic approximation formulae, *Quart. Appl. Math.* **IV** (1946), 45–99.
32* Contributions to the problem of approximation of equidistant data by analytic functions. Part B: On the problem of osculatory interpolation. A second class of analytic approximation formulae, *Quart. Appl. Math.* **IV** (1946), 112–141.
33. (With H. Rademacher), An iteration method for calculation with Laurent series, *Quart. Appl. Math.* **IV** (1946), 142–159.
34. On totally positive functions, Laplace integrals and entire functions of the Laguerre-Pólya-Schur type, *Proc. Nat. Acad. Sci.* **33** (1947), 11–17.
35. Some analytic aspects of the problem of smoothing, *Courant Anniv. Vol.* (1948), 351–370.
36. On variation diminishing integral operators of the convolution type, *Proc. Nat. Acad. Sci.* **34** (1948), 164–169.
37. (With A. Whitney), Sur la positivité des déterminants de translations des fonctions de fréquence de Pólya..., *C. R. Acad. Sci. Paris* **228** (1949), 1996–1998.
38. (With H. Rademacher), Convex domains and Chebychev's approximation problem, *Canad. J. Math.* **2** (1950), 245–256.
39* On Pólya frequency functions. II. Variation-diminishing integral operators of the convolution type, *Acta Sci. Math. (Szeged)* **12** (1950), 97–106.
40. The finite Fourier series and elementary geometry, *Amer. Math. Monthly* **57** (1950), 390–404.
41. (With M. Aissen, A. Edrei, and A. M. Whitney), On the generating functions of totally positive sequences, *Proc. Nat. Acad. Sci.* **37** (1951), 303–307.
42. (With A. M. Whitney), A theorem on polygons in n dimensions with applications to variation-diminishing and cyclic variation-diminishing linear transformations, *Compositio Math.* **9** (1951), 141–160.
43* On Pólya frequency functions. I. The totally positive functions and their Laplace transforms, *J. Analyse Math.* **1** (1951), 331–374.
44* (With T. S. Motzkin), On lineal entire functions of n complex variables, *Proc. Amer. Math. Soc.* **3** (1952), 517–526.
45* (With M. Aissen and A. M. Whitney), On the generating functions of totally positive sequences. I, *J. Analyse Math.* **2** (1952), 93–103.
46. A remark on M. M. Day's characterization of inner-product spaces and a conjecture of L. M. Blumenthal, *Proc. Amer. Math. Soc.* **3** (1952), 961–964.

47* (With Anne Whitney), On Pólya frequency functions. III. The positivity of translation determinants with an application to the interpolation problem by spline curves, *Trans. Amer. Math. Soc.* **74** (1953), 246–259.

48* On smoothing operations and their generating functions, *Bull. Amer. Math. Soc.* **59** (1953), 199–230.

49* On a theorem of Kirzbraun and Valentine, *Amer. Math. Monthly* **60** (1953), 620–622.

50* (With T. S. Motzkin), The relaxation method for linear inequalities, *Canad. J. Math.* **6** (1954), 393–404.

51* An isoperimetric inequality for closed curves convex in even-dimensional Euclidean spaces, *Acta Math.* **91** (1954), 143–164.

52. A note on multiply positive sequences and the Descartes rule of signs, *Rend. Circ. Mat. Palermo (2)* **4** (1955), 123–131.

53* On the zeros of the generating functions of multiply positive sequences and functions, *Ann. of Math.* **62** (1955), 447–471.

54. Some extremal problems for positive definite sequences and related extremal convex conformal maps of the circle, *Kon. Ned. Akad. Amsterdam* **61** (1958), 28–37.

55* (With G. Pólya), Remarks on de la Vallée Poussin means and convex conformal maps of the circle, *Pacific J. Math.* **8** (1958), 295–334.

56. Spline functions, convex curves and mechanical quadratures, *Bull. Amer. Math. Soc.* **64** (1958), 352–357.

57. On variation diminishing approximation methods, in *On Numerical Approximation*, pp. 249–274, Univ. of Wisconsin Press, Madison, Wisconsin, 1959.

58. The integrability of certain functions and related summability methods, *Amer. Math. Monthly* **66** (1959), 361–375 and 562–563.

59. On the maxima of certain Hankel determinants and the zeros of the classical orthogonal polynomials, *Kon. Ned. Akad. Amsterdam* **21** (1959), 282–290.

60* (With B. Epstein), On a conjecture concerning schlicht functions, *Bull. Amer. Math. Soc.* **65** (1959), 273–275.

61. (With A. S. Besicovitch), Sur les arcs ascendants à pente partout nulle et des problémes qui s'y rattachent, *C. R. Acad. Sci. Paris* **249** (1959), 1079–1080.

62. On the question of unicity in the theory of best approximation, *Ann. New York Acad. Sci.* **86** (1960), 682–692.

63. (With J. C. Mairhuber and R. E. Williamson), On variation diminishing transformations on the circle, *Rend. Circ. Mat. Palermo (2)* **8** (1959), 241–270.

64. (With G. Szegö), An extremum problem for polynomials, *Compositio Math.* **14** (1960), 260–268.

65. (With C. T. Yang), On the unicity of solutions of problems of best approximation, *Ann. Mat. Pura Appl. (4)* **54** (1961), 1–12.

66. (With A. S. Besicovitch), On Jordan arcs and Lipschitz classes of functions defined on them, *Acta Math.* **106** (1961), 113–136.

67. On two theorems of P. Erdös and A. Renyi, *Illinois J. Math.* **6** (1962), 53–58.

68. On two theorems of Archimedes and F. J. van den Berg, *Simon Stevin* **5** (1962), 133–138.

69. Extrema for gap power series of positve real part, *Proc. Nat. Acad. Sci.* **48** (1962), 1151–1154.

70* On the Besicovitch-Perron solution of the Kakeya problem, in *Studies in Mathematical Analysis and Related Topics*, Pólya volume, pp. 359–363, Stanford Univ. Press,

Stanford, CA, 1962.

71. On certain minima related to the Besicovitch-Kakeya problem, *Mathematica (Cluj)* **4** (1962), 145–148.

72* Spline interpolation and best quadrature formulae, *Bull. Amer. Math. Soc.* **70** (1964), 143–148.

73. Spline interpolation and the higher derivatives, *Proc. Nat. Acad. Sci.* **51** (1964), 24–28.

74. (With Ch. Pisot), Arithmetic problems concerning Cauchy's functional equation, *Illinois J. Math.* **8** (1964), 40–56.

75* On best approximations of linear operators, *Indag. Math.* **26** (1964), 155–163.

76* Spline functions and the problem of graduation, *Proc. Nat. Acad. Sci.* **52** (1964), 947–950.

77. On trigonometric spline interpolation, *J. Math. Mech.* **13** (1964), 795–826.

78. A note on the cyclotomic polynomial, *Mathématika* **11** (1964), 131–136.

79. Arithmetic problems concerning Cauchy's functional equation, *Compositio Math.* **16** (1964), 169–175.

80* On interpolation by spline functions and its minimal properties, *Internat. Schriftenreihe Numer. Math. (Birkhäuser)* **5** (1964), 109–129.

81. (With Ch. Pisot), Arithmetic problems concerning Cauchy's functional equation II, *Illinois J. Math.* **9** (1965), 129–136.

82. Extrema for gap power series of positive real part, *J. Analyse Math.* **14** (1965), 379–391.

83. On monosplines of least deviation and best quadrature formulas, *SIAM J. Numer. Anal., Ser. B.* **2** (1965), 144–170.

84. (With F. Cunningham), On the Kakeya constant, *Canad. J. Math.* **17** (1965), 946–956.

85. Extremum problems for mass-distributions in a finite interval, *Rend. Circ. Mat. Palermo (2)* **13** (1964), 1–12.

86* (With H. B. Curry), On Pólya frequency functions. IV. The fundamental spline functions and their limits, *J. Analyse Math.* **17** (1966), 71–107.

87. On monosplines of least square deviation and best quadrature formulae. II, *SIAM J. Numer. Anal.* **3** (1966), 321–328.

88. (With S. J. Einhorn), On Euclidean sets having only two distances between points, *Indag. Math.* **28** (1966), 479–488 and 489–504.

89* On Hermite-Birkhoff interpolation, *J. Math. Anal. Appl.* **16** (1966), 538–543.

90* (With M. Marsden), On variation diminishing spline approximation methods, *Mathematica (Cluj)* **8** (1966), 61–82.

91. (With S. K. Zaremba), On Cauchy's lemma concerning convex polygons, *Canad. J. Math.* **19** (1967), 1062–1071.

92. On spline functions (with a supplement by T. N. E. Greville), in *Inequalities*, O. Shisha, ed., pp. 255–291, Academic Press, NY, 1967.

93. On the Ahlberg-Nilson extension of spline interpolation: The *g*-splines and their optimal properties, *J. Math. Anal. Appl.* **21** (1968), 207–231.

94. On spline interpolation at all integer points of the real axis, *Mathematica (Cluj)* **10** (1968), 151–170. Also in *Seminaire Delange-Pisot-Poitou: 1967/68, Theorie des Nombres*, Fasc.1, Exp.1, 18pp. Secrétariat mathematique, Paris, 1969.

95. Monosplines and quadrature formulae, in *Theory and Applications of Spline Functions*, T. N. E. Greville, ed., 157–207, Academic Press, NY, 1969.

96. Spline interpolation and the higher derivatives, in *Abhandlungen zur Zahlentheorie and Analysis, zur Erinnerung an Edmund Landau (1877–1938)*, P. Turan, ed., 281–295, VEB Deutscher Verlag der Wiss., Berlin, 1968.
97. Linkages and distance geometry. I and II, *Indag. Math.* **31** (1969), 43–63.
98. Cardinal interpolation and spline functions, *J. Approximation Theory* **2** (1969), 167–206.
99. (With Zvi Ziegler), On cardinal monosplines of least L_∞-norm on the real axis, *J. Analyse Math.* **23** (1970), 409–436.
100* (With A. Cavaretta), Solution of Landau's problem concerning higher derivatives on the halfline, MRC Technical Summary Report #1050; also in *Proc. Internat. Conference on Constructive Function Theory, Varna, May 1972*, Sofia, 297–308.
101. A second look at approximate quadrature formulae and spline interpolation, *Adv. in Math.* **4** (1970), 277–300.
102. On equidistant cubic spline interpolation, *Bull. Amer. Math. Soc.* **77** (1971), 1039–1044.
103. (With A. Sharma), The interpolatory background of the Euler-MacLaurin quadrature formula, *Bull. Amer. Math. Soc.* **77** (1971), 1034–1038.
104. On polynomial spline functions on the circle. I and II, Proc. of the Conference on Constructive Function Theory, August 1969, Akademiai Kiado, Budapest, 1971, pp. 403–433.
105* The perfect B-splines and a time-optimal control problem, *Israel J. Math.* **10** (1971), 261–274.
106* Norm inequalities for a certain class of C^∞ functions, *Israel J. Math.* **10** (1971), 364–372.
107. Notes on spline functions I. The limits of the interpolating periodic spline functions as their degree tends to infinity, *Indag. Math.* **34** (1972), 412–422.
108. Cardinal interpolation and spline functions II. Interpolation of data of power growth, *J. Approximation Theory* **6** (1972), 404–420.
109. (With P. R. Lipow), Cardinal interpolation and spline functions III. Cardinal Hermite interpolation, *Linear Algebra Appl.* **6** (1973), 273–304.
110* Cardinal interpolation and spline functions IV. The exponential Euler splines, ISNM (Birkhäuser Verlag) **20** (1972), 382–404.
111. (With A. Sharma), Cardinal interpolation and spline functions V. The B-splines for cardinal Hermite interpolation, *Linear Algebra Appl.* **7** (1973), 1–42.
112. Cardinal interpolation and spline functions VI. Semicardinal interpolation and quadrature formulas, *J. Analyse Math.* **XXVII** (1974), 159–204.
113* Cardinal interpolation and spline functions VII. The behavior of cardinal spline interpolants as their degree tends to infinity, *J. Analyse Math.* **XXVII** (1974), 205–229.
114. *Cardinal Spline Interpolation*, CBMS Regional Conference Monograph No. 12, vi + 125 pages, SIAM, Philadelphia, 1973.
115. Notes on spline functions III. On the convergence of the interpolating cardinal splines as their degree tends to infinity, *Israel J. Math.* **16** (1973), 87–93.
116. (With F. B. Richards) Notes on spline functions IV. A cardinal spline analogue of the theorem of the brothers Markov, *Israel J. Math.* **16** (1973), 94–102.
117. (With S. D. Silliman) On semicardinal quadrature formulas, *Math. Comp.* **28** (1974), 483–497. See also *Approximation Theory*, G. G. Lorentz ed., Academic Press, 1973, 461–467.

118. Splines and histograms, with an Appendix by C. de Boor. In *Spline Functions and Approximation Theory*, A. Meir and A. Sharma, eds., ISNM Vol. 21, Birkhäuser Verlag, (1973), 277– 327, 329–358.
119. Spline functions and differential equations – First order equations, in *Studies in Numerical Analysis*, B.K.P. Scaife, ed., Academic Press, London (1974), 311–324.
120. Remarks concerning the numerical inversion of the Laplace transform due to Bellman, Kalaba and Lockett, *J. Math. Anal. Appl.* **43** (1973), 823–828.
121. (With C. de Boor) Unique prime factorization and lattice points, *Math. Mag.* **46** (1973), 198–203.
122. The elementary cases of Landau's problem of inequalities between derivatives, *Amer. Math. Monthly* **80** (1973), 121– 158.
123. Notes on spline functions V. Orthogonal or Legendre splines, *J. Approximation Theory* **13** (1975), 84–104.
124. Notes on spline functions VI. Extremum problems of the Landau-type for the differential operator $D^2 \pm 1$, in *Studies in Spline Functions and Approximation Theory*, S. Karlin, C.A. Micchelli, A. Pinkus and I.J. Schoenberg, eds., Academic Press (1976), 353–369.
125. Remarks on two geometric conjectures of L. Fejes Tóth, Analele ştiinţifice ale Universităţii "Al.I. Cuza" din Iaşi **XXI** (1975), 9–13.
126. (With D. J. Newman) Splines and the logarithmic function, *Pacific J. Math.* **61** (1975), 241–258.
127. On the motion of a billiard ball in two dimensions, *Delta* **5** (1975), 1–17.
128. On Micchelli's theory of cardinal L-splines, in *Studies in Spline Functions and Approximation Theory*, S. Karlin et al., eds., Academic Press (1976), 251–276.
129* On the remainders and the convergence of cardinal spline interpolation for almost periodic functions, in *Studies in Spline Functions and Approximation Theory*, S. Karlin et al., eds., Academic Press (1976), 277–303.
130. On the location of the frets on the guitar, *Amer. Math. Monthly* **83** (1976), 550–552.
131. Extremum problems for the motions of a billiard ball I. The L_p norm, $1 \leq p < \infty$, *Indag. Math.* **38** (1976), 66–75.
132. Extremum problems for the motions of a billiard ball II. The L_∞ norm, *Indag. Math.* **38** (1976), 263–279.
133. On Chebyshev and Markov-type problems for polynomials in a circular ring, in *Colloquia Mathematica Societatis Janos Bolyai 19, Fourier Analysis and Approximation Theory*, Budapest (1976), 697–712.
134. (With C. de Boor) Cardinal interpolation and spline functions VIII. The Budan-Fourier theorem for splines and applications, in *Spline Functions, Karlsruhe 1975*, K. Böhmer, G. Meinardus and W. Schempp, eds., *Lecture Notes in Mathematics* **501**, Springer-Verlag (1976), 1–79.
135. The Landau problem for motions in a ring and in bounded continua, *Amer. Math. Monthly* **84** (1977), 1–12. A somewhat expanded version appeared as MRC Technical Summary Report #1563, Oct. 1975.
136. The Landau problem for the differential operator $D^2 - a^2$ in a circular ring, in *Colloquia Mathematica Societatis Janos Bolyai, 19, Fourier Analysis and Approximation Theory*, Budapest (1976), 713–723.
137. On the zeros of successive derivatives of integral functions II, in *Complex Analysis, Kentucky 1976*, J.D. Buchholz and T.J. Suffridge eds., *Lecture Notes in Mathematics*

599, Springer-Verlag (1977), 109–116.
138. Approximating lengths, areas and volumes by polygons and polyhedra, *Delta* **7** (1977), 32–46.
139. (With T.N.E. Greville and A. Sharma) The spline interpolation of sequences satisfying a linear recurrence relation, *J. Approximation Theory* **17** (1976), 200–221. See also *Approximation Theory*, G. G. Lorentz ed., Academic Press, 1973, 365–367.
140. (With J. M. Pollin) On the matrix approach to Fibonacci numbers and Fibonacci pseudoprimes, *Fibonacci Quart.* **18** (1980), 261–268.
141* The Landau problem I. The case of motions on sets, *Indag. Math.* **40** (1978), 276–286.
142. On the arithmetic-geometric mean, *Delta* **7** (1977), 49–65.
143. On cardinal spline smoothing, in Proc. of the International Symposium on Approximation Theory, Campinas, Brazil, 1977, J. B. Prolla, ed., North-Holland Publ. Co. (1979), 383–407.
144. On a problem of Steinhaus on lattice points, *Amer. Math. Monthly* **86** (1979), 765–766.
145. On Jacobi-Bertrand's proof of a theorem of Poncelet, in *Studies in Pure Mathematics to the Memory of Paul Turán*, publ. by the Hungarian Academy of Sciences (1980), 623–627.
146. Extremum problems for the multi-dimensional case of König and Szücs of billiard ball motions, *Math. Rep. Acad. Sci. Canada* **1** (1978), 37–40.
147. Extremum problems for the motions of a billiard ball III. The multi-dimensional case of König and Szücs, *Studia Sci. Math. Hungar.* **13** (1978), 53–78.
148. Extremum problems for the motions of a billiard ball IV. A higher-dimensional analogue of Kepler's Stella octangula, *Studia Sci. Math. Hungar.* **14** (1979), 273–292.
149* The Landau problem II. The case of motions on curves, *Indag. Math.* **43** (1981), 325–335.
150* The Landau problem III. Motions on special curves and time-optimal control problems, *Indag. Math.* **43** (1981), 337–351.
151. The harmonic analysis of skew polygons as a source of outdoor sculptures, in *The Geometric Vein*, The Coxeter Festschrift, C. Davis, B. Grünbaum, F. A. Sherk eds., Springer-Verlag, New York (1982), 165–176.
152. The Landau problem for motions on curves and time-optimal control problems, in *Approximation Theory*, E.W. Cheney, ed., Academic Press (1980), 811–821.
153. (With T. S. Motzkin) On Fejér sets in linear and spherical spaces, NBS Report of August 25, 1952, 19 pages. *Colloq. Math. Soc. János Bolyai* **35** (1980), 861–875.
154. (With R. Askey and A. Sharma) Hausdorff's moment problem and expansions in Legendre polynomials, *J. Math. Anal. Appl.* **86** (1982), 237–245.
155. A direct derivation of a Jacobian identity from elliptic functions, *Amer. Math. Monthly* **88** (1981), 616–618.
156. Two applications of approximate differentiation formulae: An extremum problem for multiply monotone functions and the differentiation of asymptotic expansions, *J. Math. Anal. Appl.* **89** (1982), 251–261.
157* On polynomial interpolation at the points of a geometric progression, *Proc. Roy. Soc. Edinburgh Sect. A* **90A** (1981), 195–207.
158. Interpolating splines as limits of polynomials, *Linear Algebra Appl.* **52/53** (1983), 617–628.
159. A new approach to Euler splines, *J. Approximation Theory* **39** (1983), 324–337.

160. (With T.N.T. Goodman and A. Sharma), Piecewise smooth solutions of some difference-differential equations, *J. Approximation Theory* **48** (1986), 262–271.
161. (With T.N.E. Greville and A. Sharma) The behavior of the exponential Euler spline $S_n(x;t)$ as $n \to \infty$ for negative values of the base t, in *Approximation Theory*, CMS Conference Proceedings, Vol. 3, Amer. Math. Soc., Providence RI (1982), 185–198.
162. Euler's contributions to cardinal spline interpolation, in *Leonardt Euler 1707-1783, Beiträge zu Leben und Werk*, Birkhäuser Verlag, Basel (1984), 199–213.
163. A direct approach to the Villarceau circles of a torus, *Simon Stevin* **59** (1985), 365–372.
164. Self-reflecting skew polygons and polytopes in the 4-dimensional hypercube, *Geom. Dedicata* **14** (1983), 355–373.
165. (With A. W. Goodman) On a theorem of Szegö on univalent convex maps of the unit circle, *J. Analyse Math.* **44** (1984/5), 200–204.
166. (With D. W. Crowe) On the equidecomposability of a regular triangle and a square of equal area, *Giessener Mitteilungen* **164** (1984), 59–64.
167. On hypocycloids and their diameters, *J. College Math.* **16** (1985), 262–267.
168. A conjectured analogue of Rolle's theorem for polynomials with real and complex coefficients, *Amer. Math. Monthly* **93** (1986), 8–13.
169. On the anti-cylinder, MRC Technical Summary Report #2842, (1985). There is related material in: On the turning lozange of constant side whose vertices alternate between two fixed circles of a ring, MRC Technical Summary Report #2813, (1985).
170. (With A. W. Goodman) A proof of Grace's theorem by induction, Honam Math. J. **xx** (1987), xxx-xxx.
171. Outdoor sculptures, MRC Technical Summary Report #2900 (1986); submitted.
172. The Chinese remainder problem and polynomial interpolation, *College Math. J.* **xx** (1987), xxx-xxx.
173. On vector indices mod m, a circular slide rule mod 100, *Mathematics Magazine* **xx** (1987), xxx-xxx.

Books

Mathematical Time Exposures, 270 pages, The Mathematical Association of America, 1982.

A brief account of my life and work

This account is not as short as I wanted to make it due to my long life while my life and my work cannot be separated. I belong to a generation which was too young for active duty in the First World War and too old for the Second. To give the story some structure I divide it into significant periods,

1. **Romania 1903–1922.** I was born on April 21, 1903, in Galatz, Romania, as the youngest of four children. My father, trained as an accountant, inherited the lumber business of my grandfather Isaac Segal, on my mother's side, but my father was not a good business man. The fortunes of the family improved in 1910 when we moved to Jassy, Romania, where my father eventually became Sub-Director (Vice President) of the newly founded "Banca Moldova". The building of the Banca Moldova, at present a main Post Office of Jassy, was new in 1910, and our family occupied one-half of its second floor. We four children enjoyed the cultural interests of our parents, especially of mother's who wrote and spoke French fluently. My father was an Austro-Hungarian subject and Consul in Jassy of the Dual Monarchy. This caused difficulties during the First World War after Romania declared war on the Axis powers. After the war we became Romanian citizens.

My parents were devoted Zionists and my mother, Rachel Segal Schoenberg, was much in demand as a public speaker at Zionist meetings; she was also a fine Romanian poetess. My father Jacob Schoenberg (1864–1930) was more practically oriented and was active in establishing agricultural stations for the training of young Jewish boys and girls which were to go as farmers to the settlements in Palestine. We frequently had as house guests prominent Zionist personalities such as Ussishkin, Sokoloff, Wilensky, Colonel Wedgwood and others. In 1926 my older sister Elsa married Ussishkin's son Sama and still lives in Jerusalem as his widow.

At the age of 12 I became interested in Physics, reading mostly French textbooks. However, I soon switched to Mathematics studying by myself the French textbooks of the Frères Jesuits; also the series of textbooks founded by Hadamard of which he wrote the Geometrie. I had plenty of time for these studies, because for the last two years of High School I prepared myself at home, as was possible at the time, taking an examination at the end of the year, and obtained the High School Diploma (Baccalaureate) in 1919. In my mathematical studies I was much helped by Titu Patriciu, a good mathematician who taught High School (Liceu).

In 1919 I started my studies at the University of Jassy, founded in 1860, by attending, during 1919–1920, the course in Projective and Descriptive Geometry given by Victor Costin. This was followed during 1920–22 by the year-long courses leading to the Licenţa (M.A.) in Mathematics: Analytic Geometry (Alexandru Myller), Algebra Including Galois Theory (S. Stoilov), Analysis (Simion Sanielevici), Theory of Analytic Functions (Vera Myller - Lebedeff), Mechanics (S. Sanielevici), Astronomy (Constantin Popovici). Also outside of this program I had courses in Differential Geometry (A. Myller) and Differential Equations, in particular the theory of Fuchs (S. Sanielevici). The two Myllers were students of Hilbert, while Sanielevici was a student of Picard. I dedicated to these mathematicians my little book [**S83**]. In June 1922 I received the M.A. degree in Mathematics (very good with distinction).

2. **Germany and Romania 1922–1930.** In the fall of 1922 my father sent me to Germany. There I spent one semester in Göttingen, then three semesters in Berlin, and finally I went back to Göttingen for two semesters. One attraction of Berlin was that my sister Irma (1902–1984) was studying the piano there.

During the three years (1922–1925) spent in Germany, I attended courses given by Edmund Landau (Entire Functions, Trigonometric Series, The Big Fermat Problem, Analytic Number Theory), and Issai Schur (Algebra, Number Theory, Analytic Number Theory), and A. Ostrowski (Overconvergence of Power Series); also a Problem Seminar offered by Ostrowski and K. Granjot.

The idea for my Ph.D. thesis was suggested by Schur's course in Analytic Number Theory. He had discussed H. Weyl's theory [**Wy**] of uniform distribution of numbers mod 1 of 1916. Schur gave some new asymptotic properties of Euler's $\phi(n)$ function which suggested to me the more general theory of non-uniform distribution of numbers mod.1. This is what I initiated in my Ph.D. thesis [**1***]. Following a general theory, the main result concerned the asymptotic distribution function $F(x)$ in $[0,1]$ of the special sequence $x_n = \phi(n)/n, (n = 1, 2, ...)$. I showed that $F(x)$ is different from x and is continuous and increasing in $[0,1]$. Later Paul Erdös [**Er**] showed that $F(x)$ is a singular function, hence $F'(x) = 0$ almost everywhere. This is probably the most natural example of a singular function in the sense of Lebesgue.

The Prussian Ministry did not recognize my High School Diploma and I could not obtain my Ph.D. degree in Germany. In 1925 I returned to Jassy becoming Assistant to Victor Costin. The following remark shows the overwhelming importance for us at the time of Western culture: During 1925–1930 on the way from Romania to Germany I passed at least half a dozen times the city of Krakow; not once did I stop there to look at this interesting ancient city. On the basis of [**1***] I obtained my Ph.D. degree at the University of Jassy in June 1926 under the direction of S. Sanielevici.

Schur liked my paper and in 1928 he said to me: "Herr Doktor, in dieser Sache sind Sie allen voraus; Sie müssen diesen Vorsprung ausnützen." (Herr Doktor, in this matter you are ahead of everybody, you must utilize this advantage). H. Davenport continued my work and there were further papers by S. Bochner and B. Jessen and also by B. Jessen and A. Wintner; for references see [**18***]. I finally wrote [**18***] using the Fourier Transform.

During 1926–1927 I served one year of military service required of high school graduates: One-half year at the Field Artillery School in Timişoara (the old Hungarian Temeshvar) from which I graduated as a Corporal. The second half of that year I served in a (horse drawn) Field Artillery Regiment near Chişinau (the old Russian Kishinev) in Bessarabia. The year 1926–1927 was as peaceful as Europe experienced between the wars. In recollection and by old photographs my military service has some elements of musical comedy. As a lifelong vegetarian I could not eat at the officers' mess, but with my doctorate in Mathematics I was well treated.

The Hebrew University of Jerusalem was founded in 1925 and Edmund Landau, who was a great Zionist and had learnt Hebrew, was one of its founders. He arranged my visit to Jerusalem for the Spring Semester of 1928. There, at the Einstein Institute, I gave a course on "Higher Algebra" in which Benjamin Amira helped me with the language (Hebrew). From Heinrich Weber's book [**Wb**], pages 354–357, I learned about Felix Klein's note of 1892 on finding upper bounds for the number of real roots of an algebraic equation in a finite interval. Klein shows, for second degree equations only, that Jacobi's methods of using Descartes' rule of signs gives a bound which is less or equal to the bound furnished by the Budan-Fourier theorem. This result I generalized to arbitrary degree in [**4***] and [**15***]. See also S. Karlin's book [**K**], pp. 316-318. This was my introduction to the important notions of Total Positivity and Variation Diminishing linear transformation which were to dominate many of my papers in later years.

In his doctoral dissertation [**M**] of 1933, written under the direction of A. Ostrowski, Th. Motzkin found the necessary and sufficient conditions for a linear transformation to be variation diminishing. My own derivation of these conditions is included in my work on the zeros of polynomials which I wrote up during the late thirties and later, in several papers. Besides [**4***] and [**15***], already mentioned, they are [**19**], [**42**] (jointly with Anne Whitney), [**52**] and [**53***]. Their main results are beautifully presented in Nicola Obreschkoff's excellent book [**O**].

In Jerusalem I met the older daughter Charlotte (Dolli) of Marianne and Edmund Landau and we were married in 1930 in Berlin.

3. Chicago, Cambridge, Princeton, Waterville and Philadelphia 1930–1943. For the year 1930–1931 I was granted a Rockefeller Fellowship and we traveled to Chicago where I was to study the Calculus of Variations under the direction of Gilbert Bliss. Professor Bliss also granted me an Assistantship for 1931–1932. In Chicago I wrote the seven papers from [**5**] to [**11**], some of them jointly with Professor Bliss. With the hindsight of over half a century it is clear that the significant work in this field was being done by Marston Morse in the Calculus of Variations in the large.

Our daughter Elizabeth was born in Chicago on September 17, 1931.

For 1932-1933 we went to Cambridge MA where I attended courses at Harvard (David Widder) and M.I.T. (Dirk Struik and Jesse Douglas). In the Fall of 1933 we went to Princeton NJ where I became a Fellow of the Institute for Advanced Studies, newly founded by Abraham Flexner. I also received a small stipend for helping with the publication of the Annals of Mathematics. Under the influence of Leonard M. Blumenthal I became interested in the Distance Geometry which had been initiated by Karl Menger in Vienna in the late twenties. Stimulated by a note of M. Fréchet in the Annals of Mathematics, I wrote the series of papers from [**22**] to [**30**] on isometric imbedding of metric spaces in Hilbert space and its connection with Analysis. The dominant notions are the *Positive Definite Functions* and *Isometric Imbedding*. Some of these results are sometimes quoted in the literature only because of their purely analytic or probabilistic connections, without mention of their geometric imbedding interpretations: As examples see A. M. Yaglom [**Y**], G. Choquet [**Ch**], N. I. Akhiezer [**A**]. My main papers on the subject are [**25***], [**26***], and the joint paper [**28***] with J. von Neumann.

I recently noticed with particular pleasure that after all these years the three authors C. Berg, J. Christensen and P. Ressel discuss my work in their recent book [**BCR**]. They simplify my theory and at the same time generalize it to semigroups.

This was the worst part of the Depression and teaching positions were hard to obtain. Oswald Veblen suggested to Arnold Dresden that I might teach at Swarthmore College while Dresden went on his sabbatical leave in January 1935. We then moved to Swarthmore PA where I was Acting Assistant Professor. On Dresden's return from Cambridge University and Princeton in January 1936, I continued to teach at Swarthmore until June 1936. At that time Marston Morse, as a Colby College graduate, had recommended me for a position at Colby College.

We spent five enjoyable years (1936–1941) in Waterville ME. I usually drove by myself to Princeton for a few summer months. At Colby I taught also Mechanics and a non-technical Mathematics course open to any student. We made lifelong friends such as Mary Hatch Marshall, for many years Professor of English at Syracuse University. In 1937 our second daughter, Beatrice, was born on October 17.

In 1933 my father-in-law lost his Professorship in Göttingen due to the Nazis and moved to Berlin; there he died on February 14, 1938. His widow Marianne Landau (1886–1963), daughter of the Nobel Laureate Paul Ehrlich, the founder of Chemotherapy, left Berlin and joined our family in Waterville.

At Colby I continued working on Distance Geometry and its connection with Analysis resulting in papers already mentioned above and in [**Bl**] and [**BCR**], see [**Bl**] on pages 132–136. My one-page note [**24***], "On the Peano curve of Lebesgue" was written while proctoring a two-hour examination in Mechanics in 1937.

In 1941 John Robert Kline offered me a position at the University of Pennsylvania where I was to remain until 1965, with the interruptions of sabbaticals and leaves.

4. **Aberdeen MD 1943–1946.** Several mathematicians, among them O. Veblen, Mina Rees and Leo Zippin, were concerned with making mathematicians available for the war effort. Leo Zippin, as a Corporal, followed the development of the first electronic computer, the ENIAC, at the Moore School of the University of Pennsylvania, which was to be moved to the Ballistics Research Laboratory (the BRL) in Aberdeen MD. Leo Zippin arranged my going to the BRL for the duration.

The morning in August 1943 of my reporting for duty, Major A. A. Bennett, of Brown University, then Chief of the Computing Branch of the BRL, told me what my particular problem was to be: Trajectories of projectiles were until then computed with desk calculators by hand. Into these computations entered tables of the drag-functions of air resistance, about 24 of them, which were obtained empirically by firings of various types of projectiles. As the step of integration used in these trajectory computations was rather large and the methods of numerical integrations fairly complicated, it did not much matter that the 4-place drag-function tables were rather rough. In performing these computations on the ENIAC, which was very fast, a much simpler integration method of very small step could be used. In these methods, the accumulation of the round-off errors was unacceptable due to the rough drag-function tables; they needed to be smoothed by being approximated by analytic functions. To do this was my problem.

I solved this problem by what later I called Cardinal Spline Interpolation and Cardinal Spline Smoothing. To insure the smoothness, to 8 decimal places, of the second derivative of the approximant of the drag-function, I used the heat-flow equation on the real axis to produce approximations regular on the entire real axis.

Only the fine resources of the BRL made this work possible. We had advanced (but not electronic) punched-card machines which were well suited for this work. A full account of it is given in my papers [**31***]and [**32***]. Equidistant splines were used before me, e.g., by C. Runge [**R**] (quadratic splines, 1901) and Quade and Collatz [**QC**] (arbitrary degree, 1938) for the approximation of the Fourier coefficients of periodic functions, but not for the approximation of functions. From January 1945 to January 1946 I was Chief of the Punched Card Section of the Computing Branch.

In my work described above I was much helped by the actuarial work on so-called *Osculatory Interpolation*, and in particular by T. N. E. Greville's article [**G44**]. Greville's Selected Papers include the paper [**GV**] by Greville and Hubert Vaughan, which is a masterful exposition of the actuarial work on the subject. In [**31***] and [**32***] I also had to solve a numerical problem on Laurent series whose solution, by Hans Rademacher and me, is given in [**33**].

5. **Philadelphia, Princeton, Madison 1946–1965.** In January 1946, the family moved back to Swarthmore, while I returned to the University of Pennsylvania where I remained, with some leaves and sabbaticals, until 1965. The beneficial effect of the work

done in Aberdeen on my mathematical development soon became apparent. Work on a good applied problem becomes rewarding if one is prepared to follow purely mathematical tangents which suggest themselves.

In [**31***] and [**32***] I considered exclusively spline functions having equidistant knots. Haskel Curry recognized the possibility of defining splines with arbitrary non-equidistant knots and we wrote in 1946 essentially the first part of [**86***] which was only published in 1966. I wrote the second part of [**86***] which led to one of the four different characterizations of the so-called *Pólya frequency functions.*

In [**31***] I gave the conditions under which a convolution transformation

$$y_n = \sum_k a_{n-k} x_k \tag{1}$$

may be regarded as a smoothing formula; in the second part of [**35**] are given generating functions of totally positive sequences (a_n). A sequence (a_n) is called totally positive (tp) provided that the cyclic matrix

$$\|a_{n-k}\| \quad (-\infty < n, k < \infty)$$

has only non-negative minors; see my invited address [**48***].

The continuous analogue of (1) is the convolution

$$g(x) = \int_{-\infty}^{\infty} \Lambda(x-y) f(y)\, dy, \tag{2}$$

and the characterization of the variation diminishing transformation of the form (2) led to what is likely my best work: in [**43***] I solved a problem that was initiated by Pólya [**P**] in 1915. This problem was as follows: Let B_2 denote the class of Laguerre-Pólya-Schur entire functions of the form

$$\phi(z) = e^{-\gamma z^2} + \delta z \prod_{k=1}^{\infty} (1 + \delta_k z) e^{-\delta_k z}, \qquad (0 < \gamma + \sum_i^{\infty} \delta_k^2 < \infty). \tag{3}$$

These were shown by Laguerre and Pólya to be the only possible limits of real polynomials $P(z)$, with $P(0) = 1$, having only real zeros. Pólya [**P**] and H. Hamburger [**H**] showed in 1915 and 1920, respectively, that the reciprocal

$$1/\phi(z)$$

is represented in its vertical strip of regularity containing the origin by the Stieltjes integral

$$\frac{1}{\phi(z)} = \int_{-\infty}^{\infty} e^{-zx}\, d\alpha(x), \tag{4}$$

where $\alpha(x)$ is bounded and non-decreasing. In [**43***] I showed that (4) may be replaced by

$$\frac{1}{\phi(z)} = \int_{-\infty}^{\infty} e^{-zx} \Lambda(x)\, dx, \tag{5}$$

where $\Lambda(x)$ is summable and *totally positive.* The last condition means that if

$$(6) \qquad x_1 < x_2 < \ldots < x_n, \qquad y_1 < y_2 < \ldots < y_n \qquad (n = 1, 2, \ldots)$$

then

$$(7) \qquad \det \|\Lambda(x_i - y_k)\| \geq 0.$$

I called such $\Lambda(x)$ *Pólya frequency functions.* In a letter of October 1, 1946, Pólya wrote to me: "... Since a long time I have not seen a result in which I were more interested. ..." See also the book [**HW**] by Hirschman and Widder and particularly the important book [**K**] of Samuel Karlin. There are four distinct characterizations of Pólya frequency functions described in [**43***], [**39***], [**47***] and [**86***]. These four characterizations are conveniently stated in my monograph [**114**], Lecture 1, §3. In jest N. G. de Bruijn suggested that these frequency functions should be called Pólyamials!

The discrete analogues of Pólya frequency functions are the *totally positive sequences* (a_n) with $\Sigma a_n < \infty$. Their generating functions

$$\sum_{-\infty}^{\infty} a_n z^n$$

were determined by M. Aissen, Anne Whitney and myself in [**41**] and [**45***]. However, we left an important open problem which was elegantly solved by Albert Edrei in his paper [**Ed**]. Edrei had heard me state the problem at the International Congress of 1950 at Harvard.

My family received a heavy blow when my beloved wife Dolli died of acute Leukemia on July 2, 1949. My older daughter Elizabeth left for College (Radcliffe) in the Fall of the same year. On December 2, 1950, I married Dolly van der Hoop, of Amsterdam, who was visiting Mark Dresden and his wife in Media PA.

Following a suggestion of Th. Motzkin, John Curtiss appointed me for 1951–1952 at the Institute of Numerical Analysis of the National Bureau of Standards at U.C.L.A., where my collaboration with Motzkin produced the papers [**44***], [**50***], [**153**]. The paper [**44***] anticipated the work of W. Dahmen and C. A. Micchelli [**DM**] on multi-dimensional splines. The paper [**50***] has connections with the notion of "Complexity" of Computer Science; cf. Goffin's commentary in these *selecta.* The paper [**153**] was only published as late as 1980.

Our son Michael Jan was born in Santa Monica CA on September 12, 1951.

Variation diminishing linear transformations also led to a study of curves convex in higher dimensions and to an extension of the classical isoperimetric inequality presented in [**51***].

At Penn I gave courses in Approximation Theory, Laplace Transforms, The Moment Problem, Functional Analysis, Divergent Series, Complex Analysis, Real Variables, and shared with Rademacher the weekly Problem Seminar. I also had a few Ph.D. students: Louis Brickman, Anne Whitney (Mrs. Calloway), S. J. Einhorn, Hyman Gabai, D. S. Greenstein, R. S. Johnson, T. P. G. Liverman, J. C. Mairhuber, R. E. Williamson and Dorothy Wolfe.

In 1949 Hans Rademacher married my sister Irma who had divorced her first husband, the composer Stefan Wolpe.

During 1956–1957 I went on a sabbatical to Stanford to work with Pólya, Peter Scherk, of Toronto, replacing me at Penn. To introduce the main result of our collaboration, I first mention that I. P. Natanson in [**N**] uses the Bernstein polynomials to prove Weierstraß' approximation theorem in $[0, 1]$, and the de la Vallée Poussin means of the Fourier series to prove Weierstrass' theorem for periodic functions. In [**55***] Pólya and I showed that both these means are variation diminishing and cyclically variation diminishing, respectively. Our main "result", however, was a conjecture: *If two power series* $\sum_0^\infty a_n z^n$ *and* $\sum_0^\infty b_n z^n$ *are univalent in* $|z| < 1$ *and map this unit circle onto convex domains, then also their Hadamard product*

$$\sum_0^\infty a_n b_n z^n$$

has the same property; see [**55***], also the largely expository paper [**57**].

A group of Analysts took a fancy to this "Pólya-Schoenberg conjecture". Only after 15 years of efforts our conjecture was finally established by S. Ruscheweyh and T. Sheil-Small in their paper [**RS**].

A fine addition to our department was A. S. Besicovitch who had retired from Cambridge University in the late fifties. He used to scare students at Oral Exams with his "Tell me please" in a heavy Russian accent. For a joint paper see [**66**]; see also [**70***] and [**71**] on the Kakeya problem.

During 1960–1963 I was chairman of the Mathematics Department. For 1961–1962 I organized a Symposium on Number Theory in honor of Hans Rademacher who was to retire in June 1962. We had several guests, among them Charles Pisot of Paris, with whom I wrote the papers [**74**] and [**81**].

Following the initial papers [**31***], [**32***], I had the spline functions for myself from 1946 to about 1960, when they were rediscovered by several people. For 1963–1964 I went on sabbatical to the Institute for Advanced Studies where I developed an intensive activity writing eight papers on several subjects such as spline functions and function theory; see [**75***], [**76***], [**77**], [**78**], [**80***], [**82**], [**83**], and [**85**]. At the end of that year I felt that I needed another year's leave from Penn and I therefore accepted for 1964–1965 an appointment to the Mathematics Research Center at the University of Wisconsin-Madison.

6. Madison, Wisconsin 1965–1973. In June 1965 I resigned from Penn and joined the Mathematics Department of the UW and the MRC. During those years the main subject of my work was on *Cardinal Spline Interpolation* adequately described in my eight papers [**98**], [**108**], [**109**], [**110***], [**111**], [**112**], [**113***], and [**134**], some written jointly with P. R. Lipow, A. Sharma and C. de Boor. The results were summarized in the monograph [**114**].

Starting with the 4th paper [**110***] the Eulerian character of the theory becomes apparent. As pointed out in [**98**] the theory was initiated by Yu. N. Subbotin [**Su**] for bounded data. My series of papers pretty much solved the one-dimensional case. Generalizations to higher dimensional interpolation problems, not obtained by tensor product constructions, were obtained by C. de Boor, K. Höllig and S. Riemenschneider. In their paper [**BHR**], they write: "In this paper we carry Schoenberg's beautiful cardinal spline theory over to a two-dimensional context which is not just the tensor product of the univariate situation. We find that we must work harder, yet must be satisfied with less precise results".

For the solution of some extremum problems suggested by work of G. Glaeser and R. Louboutin see [**105***]. Long ago in [**21**] I made a peripheral contribution towards the problem of determining the so-called J. M. Whittaker constant of entire function theory. For a refinement of [**21**] see [**106***]. In 1966 I revived interest in Birkhoff's Interpolation Problem in [**89***]. Jointly with M. Marsden the variation diminishing spline approximation methods were developed in [**90***]. With S. D. Silliman I studied semi-cardinal interpolation and quadrature formulae in our paper [**117**].

In the sixties Spline Theory became one of the most active topics in Approximation Theory: already 1967 saw the appearance of the first book on the subject [**AN**] by J. H. Ahlberg, E. N. Nilson and J. L. Walsh, which was translated into Russian by Yu. N. Subbotin. This was followed by two Symposia on the subject at the MRC: 1) The Symposium of October 1968 whose Proceedings [**G69**] were edited by T. N. E. Greville, 2) The Symposium of May 1969 with its Proceedings [**S69**] edited by I. J. Schoenberg. Greville's article "Introduction to Spline Functions" in [**G69**] is still the best introduction to the subject. More recently we have the more specialized books [**Bo**] by C. de Boor and [**Sc**] by L. L. Schumaker.

My Ph.D. students in Madison were D. R. Ferguson, Alfred Cavaretta, F. R. Loscalzo, Martin Marsden, Franklin Richards, Sherwood Silliman and I. Yuan Houng.

In 1966 I attended the International Congress in Moscow where the big experience was meeting M. G. Krein after many years of correspondence.

Much traveling connected with lectures was a result of my activities. Not all these trips need be mentioned, but here are a few: During 1967 I traveled for two months in Europe visiting Romania for the first time since 1930 (Bucharest, Cluj, Jassy). I was a guest of the Romanian Academy with a chauffeured car at my disposal. In 1969, from October to December, a visit to Haifa with lectures also in Jerusalem, Tel-Aviv, and the Weizmann Institute in Rehovot. In 1971 a month's visit to the University of Dundee for a Symposium.

7. **After retiring in June 1973.** I remained at the MRC as a part-time Honorary Member and for this I warmly thank its Director, Professor John Nohel. In four consecutive years I spent one semester (or an entire year) at the following four institutions: 1) *The Weizmann Institute* in Rehovot, Israel, first semester, and [**KM**] is an outgrowth of that; 2) *The State University of California in San Diego*; 3) *The University of Pittsburgh*; 4) the entire year from June 1977 to June 1978 I was a Visiting Professor at the *United States Military Academy* at West Point NY. Out of a weekly seminar given at West Point grew my little book [**S83**] "Mathematical Time Exposures" already mentioned in Section 1.

Of the nearly fifty papers written since 1973 I will mention only a few. I begin with [**157***]. Everybody knows that Newton's general polynomial interpolation formula simplifies if the points of interpolation are in arithmetic progression. However, even Scottish mathematicians have forgotten that James Stirling found in 1730 that a similar simplification occurs if the points are in geometric progression. To remind them of it and the work of Schellbach and Runge, I wrote [**157***]. I base the discussion on the so-called q-divided differences; this led to my collaboration with D. J. Newman producing the joint paper [**126**].

The remaining papers that I wish to mention may be divided into two classes each of which deal with related problems. The first deals with

1) *The Landau-Kolmogorov inequality and related extremum problems*: They are described in the nine papers: [**100***] (jointly with A. Cavaretta), [**122**] (which received the

Lester Ford Prize), [**124**], [**135**], [**136**], [**141***], [**149***], [**150***] and [**152**].

With the paper [**100***], which solved the Landau problem for the interval $[0, \infty)$, we had some excitement in 1970. We presented it to the International Conference on Constructive Function Theory in Varma (Bulgaria), May 19-25, 1970. As I did not go to Varma, the paper was read by Penkov. I received a postcard from Paul Turan in which he wrote the following: After the presentation of [**100***] by Penkov, S. B. Stečkin announced that he had promised some years ago 4 bottles of cognac to the solver of the Landau problem for $[0, \infty)$. As I did not go to Russia after 1970, I did not receive any cognac.

2) *Extremum problems for Billiard Ball Motions in Cubes.* These problems can be traced to the paper [**KS**] by D. König and A. Szücs of 1913. Natural generalizations for higher dimensions are contained in the papers [**127**], [**131**], [**132**], [**146**], [**147**], [**148**] and [**164**]. Notice Coxeter's paper [Co]. Also notice that in [**148**] I state a conjecture in the Geometry of Numbers which I cannot establish.

During May 1977 I was in Italy (Rome, Florence, Pisa), spending three weeks at the Istituto "Mauro Picone" in Rome. There I wrote the paper [**143**] on Cardinal Spline Smoothing.

This is the end of my account. Let me thank most warmly my colleague and successor Carl de Boor for agreeing to be the Chief Editor of my Selected Papers in two volumes.

March 1986

I. J. Schoenberg

References

[**AN**] J. H. Ahlberg, E. N. Nilson and J. L. Walsh, *The Theory of Splines and Their Applications*, Academic Press, New York, 1967.

[**A**] N. I. Akhiezer, *The Classical Moment Problem*, Hafner Publishing Co., New York, 1965.

[**BCR**] Christian Berg, Christensen and Ressel, *Harmonic Analysis, The Theory of Positive Definite and Related Functions*, Springer-Verlag, Heidelberg, 1984.

[**Bl**] L. M. Blumenthal, *Theory and Applications of Distance Geometry*, Chelsea Publishing Co., New York, 1970.

[**Bo**] Carl de Boor, *A Practical Guide to Splines*, Springer-Verlag, New York, 1978.

[**BHR**] Carl de Boor, Klaus Höllig and S. Riemenschneider, Bivariate cardinal interpolation by splines on a three-direction mesh, *Illinois J. Math.* **29** (1985), 533–566,

[**Ch**] G. Choquet, *Lectures on Analysis*, W. A. Benjamin, Amsterdam, 1969.

[**Co**] H. S. M. Coxeter, The Derivation of Schoenberg's Star-Polytopes from Schoute Simplex Nets, *The Geometric Vein* (The Coxeter Festschrift), Springer-Verlag, New York, 1981.

[**DM**] W. Dahmen and C. A. Micchelli, On limits of multivariate B-splines, *J. Analyse Math.* **39**, 256–278.

[**Ed**] Albert Edrei, On the generating functions of totally positive sequences. II, *J. Analyse Math.* **2** (1952), 104–109.

[**Er**] P. Erdös, On the smoothness of the asymptotic distribution of additive arithmetical functions, *Amer. J. Math.* **61** (1939), 722–725.

[**G44**] T. N. E. Greville, The general theory of osculatory interpolation, *Trans. Actuarial Soc. America* **45** (1944), 202–265.

[G69] T. N. E. Greville, ed., *The Theory and Applications of Spline Functions*, Academic Press, New York, 1969.

[GV] T. N. E. Greville and Hubert Vaughan, Polynomial interpolation in terms of symbolic operators, in *Selected Papers of T. N. E. Greville*, T. N. E. Greville, ed., Winnipeg, Canada, 1984, 245–308.

[H] H. Hamburger, Bemerkungen zu einer Fragestellung des Herrn Pólya, *Math. Z.* **7** (1920), 302–322.

[HW] I. I. Hirschman and D. V. Widder, *The Convolution Transform*, Princeton University Press, Princeton University, 1955.

[K] S. Karlin, *Total Positivity*, vol. 1, Stanford, 1968.

[KM] S. Karlin, C. A. Micchelli, A. Pinkus and I. J. Schoenberg, eds., *Studies in Spline Functions and Approximation Theory*, Academic Press Inc., New York, 1976.

[KS] D. König and A. Szücs, Mouvement d'un Point abandonné a l'intérieur d'un cube, *Rend. Palermo* **38** (1913), 79–90.

[M] Th. Motzkin, *Beiträge zur Theorie der linearen Ungleichungen*, Doctoral Dissertation, Basel, 1933. Published in Jerusalem, 1936.

[N] I. P. Natanson, *Konstruktive Functionentheorie*, Berlin, 1955.

[P] G. Pólya, Algebraische Untersuchungen über ganze Funktionen vom Geschlechte Null und Eins, *J. Math.* **145** (1915), 224–229.

[O] N. Obreschkoff, *Verteilung und Berechnung der Nullstellen reeller Polynome*, VEB Deutscher Verlag der Wissenschaften, Berlin, 1963.

[QC] W. Quade and L. Collatz, Zur Interpolationstheorie der reellen periodischen Funktionen, *Proc. Preuss. Akad. Wiss. (Math.-Phys. Kl.)* **30** (1938), 383–429.

[R] C. Runge, Über die Darstellung willkürlicher Funktionen und die Interpolation zwischen äquidistanten Ordinaten, *Z. Angew. Math. Phys.* **46** (1901), 224–243.

[RS] S. Ruscheweyh and T. Sheil-Small, Hadamard products of schlicht functions and the Pólya-Schoenberg conjecture, *Comment. Math. Helv.* **48** (1973), 119–135.

[S69] I. J. Schoenberg, ed., *Approximations with Special Emphasis on Spline Functions*, Academic Press, New York, 1969.

[S83] I. J. Schoenberg, *Mathematical Time Exposures*, Math. Assoc. of American, 1983.

[Sc] L. L. Schumaker, *Spline Functions: Basic Theory*, John Wiley & Sons, 1981.

[Su] Yu. N. Subbotin, On the relations between finite differences and the corresponding derivatives, *Proc. Steklov Inst. Math.* **78** (1965), 24-42; *Amer. Math. Soc. Transl.* (1967), 23–42.

[Wb] H. Weber, *Lehrbuch der Algebra*, Vieweg & Sohn, Braunschweig, 1912.

[Wy] H. Weyl, Über die Gleichverteilung von Zahlen mod. Eins, *Math. Ann.* **77** (1916), 313–352.

[Y] A. M. Yaglom, *An Introduction to the Theory of Stationary Random Functions*, Prentice-Hall, Inc., 1962.

NUMBER THEORY

Über die asymptotische Verteilung reeller Zahlen mod 1.

Von

Isac Schoenberg in Jassy (Rumänien).

Die vorliegende Arbeit schließt sich an die Abhandlung von Herrn Weyl[1]) über die Gleichverteilung von Zahlen mod 1 an, und zwar werden wir hier eine Verallgemeinerung dieser Fragestellung behandeln. — Das qualitativ wesentliche an der Gleichverteilung ist das Vorhandensein einer Verteilungswahrscheinlichkeit und wir behandeln hier die Verteilung von Zahlen mod 1 nur von diesem Standpunkt aus. — Interessante Fälle einer solchen ungleichmäßigen Verteilung sind schon von den Herren Pólya[2]) und Szegö[3]) behandelt worden. Für eine eingehendere analytische Behandlung im allgemeinen Fall erscheint aber als wesentlich auch die Forderung der Stetigkeit der Verteilungswahrscheinlichkeit der Zahlen.

Unsere Aufgabe erscheint aufs engste mit dem Momentenproblem für ein endliches Intervall verknüpft, welches von Herrn Hausdorff[4]) in einer wichtigen Abhandlung behandelt worden ist.

Der Fall einer monotonen stetigen Belegungsfunktion als Lösung wurde aber darin nicht betrachtet, und da hier gerade dieser Fall erscheint, mußte ich das hier nachholen.

Diese letzte Aufgabe erscheint, im Falle des Fourierschen Momentenproblems, in enger Beziehung zu neueren Untersuchungen der Herren F. Riesz[5]), L. Neder[6]) und S. Sidon[7]) über die Fourierkoeffizienten stetiger, periodischer Funktionen von beschränkter Schwankung.

[1]) „Über die Gleichverteilung von Zahlen mod. Eins", Math. Annalen **77** (1916), S. 313–352.

[2]) „Über eine neue Weise bestimmte Integrale in der analytischen Zahlentheorie zu gebrauchen", Göttinger Nachrichten 1917, S. 149–159.

[3]) Math. és. term. tud. ért. **36** (1918), S. 531. — Vgl. Pólya und Szegö, „Aufgaben und Lehrsätze aus der Analysis", Bd. 1, II. Abschnitt, Nr. 44 und 194.

[4]) „Momentprobleme für ein endliches Intervall", Math. Zeitschr. **16**, S. 220–248.

[5]) „Über die Fourierkoeffizienten einer stetigen Funktion von beschränkter Schwankung", Math. Zeitschr. **2**, S. 312–315.

[6]) „Über die Fourierkoeffizienten der Funktionen von beschränkter Schwankung", Math. Zeitschr. **6**, S. 270–273.

[7]) „Über die Fourier-Koeffizienten einer stetigen Funktion von beschränkter Schwankung", Acta litter. ac scient. univ. Hungaricae **2**, I (1924), S. 43–46. Siehe den Zusatz am Schluß dieser Arbeit.

In den beiden letzten Paragraphen mache ich einige Anwendungen der allgemeinen Sätze.

Es sei mir noch gestattet, meinem verehrten Lehrer, Herrn Prof. S. Sanielevici in Jassy für seine freundlichen Ratschläge bei der endgültigen Abfassung dieser Arbeit meinen Dank auszusprechen.

§ 1.

Definition der asymptotischen stetigen Verteilung mod 1 und einleitende Sätze. — Die zwei Fragestellungen.

1. Es seien

$$x_{1n}, x_{2n}, \ldots, x_{nn} \qquad (n = 1, 2, 3, \ldots) \tag{1}$$

n reelle Zahlen, die wir uns schon mod 1 reduziert denken können, also alle der Strecke $0 \leqq x \leqq 1$ angehörend.

Sollte x_{in} $(i \leqq n)$ nur von i abhängen, so sind die Zahlen (1) die ersten n Zahlen einer Zahlenfolge

$$x_1, x_2, \ldots, x_n, \ldots$$

und die Reihenfolge der Glieder ist für das folgende wesentlich.

Es sei für den Augenblick x eine feste Zahl mit $0 < x < 1$ und es sei $m_x(n)$ die Anzahl derjenigen unter den Zahlen (1), welche der abgeschlossenen Teilstrecke $0 \ldots x$ angehören.

Wir sagen, daß die Zahlen (1) sich für $n \to \infty$ auf der Strecke $0 \ldots 1$ *asymptotisch stetig verteilen*, wenn für jedes feste x

$$\lim_{n \to \infty} \frac{m_x(n)}{n} = z(x) \qquad (0 < x < 1) \tag{2}$$

vorhanden ist und, $z(0) = 0$ und $z(1) = 1$ gesetzt, die Funktion $z(x)$ im abgeschlossenen Intervall $0 \leqq x \leqq 1$ stetig ist.

Die Funktion $z(x)$ wollen wir die *Verteilungsfunktion* der Zahlen (1) nennen.

Es folgt nach der Definition, daß $z(x)$ nie abnimmt. Derartige Funktionen $z(x)$, die für $0 \leqq x \leqq 1$ monoton und stetig sind, mit $z(0) = 0$, $z(1) = 1$, wollen wir fortab kurz als Verteilungsfunktionen bezeichnen.

Dann und nur dann liegt asymptotische *Gleichverteilung* vor, wenn $z(x) = x$ ist.

Wenn wir z. B. von den mod 1 reduzierten Zahlen

$$\frac{n}{1}, \frac{n}{2}, \ldots, \frac{n}{n}$$

ausgehen, so findet man, wie Herr Pólya gezeigt hat, für die Verteilungsfunktion

$$z(x) = \int_0^1 \frac{1 - t^x}{1 - t} \, dt.$$

2. Wir beweisen zunächst den folgenden

Satz I. *Es sei* $f(x)$ *für* $0 \leqq x \leqq 1$ *von beschränkter Schwankung und* $z(x)$ *sei nach der obigen Definition die Verteilungsfunktion der Zahlen* (1), *dann ist*

$$(3) \qquad \lim_{n\to\infty} \frac{f(x_{1n})+f(x_{2n})+\ldots+f(x_{nn})}{n} = \int_0^1 f(x)\, dz(x),$$

wo das Integral im Sinne von Stieltjes einen Sinn hat, weil wir $z(x)$ *stetig voraussetzen.*

Der Beweis ist ebenso leicht wie für den Spezialfall der Gleichverteilung, also $z(x) = x$[8]).

Aus der Limesgleichung (2) folgt sofort, daß (3) jedenfalls für Treppenfunktionen mit endlich vielen Treppen gültig ist.

Man kann nun, wenn $\varepsilon > 0$ gegeben wird, zwei derartige Treppenfunktionen $f_1(x)$ und $f_2(x)$ bestimmen, daß

$$f_1(x) \leqq f(x) \leqq f_2(x) \quad \text{für} \quad 0 \leqq x \leqq 1 \quad \text{und} \quad \int_0^1 f_2(x)\, dz(x) - \int_0^1 f_1(x)\, dz(x) < \varepsilon.$$

Wir wollen auch fernerhin allgemein die Abkürzung

$$M_n(F) = \frac{F(x_{1n})+F(x_{2n})+\ldots+F(x_{nn})}{n}$$

für diesen Mittelwert benutzen.

Da Formel (3) für $f_1(x)$ gilt, so ist für genügend großes n

$$M_n(f_1) > \int_0^1 f_1\, dz - \varepsilon\,.$$

Aber $\quad M_n(f) \geqq M_n(f_1) \quad$ und $\quad \int_0^1 f\,dz - \int_0^1 f_1\,dz \leqq \int_0^1 f_2\,dz - \int_0^1 f_1\,dz < \varepsilon,$

also
$$M_n(f) > \int_0^1 f\,dz - 2\varepsilon \quad \text{für großes } n.$$

Von $f_2(x)$ ausgehend, erhält man die Abschätzung nach oben

$$M_n(f) < \int_0^1 f\,dz + 2\varepsilon \quad \text{für großes } n,$$

also zusammenfassend

$$\left| M_n(f) - \int_0^1 f\,dz \right| < 2\varepsilon \quad \text{für großes } n$$

und der Satz ist bewiesen.

---

[8]) Vgl. H. Weyl a. a. O. S. 314.

Im Satz I kann natürlich die Funktion $f(x)$ auch komplex sein, wenn nur Real- und Imaginärteil den Bedingungen genügen.

Indem wir in der Formel (3) die Funktion $f(x)$ passend spezialisieren, erhalten wir, daß für $k = 1, 2, 3, \ldots$ die Grenzwerte

$$\lim_{n \to \infty} \frac{x_{1n}^k + \ldots + x_{nn}^k}{n} = \mu_k, \tag{4}$$

$$\lim_{n \to \infty} \frac{e^{2\pi i k x_{1n}} + \ldots + e^{2\pi i k x_{nn}}}{n} = \omega_k \tag{5}$$

vorhanden sind und deren Werte erhalten wir aus

$$\int_0^1 x^k \, dz(x) = \mu_k \qquad (\mu_0 = 1;\ k = 0, 1, 2, \ldots), \tag{4'}$$

$$\int_0^1 e^{2\pi i k x} \, dz(x) = \omega_k \qquad (\omega_0 = 1;\ k = 0, 1, 2, \ldots). \tag{5'}$$

Diese Grenzwerte erscheinen also als Stietjes- resp. Fouriermomente der Verteilungsfunktion $z(x)$.

3. Wir haben bisher vorausgesetzt, daß sich die Zahlen (1) asymptotisch stetig verteilen. Wir wollen nun diese Voraussetzung fallen lassen und wir beweisen, *daß aus den Gleichungen* (4) *auch die Existenz der Limites* ω_k, *also auch die Gleichungen* (5) *folgen. Allgemeiner, aus den Gleichungen* (4) *oder* (5) *folgt, daß der Limes*

$$\lim_{n \to \infty} \frac{f(x_{1n}) + f(x_{2n}) + \ldots + f(x_{nn})}{n} = L(f) \tag{6}$$

vorhanden ist, für jede für $0 \leqq x \leqq 1$ *stetige Funktion* $f(x)$.

Der Beweis ist sofort erledigt. Wir wollen etwa von den Gleichungen (4) ausgehen und zeigen, daß $L(f)$ vorhanden ist.

Es sei $\varepsilon > 0$ beliebig gegeben.

Nach dem Weierstraßschen Approximationssatz kann man ein Polynom so bestimmen, daß

$$f(x) = a_0 + a_1 x + \ldots + a_m x^m + r(x)$$

und

$$|r(x)| < \varepsilon \quad \text{für} \quad 0 \leqq x \leqq 1.$$

In der abkürzenden Bezeichnung von 2 folgt aus der letzten Gleichung

$$M_n(f) = a_0 + a_1 M_n(x) + \ldots + a_m M_n(x^m) + M_n(r)$$

und aus $M_n(x^k) \to \mu_k$ für $n \to \infty$ und $|M_n(r)| < \varepsilon$ folgt

$$|M_n(f) - (a_0 + a_1 \mu_1 + \ldots + a_m \mu_m)| < 2\varepsilon$$

für genügend großes n, also auch

$$|M_{n+p}(f) - M_n(f)| < 4\varepsilon \quad \text{für großes } n$$

und also ist Formel (6) bewiesen.

Man beweist dasselbe ebenso leicht, wenn man von den Gleichungen (5) ausgeht, indem man $f(x)$ durch ein trigonometrisches Polynom approximiert.

4. Wir benutzen jetzt den Zusammenhang mit dem Momentenproblem und beweisen den

Satz II. *Es mögen die Grenzwerte* μ_k *gemäß den Gleichungen* (4) *vorhanden sein* ($\mu_0 = 1$) *und es sei das Momentenproblem* (4′) *durch eine stetige monotone Funktion* $z(x)$ *lösbar, wobei wir* $z(0) = 0$ *voraussetzen dürfen.*

Unter diesen Umständen verteilen sich unsere Zahlen (1) *asymptotisch stetig und es ist* $z(x)$ *deren Verteilungsfunktion.*

Man kann genau dasselbe beweisen, wenn man von den Fouriermomenten ω_k *ausgeht und* $z(x)$ *den Gleichungen* (5′) *genügt.*

Es seien die Voraussetzungen des Satzes erfüllt, dann folgt nach **3** die Existenz von

$$\lim_{n\to\infty} M_n(f) = L(f) \quad \text{für jedes stetige } f(x).$$

Man sieht nun leicht, daß

$$L(f) = \int_0^1 f(x)\, dz(x),$$

denn nach (4) und (4′) ist diese Gleichung für jedes Polynom richtig und nach dem Approximationssatz folgt sie auch sofort für jede stetige Funktion.

Der Satz II besagt, daß für $0 < \vartheta < 1$, in der Bezeichnung von **1**,

$$\lim_{n\to\infty} \frac{m_\vartheta(n)}{n} = z(\vartheta),$$

für jedes solche ϑ.

Wir betrachten folgende Funktion mit einem Sprung

$$v(x) = \begin{cases} 1 & \text{für } 0 \leqq x \leqq \vartheta, \\ 0 & \text{,, } \vartheta < x \leqq 1, \end{cases}$$

und da

$$M_n(v) = \frac{m_\vartheta(n)}{n} \quad \text{und} \quad \int_0^1 v(x)\, dz(x) = z(\vartheta),$$

so genügt es zu beweisen, daß

$$\lim_{n\to\infty} M_n(v) = L(v) = \int_0^1 v(x)\, dz(x).$$

Es sei $\varepsilon > 0$ fest, genügend klein, damit $0 < \vartheta - \varepsilon$ und $\vartheta + \varepsilon < 1$. Es seien $\psi_+(x)$ und $\psi_-(x)$ die zwei stetigen Funktionen, welche die Streckenzüge $ACEF$ resp. $ABDF$ der Fig. 1 darstellen. Es ist

$$\psi_-(x) \leqq v(x) \leqq \psi_+(x), \quad \text{also} \quad M_n(\psi_-) \leqq M_n(v) \leqq M_n(\psi_+).$$

Für $\delta > 0$ fest folgt also (ψ_+ und ψ_- sind stetig!)

$$\int_0^1 \psi_-(x)\, dz(x) - \delta < M_n(v) < \int_0^1 \psi_+(x)\, dz(x) + \delta$$

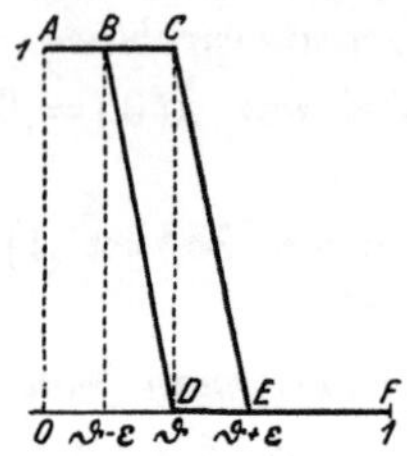

Fig. 1.

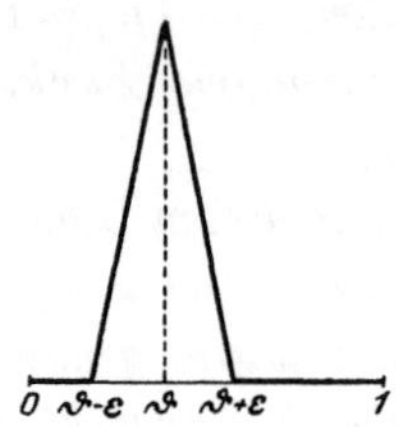

Fig. 2.

für genügend großes n und es folgt daraus die Existenz des Grenzwertes von $M_n(v)$ und aus $\int \psi_- dz \leqq \int v\, dz \leqq \int \psi_+ dz$[9]) auch sein Wert

$$\int_0^1 v\, dz(x),$$

wenn wir zeigen können, daß die Differenz

$$\Delta_\varepsilon = \int_0^1 \psi_+(x)\, dz(x) - \int_0^1 \psi_-(x)\, dz(x)$$

für genügend kleines ε auch beliebig klein wird.

Es sei $\psi(x) = \psi_+(x) - \psi_-(x)$ (Fig. 2), so folgt

$$\Delta_\varepsilon = \int_0^1 \psi(x)\, dz(x) = \int_{\vartheta-\varepsilon}^{\vartheta+\varepsilon} \psi(x)\, dz(x) = -\int_{\vartheta-\varepsilon}^{\vartheta+\varepsilon} z(x)\, d\psi(x)$$

$$= \frac{1}{\varepsilon} \int_{\vartheta}^{\vartheta-\varepsilon} z(x)\, dx + \frac{1}{\varepsilon} \int_{\vartheta}^{\vartheta+\varepsilon} z(x)\, dx$$

und dieser letzte Ausdruck strebt für $\varepsilon \to 0$ gegen $-z(\vartheta - 0) + z(\vartheta + 0)$ also gegen Null, denn $z(x)$ ist stetig. Damit ist der Satz bewiesen.

5. Man kann nun zwei verschiedene Fragestellungen verfolgen:

A. *Es sei eine Verteilungsfunktion* $z(x)$[10]) *a priori gegeben.*

---

[9]) Diese Ungleichungen gelten, weil $z(x)$ auch monoton und nie abnehmend ist.

[10]) Diese Funktion sei also für $0 \leqq x \leqq 1$ stetig, monoton $z(0) = 0$, $z(1) = 1$, siehe 1.

Es sollen Kriterien aufgestellt werden, welche die asymptotische Verteilung unserer Zahlen (1) *nach der Funktion* $z(x)$ *gewährleisten.*

B. *Man finde Kriterien, welche uns überhaupt eine asymptotische stetige Verteilung der Zahlen* (1) *erkennen lassen, wo also die Verteilungsfunktion nicht von vornhinein gegeben wird, sondern beliebig sein kann und erst hinterher durch die Zahlen* (1) *bestimmt wird.*

Es ist natürlich die zweite Fragestellung interessanter, denn wenn man von (auf irgend eine Weise) bestimmten Zahlen (1) ausgeht (wie z. B. $\frac{n}{1}, \frac{n}{2}, \ldots, \frac{n}{n}$ mod 1), so ist doch zunächst gewiß $z(x)$ ganz unbekannt.

6. Wir wollen zunächst die Fragestellung A erledigen, auf welche man mit Hilfe des Satzes II auf verschiedene Arten eingehen kann.

Für $z(x) = x$ entsteht die Frage nach der Gleichverteilung der Zahlen (1), wofür die Gleichungen

$$(7) \qquad \lim_{n\to\infty} \frac{e^{2\pi i k x_{1n}} + e^{2\pi i k x_{2n}} + \ldots + e^{2\pi i k x_{nn}}}{n} = 0 \quad \text{für} \quad k = 1, 2, 3, \ldots$$

notwendig und hinreichend sind[11]).

Im allgemeinen Fall eines gegebenen $z(x)$ sind notwendig und hinreichend für die Verteilung der Zahlen (1) nach $z(x)$, entweder die Gleichungen (4) *oder* diejenigen (5) für $k = 1, 2, 3, \ldots$, wobei die entsprechenden Zahlen μ_k und ω_k aus (4′) und (5′) zu entnehmen sind, denn alle Bedingungen des Satzes II sind damit erfüllt.

Man kann den Bedingungen auch eine andere Gestalt geben.

Es sei $$Q_0(x) \equiv 1, \; Q_1(x), \; Q_2(x), \; \ldots$$

eine Folge von Funktionen, die für $0 \leqq x \leqq 1$ stetig sind und den Orthogonalitätsrelationen

$$\int_0^1 Q_m(x) Q_n(x) \, dz(x) = 0 \qquad (m \neq n, \; m \geqq 0, \; n \geqq 0)$$

genügen. Es sei außerdem möglich, irgend eine stetige Funktion durch lineare Aggregate der $Q_n(x)$ mit konstanten Koeffizienten beliebig genau gleichmäßig zu approximieren.

Eine solche Funktionenfolge bilden z. B. die durch diese Bedingungen bis auf konstante Faktoren eindeutig bestimmten sogenannten Stieltjesschen Polynome von $z(x)$.

---

[11]) H. Weyl a. a. O. Satz 1, S. 315.

Notwendig und hinreichend für die Verteilung der Zahlen (1) *nach* $z(x)$ *sind dann die Bedingungen*

$$\lim_{n\to\infty} \frac{Q_k(x_{1n}) + Q_k(x_{2n}) + \ldots + Q_k(x_{nn})}{n} = 0 \quad (\text{für } k = 1, 2, 3, \ldots).$$

Auf den Beweis, der sofort aus dem Weierstraßschen Approximationssatz und dem Satz II folgt, will ich nicht mehr eingehen.

§ 2.

Kriterien für asymptotische stetige Verteilung mit Hilfe der Stieltjesmomente μ_k.

7. Wir haben in **3.** bewiesen, daß aus den Gleichungen (4) oder (5) auch die Gleichung (6) für jede stetige Funktion $f(x)$ folgt. Wir fassen nun die Operation $L(f)$ ins Auge unter Voraussetzung der Gleichungen (4).

Sie ist offenbar linear, das heißt $L(af + b\varphi) = aL(f) + bL(\varphi)$, und auch beschränkt. Denn aus

$$|M_n(f)| \leqq \operatorname*{Max}_{0 \leqq x \leqq 1} |f(x)| = \mathfrak{M}$$

folgt auch

$$|L(f)| \leqq \mathfrak{M}.$$

Nach einem wichtigen Satz von Herrn F. Riesz[12]) kann man nun folgern, daß es eine monotone Funktion $z(x)$ gibt, so daß für jedes stetige $f(x)$

$$L(f) = \int_0^1 f(x)\,dz(x).$$

Wir setzen $z(0) = 0$. Dann ist natürlich ($f \equiv 1$) $z(1) = 1$ und die Funktion $z(x)$ nimmt also nie ab.

Es sind also die Momentenprobleme (4′) *und* (5′) *mit Hilfe einer monotonen Funktion* $z(x)$ *lösbar*[13]).

Um aus dem Satz II zu folgern, daß sich unsere Zahlen (1) asymptotisch stetig verteilen, haben wir noch *die Stetigkeit* der auf diese Weise gewonnenen Funktion $z(x)$ nötig, und zwar muß man, um zu brauchbaren Kriterien zu gelangen, die dazu nötigen Bedingungen in den Grenzwerten

[12]) „Démonstration nouvelle d'un théorème concernant les opérations fonctionelles linéaires", Annales de l'école normale 3e série **31** (1914) S. 9–14.

Aus dem Satz von Herrn Riesz folgt zunächst, daß $z(x)$ von beschränkter Schwankung ist, aber man ersieht aus seinem Beweis a. a. O., daß, wenn immer $L(f) \geqq 0$ für $f(x) \geqq 0$ $(0 \leqq x \leqq 1)$, die Funktion $z(x)$ sogar monoton ist.

Hier ist diese Zusatzbedingung natürlich erfüllt.

[13]) Man bemerkt sofort, daß die Zahlen μ_k eine total monotone Folge bilden und dann folgt dasselbe Ergebnis auch aus einem Satz von Herrn Hausdorff.

Die Umkehrung ist auch richtig, daß heißt: wenn die total-monotone Folge μ_k $(\mu_0 = 1)$ gegeben ist, so gibt es entsprechende Zahlen (1), so daß (4) gilt.

μ_k oder in den ω_k ausdrücken. Die Untersuchung gestaltet sich ganz verschieden, je nachdem man von den μ_k oder den ω_k Gebrauch macht und wir wollen mit den ersteren beginnen.

8. Wir gehen also jetzt von der Funktion $z(t)$ aus, und zwar sei sie für $0 \leqq t \leqq 1$ monoton mit $z(0)=0$ und $z(1)=1$ und wir bilden ihre Momente

$$\int_0^1 t^k \, dz(t) = \mu_k \qquad (k = 0, 1, 2, \ldots). \tag{8}$$

Wir betrachten auch das Integral

$$f(x) = \int_0^1 \frac{dz(t)}{1 - xt} \tag{9}$$

und man sieht sofort, daß es eine, in der von 1 bis $+\infty$ längst der reellen Ache aufgeschlitzten x-Ebene, reguläre analytische Funktion darstellt.

Aus (8) und (9) folgt die Potenzreihenentwicklung

$$f(x) = \sum_{n=0}^{\infty} \mu_n x^n \qquad \text{für} \quad |x| < 1. \tag{10}$$

Wir verwerten die durch (10) gegebene Funktion und beweisen den

Hilfssatz I. *Damit die monotone Funktion $z(t)$ für $0 \leqq t \leqq 1$ stetig sei, ist notwendig und hinreichend, daß die Funktion $f(x)$ folgenden Randbedingungen genügt:*

(α) *Für jedes feste r mit $1 \leqq r < \infty$ sei*

$$\lim_{x \to r} (x - r) f(x) = 0,$$

wenn x auf irgend einem Weg[14]) *mit bestimmter Tangente, welche mit der Strecke $1 - +\infty$ nicht zusammenfällt, in r einläuft.*

(β) *Es sei*

$$\lim_{x \to \infty} f(x) = 0$$

wenn x in irgend einer, von derjenigen der positiven reellen Achse verschiedenen Richtung[14]) *nach ∞ strebt.*

Es sei
$$\vartheta = \frac{1}{r}.$$
Nach (9) ist

$$(1 - x\vartheta) f(x) = \int_0^1 \frac{1 - x\vartheta}{1 - xt} \, dz(t).$$

[14]) Ein Weg genügt, dasselbe folgt schon für die anderen.

12*

Die Kurve, auf der $\frac{1}{x}$ in ϑ einläuft, bildet auch einen endlichen Winkel mit der Strecke $0 - 1$ und es folgt daraus, daß

$$\left| \frac{1 - x\vartheta}{1 - x t} \right| < M$$

also beschränkt, für $0 \leqq t \leqq 1$ und $\frac{1}{x}$ auf jener festen Kurve. *Es sei* $z(t)$ *an der Stelle* $t = \vartheta$ *stetig.* Es sei auch $\vartheta < 1$.

Wenn wir die Umgebung $\vartheta - \varepsilon \ldots \vartheta + \varepsilon$ von ϑ herausschneiden ($\varepsilon > 0$ genügend klein), so strebt für $\frac{1}{x} \to \vartheta$ (auf der Kurve!) der Ausdruck

$$\left| \frac{1 - x\vartheta}{1 - x t} \right| = \left| \frac{\frac{1}{x} - \vartheta}{\frac{1}{x} - t} \right| \to 0$$

und zwar gleichmäßig für alle Werte von t mit $|t - \delta| \geqq \varepsilon$.

Für $\frac{1}{x}$ genügend nahe an ϑ ist also

$$|1 - x\vartheta| \cdot |f(x)| < \int\limits_{\vartheta-\varepsilon}^{\vartheta+\varepsilon} \left| \frac{1 - x\vartheta}{1 - x t} \right| d z(t) + \varepsilon < M(z(\vartheta + \varepsilon) - z(\vartheta - \varepsilon)) + \varepsilon$$

und wird also unendlich klein für ε genügend klein da wir $z(t)$ in ϑ stetig voraussetzen. Es ist also (α) erfüllt.

Derselbe Schluß gilt auch für $\vartheta = 1$.

Es sei $z(t)$ *an der Stelle* $t = \vartheta$ *unstetig mit dem Sprung* σ.

Wir betrachten die Funktion $z_1(t)$, die für $t = \vartheta$ stetig ist:

$$z_1(t) = z(t) \quad \text{für} \quad 0 \leqq t < \vartheta \quad \text{und} \quad z_1(t) = z(t) - \sigma \quad \text{für} \quad \vartheta < t \leqq 1.$$

Es ist dann

$$(1 - x\vartheta) f(x) = \int\limits_0^1 \frac{1 - x\vartheta}{1 - x t} d z(t) = \int\limits_0^1 \frac{1 - x\vartheta}{1 - x t} d z_1(t) + \sigma,$$

denn das Integral ändert sich um den Sprung multipliziert mit dem Wert des Integranden an der Sprungstelle, hier gleich 1.

Da z_1 in ϑ stetig ist, so strebt, wie schon bewiesen, das Integral der rechten Seite gegen Null und $(1 - x\vartheta) f(x) \to \sigma \neq 0$ für $\frac{1}{x} \to \vartheta$. Damit ist (α) bewiesen und der Beweis von (β) ist auch schon geführt, wenn man so wie oben mit $\vartheta = 0$ operiert.

Wenn z. B. $z(t) = t$, so ist

$$\mu_k = \int\limits_0^1 t^k \, d t = \frac{1}{k+1}$$

und

$$f(x)=1+\frac{x}{2}+\frac{x^2}{3}+\ldots \quad =-\frac{\log(1-x)}{x}$$

genügt offenbar den Bedingungen (α) und (β).

9. Um diesen Hilfssatz zu verwerten, benötigen wir auch die Funktion

(11) $$\Phi(s)=\int_0^1 t^s\,dz(t) \quad \text{für} \quad \Re s>0 \text{ [15]}.$$

Wir beweisen zunächst kurz, daß $\Phi(s)$ *in der Halbebene* $\Re s>0$ *analytisch und regulär ist.*

In der Tat sei $s=\sigma+i\lambda$ fest $(\sigma>0)$ und $0<|h|<\frac{\sigma}{2}$.

Es ist

$$\frac{\Phi(s+h)-\Phi(s)}{h}-\int_0^1 t^s \lg t\,dz(t)=\int_0^1\left(\frac{t^h-1}{h}-\lg t\right)t^{\sigma+i\lambda}\,dz(t),$$

also

$$\left|\frac{\Phi(s+h)-\Phi(s)}{h}-\int_0^1 t^s \lg t\,dz(t)\right|\leqq\int_0^1\left|\frac{t^h-1}{h}-\lg t\right|\cdot t^\sigma\,dz(t).$$

Für $0<t\leqq 1$ ist aber

$$\left|\frac{t^h-1}{h}-\lg t\right|=\left|\frac{e^{h\lg t}-1}{h}-\lg t\right|=\left|\frac{h}{2!}\lg^2 t+\frac{h^2}{3!}\lg^3 t+\ldots\right|$$

$$\leqq|h|\lg^2 t\left(1+\frac{|h|}{1!}\lg\frac{1}{t}+\frac{|h|^2}{2!}\lg^2\frac{1}{t}+\ldots\right)=|h|\lg^2 t\, e^{|h|\lg\frac{1}{t}}$$

$$=|h|\cdot t^{-|h|}\lg^2 t\leqq|h|\cdot t^{-\frac{\sigma}{2}}\lg^2 t,$$

weil

$$|h|<\frac{\sigma}{2}.$$

Also ist

$$\left|\frac{\Phi(s+h)-\Phi(s)}{h}-\int_0^1 t^s\lg t\,dz(t)\right|\leqq|h|\cdot\int_0^1 t^{\frac{\sigma}{2}}\lg^2 t\,dz(t)$$

und für $h\to 0$ erhält man $\Phi'(s)=\int_0^1 t^s\lg t\,dz(t)$ und $\Phi(s)$ ist regulär nach dem Satz von Goursat.

Aus (11) *folgt auch sofort* $|\Phi(s)|\leqq 1$.

Diese Funktion besitzt auch noch folgende Eigenschaft:

---

[15]) $t^s=e^{s\lg t}$, $\lg t$ reell.

Für $\sigma \to +0$ strebt $\Phi(\sigma + i\lambda)$ gegen einen Grenzwert, den wir mit $\Phi(i\lambda)$ bezeichnen, und zwar gleichmäßig für $-\infty < \lambda < +\infty$, $\Phi(i\lambda)$ ist also eine stetige Funktion von λ.

Zum Beweis definieren wir

$$\Phi(i\lambda) = \lim_{\varepsilon \to 0} \int_\varepsilon^1 t^{i\lambda}\, dz(t) = \int_0^1 t^{i\lambda}\, dz(t) \tag{11'}$$

als konvergentes uneigentliches Stieltjesintegral. Für den Augenblick wollen wir mit $z_1(t)$ die Funktion bezeichnen, die gleich $z(t)$ für $0 < t \leqq 1$ aber $z_1(0) = z(+0)$ (Abkürzung für $\lim\limits_{\varepsilon \to +0} z(\varepsilon)$). Es ist dann auch

$$\Phi(i\lambda) = \int_0^1 t^{i\lambda}\, dz_1(t), \qquad \Phi(\sigma + i\lambda) = \int_0^1 t^{\sigma + i\lambda}\, dz_1(t) \quad \text{für} \quad \sigma > 0$$

und also

$$|\Phi(\sigma + i\lambda) - \Phi(i\lambda)| = \left| \int_0^1 t^{\sigma+i\lambda}\, dz_1(t) - \int_0^1 t^{i\lambda}\, dz_1(t) \right|$$

$$= \left| \int_0^1 (t^\sigma - 1)\, t^{i\lambda}\, dz_1(t) \right| \leqq \int_0^1 (1 - t^\sigma)\, dz_1(t)$$

eine von λ unabhängige Schranke, welche für $\sigma \to 0$ auch gegen Null strebt, weil $z_1(t)$ für $z = 0$ stetig ist.

Es gilt also die Darstellung (11) in der ganzen Halbebene auch für $\sigma = 0$, aber in diesem Fall ist das Integral im Sinne $\lim\limits_{\varepsilon \to 0} \int\limits_\varepsilon$ zu nehmen.

Für $s = 0$ ist also

$$\Phi(0) = z(1) - z(+0) = 1 - z(+0),$$

so daß $\Phi(0) = 1$ notwendig und hinreichend für die Stetigkeit von $z(t)$ an der Stelle $t = 0$ ist.

10. Wir wollen auch die Bedingungen für die Stetigkeit von $z(t)$ für $0 < t \leqq 1$ mit Hilfe der Funktion $\Phi(s)$ ausdrücken.

Die Bedingung (α) des Hilfssatzes I benutzt die Werte von $f(x)$ für Argumente x, die außerhalb des Konvergenzbereiches $|x| < 1$ von (10) liegen, und wir werden zum Ziel gelangen, wenn wir die analytische Fortsetzung des Funktionselementes (10) mit Hilfe von $\Phi(s)$ ausführen.

Nach (10) und (11) ist

$$f(x) = 1 + \sum_{n=1}^{\infty} \Phi(n)\, x^n$$

und da $\Phi(s)$ für $\Re s > 0$ regulär und daselbst absolut $\leqq 1$ ist, so kann man hier einen allgemeineren Satz von Le Roy und Lindelöf benützen, den ich dem Buch „Calcul des Résidus"[16]) von Herrn Lindelöf entnehme

---

[16]) Paris 1905, S. 109—112, insbesondere Formel (4) auf S. 111. — Was dort $\varphi(z)$ ist, heißt hier $\Phi(z)$ und in unserem Spezialfall ist $\vartheta = 0$ und $m = 1$ zu nehmen.

und erhält sofort die Darstellung

$$f(x) = 1 + \int_{\alpha - i\infty}^{\alpha + i\infty} \frac{\Phi(z)\, x^z}{1 - e^{2\pi i z}} dz \qquad (0 < \alpha < 1)$$

für $|x| > 0$ und $0 < \operatorname{arc} x < 2\pi$, $(x^z = e^{z \lg x},\ \lg x = \lg|x| + i \operatorname{arc} x)$.

Da $\Phi(\sigma + i\lambda)$ gleichmäßig in λ gegen $\Phi(i\lambda)$ strebt, für $\sigma \to 0$, so dürfen wir in dieser Formel den Integrationsweg $\Re z = \alpha$ mit der imaginären Achse vertauschen und erhalten

$$(12) \qquad f(x) = 1 + \int_{-\infty i}^{\infty i}{}' \frac{\Phi(z)\, x^z}{1 - e^{2\pi i z}} dz$$

in demselben Bereich für x, wobei der Strich am Integral andeuten soll, daß wir den Pol im Nullpunkt längs des Halbkreises

$$z = \varkappa e^{i\omega}, \qquad -\frac{\pi}{2} \leqq \omega \leqq \frac{\pi}{2} \qquad (0 < \varkappa < 1)$$

umgehen.

Das Integral (12) konvergiert natürlich absolut, wie man sofort einsieht, weil wir $0 < \operatorname{arc} x < 2\pi$ festgesetzt haben.

11. Es sei nun $x = r e^{i\vartheta}$, r fest $1 \leqq r < \infty$ und $0 < \vartheta < 2\pi$. Dann ist

$$\lim_{\vartheta \to +0} \frac{|x - r|}{\vartheta} = r$$

und die Bedingung (α) von Hilfssatz I ist mit

$$(13) \qquad \lim_{\vartheta \to +0} \vartheta \cdot f(r e^{i\vartheta}) = 0$$

gleichwertig.

Wir setzen zur Abkürzung

$$\varphi(z) = \frac{\Phi(z)\, x^z}{1 - e^{2\pi i z}}$$

und dann folgt aus (12)

$$(14) \quad \vartheta f(r e^{i\vartheta}) = \vartheta + \vartheta \cdot \int_{-\infty i}^{-\varkappa i} \varphi(z)\, dz + \vartheta \cdot \int_{\text{n.d.Halbkreis}} \varphi(z)\, dz + \vartheta \cdot \int_{\varkappa i}^{\infty i} \varphi(z)\, dz.$$

Wir wollen die zwei ersten Integrale abschätzen:

$$\left| \int_{-\infty i}^{-\varkappa i} \varphi(z)\, dz \right| = \left| \int_{-\infty}^{-\varkappa} \frac{\Phi(i\lambda)\, e^{-\lambda\vartheta + i\lambda \lg r}}{1 - e^{-2\pi\lambda}} d\lambda \right| = \left| \int_{\varkappa}^{\infty} \frac{\Phi(-i\lambda)\, e^{\lambda\vartheta - i\lambda \lg r}}{1 - e^{2\pi\lambda}} d\lambda \right|$$

$$\leqq \int_{\varkappa}^{\infty} \frac{e^{\vartheta\lambda}}{e^{2\pi\lambda} - 1} d\lambda < \int_{\varkappa}^{\infty} \frac{e^{\frac{\pi}{2}\lambda}}{e^{2\pi\lambda} - 1} d\lambda \quad \text{für} \quad 0 < \vartheta < \frac{\pi}{2},$$

also eine von ϑ unabhängige Schranke.

Für $x = r\,e^{i\vartheta}$, $z = \varkappa\,e^{i\omega}$ ist $|x^z| = e^{\varkappa(\lg r\cdot\cos\omega - \vartheta\sin\omega)}$ also, für unsere Werte von r und ϑ ist

$$|x^z| \leqq e^{\varkappa(\lg r + \vartheta)}.$$

Folglich ist

$$\left|\int\limits_{\text{n. d. Hkr.}} \varphi(z)\,dz\right| = \int\limits_{\text{n. d. Hkr.}} \frac{e^{\varkappa\left(\lg r + \frac{\pi}{2}\right)}}{\left|1 - e^{2\pi i z}\right|}\cdot|dz| \quad \text{für} \quad 0 < \vartheta < \frac{\pi}{2},$$

wieder eine von ϑ unabhängige Schranke.

Aus (14) folgt also (13), wenn auch das letzte Glied auf der rechten Seite von (14) für $\vartheta \to 0$ gegen Null strebt, also

$$\lim_{\vartheta\to+0} \vartheta \int\limits_{\varkappa}^{\infty} \frac{\Phi(i\lambda)\,e^{-\lambda\vartheta + i\lambda\lg r}}{1 - e^{-2\pi\lambda}}\,d\lambda = 0.$$

Wir ersetzen λ durch $\frac{\lambda}{\vartheta}$ und haben

$$\left|\vartheta \int\limits_{\varkappa}^{\infty} \frac{\Phi(i\lambda)\,e^{-\lambda\vartheta + i\lambda\lg r}}{1 - e^{-2\pi\lambda}}\,d\lambda\right| = \left|\int\limits_{\vartheta\varkappa}^{\infty} \frac{\Phi\left(i\frac{\lambda}{\vartheta}\right)e^{-\lambda + i\frac{\lambda}{\vartheta}\lg r}}{1 - e^{-2\pi\frac{\lambda}{\vartheta}}}\,d\lambda\right| \leqq \int\limits_{\varkappa\vartheta}^{\infty} \frac{\left|\Phi\left(i\frac{\lambda}{\vartheta}\right)\right| e^{-\lambda}}{1 - e^{-2\pi\varkappa}}\,d\lambda$$

$$\leqq \frac{1}{1 - e^{-2\pi\varkappa}} \int\limits_0^{\infty} \left|\Phi\left(i\frac{\lambda}{\vartheta}\right)\right| e^{-\lambda}\,d\lambda.$$

In dieser letzten Schranke ist r ganz fortgefallen, so daß die Bedingung (α) des Hilfssatzes gewiß erfüllt ist, wenn $\left(k = \frac{1}{\vartheta}\right)$

$$\lim_{k\to\infty} \int\limits_0^{\infty} |\Phi(i\lambda k)|\,e^{-\lambda}\,d\lambda = 0. \tag{15}$$

Wir wollen der Gleichung (15) eine bequemere Form geben.

Aus $|\Phi(i\lambda)| \leqq 1$ folgt, daß das uneigentliche Integral in (15) gleichmäßig in bezug auf k konvergiert, so daß zum Bestehen von (15) hinreichend ist, daß

$$\int\limits_0^{l} |\Phi(i\lambda k)|\,e^{-\lambda}\,d\lambda \to 0 \quad \text{für} \quad k \to \infty,$$

für jeden festen Wert von $l > 0$ und diese letzte Bedingung ist auch mit

$$\int\limits_0^{l} |\Phi(i\lambda k)|\,d\lambda \to 0 \quad \text{für} \quad k \to \infty,$$

gleichwertig.

Wir ersetzen im letzten Integral λ durch $\frac{\lambda}{k}$ und erhalten

$$\frac{1}{k}\int\limits_0^{lk} |\Phi(i\lambda)|\,d\lambda \to 0$$

und wenn wir $l k = x$ setzen

$$\frac{l}{x}\int_0^x |\Phi(i\lambda)|\, d\lambda \to 0,$$

also endlich

$$\lim_{x\to\infty} \frac{1}{x}\int_0^x |\Phi(i\lambda)|\, d\lambda = 0, \tag{16}$$

eine Bedingung, welche verlangt, daß $|\Phi(i\lambda)|$ im Mittel gegen Null konvergiere.

Wir bemerken nun, daß die Funktion $\Phi(s)$ nach (11) den Gleichungen

$$\Phi(n) = \mu_n \qquad (n = 1, 2, 3, \ldots)$$

genügt und da noch

$$|\Phi(s)| \leqq 1 \quad \text{für} \quad \Re s > 0,$$

so folgt nach einem allgemeinen Satze[17]), daß die analytische Funktion $\Phi(s)$ durch diese Bedingungen eindeutig festgelegt ist. Wir erhalten zusammenfassend, indem wir Satz II heranziehen, den

Satz III. *Für die Zahlen* $x_{1n}, x_{2n}, \ldots, x_{nn}$ (1) *der Strecke* $0 \ldots, 1$ *mögen die Grenzwerte*

$$\lim_{n\to\infty} \frac{x_{1n}^k + x_{2n}^k + \ldots + x_{nn}^k}{n} = \mu_k \tag{4}$$

für $k = 1, 2, 3, \ldots$ *vorhanden sein.*

Es existiert eine eindeutig bestimmte für $\Re s > 0$ *reguläre und beschränkte Funktion* $\Phi(s)$, *welche den Bedingungen*

$$\Phi(n) = \mu_n \quad \text{für} \quad n = 1, 2, 3, \ldots$$

genügt.

Wenn $z(t)$ *die Lösung* $(z(0) = 0)$ *des Momentenproblems*

$$\int_0^1 t^k\, dz(t) = \mu_k \qquad (k = 0, 1, 2, \ldots;\ \mu_0 = 1) \tag{4'}$$

bedeutet, so ist

$$\Phi(s) = \int_0^1 t^s\, dz(t), \qquad \Re s > 0$$

eine Darstellung der Funktion $\Phi(s)$.

Es existiert

$$\lim_{\sigma\to+0} \Phi(\sigma + i\lambda) = \Phi(i\lambda) \qquad (s = \sigma + i\lambda)$$

gleichmäßig für $-\infty < \lambda < \infty$.

---

[17]) Vgl. G. H. Hardy, „On two theorems of F. Carlson und S. Wiegert", Acta Math. 42, S. 327—339.

Die Zahlen (1) *verteilen sich gewiß asymptotisch stetig, wenn*

$$(17) \qquad \Phi(0)=1 \quad und \quad \lim_{x\to\infty}\frac{1}{x}\int_0^x |\Phi(i\lambda)|\,d\lambda=0.$$

Die entsprechende Verteilungsfunktion ist $z(t)$, *welches nach den Bedingungen* (17) *notwendig stetig ist.*

Wenn wir in der Formel (11′) von **9** die Substitution $x=\lg\frac{1}{t}$ ausführen, so erhalten wir ein Integral, mit dessen Hilfe wir das Ergebnis (16) auch so ausdrücken können:

Es sei $\psi(x)$ *eine für* $x\geqq 0$ *monotone Funktion, welche für* $x\to\infty$ *endlich bleibt, und*

$$\varrho(\lambda)=\left|\int_0^\infty e^{i\lambda x}\,d\psi(x)\right|,$$

wenn

$$\lim_{x\to\infty}\frac{1}{x}\int_0^x \varrho(\lambda)\,d\lambda=0,$$

dann ist die Funktion $\varphi(x)$ *gewiß für* $x\geqq 0$ *auch stetig*[18]).

§ 3.

Kriterien für asymptotische stetige Verteilung mit Hilfe der Fouriermomente ω_k.

12. In diesem Paragraphen beweisen wir den weiter unten angeführten Satz IV.

Wir beginnen mit dem Beweis eines Hilfssatzes.

Es sei $z_0(t)$ eine für $0\leqq t\leqq 2\pi$ nie abnehmende Funktion und

$$(18) \qquad \omega_k'=\frac{1}{2\pi}\int_0^{2\pi} e^{-ikt}\,dz_0(t).$$

Wir betrachten die schon von Herrn G. Herglotz[19]) benutzte Funktion

$$(19) \qquad F(x)=\omega_0'+2\omega_1' x+2\omega_2' x^2+\dots.$$

Diese Potenzreihe konvergiert für $|x|<1$, und man findet leicht aus (18) und (19) die Darstellung

[18]) Unter leicht einschränkenden Nebenbedingungen kann man diesen Satz auch für die ganze reelle Achse ($-\infty<x<\infty$) ausdehnen.

[19]) „Über Potenzreihen mit positivem, reellen Teil im Einheitskreis", Sächsische Sitzungsberichte **63** (1911), S. 501–511.

(20) $$F(x) = \frac{1}{2\pi} \int_0^{2\pi} \frac{e^{it}+x}{e^{it}-x}\, dz_0(t) \qquad |x| < 1,$$

ein verallgemeinertes Poissonsches Integral.

Hilfssatz II. *Für* $0 < \vartheta < 2\pi$ *und fest ist*

$$\lim_{r \to 1} (1-r)\, F(r e^{i\vartheta}) = \frac{1}{\pi} \{z_0(\vartheta + 0) - z_0(\vartheta - 0)\}$$

und $(\vartheta = 0)$

$$\lim_{r \to 1} (1-r)\, F(r) = \frac{1}{\pi} \{z_0(+0) - z_0(0) + z_0(2\pi) - z_0(2\pi - 0)\}.$$

Aus (20) folgt

$$(1-r)\, F(r e^{i\vartheta}) = \frac{1}{2\pi} \int_0^{2\pi} \frac{e^{i(t-\vartheta)}+r}{e^{i(t-\vartheta)}-r} \cdot (1-r)\, dz_0(t),$$

und daraus folgen die Gleichungen des Hilfssatzes II nach ähnlichen Schlüssen, wie sie in 8 zum Beweis des Hilfssatzes I verwendet wurden. Die Sache ist derart einfach, daß wir dabei nicht mehr verweilen.

Insbesondere gilt:

Damit $z_0(t)$ *für* $0 \leqq t \leqq 2\pi$ *auch stetig sei, ist notwendig und hinreichend, daß*

(21) $$\lim_{r \to 1} (1-r)\, F(r e^{i\vartheta}) = 0$$

für jedes feste ϑ *mit* $0 \leqq \vartheta < 2\pi$.

Ein bekannter Satz über Potenzreihen lautet folgendermaßen:

Die Koeffizienten der Potenzreihe

$$\varphi(x) = \sum_{\nu=1}^{\infty} c_\nu x^\nu$$

mögen der Gleichung $\lim_{n \to \infty} \frac{c_1 + c_2 + \dots + c_n}{n} = 0$ *genügen. Dann ist dieselbe für* $|x| < 1$ *gewiß konvergent* und

$$\lim_{r \to 1} (1-r)\, \varphi(r) = 0.$$

Indem wir diesen Satz auf die Potenzreihe (19) anwenden, folgt, daß (21) gewiß dann erfüllt ist, wenn

(22) $$\lim_{n \to \infty} \frac{|\omega_1'| + |\omega_2'| + \dots + |\omega_n'|}{n} = 0,$$

und *aus* (22) *folgt also a fortiori die Stetigkeit von* $z_0(t)$ *für* $0 \leqq t \leqq 2\pi$.

Um zu unserer ursprünglichen Aufgabe zu gelangen, gehen wir wieder von den Zahlen (1) und den Gleichungen (5) in **2** aus. Wir sahen in **7**, daß, wenn (5) vorhanden sind, sich das Momentenproblem (4′) und also auch (5′) durch eine monotone (nie abnehmende) Funktion $z(t)$ lösen läßt. Aus (5′) folgt

$$\frac{\overline{\omega}_k}{2\pi} = \frac{1}{2\pi}\int_0^{2\pi} e^{-ikt}\, dz\left(\frac{t}{2\pi}\right) = \omega_k',$$

wenn wir $z_0(t) = z\left(\frac{t}{2\pi}\right)$ setzen, und wenn wir nun das Ergebnis (22) mit dem Satz II verbinden, so folgt, *daß*

$$(23) \qquad \lim_{n\to\infty} \frac{|\omega_1| + |\omega_2| + \ldots + |\omega_n|}{n} = 0$$

hinreichend für die asymptotische stetige Verteilung der Zahlen (1) *ist.*

Wir werden weiter noch beweisen, daß die Bedingung (23) auch notwendig ist.

13. Es sei $f(t)$ eine für $-\infty < t < \infty$ erklärte reelle Funktion mit der Periode 2π und in $0 \leqq t \leqq 2\pi$ von beschränkter Schwankung.

Es seien a_k und b_k die Fourierkoeffizienten von $f(t)$. Herr F. Riesz[20]) zeigt an einem Beispiel, daß für eine solche auch noch überall stetige Funktion $f(t)$ nicht notwendig

$$(24) \qquad \lim_{n\to\infty} n a_n = 0, \qquad \lim_{n\to\infty} n b_n = 0$$

zugleich gelten.

Herr L. Neder[21]) beweist aber umgekehrt, daß aus unseren ersten Voraussetzungen für $f(t)$ und (24) auch die Stetigkeit von $f(t)$ folgt, oder genauer, daß $f(t)$ nur hebbare Unstetigkeiten aufweisen kann.

In einer neueren Arbeit hat Herr S. Sidon[22]) bewiesen, daß, wenn unser $f(t)$ auch stetig ist, zwar nicht (24), aber gewiß

$$(25) \quad \lim_{n\to\infty} \frac{|a_1| + 2|a_2| + \ldots + n|a_n|}{n} = 0, \qquad \lim_{n\to\infty} \frac{|b_1| + 2|b_2| + \ldots + n|b_n|}{n} = 0$$

gelten müssen.

Wir wollen in diesem Zusammenhang zunächst beweisen, daß auch umgekehrt die Bedingungen (25) hinreichend für die Stetigkeit (hebbare Unstetigkeiten ausgenommen) von $f(t)$ sind.

---

[20]) a. a. O.
[21]) a. a. O.
[22]) a. a. O.

In der Tat sei

$$\omega'_k = \frac{1}{2\pi}\int_0^{2\pi} e^{-ikt}\, d f(t).$$

Es folgt durch partielle Integration $(f(0) = f(2\pi)!)$

(26) $$\omega'_k = \frac{ik}{2\pi}\int_0^{2\pi} f(t)\, e^{-ikt}\, dt = \frac{i}{2}(k a_k - i k b_k)$$

und

$$|\omega'_k| \leqq |k a_k| + |k b_k|,$$

so daß aus (25) auch

$$\lim_{n\to\infty} \frac{|\omega'_1| + \ldots + |\omega'_n|}{n} = 0$$

folgt. Wenn wir

$$F(x) = \omega'_0 + 2\omega'_1 x + 2\omega'_2 x^2 + \ldots$$

setzen, so folgt aus der letzten Limesgleichung und einem schon benutzten Satz, daß

(27) $$\lim_{r\to 1}(1-r)\, F(r e^{i\vartheta}) = 0$$

für $0 \leqq \vartheta < 2\pi$. Es ist aber $f(t)$ von beschränkter Schwankung, also können wir setzen

(28) $$f(t) = z_1(t) - z_2(t),$$

wo z_1 und z_2 für $0 \leqq t \leqq 2\pi$ nie abnehmen.

Es seien $F_1(x)$ und $F_2(x)$ die der Funktion $F(x)$ analogen, mit den Momenten von $z_1(t)$ bzw. $z_2(t)$ gebildeten Funktionen. Nach (28) folgt

$$F(x) = F_1(x) - F_2(x),$$

und aus (27) erhält man

$$\lim_{r\to 1}(1-r)\, F_1(r e^{i\vartheta}) = \lim_{r\to 1}(1-r)\, F_2(r e^{i\vartheta}) \quad \text{für} \quad 0 \leqq \vartheta < 2\pi.$$

Nach dem Hilfssatz II ist also für $0 < \vartheta < 2\pi$

$$z_1(\vartheta+0) - z_1(\vartheta-0) = z_2(\vartheta+0) - z_2(\vartheta-0),$$

also nach (28)

$$f(\vartheta+0) = f(\vartheta-0).$$

Für $\vartheta = 0$ folgt

$$\begin{aligned} &z_1(+0) - z_1(0) + z_1(2\pi) - z_1(2\pi-0) \\ &= z_2(+0) - z_2(0) + z_2(2\pi) - z_2(2\pi-0), \end{aligned}$$

was nach $f(0) = f(2\pi)$ und (28) auf

$$f(+0) = f(2\pi - 0)$$

herauskommt.

Die Funktion $f(t)$ kann also nur hebbare Unstetigkeiten aufweisen.

14. Wir können aus dem Satze von Herrn Sidon sehr leicht die Behauptung von **12** ableiten, welche aussagt, daß (23) nicht nur hinreichend, sondern auch notwendig für die asymptotische stetige Verteilung der Zahlen (1) ist.

In der Tat, wenn $z(t)$ die stetige Verteilungsfunktion ist, so ist auch das dortige $z_0(t) = z\left(\frac{t}{2\pi}\right)$ stetig für $0 \leqq t \leqq 2\pi$.

$$f(t) = z_0(t) - \frac{1}{2\pi} t$$

ist stetig und von beschränkter Schwankung für $0 \leqq t \leqq 2\pi$. Außerdem ist $f(0) = f(2\pi)$. Unter diesen Voraussetzungen ist ω_k' in (18) mit dem ω_k' in (26) identisch, also

$$|\omega_k'| \leqq |k a_k| + |k b_k| \qquad (k = 1, 2, 3, \ldots)$$

und da $f(t)$ stetig ist, so folgt nach den Sidonschen Gleichungen (25) auch (22) und folglich auch (23).

Zusammenfassend erhalten wir den folgenden

Satz IV. *Notwendig und hinreichend für die asymptotische stetige Verteilung* mod 1 *der Zahlen* $x_{1n}, x_{2n}, \ldots, x_{nn}$ $(n = 1, 2, 3, \ldots)$ *ist die Existenz der Grenzwerte*

$$\lim_{n\to\infty} \frac{e^{2\pi i k x_{1n}} + e^{2\pi i k x_{2n}} + \ldots + e^{2\pi i k x_{nn}}}{n} = \omega_k \qquad (k = 1, 2, 3, \ldots)$$

mit der Bedingung

$$\lim_{n\to\infty} \frac{|\omega_1| + |\omega_2| + \ldots + |\omega_n|}{n} = 0.$$

§ 4.

Eine Anwendung des Satzes

15. Es seien

$$x_{1n}, x_{2n}, \ldots, x_{nn}$$

n Zahlen der Strecke $0 \ldots 1$, die sich für $n \to \infty$ asymptotisch stetig nach $z(x)$ verteilen mögen.

Wenn $\psi(x)$ irgendeine Verteilungsfunktion ist, so folgt sofort aus Satz I, daß sich die Zahlen

$$\psi(x_{1n}), \psi(x_{2n}), \ldots, \psi(x_{nn})$$

asymptotisch stetig nach $z(\psi_{-1}(x))$ verteilen, wenn $\psi_{-1}(x)$ die Umkehrungsfunktion von $\psi(x)$ bedeutet. Insbesondere sind die Zahlen $z(x_{1n}), z(x_{2n}), \ldots, z(x_{nn})$ asymptotisch gleichverteilt für $n \to \infty$[23]).

Nicht ganz so selbstverständlich ist die Antwort auf die Frage: *Wie verteilen sich die Zahlen* $\frac{1}{x_{1n}}, \frac{1}{x_{2n}}, \ldots, \frac{1}{x_{nn}}$ mod 1? Man muß natürlich dabei voraussetzen, daß keines der $x_{\nu n}$ verschwindet.

Wir werden zeigen, daß sich diese Zahlen asymptotisch stetig mod 1 *nach der Verteilungsfunktion*

$$z^*(x) = \sum_{n=1}^{\infty} \left\{ z\left(\frac{1}{n}\right) - z\left(\frac{1}{x+n}\right) \right\}$$

verteilen.

Der allgemeinen Methode folgend, wollen wir zunächst zeigen, daß

$$(29) \qquad \left\{ \begin{array}{l} \omega_k^* = \displaystyle\int_0^1 e^{2\pi i \frac{k}{x}} dz(x) \\ \text{gesetzt, dann} \\ \displaystyle\lim_{n\to\infty} \frac{e^{2\pi i \frac{k}{x_{1n}}} + e^{2\pi i \frac{k}{x_{2n}}} + \cdots + e^{2\pi i \frac{k}{x_{nn}}}}{n} = \omega_k^* \end{array} \right.$$

ist. Wir bemerken, daß diese Behauptung nicht direkt aus Satz I für $f(x) = e^{2\pi i \frac{k}{x}}$ folgt, denn diese Funktion ist in ihrem Real- und Imaginärteil von unbeschränkter Schwankung. Darum muß man etwas vorsichtig sein.

Es sei $\delta > 0$ und < 1, genügend klein, so daß $z(\delta) < 1$. Es sei m_δ die Anzahl derjenigen Zahlen $x_{1n}, x_{2n}, \ldots, x_{nn}$, welche der Strecke $0 \ldots \delta$ angehören, und wir setzen $m'_\delta = n - m_\delta$ für die Anzahl der übrigen Zahlen.

Es sei $\varepsilon > 0$ genügend klein, so daß $z(\delta) + \varepsilon < 1$, dann folgt aus $\lim\limits_{n\to\infty} \frac{m_\delta}{n} = z(\delta)$, daß $\frac{m_\delta}{n} < z(\delta) + \varepsilon$ für $n > n_0(\delta, \varepsilon)$.

Da $z(\delta) + \varepsilon < 1$, so folgt $\frac{m_\delta}{n} < 1$, also $m'_\delta > 0$ für diese Werte von n. Wir haben ferner

---

[23]) Ich möchte in diesem Zusammenhang bemerken, daß wenn eine Folge $x_1, x_2, x_3, \ldots$ von Zahlen der Strecke $0 \ldots 1$ gegeben ist, welche diese Strecke überall dicht bedeckt, man immer in passender Weise die Zahlen dieser Folge umordnen kann, so daß die umgeordnete Folge sich nach einer a priori beliebig gegebenen Verteilungsfunktion verteilt. Vgl. hierzu J. v. Neumann, Gleichmäßig dichte Zahlenfolgen (ungarisch), Mathematikai és Physikai Lapok **32**.

$$\frac{1}{n}\sum_{\nu=1}^{n} e^{2\pi i\frac{k}{x_{\nu n}}} - \frac{1}{m'_\delta}\sum_{\delta<x_{\nu n}} e^{2\pi i\frac{k}{x_{\nu n}}} = \frac{1}{n}\sum_{x_{\nu n}\leqq\delta} + \frac{1}{n}\sum_{\delta<x_{\nu n}} - \frac{1}{m'_\delta}\sum_{\delta<x_{\nu n}}$$

$$= \frac{1}{n}\sum_{x_{\nu n}\leqq\delta} - \frac{m_\delta}{n}\cdot\frac{1}{m'_\delta}\sum_{\delta<x_{\nu n}},$$

also

$$(30)\qquad \left|\frac{1}{n}\sum_{\nu=1}^{n} e^{2\pi i\frac{k}{x_{\nu n}}} - \frac{1}{m'_\delta}\sum_{\delta<x_{\nu n}} e^{2\pi i\frac{k}{x_{\nu n}}}\right| \leqq \frac{m_\delta}{n} + \frac{m_\delta}{n} < 2(z(\delta)+\varepsilon)$$

$$\text{für } n > n_0(\delta, \varepsilon).$$

Man sieht nun, daß sich die Zahlen $x_{1n}, \ldots, x_{nn}$, welche der Strecke $\delta \ldots 1$ angehören, auf dieser Strecke asymptotisch nach der Funktion $\frac{z(x)-z(\delta)}{1-z(\delta)}$ verteilen.

Es ist klar, was wir darunter verstehen, und daß der Satz I auch für ein solches von $0 \ldots 1$ verschiedenes Intervall gilt. Daraus folgt

$$(31)\qquad \left|\frac{1}{m'_\delta}\sum_{\delta<x_{\nu n}} e^{2\pi i\frac{k}{x_{\nu n}}} - \frac{1}{1-z(\delta)}\int_\delta^1 e^{2\pi i\frac{k}{x}}\,dz(x)\right| < \varepsilon$$

$$\text{für } n > n_1(\delta, \varepsilon) \quad (n_1 > n_0).$$

Endlich ist für ein genügend kleines $\delta < \delta_0(\varepsilon)$

$$\left|\int_0^1 e^{2\pi i\frac{k}{x}}\,dz(x) - \frac{1}{1-z(\delta)}\int_\delta^1 e^{2\pi i\frac{k}{x}}\,dz(x)\right| < \varepsilon.$$

Aus dieser Ungleichung in Verbindung mit (30) und (31) folgt

$$\left|\frac{1}{n}\sum_{\nu=1}^{n} e^{2\pi i\frac{k}{x_{\nu n}}} - \int_0^1 e^{2\pi i\frac{k}{x}}\,dz(x)\right| < 2(z(\delta)+\varepsilon) + 2\varepsilon$$

$$\text{für } \delta < \delta_0(\varepsilon) \text{ und } n > n_1(\delta, \varepsilon)$$

und (29) ist bewiesen.

16. Wir müssen also jetzt aus den Gleichungen

$$(32)\qquad \int_0^1 e^{2\pi i k x}\,dz^*(x) = \int_0^1 e^{2\pi i\frac{k}{x}}\,dz(x) \qquad (k = 0, 1, 2, \ldots)$$

die Funktion $z^*(x)$ bestimmen.

Wir könnten hier nach Satz IV beweisen, daß $z^*(x)$ stetig ist, aber alles ist hier überflüssig, denn man kann leicht nachrechnen, daß

$$(33)\qquad z^*(x)=\sum_{n=1}^{\infty}\left\{z\left(\frac{1}{n}\right)-z\left(\frac{1}{n+x}\right)\right\}\ ^{24)}$$

diesen Gleichungen genügt. In der Tat ist

$$\int_0^1 e^{2\pi i\frac{k}{x}}dz(x)=\sum_{n=1}^{\infty}\int_{\frac{1}{n+1}}^{\frac{1}{n}} e^{2\pi i\frac{k}{x}}dz(x)=-\sum_{n=1}^{\infty}\int_0^1 e^{2\pi iky}dz\left(\frac{1}{n+y}\right)$$

$$=\sum_{n=1}^{\infty}\int_0^1 e^{2\pi iky}d\left\{z\left(\frac{1}{n}\right)-z\left(\frac{1}{n+y}\right)\right\}=\int_0^1 e^{2\pi iky}dz^*(y)\ ^{25)},$$

also (32) ist bewiesen.

Wenn wir insbesondere von asymptotisch gleichverteilten Zahlen ausgehen, also $z(x)=x$, so folgt aus (33)

$$z^*(x)=\sum_{n=1}^{\infty}\left(\frac{1}{n}-\frac{1}{n+x}\right)=c+\frac{\Gamma'(1+x)}{\Gamma(1+x)}=\int_0^1\frac{1-t^x}{1-t}dt.$$

Nach dieser Funktion verteilen sich also die Zahlen

$$\frac{n}{1},\frac{n}{2},\ldots,\frac{n}{n}\ \mathrm{mod}\ 1\ ^{26)},$$

denn die inversen Zahlen $\frac{1}{n},\frac{2}{n},\ldots,\frac{n}{n}$ sind asymptotisch gleichverteilt.

Für $z(x)=x^k$ (k ganz >1) folgt, daß sich die Zahlen

$$\sqrt[k]{\frac{n}{1}},\sqrt[k]{\frac{n}{2}},\ldots,\sqrt[k]{\frac{n}{n}}\ \mathrm{mod}\ 1$$

nach der Funktion

$$z_k^*(x)=\sum_{n=1}^{\infty}\frac{1}{n^k}+\frac{(-1)^{k-1}}{(k-1)!}\frac{d^k}{dx^k}\log\Gamma(x+1)$$

verteilen.

§ 5.

Anwendung auf die Verteilung der Zahlen $\frac{\varphi(1)}{1},\frac{\varphi(2)}{2},\frac{\varphi(3)}{3},\ldots$. [27]

17. In diesem letzten Teil wollen wir als Anwendung des Satzes III beweisen, *daß sich diese Zahlenfolge asymptotisch stetig auf der Strecke* 0 ... 1 *verteilt.*

---

[24] Diese Reihe hat nicht negative Glieder und konvergiert gleichmäßig für $0\leq x\leq 1$; denn $1=\sum\left\{z\left(\frac{1}{n}\right)-z\left(\frac{1}{n+1}\right)\right\}$ ist eine Majorante.

[25] Man darf Summation und Integration vertauschen, was nach partieller Integration sofort folgt.

[26] Vgl. G. Pólya a. a. O.

[27] Das Symbol $\varphi(n)$ bedeutet die Eulersche Funktion.

Wir beginnen mit einer ersten Verteilungseigenschaft dieser Folge: *Sie bedeckt überall dicht die Strecke* 0 ... 1.

Der Beweis ist ganz einfach. Es ist $\frac{\varphi(n)}{n} = \prod_{p\mid n}\left(1-\frac{1}{p}\right)$ und es genügt also zu zeigen, daß die Zahlen der Gestalt

$$\left(1-\frac{1}{p_1}\right)\left(1-\frac{1}{p_2}\right)\dots\left(1-\frac{1}{p_r}\right),$$

($p_1, p_2, \dots, p_r$ sind verschiedene Primzahlen in irgendwelcher Anzahl), die Strecke 0 ... 1 überall dicht bedecken.

Die Funktion $\log x$ ist für $x \geqq 1$ stetig und darum genügt es zu diesem Zweck zu zeigen, daß man mit den beliebigen endlichen Teilsummen, die mit Gliedern aus der Folge

$$\log\frac{2}{1},\ \log\frac{3}{2},\ \log\frac{5}{4},\ \log\frac{7}{6},\dots,\ \log\frac{p}{p-1},\dots$$

gebildet sind, die positive Halbachse überall dicht bedecken kann. Das folgt aber sofort aus der Divergenz der Reihe $\sum_p \log\frac{p}{p-1}$, denn wenn $x > \varepsilon > 0$ fest gegeben sind, so ist $0 < \log\frac{p}{p-1} < \varepsilon$ für $p \geqq q$ und mindestens eine Teilsumme der Reihe

$$\sum_{p\geqq q}\log\frac{p}{p-1}$$

fällt zwischen $x-\varepsilon$ und $x+\varepsilon$, denn die Reihe divergiert und ihre Glieder sind positiv und kleiner als ε.

18. Diese Eigenschaft benutzen wir im folgenden nicht, wohl aber einen Satz von Herrn Prof. J. Schur, welchen ich aus seinen Vorlesungen vom W.-S. 1923/24 kenne und der aussagt: *Für jeden komplexen Wert von s ist*

$$\lim_{n\to\infty}\frac{\left(\frac{\varphi(1)}{1}\right)^s+\left(\frac{\varphi(2)}{2}\right)^s+\dots+\left(\frac{\varphi(n)}{n}\right)^s}{n} = \Phi(s) \tag{34}$$

und $\Phi(s)$ ist ganz transzendent mit der Produktdarstellung

$$\Phi(s) = \prod_p\left\{1-\frac{1}{p}+\frac{1}{p}\left(1-\frac{1}{p}\right)^s\right\}. \tag{35}$$

p durchläuft alle Primzahlen und das Produkt (35) konvergiert absolut und gleichmäßig in jedem endlichen Gebiet der s-Ebene.

Wir benutzen diesen Satz nur für positives ganzes $s = k$ und er zeigt uns, daß die Grenzwerte μ_k vorhanden sind und

$$\Phi(k) = \mu_k \qquad (k = 1, 2, 3, \dots).$$

Außerdem ist für $s=\sigma+i\lambda$ und $\sigma\geqq 0$

$$|\Phi(s)|=\prod_p\left|1-\frac{1}{p}+\frac{1}{p}\left(1-\frac{1}{p}\right)^s\right|\leqq\prod_p\left\{1-\frac{1}{p}+\frac{1}{p}\left(1-\frac{1}{p}\right)^\sigma\right\}\leqq 1$$

und $\Phi(s)$ ist also genau die Funktion des Satzes III.

Die erste Bedingung (17) $\Phi(0)=1$ ist offenbar erfüllt und man muß nun auch die zweite nachweisen.

Aus

$$\Phi(i\lambda)=\prod_p\left(1-\frac{1}{p}+\frac{1}{p}e^{-i\lambda\log\frac{p}{p-1}}\right)$$

folgt

$$\log|\Phi(i\lambda)|=\sum_p\log\left|1-\frac{1}{p}+\frac{1}{p}e^{i\lambda\lg\frac{p}{p-1}}\right|$$

$$=\Re\sum_p\log\left(1-\frac{1}{p}+\frac{1}{p}e^{i\lambda\lg\frac{p}{p-1}}\right).$$

Für $|x|<1$ ist $\log(1-x)=-x+x^2\cdot\alpha(x)$ $(\lg 1=0)$ und für $|x|\leqq\frac{2}{3}$ ist $|\alpha(x)|<2$. Daraus folgt

$$\Re\sum_p\log\left(1-\frac{1}{p}+\frac{1}{p}e^{i\lambda\lg\frac{p}{p-1}}\right)=\log\left|\frac{1}{2}+\frac{1}{2}e^{i\lambda\lg 2}\right|-\Re\sum_{p\geqq 3}\frac{1-e^{i\lambda\lg\frac{p}{p-1}}}{p}$$

$$+\Re\sum_{p\geqq 3}\frac{\left(1-e^{i\lambda\lg\frac{p}{p-1}}\right)^2}{p^2}\cdot\alpha\left(\frac{1-e^{i\lambda\lg\frac{p}{p-1}}}{p}\right).$$

Für $p\geqq 3$ ist $\left|\frac{1-e^{i\lambda\lg\frac{p}{p-1}}}{p}\right|\leqq\frac{2}{3}$ und die letzte Reihe auf der rechten Seite besitzt die von λ unabhängige Majorante $\sum\frac{8}{p^2}$. Also ist

$$\log|\Phi(i\lambda)|=\log\left|\cos\frac{\lambda\lg 2}{2}\right|-2\sum_p\frac{1}{p}\sin^2\left(\frac{\lambda}{2}\lg\frac{p}{p-1}\right)+0(1)$$

und endlich

$$|\Phi(i\lambda)|=\left|\cos\frac{\lambda\lg 2}{2}\right|\cdot e^{-2\sum\limits_p\frac{1}{p}\sin^2\left(\frac{\lambda}{2}\lg\frac{p}{p-1}\right)+0(1)} \tag{36}$$

für $-\infty<\lambda<+\infty$.

Die Bedingung (17) des Satzes III ist erfüllt, wenn

$$\lim_{x\to\infty}\frac{1}{x}\int_0^x e^{-2\sum\frac{1}{p}\sin^2\left(\frac{\lambda}{2}\lg\frac{p}{p-1}\right)}d\lambda=0.$$

13*

Wir ersetzen λ durch $2x$ und dieze Gleichung ist à fortiori erfüllt, wenn

$$(37) \qquad \lim_{x\to\infty} \frac{1}{x}\int_0^x e^{-\sum_p \frac{1}{p}\sin^2\left(x\lg\frac{p}{p-1}\right)} dx = 0 .$$

19. Wir beweisen zunächst: *Die Zahlen der Folge*

$$\lg\frac{2}{1},\ \lg\frac{3}{2},\ \lg\frac{5}{4},\ \ldots,\ \lg\frac{p}{p-1},\ \ldots$$

sind linear unabhängig, das heißt: eine Gleichung

$$\sum c_p \log\frac{p}{p-1} = 0$$

mit ganzen rationalen c_p und wo p endlich viele verschiedene Primzahlwerte durchläuft, ist nur für $c_p = 0$ möglich.

Angenommen es wären nicht alle $c_p =$ Null, dann darf man $c_p \neq 0$ für alle c_p voraussetzen. Die Gleichung der Behauptung kommt auf

$$\Pi\left(\frac{p}{p-1}\right)^{c_p} = 1 \quad \text{oder} \quad \Pi p^{c_p} = \Pi(p-1)^{c_p}$$

hinaus.

Es sei q der größte Primzahlwert von p. Beide Glieder der letzten Gleichung sind rationale Brüche, der erste sogar irreduzibel und q teilt seinen Nenner oder den Zähler.

Dies trifft nicht auch für den rechten Bruch ein und darum ist unsere Annahme absurd.

20. Wir können nun (37) beweisen. Es sei P eine feste Primzahl. Es ist

$$(38) \qquad \frac{1}{x}\int_0^x e^{-\sum_p \frac{1}{p}\sin^2\left(x\lg\frac{p}{p-1}\right)} dx \leqq \frac{1}{x}\int_0^x e^{-\sum_{p\leqq P} \frac{1}{p}\sin^2\left(x\lg\frac{p}{p-1}\right)} dx$$

und wir beweisen folgende zwei Behauptungen

1°. *Der Ausdruck auf der rechten Seite von* (38) *strebt für* $x \to \infty$ *gegen einen positiven Grenzwert* $A^{(P)}$.

2°. *Es ist* $\lim_{P\to\infty} A^{(P)} = 0$.

Daraus folgt auch (37), denn wenn L den lim sup des Ausdruckes auf der linken Seite von (38) bedeutet, so folgt aus (38) und 1°.

$$L \leqq A^{(P)}$$

und P war beliebig, also nach 2° für $P \to \infty$

$$L = 0$$

also (37).

Wir beginnen mit 1°. Wir betrachten die Fourierentwicklung der Funktion

$$e^{-\frac{1}{p}\sin^2 z} = a_0^{(p)} + 2a_1^{(p)}\cos z + 2a_2^{(p)}\cos 2z + \ldots = \sum_{-\infty}^{+\infty} a_n^{(p)} e^{izn} \quad (a_n^{(p)} = a_{-n}^{(p)})$$

und $\sum_{-\infty}^{\infty} |a_n^{(p)}|$ konvergiert, weil die dargestellte Funktion eine stetige zweite Ableitung besitzt[28]).

Wir ersetzen in dieser Reihenentwickelung z durch $x \log \frac{p}{p-1}$, lassen p die Werte 2, 3, 5, 7, ..., P durchlaufen und multiplizieren miteinander die herauskommenden Gleichungen.

Man erhält auf diese Weise die verallgemeinerte Fourierentwickelung

$$(39) \qquad \psi(x) = e^{-\sum\limits_{p \leqq P} \frac{1}{p} \sin^2\left(x \lg \frac{p}{p-1}\right)} = A_0 + \sum_{\nu=1}^{\infty} A_\nu e^{i\lambda_\nu x},$$

wo alle λ_ν untereinander und von 0 verschieden sind[29]).

Aus der Behauptung in **19** über die Unabhängigkeit der Zahlen $\log \frac{p}{p-1}$ folgt

$$(40) \qquad A_0 = a_0^{(2)} a_0^{(3)} a_0^{(5)} \ldots a_0^{(P)}$$

und wir wollen jetzt zeigen, daß gemäß 1°.

$$(41) \qquad \lim_{x \to \infty} \frac{1}{x} \int_0^x \psi(x)\, dx = A_0.$$

In der Tat, $\sum_{\nu=1}^{\infty} |A_\nu|$ ist gewiß konvergent. Es sei $\varepsilon > 0$ gegeben, dann bestimmen wir m mit $\sum_{\nu=m+1}^{\infty} |A_\nu| < \frac{\varepsilon}{2}$.

Es folgt dann aus (39)

$$\left|\psi(x) - A_0 - \sum_{\nu=1}^{m} A_\nu e^{i\lambda_\nu x}\right| < \frac{\varepsilon}{2}$$

und also

$$\left|\frac{1}{x}\int_0^x \psi(x)\, dx - A_0 - \sum_{\nu=1}^{m} A_\nu \frac{1}{x}\int_0^x e^{i\lambda_\nu x}\, dx\right| < \frac{\varepsilon}{2}$$

oder

$$\left|\frac{1}{x}\int_0^x \psi(x)\, dx - A_0 - \sum_{\nu=1}^{m} A_\nu \frac{e^{i\lambda_\nu x} - 1}{i\lambda_\nu x}\right| < \frac{\varepsilon}{2}$$

---

[28]) Vgl. L. Bieberbach, Differential- und Integralrechnung II, S. 81.

[29]) Die Funktion $\psi(x)$ ist fastperiodisch.

und also für genügend großes x $\left|\frac{1}{x}\int\limits_0^x \psi(x)\,dx - A_0\right| < \varepsilon$ und (41) ist bewiesen. Also ist A_0 das $A^{(P)}$ von 1°. Aus (40) wird nun auch 2°. folgen.

Es ist

$$a_0^{(p)} = \frac{1}{2\pi}\int\limits_0^{2\pi} e^{-\frac{1}{p}\sin^2 z}\,dz = \frac{1}{\pi}\int\limits_0^{\pi} e^{-\frac{1}{p}\sin^2 z}\,dz < 1$$

und die Behauptung 2° ist bewiesen, wenn wir zeigen, daß das Produkt $\prod_p a_0^{(p)}$ gegen Null divergiert, oder daß die Reihe mit positiven Gliedern

$$\sum_p (1 - a_0^{(p)})$$

divergiert.

Dies folgt nun sofort, denn für $0 \leqq x \leqq 1$ ist $1 - e^{-x} \geqq \frac{x}{2}$ und also

$$\sum_{p \leqq P}(1 - a_0^{(p)}) = \frac{1}{\pi}\sum_{p \leqq P}\int\limits_0^{\pi}\left(1 - e^{-\frac{1}{p}\sin^2 z}\right)dz \geqq \frac{1}{\pi}\sum_{p \leqq P}\int\limits_0^{\pi}\frac{\sin^2 z}{2p}\,dz,$$

also

$$\sum_{p \leqq P}(1 - a_0^{(p)}) \geqq \frac{1}{2\pi}\int\limits_0^{\pi}\sin^2 z\,dz \cdot \sum_{p \leqq P}\frac{1}{p}$$

und da $\sum_p \frac{1}{p}$ divergiert, ist alles bewiesen.

(Eingegangen am 7. Juli 1926.)

Zusatz bei der Korrektur (Februar 1928).

Ich nahm erst letztens davon Kenntnis, daß Herr Norbert Wiener[30]) in einer neueren Arbeit den Satz von § 3 schon bewiesen hat, und zwar in einer wenig abweichenden Gestalt.

Er benutzt statt der Sidonschen Bedingungen (25) von § 3 die Limesgleichung

$$\lim_{n \to \infty}\frac{1}{n}\sum_{\nu=1}^{\nu=n}\nu^2(a_\nu^2 + b_\nu^2) = 0, \qquad (42)$$

[30]) *"The quadratic variation of a function and its Fourier coefficients"*, Journal of Math. and Phys. Massachusetts **3** (1924), pp. 72–94, insbesondere p. 81.

und zeigt, daß dieselbe *notwendig und hinreichend ist für die Stetigkeit der periodischen Funktion $f(x)$ von beschränkter Schwankung, deren Fourierkoeffizienten a_n und b_n sind.*

Es sind aber $n|a_n|$ und $n|b_n|$ beschränkt, und die Bedingung (42) erscheint gleichbedeutend mit den Bedingungen (25) infolge des einfachen Satzes:

Es sei

$$c_1, c_2, c_3, \ldots$$

eine beschränkte Folge nichtnegativer Zahlen.

Aus

$$\lim_{n\to\infty} \frac{c_1 + c_2 + \ldots + c_n}{n} = 0 \tag{43}$$

folgt

$$\lim_{n\to\infty} \frac{c_1^2 + c_2^2 + \ldots + c_n^2}{n} = 0 \tag{44}$$

und umgekehrt.

Beweis. Wir dürfen natürlich $0 \leqq c_n \leqq 1$ annehmen und dann folgt selbstverständlich (44) aus (43).

Die Umkehrung folgt aus der Ungleichung

$$\frac{c_1 + c_2 + \ldots + c_n}{n} \leqq \sqrt{\frac{c_1^2 + c_2^2 + \ldots + c_n^2}{n}},$$

einer Anwendung der Schwarzschen Ungleichung.

Es genügt nacheinander $n|a_n|$ und $n|b_n|$ für c_n zu setzen, um unsere obige Behauptung zu bekommen.

Aus demselben Grunde ist die Bedingung des Satzes IV mit der Bedingung

$$\lim_{n\to\infty} \frac{|\omega_1|^2 + |\omega_2|^2 + \ldots + |\omega_n|^2}{n} = 0$$

gleichbedeutend.

Reprinted from
Math. Z. **28** (1928), pp. 171–199

ON ASYMPTOTIC DISTRIBUTIONS OF ARITHMETICAL FUNCTIONS*

BY

I. J. SCHOENBERG

INTRODUCTION

1. The present note was suggested by recent work of H. Davenport, [3],† S. Bochner and B. Jessen, [2], and A. Wintner and B. Jessen, [6]. Davenport established the existence of asymptotic distribution functions for a certain class of arithmetical functions by an extension of a method previously used by the author, [8], [9], in a similar investigation. This method was based on the consideration of the moments of the distribution functions. In questions of asymptotic distribution, however, Bochner and Jessen have shown the great advantage of dealing directly with the Fourier transforms of the distribution functions. This advantage becomes again apparent if the method of Fourier transforms, whose adaptation to sequences is fully developed in §I, is applied to Davenport's problem. This is precisely what we shall do in §II; the result thus obtained (Theorem 1) insures the existence of the asymptotic distribution function for a very large class of (positive and multiplicative) arithmetical functions. It includes Davenport's and the author's previous results and yields readily (by suitable specializations of the arithmetical function involved) the frequencies of certain classes of integers investigated by W. Feller and E. Tornier, [4], in an entirely different way.

The connection with the work of Wintner and Jessen, [6], is as follows. The distribution function $\omega(x)=\chi(e^u)$ of Theorem 1 is a special example of the infinite convolutions of purely discontinuous distribution functions investigated by these authors. They have shown ([6], Theorem 35) that such infinite convolutions can be only either purely discontinuous or else everywhere continuous, and in the latter case either singular functions or else absolutely continuous functions. These general results apply immediately to our special situation, but new and probably difficult problems arise which may be mentioned here. Theorem 1 gives simple sufficient conditions to insure continuity or discontinuity of $\omega(x)$; the problem of finding similar necessary and sufficient conditions for continuity remains unsolved. The more delicate problem of deciding whether a continuous $\omega(x)$ is singular or absolutely

* Presented to the Society, March 31, 1934; received by the editors July 25, 1935.

† Numbers in brackets refer to the Bibliography at the end of this paper.

continuous is likewise unsolved. It can be shown that a continuous $\omega(x)$ is necessarily singular if $\log f(p_n)=O(k^{-n})$ ($p_n=n$th prime, $k>1$). However, I do not have any example of an absolutely continuous $\omega(x)$.

I. Asymptotic distributions of real sequences

2. Let us recall first a few well known definitions. A finite or infinite class C of increasing positive integers $m_1, m_2, m_3, \cdots$ is said to have a *frequency* (or density) $F\{C\}$ if

$$\lim_{n\to\infty} (1/n) \sum_{m_\nu \leqq n} 1 = F\{C\}.$$

In case this limit does not exist then the upper limit of the same expression is the upper frequency $\overline{F}\{C\}$ and the lower limit is the lower frequency $\underline{F}\{C\}$ of the class C. A function $\omega(x)$ defined for $-\infty<x<\infty$, which is monotonic with $\omega(-\infty)=0$, $\omega(\infty)=1$, is called a *distribution function* (d.f.). A real sequence $x_1, x_2, x_3, \cdots$ is said to have an *asymptotic distribution function* (abbreviated: a.d.f.) $\omega(x)$ if for every point of continuity $x=\xi$ of $\omega(x)$ we have

(1)
$$F\{x_n \leqq \xi\} = \omega(\xi).$$

In this relation $F\{x_n\leqq\xi\}$ means the frequency of the class of integers n for which $x_n\leqq\xi$. For example we have so called equi-distribution* in the interval (0, 1) if the above definition holds with $\omega(x)=0$ for $x<0$, $=x$ for $0\leqq x\leqq 1$, $=1$ for $x>1$.

Let $N(x_\nu\leqq x)_n$ denote the number of those elements among the first n elements of the sequence $\{x_\nu\}$ which are $\leqq x$. With our sequence we may connect a sequence of distribution functions (step functions)

$$\omega_n(x) = (1/n)N(x_\nu \leqq x)_n \qquad (-\infty < x < \infty;\, n = 1, 2, 3, \cdots).$$

A comparison with our previous definition shows that *the d.f. $\omega(x)$ is the a.d.f. of our sequence $\{x_n\}$ if and only if the relation*

(2)
$$\lim_{n\to\infty} \omega_n(\xi) = \omega(\xi)$$

holds for every continuity point $x=\xi$ of $\omega(x)$. The limiting relation (2) is usually described by saying that the sequence of d.f. $\omega_n(x)$ converges *essentially* to the d.f. $\omega(x)$.

Throughout this note we write

$$\frac{g(x_1) + g(x_2) + \cdots + g(x_n)}{n} = M_n\{g(x)\}, \qquad \lim_{n\to\infty} M_n\{g(x)\} = M\{g(x)\},$$

---

* H. Weyl, [10].

provided the last limit exists. An important property of asymptotic distributions of sequences is contained in the following

LEMMA 1. *Let the sequence* $\{x_n\}$ *admit the a.d.f.* $\omega(x)$ *and let* $g(x)$ *be a bounded continuous complex-valued function for* $-\infty < x < \infty$. *Then*

$$M\{g(x)\} = \int_{-\infty}^{\infty} g(x)d\omega(x). \tag{3}$$

Since

$$M_n\{g(x)\} = \int_{-\infty}^{\infty} g(x)d\omega_n(x),$$

the relation (3) is a special case of a theorem of P. Lévy.*

3. The following theorem gives a criterion for asymptotic distributions of sequences.

LEMMA 2. *Necessary and sufficient conditions that a sequence* $\{x_n\}$ *shall have an asymptotic distribution are as follows: The mean value*

$$M\{e^{itx}\} = \lim_{n\to\infty} \frac{1}{n}(e^{itx_1} + \cdots + e^{itx_n}) = L(t) \tag{4}$$

shall exist for every real t and be an everywhere continuous function of t. If these conditions hold then $L(t)$ *is of the form*

$$L(t) = \int_{-\infty}^{\infty} e^{itx}d\omega(x) \tag{5}$$

where $\omega(x)$ *is the a.d.f. of our sequence* $\{x_n\}$.

The necessity of these conditions is a consequence of Lemma 1. In view of (4) we have

$$\sum_{\mu,\nu=1}^{m} L(t_\mu - t_\nu)\rho_\mu\bar{\rho}_\nu = \sum_{\mu,\nu} M\{e^{i(t_\mu - t_\nu)x}\}\rho_\mu\bar{\rho}_\nu = M\left\{\sum_{\mu,\nu} e^{i(t_\mu - t_\nu)x}\rho_\mu\bar{\rho}_\nu\right\}$$
$$= M\left\{\left|\sum_{\mu=1}^{m} e^{it_\mu x}\rho_\mu\right|^2\right\} \geqq 0,$$

hence $L(t)$ is a positive-definite function which, being assumed continuous,

*P. Lévy, [7], pp. 195–196. Lévy's theorem is as follows: If a sequence of d.f. $\omega_n(x)$ converges essentially to a d.f. $\omega(x)$, then

$$\int_{-\infty}^{\infty} g(x)d\omega_n(x) \to \int_{-\infty}^{\infty} g(x)d\omega(x),$$

for every bounded continuous $g(x)$.

is of the form (5) with a non-decreasing $\omega(x)$ uniquely defined by $\omega(-\infty)=0$.* For $t=0$ we get $\omega(+\infty)=L(0)=1$, hence $\omega(x)$ is a d.f. From (4) we infer

$$\lim_{n\to\infty}\int_{-\infty}^{\infty} e^{itx}d\omega_n(x) = \int_{-\infty}^{\infty} e^{itx}d\omega(x) \text{ for all real } t, \tag{6}$$

and a theorem of P. Lévy† insures the relation (2), i.e., $\omega(x)$ is the a.d.f. of our sequence.

II. Asymptotic distributions of multiplicative arithmetical functions

4. Let $f(n)$ be a multiplicative arithmetical function, that is, a function defined for $n=1, 2, 3, \cdots$ and satisfying the relations

$$f(mn) = f(m)f(n) \text{ if } (m, n) = 1, \qquad f(1) = 1. \tag{7}$$

As an immediate consequence of the unique factorization of integers into powers of primes, a multiplicative function is completely defined by prescribing arbitrarily the values of $f(p^\alpha)$ for all primes p and integers $\alpha \geqq 1$. In describing such functions we therefore need to consider only the $f(p^\alpha)$.

Our problem is as follows: Under what assumptions does the sequence $u_n = f(n)$ have an asymptotic distribution function, and how is this function connected with the $f(p^\alpha)$? The results of this note in this direction are contained in the following

Theorem 1. *Let a multiplicative arithmetical function $f(n)$ satisfy the conditions*

(i) $f(p^\alpha) > 0$,

(ii) *the series*

---

* This follows from an important theorem of Bochner: A continuous positive-definite function is of the form (5) with a non-decreasing $\omega(x)$. The converse is obviously true. See Bochner [1], p. 76.

† In fact Lévy, [7], p. 197, in deriving (2) from (6), assumes that (6) holds uniformly in every finite t-interval. That this additional assumption is not necessary was shown by Bochner, [1], p. 72, Theorem 21. Bochner's statement proves that we can add suitable constants to our functions $\omega_n(x)$ so as to make them tend essentially to $\omega(x)$, i.e., there is a sequence of constants c_n and a sequence of functions $\psi_n(x)$ such that

$$\omega_n(x) + c_n = \omega(x) + \psi_n(x) \text{ and } \lim_{n\to\infty} \psi_n(x) = 0$$

at every point of continuity of $\omega(x)$. From $c_n = \omega(x) - \omega_n(x) + \psi_n(x)$ we derive for every point of continuity of $\omega(x)$ the inequalities

$$\omega(x) - 1 \leqq \omega(x) - \limsup \omega_n(x) \leqq \liminf c_n \leqq \limsup c_n \leqq \omega(x) - \liminf \omega_n(x) \leqq \omega(x);$$

allowing here $x\to\infty$ and $x\to-\infty$, we derive the result $\liminf c_n = \limsup c_n = 0$, hence (2) holds. For our particular purpose (Theorem 1) Lévy's restricted statement would suffice, for without more trouble we can prove that (16) (the analogue of (4)) holds uniformly in every finite t-interval.

(8) $$\sum_p \frac{1}{p}\|\log f(p)\| \textit{ converges},$$

where, as a matter of notation, $\|x\| = \min(1, |x|)$. *Then* $f(n)$ *has the following properties:*

1. *The sequence* $u_n = f(n)$ *has an asymptotic distribution function* $\chi(u)$ *with*

$$\chi(u) = \chi(+0) = 0 \textit{ for } u \leqq 0,$$

and the Fourier transform of $\chi(e^x)(-\infty < x < \infty)$ *is*

(9) $$\int_{-\infty}^{\infty} e^{itx}d\chi(e^x) = L(t) = \prod_p \left\{\left(1-\frac{1}{p}\right)\left(1+\frac{1}{p}\exp\left[it\log f(p)\right] + \frac{1}{p^2}\exp\left[it\log f(p^2)\right]+\cdots\right)\right\},$$

the infinite product being absolutely and uniformly convergent in every finite t-interval.

2. *The set of points of increase of the distribution function* $\chi(u)$ *is identical with the sequence of points* $u_n = f(n)$ *together with the limit points of this sequence.*

3. *The distribution function* $\chi(u)$ *is purely discontinuous if the series*

(10) $$\sum_{f(p)\neq 1} \frac{1}{p} \textit{ converges}.$$

The function $\chi(u)$ *is everywhere continuous if there exists a sequence of increasing primes,* $q_1, q_2, q_3, \cdots$, *with*

(11) $$f(q_\mu) \neq f(q_\nu) \textit{ for } \mu \neq \nu,$$

and such that

(12) $$\sum_{\nu=1}^{\infty} \frac{1}{q_\nu} \textit{ diverges.}*$$

---

* Davenport's conditions (see [3], p. 10) for the existence of $\chi(u)$ are as follows:

(i') $0 < f(n) \leqq 1$,

(ii') there are two positive constants C and c such that

(8') $$0 \leqq f(p^{\alpha-1}) - f(p^\alpha) \leqq Cp^{-c\alpha}$$

for $\alpha \geqq 1$ and all primes p.

One should remark first that Theorem 1 imposes no conditions whatever on the values of $f(p^\alpha)$ for $\alpha > 1$, except condition (i). Moreover, for $\alpha = 1$ (ii') gives $0 \leqq 1 - f(p) \leqq Cp^{-c}$ and this implies already the convergence of our series (8). Hence all of the inequalities (8') for $\alpha > 1$ are superfluous as far as the existence of the d.f. $\chi(u)$ is concerned.

5. In order to prove the first part of Theorem 1 let us consider the infinite product

$$
\begin{aligned}
L(t) = \prod_p \{1 &+ p^{-1}(\exp [it \log f(p)] - 1) \\
&+ p^{-2}(\exp [it \log f(p^2)] - \exp [it \log f(p)]) + \cdots \} = \prod_p (1 + a_p).
\end{aligned} \tag{13}
$$

By means of the inequalities

$$| \sin x | \leqq \|x\| = \min (1, | x |), \qquad \|xt\| \leqq \|x\| \max (1, | t |),$$

we have

$$
\begin{aligned}
| a_p | &\leqq p^{-1} | \exp [it \log f(p)] - 1 | \\
&\quad + p^{-2} | \exp [it \log f(p^2)] - \exp [it \log f(p)] | + \cdots \\
&\leqq 2p^{-1} \left| \sin \frac{t \log f(p)}{2} \right| + 2(p^{-2} + p^{-3} + \cdots) \\
&\leqq 2p^{-1} \left\| \frac{t}{2} \log f(p) \right\| + 2p^{-1}(p-1)^{-1} \\
&< 2p^{-1} \|\log f(p)\| \max \left(1, \frac{|t|}{2}\right) + 2(p-1)^{-2} \\
&= p^{-1} \|\log f(p)\| \max (2, | t |) + 2(p-1)^{-2}.
\end{aligned}
$$

For $|t| \leqq T$, the series $\sum a_p$ is therefore dominated by the convergent series

$$\sum_p (p^{-1}\|\log f(p)\| \max (2, T) + 2(p-1)^{-2})$$

and the infinite product (13) converges absolutely and uniformly in every finite t-interval. As a further consequence of the last result we have

$$\prod_p \left\{1 + \sum_{\alpha=1}^{\infty} p^{-\alpha} | \exp [it \log f(p^\alpha)] - \exp [it \log f(p^{\alpha-1})] | \right\} = \sum_{m=1}^{\infty} \frac{| \rho_t(m) |}{m}$$

with

$$
\begin{aligned}
\rho_t(m) &= \prod_{p^\alpha | m} (\exp [it \log f(p^\alpha)] - \exp [it \log f(p^{\alpha-1})]) \\
&= \sum_{d | m} \mu(d) \exp \left[it \log f\left(\frac{m}{d}\right) \right],
\end{aligned} \tag{14}
$$

where p^α are the powers of different primes in the canonical decomposition of m and where $\mu(m)$ is the Möbius function. From the convergence of the last series one readily derives the relation

$$\lim_{n\to\infty} \frac{1}{n} \sum_{m=1}^{n} | \rho_t(m) | = 0. \tag{15}$$

By the inversion formula of Möbius, (14) implies

$$\exp [it \log f(m)] = \sum_{d|m} \rho_t(d),$$

and therefore

$$\begin{aligned} \frac{1}{n} \sum_{m=1}^{n} \exp [it \log f(m)] &= \frac{1}{n} \sum_{m=1}^{n} \sum_{d|m} \rho_t(d) = \frac{1}{n} \sum_{m=1}^{n} \left[\frac{n}{m}\right] \rho_t(m) \\ &= \sum_{m=1}^{\infty} \frac{\rho_t(m)}{m} + \frac{1}{n} \sum_{m=1}^{n} \left\{\frac{n}{m} - R\left(\frac{n}{m}\right)\right\} \rho_t(m) - \sum_{m=1}^{\infty} \frac{\rho_t(m)}{m} \\ &= \sum_{m=1}^{\infty} \frac{\rho_t(m)}{m} - \frac{1}{n} \sum_{m=1}^{n} R\left(\frac{n}{m}\right) \rho_t(m) - \sum_{m=n+1}^{\infty} \frac{\rho_t(m)}{m}, \end{aligned}$$

where $R(x) = x - [x]$. Using (15) we derive

$$\lim_{n\to\infty} \frac{1}{n} \sum_{m=1}^{n} e^{it \log f(m)} = \sum_{m=1}^{\infty} \frac{\rho_t(m)}{m} = L(t), \tag{16}$$

where $L(t)$ is the infinite product (13). Since $L(t)$ is continuous, (16) and Lemma 2 show that the sequence $\{\log f(n)\}$ admits an a.d.f. $\omega(x)$ whose Fourier transform is (5), hence the original sequence $\{f(n)\}$ admits the a.d.f. $\chi(u)$ defined as follows:

$$\chi(u) = \begin{cases} 0 & \text{for } u \leqq 0, \\ \omega(\log u) & \text{for } u > 0, \end{cases}$$

with $\chi(+0) = \omega(-\infty) = 0$. The remark that $\omega(x) = \chi(e^x)$ completes the proof of the first part of Theorem 1.

6. Let us pass to the proof of the second part of Theorem 1. Jessen and Wintner ([6], Theorem 3) have proved the following general result. Let $\sigma_1(x), \sigma_2(x), \sigma_3(x), \cdots$ be a sequence of d.f. such that the convolution $\omega_n(x) = \sigma_1(x) * \sigma_2(x) * \cdots * \sigma_n(x)$ converges essentially to a d.f. $\omega(x)$; this d.f. $\omega(x)$ is called the infinite convolution of the sequence $\{\sigma_\nu(x)\}$, and we write

$$\omega(x) = \sigma_1(x) * \sigma_2(x) * \sigma_3(x) * \cdots . \tag{17}$$

Let generally $S(\phi)$ denote the set of points of increase of a d.f. ϕ. Then $S(\omega_n)$ is the *vectorial* sum of sets (in the sense of Bohr)

$$S(\omega_n) = S(\sigma_1) + S(\sigma_2) + \cdots + S(\sigma_n), \tag{18}$$

and $S(\omega)$ is the *limit* of $S(\omega_n)$ in the following sense: A point ξ belongs to $S(\omega)$ if and only if it is the limit of a sequence of points x_n with $x_n \subset S(\omega_n)$.

If the origin O belongs to all $S(\sigma_n)$, then $S(\omega_1) \subset S(\omega_2) \subset S(\omega_3) \subset \cdots$ in virtue of (18), and now $S(\omega)$ is identical with the closure of the ordinary limit of $S(\omega_n)$, i.e., *a point belongs to* $S(\omega)$ *if it belongs to some* $S(\omega_n)$ *or else is a limit of such points.* Formula (9) shows that our d.f. $\omega(x) = \chi(e^x)$ is the infinite convolution of the sequence of d.f. $\sigma_\nu(x)$ of Fourier transforms

$$\int_{-\infty}^{\infty} e^{itx} d\sigma_\nu(x) = (1 - p_\nu^{-1})(1 + p_\nu^{-1} \exp [it \log f(p_\nu)] + p_\nu^{-2} \exp [it \log f(p_\nu^2)] + \cdots) \tag{19}$$

where p_ν stands for the νth prime; hence $S(\omega_n)$ is identical with the set of points $\log f(p_1^{\alpha_1} p_2^{\alpha_2} \cdots p_n^{\alpha_n})$ $(\alpha_\nu \geqq 0)$ and the second part of Theorem 1 is established.

7. Passing to the third part of Theorem 1 we remark that (10) implies (8). However, if (10) converges we need not consider (8) at all, for now the series

$$\sum_p p^{-1}(\exp [it \log f(p)] - 1)$$

is dominated by the convergent series

$$\sum_{f(p) \neq 1} 2p^{-1}$$

for all values of t, which implies the uniform convergence of product (13) for all real t. The transform $L(t)$ is therefore almost periodic and the d.f. $\omega(x)$ is necessarily purely discontinuous.

Let us now assume (8), (11) *and* (12) *to hold.* By (17) and (19) we have

$$\omega = \sigma_1 * \sigma_2 * \sigma_3 * \cdots = (\sigma_1 * \cdots * \sigma_{n-1} * \sigma_{n+1} * \cdots) * \sigma_n = \phi_n * \sigma_n.$$

Denoting by $\tilde{\omega}(E)$, $\tilde{\phi}_n(E)$, $\tilde{\sigma}_n(E)$ the set functions corresponding to the point functions ω, ϕ_n, σ_n, we have

$$\tilde{\omega}(E) = \int \tilde{\phi}_n(E - y)\tilde{\sigma}_n(dy).†$$

If $\omega(x)$ is not everywhere continuous, there is a point x with $\tilde{\omega}(x) = c > 0$. Now by (19), writing $\log f(p_n) = \lambda_n$,

$$\tilde{\omega}(x) = \int \tilde{\phi}_n(x - y)\tilde{\sigma}_n(dy)$$

$$= \tilde{\phi}_n(x)\left(1 - \frac{1}{p_n}\right) + \tilde{\phi}_n(x - \lambda_n)\left(\frac{1}{p_n} - \frac{1}{p_n^2}\right) + \cdots = c > 0.$$

† A Lebesgue-Radon integral. For notation and references see [6], §2.

Since $0 \leqq \tilde{\phi}_n(E) \leqq 1$, we may infer that

$$\tilde{\phi}_n(x) > c/2 \text{ for } n \text{ sufficiently large,}$$

hence

$$\tilde{\omega}(x+\lambda_n) = \tilde{\phi}_n(x+\lambda_n)\left(1-\frac{1}{p_n}\right) + \tilde{\phi}_n(x)\left(\frac{1}{p_n}-\frac{1}{p_n^2}\right) + \cdots$$
$$> \frac{c}{2}\left(\frac{1}{p_n}-\frac{1}{p_n^2}\right).$$

This result, however, leads to a contradiction, for $\sum\tilde{\omega}(x+\lambda_n)$, summed over all different λ_n (with n large), is on one hand $\leqq 1$, on the other hand it is $>(c/2)\sum(p_n^{-1}-p_n^{-2})$, and this last series diverges by our assumptions (11) and (12).*

III. Frequencies of certain classes of integers

8. Theorem 1 applies with great ease to the multiplicative arithmetical functions

$$f(n) = \phi(n)/n \qquad (\phi(n) \text{ is the Euler function}),†$$
$$f(n) = n/\sigma(n) \qquad (\sigma(n) = \text{sum of divisors of } n),‡$$

for $f(p) = 1 - 1/p$ and $p/(1+p)$ respectively, and both series

$$\sum_p \frac{1}{p}\left|\log\left(1-\frac{1}{p}\right)\right|, \qquad \sum_p \frac{1}{p}\log\left(1+\frac{1}{p}\right)$$

are convergent. Moreover both functions are everywhere dense in the interval (0, 1), as even their values of the form

$$f(q_1q_2\cdots q_m) = \left(1-\frac{1}{q_1}\right)\cdots\left(1-\frac{1}{q_m}\right) \text{ and } \frac{q_1}{1+q_1}\cdots\frac{q_m}{1+q_m} \text{ respectively,}$$

where $q_1, \cdots, q_m$ are different primes, have this property.§ Since the $f(p)$ are all different, the distribution functions are continuous. We can therefore conclude that the frequency $F\{\sigma(n) > kn\}$ of k-abundant numbers|| is continuous and *strictly decreasing* for increasing $k \geqq 1$.

* The conditions (11), (12), and the above proof of their sufficiency are due to Dr. Jessen. My original conditions were more stringent.

† Schoenberg [8].

‡ Davenport [3].

§ See [8], p. 194.

|| Davenport [3], p. 830.

These results may be extended to various generalizations of Euler's function as well as to the function

$$f(n) = n^s/\sigma_s(n) = n^s/\sum_{d|n} d^s \qquad (s > 0).$$

For $0<s\leqq 1$, this last function is everywhere dense in $(0, 1)$; if $s>1$ we have

$$\liminf f(n) = 1/\zeta(s), \qquad \limsup f(n) = 1.\dagger$$

9. We now shall apply Theorem 1 to certain arithmetical functions leading to purely discontinuous distribution functions. A few preliminary considerations are necessary.

Let $x_1, x_2, x_3, \cdots$ be a sequence of real elements. Let $\lambda_1, \lambda_2, \lambda_3, \cdots$ be all the *different* values of the elements of the sequence $\{x_n\}$, i.e., $\lambda_\mu \neq \lambda_\nu$ if $\mu \neq \nu$ while any x_n is equal to some λ_μ and vice versa. Let us further assume that the sequence $\{x_n\}$ has an a.d.f. $\omega(x)$ whose Fourier transform is

$$\int_{-\infty}^{\infty} e^{itx} d\omega(x) = \sum_{m=1}^{\infty} A_m{}^{it\lambda_m}, \tag{20}$$

hence

$$A_m \geqq 0, \qquad \sum_{m=1}^{\infty} A_m = 1. \tag{21}$$

We shall need the following

LEMMA 3. *If the sequence $\{\lambda_\mu\}$ has no finite limit point, then*

$$F\{x_\nu = \lambda_m\} = A_m, \tag{22}$$

$$F\{x_\nu = \lambda_{m_1}, \lambda_{m_2}, \lambda_{m_3}, \cdots\} = A_{m_1} + A_{m_2} + A_{m_3} + \cdots \tag{23}$$
$$(m_1 < m_2 < m_3 < \cdots).$$

For if the interval $\lambda_m - \epsilon \cdots \lambda_m + \epsilon$ is free of values λ_μ $(\mu \neq m)$, then $\lambda_m \pm \epsilon$ are points of continuity (in fact points of constancy) of $\omega(x)$, hence

$$\lim_{n\to\infty} \frac{1}{n} N(x_\nu \leqq \lambda_m \pm \epsilon)_n = \omega(\lambda_m \pm \epsilon) = \omega(\lambda_m \pm 0)$$

and therefore

$$\lim_{n\to\infty} \frac{1}{n}\{N(x_\nu \leqq \lambda_m + \epsilon)_n - N(x_\nu \leqq \lambda_m - \epsilon)_n\} = \lim_{n\to\infty} \frac{1}{n} N(x_\nu = \lambda_m)_n$$
$$= \omega(\lambda_m + 0) - \omega(\lambda_m - 0) = A_m,$$

which proves (22).

† Gronwall [5]; presumably $f(n)$ is everywhere dense in $(\zeta^{-1}(s), 1)$.

Let C_m be the class of integers ν for which $x_\nu = \lambda_m$. The totality T of all positive integers is thus decomposed as a sum of classes

$$T = C_1 + C_2 + C_3 + \cdots, \text{ with } F\{C_m\} = A_m,$$

and (23) is equivalent to

$$(24) \qquad F\{C_{m_1} + C_{m_2} + \cdots\} = F\{C_{m_1}\} + F\{C_{m_2}\} + \cdots.†$$

From

$$C = C_{m_1} + C_{m_2} + \cdots + C_{m_{k-1}} + (C_{m_k} + C_{m_{k+1}} + \cdots)$$

we get

$$(25) \quad \overline{F}\{C\} \leqq F\{C_{m_1}\} + \cdots + F\{C_{m_{k-1}}\} + \overline{F}\{C_{m_k} + C_{m_{k+1}} + \cdots\}.$$

On the other hand

$$\begin{aligned}\overline{F}\{C_{m_k} + C_{m_{k+1}} + \cdots\} &\leqq \overline{F}\{C_{m_k} + C_{m_k+1} + C_{m_k+2} + \cdots\}\\ &= 1 - F\{C_1 + C_2 + \cdots + C_{m_k-1}\}\\ &= 1 - F\{C_1\} - \cdots - F\{C_{m_k-1}\} \to 0 \text{ as } k \to \infty\end{aligned}$$

in view of (21). Hence, as $k \to \infty$, (25) gives

$$\overline{F}\{C\} \leqq F\{C_{m_1}\} + F\{C_{m_2}\} + \cdots.$$

This together with the obvious relation

$$\underline{F}\{C\} \geqq F\{C_{m_1}\} + F\{C_{m_2}\} + \cdots$$

proves (24).

10. Let us devide the totality T of positive integers into various classes as follows. Call $C(1)$ the class of square-free numbers. If n is not square-free let $q_1^{\alpha_1} q_2^{\alpha_2} \cdots q_r^{\alpha_r}$ $(\alpha_1 > 1, \cdots, \alpha_r > 1)$ be the product of *all* the powers of primes in its canonical decomposition and call $C(q_1^{\alpha_1} q_2^{\alpha_2} \cdots q_r^{\alpha_r})$ the class of all numbers n having the same product $q_1^{\alpha_1} \cdots q_r^{\alpha_r}$ of powers of primes in their canonical decomposition.

What is the frequency $F\{C(q_1^{\alpha_1} q_2^{\alpha_2} \cdots q_r^{\alpha_r})\}$ *of the class* $C(q_1^{\alpha_1} q_2^{\alpha_2} \cdots q_r^{\alpha_r})$? It can be immediately computed from (9) by specializing conveniently the function $f(n)$. For all primes p let

$$(26) \qquad f(p) = 1, \; f(p^\alpha) = p^\alpha \text{ for } \alpha > 1.$$

The series (10) is void and the a.d.f. $\omega(x)$ of the sequence $\{\log f(n)\}$ is purely discontinuous. Since $\Pi(1 - p^{-2}) = \zeta(2)^{-1} = 6/\pi^2$, its Fourier transform given by (9) becomes

† This relation, which I owe to Dr. Jessen, is not true for every decomposition $T = C_1 + C_2 + \cdots$, for even the relation $F\{C_1 + C_2 + \cdots\} = F\{C_1\} + F\{C_2\} + \cdots$ breaks down in the case of the decomposition $T = (1) + (2) + (3) + \cdots$.

$$
(27)\qquad \begin{aligned}
\int_{-\infty}^{\infty} e^{itx} d\omega(x) = L(t) &= \frac{6}{\pi^2}\prod_p \left\{1 + p^{-2}\left(1+\frac{1}{p}\right)^{-1} \exp\,[it\log p^2] \right. \\
&\qquad \left. + p^{-3}\left(1+\frac{1}{p}\right)^{-1}\exp\,[it\log p^3] + \cdots\right\} \\
&= \frac{6}{\pi^2}\left\{1 + \sum_{(q_1^{\alpha_1}\cdots q_r^{\alpha_r})}^{(\alpha>1)} q_1^{-\alpha_1}q_2^{-\alpha_2}\cdots q_r^{-\alpha_r}\left(1+\frac{1}{q_1}\right)^{-1} \right. \\
&\qquad \left. \cdots\left(1+\frac{1}{q_r}\right)^{-1}\exp\,[it\log(q_1^{\alpha_1}\cdots q_r^{\alpha_r})]\right\}.
\end{aligned}
$$

Since $\log f(n) = \log(q_1^{\alpha_1}q_2^{\alpha_2}\cdots q_r^{\alpha_r})$ $(\alpha>1)$ if and only if n belongs to the class $C(q_1^{\alpha_1}q_2^{\alpha_2}\cdots q_r^{\alpha_r})$ and since the various values of $\log(q_1^{\alpha_1}\cdots q_r^{\alpha_r})$ have no finite limit point (as logarithms of different integers) Lemma 3 shows that

$$
(28)\qquad \begin{aligned}
&F\{C(q_1^{\alpha_1}q_2^{\alpha_2}\cdots q_r^{\alpha_r})\} \\
&\quad = \frac{6}{\pi^2} q_1^{-\alpha_1}q_2^{-\alpha_2}\cdots q_r^{-\alpha_r}\left(1+\frac{1}{q_1}\right)^{-1}\cdots\left(1+\frac{1}{q_r}\right)^{-1}.
\end{aligned}
$$

By the same lemma we obtain the frequency of a sum of classes $C(q_1^{\alpha_1}\cdots q_r^{\alpha_r})$ by simply adding the frequencies of the individual classes.† Feller and Tornier determine the frequency of the class of numbers n which have an even number of *powers* of primes in their canonical decomposition (loc. cit., p. 229). We may obtain their result directly from (28), for

$$
\begin{aligned}
F\left\{\sum_{r\ \text{even}} C(q_1^{\alpha_1}\cdots q_r^{\alpha_r})\right\} &= \frac{6}{\pi^2}\sum_{r\ \text{even}} \frac{1}{(q_1^2-1)\cdots(q_r^2-1)} \\
&= \frac{6}{2\pi^2}\left\{\prod_p\left(1+\frac{1}{p^2-1}\right) + \prod_p\left(1-\frac{1}{p^2-1}\right)\right\} \\
&= \frac{1}{2}\prod_p(1-p^{-2})\left\{\prod_p \frac{1}{1-p^{-2}} + \prod_p \frac{1-2p^{-2}}{1-p^{-2}}\right\} \\
&= \frac{1}{2} + \frac{1}{2}\prod_p(1-2p^{-2}).
\end{aligned}
$$

11. The derivation of (28) was essentially based on the fact that the sequence $\{\log f(n)\}$ defined by (26) has an a.d.f. $\omega(x)$ with the transform (27). It is of some interest to point out that this result may be derived by means of elementary properties of trigonometric polynomials only, without reference to Stieltjes integrals or the more refined theorems used in §1. To discuss a

† This is Theorem 19 of Feller and Tornier, [4], p. 228.

somewhat more general situation than (26) let us assume (a) that the series (10) converges, (b) that the sequence $\{\lambda_m\}$ of the different values of the elements of the sequence $\{\log f(n)\}$ has $+\infty$ as its only limiting point, in which case by renumbering the λ_m we may assume $\lambda_1<\lambda_2<\lambda_3<\cdots<\lambda_m\rightarrow\infty$.

We know that (10) implies that $L(t)$ is an almost periodic function with a Fourier expansion of the type

$$(29) \qquad L(t) = \sum_{m=1}^{\infty} A_m e^{it\lambda_m} \qquad \left(A_m \geqq 0, \sum_{1}^{\infty} A_m = 1\right).$$

Another immediate consequence of (10) is that the limiting relation (16) holds uniformly for all real t. Hence if we write

$$\frac{1}{n}\sum_{m=1}^{n} e^{it\log f(m)} = \sum{}^{*} a_m^{(n)} e^{it\lambda_m},$$

we know that these polynomials converge uniformly to $L(t)$. From $A_m = \mathfrak{M}_t\{L(t)e^{-it\lambda_m}\}$ and the similar formulas for our polynomials we get

$$\lim_{n\rightarrow\infty} a_m^{(n)} = A_m \qquad (m = 1, 2, 3, \cdots),$$

and therefore (in the notation of §1)

$$\omega_n(x) = \sum_{\lambda_m<x} a_m^{(n)} \rightarrow \sum_{\lambda_m<x} A_m = \omega(x)$$

for every $x \neq \lambda_1, \lambda_2, \cdots$, for both of the sums involved in the last relation contain a finite number of terms only. But this relation means precisely that $\omega(x)$ is the a.d.f. of the sequence $\{\log f(n)\}$.

12. A great number of examples of classes of integers could be indicated whose frequencies can be computed by the method used above. We shall discuss only two more examples already considered by Feller and Tornier ([4], pp. 215 and 224).

Let Γ be the class of numbers of the form $n = q_1^{\alpha_1} q_2^{\alpha_2} \cdots q_r^{\alpha_r}$ $(\alpha>1)$, i.e., if $p|n$ then also $p^2|n$. To compute $F\{\Gamma\}$ let Γ_m be the class of numbers for which the above property ($p|n$ implies $p^2|n$) is required only for the first m primes $p_1, p_2, \cdots, p_m$. Obviously $\Gamma\subset\Gamma_m$. Consider the multiplicative function $f(n)$ defined by

$$f(p_1) = p_1,\ f(p_2) = p_2, \cdots, f(p_m) = p_m,\ f(p_{m+1}) = f(p_{m+2}) = \cdots = 1,$$
$$f(p^\alpha) = 1 \text{ for } \alpha > 1.$$

(10) is fulfilled and (9) becomes

$$L(t) = \prod_{\nu=1}^{m} \left\{1 - p_\nu^{-1} + p_\nu^{-2} + p_\nu^{-1}(1 - p_\nu^{-1}) \exp \left[it \log p_\nu\right]\right\}.$$

Since $\log f(n) = 0$ if and only if $n \subset \Gamma_m$, (22) gives

$$\overline{F}\{\Gamma\} \leqq F\{\Gamma_m\} = \prod_{\nu=1}^{m} (1 - p_\nu^{-1} + p_\nu^{-2}) \text{ which } \to 0 \text{ as } m \to \infty,$$

hence $F\{\Gamma\} = 0$.

Let $k_1, k_2, k_3, \cdots$ be a sequence of positive integers some of which may be infinite and let K be the class of integers not divisible by any of the prime powers $p_1^{k_1}, p_2^{k_2}, p_3^{k_3}, \cdots$ ($k_m = \infty$ means that there is no restriction at all with respect to p_m). To determine $F\{K\}$ let us assume first that the series

$$\sum_{\nu=1}^{\infty} p_\nu^{-k_\nu} \tag{30}$$

converges. Define a multiplicative function $f(n)$ by

$$f(p_\nu) = f(p_\nu^2) = \cdots = f(p_\nu^{k_\nu - 1}) = 1, \; f(p_\nu^\alpha) = p_\nu^\alpha \text{ for } \alpha \geqq k_\nu \qquad (\nu = 1, 2, 3, \cdots).$$

Since

$$\sum_{f(p) \neq 1} p^{-1} = \sum_{(k_\nu = 1)} p_\nu^{-1} \leqq \sum_{\nu=1}^{\infty} p_\nu^{-k_\nu}$$

converges, the a.d.f. of $\{\log f(n)\}$ is a step function with the transform

$$L(t) = \prod_{\nu=1}^{\infty} \{(1 - p_\nu^{-1})(1 + p_\nu^{-1} + \cdots + p_\nu^{-(k_\nu - 1)} + p_\nu^{-k_\nu} \exp \left[it \log p_\nu^{k_\nu}\right] + \cdots)\}.$$

Since $\log f(n) = 0$ if and only if $n \subset K$, we have

$$F\{K\} = \prod_{\nu=1}^{\infty} \{(1 - p_\nu^{-1})(1 + p_\nu^{-1} + \cdots + p_\nu^{-(k_\nu - 1)})\} = \prod_{\nu=1}^{\infty} (1 - p_\nu^{-k_\nu}). \tag{31}$$

If (30) diverges let K_m be the class of integers similar to K but for the new sequence of exponents $k_1, k_2, \cdots, k_m, \infty, \infty, \cdots$; then $K \subset K_m$ and (31) applied for the new sequence gives

$$\overline{F}\{K\} \leqq F\{K_m\} = \prod_{\nu=1}^{m} (1 - p_\nu^{-k_\nu}) \text{ which } \to 0 \text{ as } m \to \infty.$$

Hence $F\{K\} = 0$ and the formula (31) is again valid.

Obviously our last two examples are also particular cases of the elementary scheme discussed above.

Added in proof, January, 1936. In a recent note *On the density of some sequences of numbers*, Journal of the London Mathematical Society, vol. 10 (1935), pp. 120–125, P. Erdös proves without using Fourier analysis two theorems which in our notation are as follows.

1. *If $f(n)$ is a multiplicative function satisfying the conditions*

$$f(n) \geqq 1, \tag{32}$$

$$\sum_p \|\log f(p)\|/p \textit{ converges}, \tag{33}$$

$$f(p_1) \neq f(p_2) \textit{ if } p_1, p_2 \textit{ are different primes}, \tag{34}$$

the sequence $\{f(n)\}$ has a continuous asymptotic distribution function.

2. *If the multiplicative function $f(n)$ satisfying* (32) *is such that*

$$\sum_p \|\log f(p)\|/p \textit{ diverges}, \tag{35}$$

then

$$F\{f(n) \geqq \tau\} = 1 \textit{ for any real } \tau \geqq 1. \tag{36}$$

The first theorem of Erdös is obviously a consequence of Theorem 1. This is not true for the second theorem; I want to show, however, how it can be derived from Theorem 1 by a simple additional argument involving moments rather than Fourier transforms.

Let $f_k(n)$ be an auxiliary multiplicative function defined as follows:

$$\begin{aligned} f_k(p_\nu^\alpha) &= f(p_\nu^\alpha) \qquad (\nu = 1, 2, \cdots, k), \\ f_k(p_\nu^\alpha) &= 1 \quad (\nu = k+1, k+2, \cdots; \alpha \geqq 1). \end{aligned} \tag{37}$$

The sequence $\{\log f_k(n)\}$ has an a.d.f. $\omega_k(x)$ and the a.d.f. of the sequence $\{f_k^{-1}(n)\}$ (contained within $0 < t \leqq 1$) is therefore $\chi_k(t) = \omega_k(-\log t)$. For $s > 0$ we have

$$\begin{aligned} &\prod_{\nu=1}^{k} \{(1 - p_\nu^{-1})(1 + p_\nu^{-1} \exp[-s \log f(p_\nu)] + \cdots)\} \\ &= \int_0^\infty e^{-sx} d\omega_k(x) = \int_0^1 t^s d\chi_k(t). \end{aligned} \tag{38}$$

The product (38) tends to zero as $k \to \infty$ on account of (35). But

$$\lim_{k\to\infty} \int_0^1 t^s d\chi_k(t) = 0, \text{ for } s > 0,$$

implies (see [8], pp. 175–176) $\chi_k(t) \to 1$ for $0 < t \leqq 1$. Hence

$$\underline{F}\{f(n) \geqq \tau\} = \underline{F}\{f^{-1}(n) \leqq \tau^{-1}\} \geqq F\{f_k^{-1}(n) \leqq \tau^{-1}\}$$
$$= \chi_k(\tau^{-1}) \to 1 \text{ as } k \to \infty$$

and (36) is proved. The last inequality for frequencies follows from $f^{-1}(n) \leqq f_k^{-1}(n)$ which is due to (32) and (37).

Bibliography

1. S. Bochner, *Vorlesungen über Fouriersche Integrale*, Leipzig, 1932.

2. S. Bochner and B. Jessen, *Distribution functions and positive-definite functions*, Annals of Mathematics, (2), vol. 35 (1934), pp. 252–257.

3. H. Davenport, *Über numeri abundantes*, Sitzungsberichte der Preussischen Akademie, Physikalisch-Mathematische Klasse, 1933, pp. 830–837.

4. W. Feller und E. Tornier, *Mengentheoretische Untersuchungen von Eigenschaften der Zahlenreihe*, Mathematische Annalen, vol. 107 (1932), pp. 188–232.

5. T. H. Gronwall, *Some asymptotic expressions in the theory of numbers*, these Transactions, vol. 14 (1913), pp. 113–122.

8. B. Jessen and A. Wintner, *Distribution functions and the Riemann zeta function*, these Transactions, vol. 38 (1935), pp. 48–88.

7. P. Lévy, *Calcul des Probabilités*, Paris, 1925.

8. I. J. Schoenberg, *Über die asymptotische Verteilung reeller Zahlen* mod 1, Mathematische Zeitschrift, vol. 28 (1928), pp. 171–200.

9. I. J. Schoenberg, *Über total monotone Folgen mit stetiger Belegungsfunktion*, vol. 30 (1929), pp. 761–767.

10. H. Weyl, *Über die Gleichverteilung von Zahlen* mod 1, Mathematische Annalen, vol. 77 (1916), pp. 313-352.

Institute for Advanced Study,
Princeton, N. J., and
Swarthmore College,
Swarthmore, Pa.

Reprinted by TRUEX*press* 53 Holywell Oxford England

Reprinted from
Trans. Amer. Math. Soc. **39**(1936), pp. 315–330

REGULAR SIMPLICES AND QUADRATIC FORMS

I. J. SCHOENBERG*.

[*Extracted from the Journal of the London Mathematical Society, Vol.* 12, 1937.]

1. R. E. A. C. Paley has investigated a certain type of orthogonal matrices which he called U-matrices (Paley [6]). The matrix

$$\|a_{ik}\| \quad (i, k = 0, 1, \ldots, n)$$

is called a U-matrix if $a_{ik} = \pm 1$ and if any two rows (and therefore any two columns) are orthogonal to each other. Apart from the trivial cases $n = 0$ or 1, U-matrices of order $n+1$ can exist only if $n+1$ is a multiple of 4. After preliminary work by Sylvester, Hadamard ([4]), Scarpis, and Gilman†, Paley has indicated certain classes of numbers $n+1$ for which there exist U-matrices of order $n+1$; these classes cover, in particular, all multiples of 4 not exceeding 200, with the exception of six such multiples. The analytical interest of U-matrices, apart from their combinatorial and geometrical aspects discussed by J. A. Todd and H. S. M. Coxeter, lies chiefly in the fact, discovered by Hadamard, that there exist U-matrices of order $n+1$ if and only if

$$\max_{-1 \leqslant x_{ik} \leqslant 1} \left| \det \|x_{ik}\|_0^n \right| = (n+1)^{\frac{1}{2}(n+1)},$$

in which case this maximum is reached for a U-matrix $\|x_{ik}\| = \|a_{ik}\|$.

Let $U_{n+1} = \|a_{ik}\|$ be a U-matrix. By obvious changes of signs in rows we may assume that $a_{i0} = +1$. Denote by E_n the n-dimensional Euclidean space referred to rectangular coordinates. Coxeter ([2]) has noticed the readily established fact that the $n+1$ points of E_n,

$$P_i = (a_{i1}, a_{i2}, \ldots, a_{in}) \quad (i = 0, 1, \ldots, n), \tag{1}$$

are the vertices of a regular simplex σ_n (*i.e.* all distances $P_i P_k$ are equal) which is, of course, inscribed in the hyper-cube γ_n of vertices $(\pm 1, \ldots, \pm 1)$ and conversely, if (1) are the vertices of a regular σ_n inscribed in γ_n $(\sigma_n \subset \gamma_n)$, then $\|a_{ik}\|_0^n$, with $a_{i0} = +1$, is a U-matrix of order $n+1$. Thus the problem of the existence of U-matrices of order $n+1$ is equivalent to the problem of finding the dimensions n for which a σ_n may be inscribed in γ_n.

---

* Received 10 August, 1936; read 12 November, 1936.

† For references to the work of Sylvester, Scarpis, and Gilman, see Paley [6].

2. The present note does not contribute to a solution of this difficult problem. It is concerned with the following less restrictive question. Denote by L_n the n-dimensional lattice of points of E_n with integral coordinates. *For what dimensions n can a σ_n be inscribed in L_n?* Thus instead of requiring, as in the problem of Coxeter and Paley, that the vertices of σ_n have coordinates ± 1, we merely require these coordinates to be rational integers*. A complete answer to this question is given by the following theorem.

THEOREM 1. *It is possible to inscribe a regular simplex σ_n in the ordinary rectangular lattice L_n only in the following cases:*

1. *If n is even, the construction is possible if and only if $n+1$ is a perfect square, i.e. for $n = 4h(h+1)$ $(h = 1, 2, 3, \dots)$.*

2. *If $n \equiv 3 \pmod 4$, the construction is always possible.*

3. *If $n \equiv 1 \pmod 4$, the construction is possible if and only if $n+1$ is not divisible to an odd exponent by a prime number of the form $4k+3$, i.e. for*

$$n = (2h+1)^2 + (2k+1)^2 - 1 \quad (h, k = 0, 1, 2, \dots)\dagger.$$

The construction is therefore possible for

$$n = 1, 3, 7, 8, 9, 11, 15, 17, 19, 23, 24, 25, 27, 31, 33, \dots,$$

and impossible for the dimensions omitted. The yet unproved conjecture of Paley is that $\sigma_n \subset \gamma_n$ is always possible whenever $4 \mid n+1$, *i.e.*

$$n \equiv 3 \pmod 4.$$

The second part of our theorem ensures the possibility of $\sigma_n \subset L_n$ for these values of n.

3. Concluding this introduction, I should like to point out that the problem $\sigma_n \subset L_n$ is to a certain extent the only non-trivial question of this kind. Note first that $\sigma_n \subset L_{n+1}$ is always possible, for the end points of the coordinate unit-vectors of L_{n+1} are the vertices of such a σ_n. Consider next a regular polygon $\Pi_n = M_1 M_2 \dots M_n$ of n sides and let us ask in what

* For the elementary problem of disproving the possibility of $\sigma_2 \subset L_2$, see Pólya and Szegö [7], 156; during its discussion in a problem-seminar at Swarthmore College, one of my students, Mr. Alan Bloch, asked whether $\sigma_2 \subset L_3$ is possible. His question gave the initial impulse to the present investigation.

† This parametric representation follows from the remark that the numbers here described are precisely the numbers which are the sum of two squares.

general (oblique) lattice L of m dimensions it might be inscribed. If $\Pi_n \subset L$, we may obviously assume (since O has rational coordinates in L) that the centre O of Π_n is also a point of L. Let $\overrightarrow{ON} = \overrightarrow{OM_1} + \overrightarrow{OM_3}$. Notice that $N \subset L$ and $ON/OM_2 = 2\cos(2\pi/n)$. Hence $\cos(2\pi/n)$ must be rational, which is the case only for $n = 3$, 4, and 6. We have thus proved: *No regular polygon, except the triangle, square, and hexagon, can be inscribed in any general lattice of any number of dimensions.* This implies similar negative results for all the regular polyhedra and polytopes containing regular pentagons.

I. *The connection with the theory of quadratic forms and the case in which n is even.*

4. Let $\sigma = P_0 P_1 \dots P_n$ be a simplex in E_n and let $a_{ik} = P_i P_k$ $(a_{ii} = 0)$ be the lengths of its edges. Denote by $L(\sigma)$ the n-dimensional lattice of points generated by linear combinations of the vectors $\mathfrak{p}_i = \overrightarrow{P_0 P_i}$ with integral coefficients. The connection between the lattice $L(\sigma)$ and the positive definite quadratic form

$$(2) \qquad F(x) = \tfrac{1}{2} \sum_{i,k=1}^{n} (a_{0i}^2 + a_{0k}^2 - a_{ik}^2)\, x_i x_k = \sum_{i,k=1}^{n} (\mathfrak{p}_i \mathfrak{p}_k)\, x_i x_k$$

$$= |\mathfrak{p}_1 x_1 + \mathfrak{p}_2 x_2 + \dots + \mathfrak{p}_n x_n|^2$$

is simple and well-known*. To a different system of vectors like $\mathfrak{p}_1, \dots, \mathfrak{p}_n$, which generate the same lattice $L(\sigma)$, there corresponds a quadratic form which is arithmetically equivalent to F, *i.e.* derivable from F by a linear transformation with integral coefficients and determinant ± 1, and conversely. A given positive definite quadratic form thus completely determines a corresponding lattice apart from a rigid displacement and a reflexion in a hyperplane, that is, determines its size.

In addition to σ, let us now consider a second given simplex $\sigma' = P_0' \dots P_n'$ of E_n generating a lattice $L(\sigma')$, and let $F'(y_1, \dots, y_n)$ be its corresponding quadratic form. When can a certain convenient multiple of $L(\sigma)$, say $\lambda L(\sigma)$ [obtained from $L(\sigma)$ by a similitude followed, perhaps, by a reflexion], be inscribed in $L(\sigma')$?

* See H. Minkowski [5], 243–254, where also references to Gauss and Dirichlet are to be found. The author first noticed the identity (2) in an article of Fréchet [3], and recognized its importance for certain problems of metric geometry (Schoenberg [8]).

We need the following definitions. Two quadratic forms $G(x)$ and $G'(y)$ are said to be *rationally equivalent* if there exists a non-singular linear transformation with rational coefficients $(y) = R(x)$, such that $G(x) = G'(y)$. The forms $G(x)$ and $G'(y)$ are said to be *rationally equivalent save for a factor* if, for a suitable constant λ, λG and G' are rationally equivalent. On the basis of the correspondence between lattices and quadratic forms the following lemma, whose proof we may omit, is readily established.

LEMMA 1. *Let $\sigma = P_0 P_1 \dots P_n$ and $\sigma' = P_0' P_1' \dots P_n'$ be two given n-dimensional simplices, and let*

$$F(x) = \sum_{i,k=1}^{n} \left(\overrightarrow{P_0 P_i} \cdot \overrightarrow{P_0 P_k}\right) x_i x_k, \qquad F'(y) = \sum_{i,k=1}^{n} \left(\overrightarrow{P_0' P_i'} \cdot \overrightarrow{P_0' P_k'}\right) y_i y_k$$

*be their corresponding quadratic forms. Let $L(\sigma)$ and $L(\sigma')$ be the lattices generated by the sets of vectors $\{\overrightarrow{P_0 P_i}\}$ and $\{\overrightarrow{P_0' P_i'}\}$ respectively. A suitable multiple $\lambda L(\sigma)$ of the lattice $L(\sigma)$ can be inscribed in $L(\sigma')$ if and only if the quadratic forms $F(x)$ and $F'(y)$ are rationally equivalent up to a factor**.

5. The particular case with which we are concerned in Theorem 1 is a regular simplex $\sigma = \sigma_n$ with all its edges equal to $\sqrt{2}$, say, while $\sigma' = \sigma_n'$ is made up of n mutually orthogonal unit-vectors with common origin. In view of (2), the corresponding forms are

$$F(x) = 2 \sum_{i=1}^{n} x_i^2 + \sum_{i \neq k} x_i x_k, \quad F'(y) = y_1^2 + \dots + y_n^2. \tag{3}$$

Now $F(x)$ may be replaced for our purposes by a considerably simpler form. Indeed, the identity

$$2 \sum_{i=1}^{n} x_i^2 + \sum_{i \neq k} x_i x_k = \sum_{i=1}^{n} \frac{i+1}{i} \left(x_i + \frac{x_{i+1}}{i+1} + \frac{x_{i+2}}{i+1} + \dots\right)^2 \quad (x_i = 0 \text{ if } i > n) \tag{4}$$

shows that $F(x)$ is rationally equivalent to the form

$$f_n(x) = \sum_{i=1}^{n} i(i+1) x_i^2 = \sum_{i=1}^{n} a_i x_i^2. \tag{5}$$

Our problem is therefore to investigate the values of n for which an identity

$$\lambda \sum_{i=1}^{n} a_i x_i^2 = y_1^2 + \dots + y_n^2 \tag{6}$$

---

* Minkowski does not seem to have pointed out this geometrical interpretation of rational equivalence save for a factor, although it probably was known to him.

is a consequence of a certain linear transformation

$$(y) = R(x) \quad (|R| \neq 0) \tag{7}$$

with rational coefficients. Notice that λ is necessarily rational.

5. Let now n be even. From (6) and (7) we have

$$\lambda^n (n!)^2 (n+1) = |R|^2.$$

Since λ is rational and n is even, we conclude that $n+1$ *is a perfect square.* In order to prove also the direct statement of the first part of Theorem 1 we perform the following simple construction.

Let $\sigma' = OA_1A_2 \dots A_n$ be the simplex generating L_n, *i.e.* let $\overrightarrow{OA_i}$ be mutually orthogonal unit-vectors, and consider the regular simplex $\sigma = A_0A_1 \dots A_n$ having with σ' a common base, with A_0 and O on opposite sides of this base. The coordinates of A_0 are readily found. The coordinates of the centroid C of the base $A_1A_2 \dots A_n$ are all equal to $1/n$. Since the altitude of a regular simplex of side 1 is $(n+1)^{\frac{1}{2}}/(2n)^{\frac{1}{2}}$, we have $A_0C = 2^{\frac{1}{2}}(n+1)^{\frac{1}{2}}/(2n)^{\frac{1}{2}} = (n+1)^{\frac{1}{2}} n^{-\frac{1}{2}}$. The points O, C, A_0 are collinear and the direction cosines of their line are all equal to $n^{-\frac{1}{2}}$; hence the coordinates of A_0 are all equal to

$$\frac{1}{n} + \frac{1}{\sqrt{n}} \sqrt{\left(\frac{n+1}{n}\right)} = \frac{1+\sqrt{(n+1)}}{n}.$$

This is a rational number if $n+1$ is a square, and then the regular simplex $n\sigma$ can be inscribed in L_n. This proves that *it is always possible to inscribe a regular simplex in L_n if $n+1$ is a perfect square.* This completes the proof of the first part of Theorem 1.

II. *Application of Minkowski's theory of rational equivalence of quadratic forms for the case of n odd.*

7. Minkowski has set up a complete set of arithmetical invariants of quadratic forms with rational coefficients with respect to transformations of type (7)*; foremost among them is a certain sequence of units C_p $(=\pm 1)$ defined for all odd primes p. We start by determining these units of the quadratic form (5).

We make use of the Gauss sums

$$\phi(h;\, n) = \sum_{s=1}^{n} e^{2\pi i h s^2/n}, \tag{8}$$

* H. Minkowski, "Über die Bedingungen, unter welchen zwei quadratische Formen mit rationalen Koeffizienten ineinander rational transformiert werden können" [5], 219–239. See also the work of H. Hasse in the *Journal für Math.*, 152 (1923), 129–148, 205–224.

recalling that

(8′) $$\phi(h;\, n) = \sqrt[+]{n}(h|n)i^{\{\frac{1}{2}(n-1)\}^2} \quad \text{if} \quad (h,\, n) = 1, \quad n \text{ odd},$$

and

(8″) $$\phi(h;\, n) = d\phi(h/d;\, n/d) \quad \text{if} \quad (h,\, n) = d,$$

where $(h|n)$ denotes the Legendre-Jacobi symbol*.

The determination of Minkowski's units C_p depends on a wide generalization of the formula (8′). Let $f(x_1, \ldots, x_n)$ be a non-degenerate form with *integral* coefficients and determinant Δ, and let

$$f(h;\, N) = \sum_{x_1, \ldots, x_n = 1}^{N} e^{2\pi i h f(x_1, \ldots, x_n)/N} \quad [(h,\, N) = 1].$$

If p is an odd prime and $p^\partial \| \Delta$†, Minkowski's extension of (8′) (*loc. cit.*, page 229) is as follows:

(M) $$f(h;\, p^t) = (h\,|\,p^{\partial+nt})\, C_p\, c_p{}^t i^{\{\frac{1}{2}(p^\partial+nt-1)\}^2}\, p^{\frac{1}{2}(\partial+nt)}.$$

Here c_p stand for a second sequence of units ± 1 and (M) is valid for sufficiently large values of t (*e.g.* $t \geqslant \partial$ will do). If $p \nmid \Delta$ ($\partial = 0$), then $C_p = +1$, as is seen from (M) for $h = 1$, $t = 0$.

The determinant of (5) is $\Delta = (n!)^2(n+1)$; denote by A its largest squarefree divisor, so that A is the squarefree part of $n+1$. Finally, let

$$p^\partial \| \Delta, \quad p^{\delta_\nu} \| a_\nu \{= \nu(\nu+1)\}, \quad p^\delta \| A \quad [\Sigma\delta_\nu = \partial,\ \delta = 0 \text{ or } 1,\ \partial \equiv \delta \pmod 2].$$

If $p \nmid \Delta$, we know that $C_p = +1$; if $p\,|\,\Delta$ ($\partial > 0$), apply Minkowski's formula (M) for $h = 1$ and t even and greater than ∂. We get

$$\begin{aligned} f(1;\, p^t) &= \sum_{x_1, \ldots, x_n = 1}^{p^t} e^{2\pi i (a_1 x_1^2 + \ldots + a_n x_n^2)/p^t} = \prod_{\nu=1}^{n} \phi(a_\nu;\, p^t) \\ &= \prod_{\nu=1}^{n} p^{\delta_\nu} p^{\frac{1}{2}(t-\delta_\nu)} i^{\{\frac{1}{2}(p^{t-\delta_\nu}-1)\}^2} (a_\nu p^{-\delta_\nu}\,|\,p^{t-\delta_\nu}) \\ &= p^{\frac{1}{2}(nt+\partial)}\, i^{\sum_\nu \{\frac{1}{2}(p^{t-\delta_\nu}-1)\}^2} \prod_{\delta_\nu \text{ odd}} (a_\nu p^{-\delta_\nu}\,|\,p) \\ &= C_p i^{\{\frac{1}{2}(p^\partial+nt-1)\}^2} p^{\frac{1}{2}(nt+\partial)}; \end{aligned}$$

hence

(9) $$C_p = i^{\sum_\nu \{\frac{1}{2}(p^{t-\delta_\nu}-1)\}^2 - \{\frac{1}{2}(p^{nt+\partial}-1)\}^2} \Big(\prod_{\delta_\nu \text{ odd}} a_\nu p^{-\delta_\nu}\,\Big|\,p\Big).$$

---

* See P. Bachmann [1], 165.

† $p^\partial \| \Delta$ indicates that p^∂ is the highest power of p ($\partial \geqslant 0$) which divides Δ.

To abbreviate, let

(10) $$\{n;\ p\} = \Big(\prod_{\delta_\nu \text{ odd}} a_\nu p^{-\delta_\nu} \,|\, p\Big).$$

If $p \equiv 1 \pmod 4$, (9) gives

(11) $$C_p = \{n;\ p\}.$$

If $p \equiv -1 \pmod 4$, the exponent of i in (9) is modulo 4 congruent to

$$\sum_{\nu=1}^{n} \left(\frac{1-(-1)^{\delta_\nu}}{2}\right)^2 - \frac{1-(-1)^{\delta}}{2} = 2\sigma,$$

where $\sigma = \sum\limits_{a_\nu \text{ odd}} 1$, with $p^{a_\nu} \| \nu$ $(\nu = 1, 2, \ldots, n)$. In both cases (9) reduces to

(12) $$C_p = (-1)^{\sigma \frac{1}{2}(p-1)} \{n;\ p\},$$

where σ stands for the number of those numbers $\nu \leqslant n$ which are divisible by p to an odd exponent.

Let us now find a convenient expression for $\{n;\ p\}$. If $p+A\,(\delta = 0)$, we get, from (10),

$$\{n;\ p\} = \Big(\prod_{a_\nu \text{ odd}} \{(\nu-1)\,\nu p^{-a_\nu} \nu(\nu+1)\, p^{-a_\nu}\} \,|\, p\Big)$$

$$= \Big(\prod_{a_\nu \text{ odd}} (\nu^2-1) \,|\, p\Big) = (-1\,|\,p)^{\sigma} = (-1)^{\sigma \frac{1}{2}(p-1)}.$$

If, however, $p\,|\,A\,\{\delta = 1,\ n \equiv -1 \pmod p\}$, then we have in $\{n\,|\,p\}$ one more factor, namely

$$\{n(n+1)p^{-\delta_n} \,|\, p\} = (nAp^{-\delta} \,|\, p) = (-1\,|\,p)(Ap^{-\delta}\,|\,p) = (-1)^{\frac{1}{2}(p-1)}(Ap^{-\delta}\,|\,p).$$

In both cases we therefore have

(13) $$\{n;\ p\} = (-1)^{\sigma \frac{1}{2}(p-1) + \delta \frac{1}{2}(p-1)} (Ap^{-\delta}\,|\,p)^{\delta}.$$

Combining (12) and (13), we get

(14) $$C_p = (-1)^{\delta \frac{1}{2}(p-1)} (Ap^{-\delta}\,|\,p)^{\delta},$$

and this formula is valid for all odd primes p.

8. Let a be a positive squarefree integer. Our problem is to find when the quadratic form

$$af_n'(x) = ax_1^2 + ax_2^2 + \ldots + ax_n^2 \quad (n \text{ odd})$$

is rationally equivalent to the form (5). One of Minkowski's trivial invariants of a form is the squarefree part of its determinant; hence necessarily $a = A$. Let us now compute the units C_p' of the form Af_n'. We have $C_p' = +1$ if $p \nmid A$. If $p \mid A$, we apply Minkowski's formula (M) again, with $h = 1$, $t = 2$, $(\partial' = n)$, and obtain

$$f'(1\,;\, p^2) = \{\phi(A\,;\, p^2)\}^n = p^{n+\frac{1}{2}n}(Ap^{-1} \mid p)^n\, i^{n\{\frac{1}{2}(p-1)\}^2}$$

$$= C_p'\, i^{\{\frac{1}{2}(p^{3n}-1)\}^2}\, p^{\frac{1}{2}3n},$$

whence

$$C_p' = i^{n\{\frac{1}{2}(p-1)\}^2 - \{\frac{1}{2}(p^{3n}-1)\}^2}(Ap^{-1} \mid p).$$

If $p \equiv 1 \pmod 4$ this gives $C_p' = (Ap^{-1} \mid p)$. If $p \equiv 3 \pmod 4$ we have likewise

$$C_p' = i^{n-1}(Ap^{-1} \mid p) = (-1)^{\frac{1}{2}(n-1)}(Ap^{-1} \mid p).$$

In each case we have

$$C_p' = (-1)^{\delta\,\frac{1}{2}(n-1)\,\frac{1}{2}(p-1)}(Ap^{-\delta} \mid p)^{\delta} \tag{15}$$

for all odd primes p.

In order to see for what values of n the units C_p and C_p' are equal for all p, let us form their product; (14), (15) give

$$C_p\, C_p' = (-1)^{\delta\,\frac{1}{2}(p-1)\,\frac{1}{2}(n+1)}.$$

Now, if $n \equiv 3 \pmod 4$, we always have $C_p\, C_p' = +1$, *i.e.* $C_p = C_p'$, and therefore f_n and Af_n' are rationally equivalent by Minkowski's fundamental Theorem 1 (*loc. cit.*, p. 222). This proves the second part of Theorem 1. If, however, $n \equiv 1 \pmod 4$, *i.e.* $(n+1)/2$ odd, we shall have $C_p\, C_p' = +1$ for all odd primes p, if and only if $\delta = 0$, whenever $p \equiv 3 \pmod 4$. This proves the last part of our theorem.

References.

1. P. Bachmann, *Zahlentheorie*, 2 (Leipzig, 1894).
2. H. S. M. Coxeter, "Regular compound polytopes in more than four dimensions", *Journal of Math. and Phys.*, 12 (1933), 334–345.
3. M. Fréchet, "Sur la définition axiomatique d'une classe d'espaces . . .", *Annals of Math.* (2), 36 (1935), 705–718.
4. J. Hadamard, "Résolution d'une question relative aux déterminants", *Bulletin des Sciences Math.* (2), 17 (1893), 240–246.
5. H. Minkowski, *Gesammelte Abhandlungen*, 1 (Leipzig und Berlin, 1911).
6. R. E. A. C. Paley, "On orthogonal matrices", *Journal of Math. and Phys.*, 12 (1933), 311–320.
7. G. Pólya und G. Szegö, *Aufgaben aus der Analysis*, 2 (Berlin, 1925).
8. I. J. Schoenberg, "Remarks to Maurice Fréchet's article . . .", *Annals of Math.* (2), 36 (1935), 724–732.

Colby College,

Waterville, Maine.

Printed by C. F. Hodgson & Son, Ltd., Newton St., London, W.C.2.

Commentary

Schoenberg's first paper, "Über die asymptotische Verteilung reeller Zahlen *mod* 1" [**1***], was an important paper. It presents a general theory of non-uniform distributions of numbers in [0,1]. The main result is that $\varphi(n)/n$ has a continuous distribution function, i.e., for each $z \in [0,1]$, the limit

$$\lim_{n\to\infty} \#\{m \le n : \varphi(m) \le zm\}/n$$

exists and depends continuously on z. Here, $\varphi(n)$ is Euler's totient function, i.e., $\varphi(n)$ gives the number of nonnegative integers less than n and relatively prime to it. Schoenberg used methods of Fourier analysis. Subsequent authors have given elementary proofs and many generalizations; the subject matter eventually developed into probabilistic number theory [**E**]. Very soon after [**1***] appeared, Behrend, Chowla [**C**], and Davenport [**D**] each proved independently that the density of abundant numbers exists by showing that $\sigma(n)/n$ has a continuous distribution function. Here, $\sigma(n)$ is the sum of the divisors of n, and n is called 'abundant' if $\sigma(n) > 2n$. A little later I proved this in [**Er1**] by elementary methods by showing that the sum of the reciprocals of the primitive abundant numbers converges. For further references and details, see Elliott's book [**E**].

Stimulated by [**D**] and other papers, Schoenberg returned to this subject in his paper [**18***], "On Asymptotic Distributions of Arithmetical Functions", in which he proved that, under fairly general conditions, the distribution function of a multiplicative function exists. This result was ultimately subsumed in the Erdős-Wintner theorem which gives a necessary and sufficient condition for the existence and continuity of the distribution function of an additive arithmetical function; see [**E**], especially Chapter 5. Schoenberg also gave a sufficient condition for the distribution function to be purely singular, and a sufficient condition for it to be continuous. I later proved (in [**Er2**]) that the negation of Schoenberg's sufficient condition for the singularity of the distribution function is necessary and sufficient for the continuity of the distribution function.

Schoenberg also asks for a necessary and sufficient condition for the absolute continuity of the distribution function. This problem is still unsolved and seems very difficult. A few years later I proved (in [**Er3**]) that the distribution function of $\sigma(n)/n$, and in fact of most of the usual arithmetic functions, is purely singular, but I gave examples of arithmetic functions whose distribution function is absolutely continuous and in fact is an entire function. Some of my conjectures were settled by G.J. Babu (see, e.g., [**B**]), but a necessary and sufficient condition for the absolute continuity of the distribution function is nowhere in sight and perhaps there is no simple condition.

Schoenberg's paper [**79**] on "Arithmetic problems concerning Cauchy's functional equation" is a brief report on the material in his joint paper [**81**] with Ch. Pisot with the same title. The authors consider the following problem. Let P be a set of k distinct primes and let A be the set of all integers composed of the p's in P. Assume that the function f is strictly monotone on A and satisfies

$$f(a) = \sum_{p^\alpha \| a} f(p^\alpha),$$

where $p^\alpha \| a$ means that p^α divides a but $p^{\alpha+1}$ does not. Does it then follow that $f(n) = c\log n$? Yes if $k > 2$ but No if $k = 2$. The authors also ask this question when the strict monotonicity of f is replaced by the condition that

$$f(a_{i+1}) - f(a_i) \to 0, \tag{1}$$

where $a_1 < a_2 < \dots$ gives the elements of A in order. As far as I know this problem is still open. I conjectured and Wissing proved that if $f(n+1) - f(n) < c$, then $f(n) = \alpha \log n + g(n)$ for some bounded g. Perhaps (1) could be replaced by

$$f(a_{i+1}) - f(a_i) < c \tag{1$'$}$$

and this might imply that $f(n) = c \log n + g(n)$ for some bounded g.

References

[**B**] G.J. Babu, Absolutely continuous distribution functions of additive functions $f(p) = (\log p)^{-a}, a > 0$, *Acta Arith.* **26** (1974/5), 401–403.

[**C**] S. Chowla, On abundant numbers, *J. Indian Math. Soc.* **2** (1934), 30–34.

[**D**] H. Davenport, Über numeri abundantes, *Sitzungsber. Preuss. Akad. Wiss. Berlin* **27** (1933), 830–837.

[**E**] P.D.T.A. Elliott, *Probabilistic Number Theory*, Vol. 1 and 2, Springer-Verlag, 1980.

[**Er1**] P. Erdős, On the density of abundant numbers, *J. London Math. Soc.* **9** (1934), 278–282.

[**Er2**] P. Erdős, On the density of some sequence of numbers. III, *J. London Math. Soc.* **13** (1938), 119–127.

[**Er3**] P. Erdős, On the smoothness of the asymptotic distribution of additive arithmetical functions, *Amer. J. Math.* **61** (1939), 722–725.

Paul Erdős
Hungarian Academy of Sciences

LINEAR INEQUALITIES

ON FINITE AND INFINITE COMPLETELY MONOTONIC SEQUENCES*

BY I. J. SCHOENBERG

1. *Introduction.* As one of several important results concerning the problem of moments for a finite interval, Hausdorff† has proved the following theorem.

The problem of moments, which is to find a real function $\chi(t)$ *for* $0 \leqq t \leqq 1$, *such that*

$$(1) \qquad \mu_n = \int_0^1 t^n d\chi(t), \qquad (n = 0, 1, 2, \cdots),$$

in which the sequence $\mu_0, \mu_1, \mu_2, \cdots$ *is given in advance, has a monotonic non-decreasing solution, if and only if*

$$(2) \quad \Delta^l \mu_m = \sum_{\nu=0}^{l} (-1)^\nu \binom{l}{\nu} \mu_{m+\nu} \geqq 0, \qquad \textit{for } l, m = 0, 1, 2, \cdots,$$

in which case the sequence is said to be completely monotonic.

The same problem (1) *has a solution* $\chi(t)$ *which is of bounded variation, if and only if*

---

* Presented to the Society, November 28, 1931.

† *Ueber das Momentenproblem für ein endliches Interval*, Mathematische Zeitschrift, vol. 16 (1923), pp. 220–248. We refer to §§1 and 2.

$$\sum_{m=0}^{p} \binom{p}{m} | \Delta^{p-m}\mu_m | = O(1) \text{ for } p \to \infty.\text{*} \tag{3}$$

Hausdorff proves this theorem starting from the system (1), which defines a positive linear functional operation by interpolation in function space, and proves that the conditions (2) are necessary and sufficient for the existence of such a functional operation defined by the function $\chi(t)$. In this note we shall proceed in the opposite way. We shall solve the set (2) of linear inequalities in infinitely many variables and shall find that its most general solution may be represented parametrically in the form (1). In another paper the same point of view will be applied to similar but more general problems considered by Carathéodory, F. Riesz, and Hausdorff.

2. *Finite Sequences.* It seems natural to start first with the following problem. Let us call the finite sequence $\mu_0, \mu_1, \cdots, \mu_p$ *completely monotonic if*

$$\Delta^l \mu_m \geqq 0 \quad \text{for } l, m = 0, 1, 2, \cdots, p, \text{ with } l + m \leqq p.\dagger \tag{4}$$

We seek a parametric representation for the most general solution of the set (4) *of linear inequalities.*

Such a representation is readily found. The identity $\Delta^l \mu_m = \Delta^l \mu_{m+1} + \Delta^{l+1}\mu_m$ shows that the set (4) is a consequence of its sub-set

$$\Delta^{p-m}\mu_m \geqq 0, \quad \text{for } m = 0, 1, 2, \cdots, p, \tag{5}$$

which means that the $p+1$ linear forms

$$\begin{aligned}
\rho_{p,0} &= \mu_0 - \binom{p}{1}\mu_1 + \binom{p}{2}\mu_2 - \cdots + (-1)^p \mu_p,\\
\rho_{p,1} &= \mu_1 - \binom{p-1}{1}\mu_2 + \cdots + (-1)^{p-1}\mu_p,\\
\rho_{p,2} &= \mu_2 - \cdots + (-1)^{p-2}\mu_p,\\
&\;\;\vdots\\
\rho_{p,p-1} &= \mu_{p-1} - \mu_p,\\
\rho_{p,p} &= \mu_p,
\end{aligned} \tag{6}$$

* This statement is certainly verified for the previous case, because (2) gives $\sum_{m=0}^{p}\binom{p}{m}|\Delta^{p-m}\mu_m| = \sum_{m=0}^{p}\binom{p}{m}\Delta^{p-m}\mu_m = \mu_0$.

† This definition is justified by the fact that these are all the higher differences which may be derived from our finite sequence.

should be non-negative. The linear transformation inverse to (6) is

$$
\begin{aligned}
\mu_0 &= \rho_{p,0} + \binom{p}{1}\rho_{p,1} + \binom{p}{2}\rho_{p,2} + \cdots + \rho_{p,p}, \\
\mu_1 &= \rho_{p,1} + \binom{p-1}{1}\rho_{p,2} + \cdots + \rho_{p,p}, \\
\mu_2 &= \rho_{p,2} + \cdots + \rho_{p,p}, \\
&\vdots \\
\mu_{p-1} &= \rho_{p,p-1} + \rho_{p,p}, \\
\mu_p &= \rho_{p,p}.
\end{aligned} \tag{7}
$$

Our problem is solved by the set (7) which yields for

$$\rho_{p,0} \geqq 0, \quad \rho_{p,1} \geqq 0, \cdots, \rho_{p,p} \geqq 0, \tag{8}$$

a parametric representation for the most general finite sequence $\mu_0, \mu_1, \cdots, \mu_p$, which is completely monotonic.

3. *Infinite Sequences.* We return to our original problem. The inequalities (2) are certainly necessary for a parametric representation (1) with $\chi(t)$ monotonic, since

$$\Delta^l \mu_m = \int_0^1 t^m(1-t)^l d\chi(t) \geqq 0.$$

Conversely, suppose that now the inequalities (2) hold. The set of inequalities (4) holds for every value of $p = 1, 2, 3, \cdots$ and therefore also (6), (7), and (8) are true for these values. We shall obtain the set of equations (1) from the set (7) by making $p \to \infty$.

The nth equation of the set (7) gives for $p \geqq n$

$$\mu_n = \sum_{m=n}^{p} \binom{p-n}{m-n} \rho_{p,m} = \sum_{m=n}^{p} \frac{\binom{p-n}{m-n}}{\binom{p}{m}} \lambda_{p,m},$$

and finally

$$\mu_n = \sum_{m=0}^{p} \frac{m(m-1)(m-2)\cdots(m-n+1)}{p(p-1)(p-2)\cdots(p-n+1)} \lambda_{p,m}, \tag{9}$$

if we put

$$\lambda_{p,m} = \binom{p}{m}\rho_{p,m} = \binom{p}{m}\Delta^{p-m}\mu_m.$$

The right side of equation (9) may be written as a Stieltjes integral if we introduce with Hausdorff the function $\chi_p(t)$ defined as follows: $\chi_p(t)$ is a step-function defined for $0 \leqq t \leqq 1$, with $\chi_p(0)=0$, its points of discontinuity being for the values $t=m/p$ $(m=0, 1, 2, \cdots, p)$, where its jump has the value $\lambda_{p,m}$. This function will be completely defined by the additional condition $2\chi_p(t)=\chi_p(t+0)+\chi_p(t-0)$ for $0<t<1$. From the first equation (7) we get $\chi_p(1)=\mu_0$ and the conditions (8) show that $\chi_p(t)$ is a non-decreasing function. From (9) we obtain

$$(10)\qquad \mu_n = \int_0^1 \frac{t\left(t-\frac{1}{p}\right)\left(t-\frac{2}{p}\right)\cdots\left(t-\frac{n-1}{p}\right)}{\left(1-\frac{1}{p}\right)\left(1-\frac{2}{p}\right)\cdots\left(1-\frac{n-1}{p}\right)}d\chi_p(t)$$
$$= \int_0^1 t^n d\chi_p(t) + O\left(\frac{1}{p}\right).$$

The remainder of the proof is similar to Hausdorff's. A theorem of E. Helly* insures the existence of a sub-sequence $\chi_q(t)$ which converges to a monotonic function $\bar{\chi}(t)$. When $p=q\rightarrow\infty$, equation (10) reduces to (1) with $\chi(t)=\bar{\chi}(t)$. The relation $\int_0^1 t^n d\chi_q(t)\rightarrow\int_0^1 t^n d\bar{\chi}(t)$ follows from a theorem of Helly (loc. cit. pp. 288–289) or just as well from a known theorem on Riemann integrals as follows:

$$\int_0^1 t^n d\chi_q(t) = \chi_q(1) - n\int_0^1 \chi_q(t)t^{n-1}dt$$
$$\rightarrow \bar{\chi}(1) - n\int_0^1 \bar{\chi}(t)t^{n-1}dt = \int_0^1 t^n d\bar{\chi}(t).$$

This proves the first part of the theorem. Using the fact that $\chi(t)$ is uniquely defined at all its points of continuity by the set (1) and the condition $\chi(0)=0$, Hausdorff proved (loc. cit., pp. 225–226) that the whole sequence $\chi_p(t)$ converges to a monotonic function $\chi(t)$ which solves the problem of moments.

* E. Helly, *Über lineare Funktionaloperationen*, Sitzungsberichte der Wiener Akademie, vol. 121 IIa (1912), p. 286.

The second statement of the theorem concerning the solution of bounded variation is also readily proved. If we suppose the function $\chi(t)$ of equations (1) to be of bounded variation, then $\chi(t)=\chi^1(t)-\chi^2(t)$, with $\chi^1(t)$ and $\chi^2(t)$ both non-decreasing functions. In Hausdorff's notation (loc. cit., p. 231) we have $\lambda_{p,m}=\lambda_{p,m}^1-\lambda_{p,m}^2$ and hence

$$\sum_{m=0}^{p}|\lambda_{p,m}| \leqq \sum_{m=0}^{p}\lambda_{p,m}^{1} + \sum_{m=0}^{p}\lambda_{p,m}^{2} = \mu_0^1 + \mu_0^2,$$

which proves the necessity of the condition (3). Conversely, if (3) or $\sum_{m=0}^{p}|\lambda_{p,m}|<K$ holds for $p\to\infty$, then the sequence $\chi_p(t)$ is a sequence of uniformly bounded functions of uniformly bounded variations. Another theorem of E. Helly* insures the existence of a sub-sequence $\chi_q(t)$ converging to a function $\bar{\chi}(t)$ of bounded variation. Just as above, on making $p=q\to\infty$, equation (10) becomes (1) with $\chi(t)=\bar{\chi}(t)$ and the second part of the theorem is proved. Moreover, formula (10) and the same arguments used by Hausdorff for the previous case permit us to prove that the whole sequence $\chi_p(t)$ converges to a function $\chi(t)$ of bounded variation satisfying the equations (1).

THE UNIVERSITY OF CHICAGO

---

* E. Helly, loc. cit., §6, Theorem 7 on p. 283. This theorem is a corollary of his first theorem used above.

Reprinted from
Bull. Amer. Math. Soc. (Feb. 1932), pp. 72–76

THE RELAXATION METHOD FOR LINEAR INEQUALITIES

T. S. MOTZKIN AND I. J. SCHOENBERG

I. Statement of Problem and Main Results

1. The relaxation method. Let A be a closed set of points in the n-dimensional euclidean space E_n. If p and p_1 are points of E_n such that

$$|p - a| > |p_1 - a|, \text{ for every } a \in A, \tag{1.1}$$

then p_1 is said to be *point-wise closer* than p to the set A. If p is such that there is no point p_1 which is point-wise closer than p to A, then p is called a *closest point* to the set A. In 1922 Fejér **(2)** made the interesting observation that the set of closest points to A is identical with the convex hull $K(A)$ of the set A. We have mentioned this remark because it will suggest a way of dealing with our main problem, to which we now turn.

We are given a consistent system of m linear inequalities

$$\sum_{j=1}^{n} a_{ij}x_j + b_i \geqslant 0 \qquad (i = 1, \ldots, m). \tag{1.2}$$

The coefficients a_{ij} and b_i being given numerically, the problem is to devise a numerical procedure which will furnish a solution $(x_1, \ldots, x_n)$ of the system (1.2). In the case of a homogeneous system, i.e. when all $b_i = 0$, we add the obvious requirement that the solution $(x_1, \ldots, x_n)$ obtained be different from the trivial solution $(0, \ldots, 0)$.

A natural approach to this problem will be suggested by Fejér's idea as soon as we place the problem in its customary geometric setting. Each of the inequalities (1.2) defines a closed half-space

$$H_i\colon \sum_{j=1}^{n} a_{ij}x_j + b_i \geqslant 0, \tag{1.3}$$

in terms of which the set of points corresponding to the solutions of (1.2) is identical with the convex polytope

$$A = \bigcap_{i=1}^{m} H_i, \tag{1.4}$$

which is assumed from the outset not to be void. Let $p \notin A$ be given. The following simple construction furnishes a point p_1 which is point-wise closer than p to A: Clearly $p \notin H_j$ for some j. Let p' be symmetric to p with respect to the

Received May 11, 1953. Presented to the American Mathematical Society, September, 1952. This work was performed on a National Bureau of Standards contract with the University of California, Los Angeles, and was sponsored (in part) by the Office of Scientific Research, USAF.

boundary π_j of H_j. If p_1 is on the segment joining p to p' and $p_1 \neq p, p'$, then clearly if $a \in H_j$ then

$$|p - a| > |p_1 - a|$$

holds. As $A \subset H_j$, we see that (1.1) is verified, i.e. p_1 *is point-wise closer than* p *to* A. The numerical "construction" of p_1 is easily done as follows. Let q be the projection of p on the hyperplane π_j, choose a number λ such that $0 < \lambda < 2$ and set

$$p_1 = p + \lambda(q - p).$$

In passing from p to p_1, the point-wise approach to A would seem to be strongest if among the H_j, not containing p, we select the one which is furthest away from p. If $\lambda = 2$ then $p_1 = p'$ and then (1.1) again holds, with the exception that we have the equality sign for the points of A which are on the boundary of H_j, if such points exist.

These remarks suggest the following systematic search for a point of A: Choose a point p at will. If $p \in A$, i.e. its coordinates satisfy (1.2), our quest has ended. If $p \notin A$, let H_j be such that

$$\text{dist}(p, H_j) = \max_i \text{dist}(p, H_i), \tag{1.5}$$

where dist denotes euclidean distance. Let q be such that

$$q \in H_j, \quad |p - q| = \text{dist}(p, H_j). \tag{1.6}$$

If λ is a constant, $0 < \lambda \leqslant 2$, we define

$$p_1 = p + \lambda(q - p). \tag{1.7}$$

For convenience we abbreviate this construction by writing

$$p_1 = F_\lambda(p). \tag{1.8}$$

where $F_\lambda(p)$, defined for $p \notin A$, has been made single-valued by some preassigned rule for choosing j in case that (1.5) should not define j uniquely.[1] If $p_1 \in A$ our process has terminated. If $p_1 \notin A$ we iterate (1.8), obtaining

$$p_2 = F_\lambda(p_1),$$

and we continue in like manner deriving a sequence of points $p = p_0, p_1, p_2, \ldots$, all outside A and connected by the relation

$$p_{\nu+1} = F_\lambda(p_\nu) \qquad (\nu = 0, 1, \ldots). \tag{1.9}$$

There are two alternatives: (1) The process terminates after N steps with a point $p_N \in A$; (2) The process continues indefinitely producing an infinite sequence $\{p_\nu\}$.

2. Statement of the main theorems. S. Agmon **(1)** has recently shown that if $0 < \lambda < 2$ and the sequence $\{p_\nu\}$ is infinite, then p_ν converges, as $\nu \to \infty$,

[1]For instance the smallest j satisfying (1.5).

to a point on the boundary of A. We give a new proof of this result (Theorem 1, Case 1, and Theorem 2, Case 1, below). Our main contribution, however, is the investigation of the case when $\lambda = 2$. Throughout this paper we denote by r the dimensionality of the polytope A defined by (1.4). As we assume that A is not void, r may have any value from zero to n. We denote by L_r the r-flat which contains A.

THEOREM 1. *We assume that $r = n$, i.e. A is not contained in any hyperplane of E_n. Let $\{p_\nu\}$ be a sequence of points obtained by the process described in* § 1. *There are two cases:*

CASE 1. *If $0 < \lambda < 2$ then either $\{p_\nu\}$ terminates or else p_ν converges to a point l on the boundary of A.*

CASE 2. *If $\lambda = 2$ then the sequence $\{p_\nu\}$ always terminates.*

The formulation of our results for the case when $r < n$ requires the following remarks concerning spherical surfaces. Let L_r be a given r-flat in E_n, $0 \leqslant r \leqslant n - 1$. We are also given a point p, $p \notin L_r$. *Let X be the locus of points x such that*

$$|x - a| = |p - a|, \qquad \textit{for every } a \in L_r. \tag{1.10}$$

We claim that X is a spherical surface S_{n-r-1} of dimension $n - r - 1$. Thus if $r = 0$, then L_r reduces to a single point a and X is evidently the S_{n-1} with center at a passing through p. In the other extreme case when $r = n - 1$, L_r is a hyperplane and the locus X contains exactly two points: the point p and its symmetric image with respect to L_r. These two points form a S_0 located on the line through p which is normal to L_r. A general proof of our assertion is as follows: Let b be the orthogonal projection of p onto L_r. Erect at b the $(n - r)$-flat L'_{n-r} which is normal to L_r. Evidently $p \in L'_{n-r}$. *Then $x \in X$ if and only if $x \in L'_{n-r}$ and $|x - b| = |p - b|$.* Indeed, assume that $x \in X$. By (1.10), for $a = b$, we obtain that $|x - b| = |p - b|$. This last equality and (1.10) show that for every $a \in L_r$ the two triangles xba and pba are congruent. Since $\angle pba = 90°$ we conclude that $\angle xba = 90°$. Hence the line joining b and x is normal to L_r and $x \in L'_{n-r}$. Conversely, if $x \in L'_{n-r}$ and $|x - b| = |p - b|$, let us show that $x \in X$. This is now clear because the two triangles pba and xba $(a \in L_r)$ are right-angled at b and have equal legs respectively. This implies (1.10), hence $x \in X$. The locus X may accordingly be defined by the two conditions

$$x \in L'_{n-r}, \quad |x - b| = |p - b|,$$

and is therefore seen to be identical with the spherical surface S_{n-r-1} of L'_{n-r} having its center at b and passing through p. We shall refer to S_{n-r-1} as a spherical surface having L_r as its axis, for indeed, by (1.10), L_r is precisely the locus of points a with the property of being equidistant from all points of S_{n-r-1}.

THEOREM 2. *We assume that* $r < n$, $A \subset L_r$. *Let* $\{p_\nu\}$ *be a sequence of points obtained by the process of* § 1.

CASE 1. *If* $0 < \lambda < 2$, *then* $\{p_\nu\}$ *either terminates or else* p_ν *converges to a point* l *of* A.

CASE 2. *If* $\lambda = 2$, *then* $\{p_\nu\}$ *either terminates or else there is a number* ν_0 *such that the points* p_ν, *for* $\nu \geqslant \nu_0$, *are on a spherical surface* S_{n-r-1} *having* L_r *as its axis.*

3. Remarks. (a) The procedure here described for finding a solution of (1.2) is called the *relaxation method*, especially if $\lambda = 1$, when it may also be called the *projection method*. We speak of *under-relaxation* or *over-relaxation* depending on whether $0 < \lambda < 1$ or $1 < \lambda < 2$. The case when $\lambda = 2$ is an extreme case of over-relaxation which may also be called the *reflexion method.*

(b) Theorem 1, Case 2, describes the main advantage of the reflexion method ($\lambda = 2$). No other value of λ, $0 < \lambda < 2$, has the property of always leading to a terminating sequence if $r = n$. If $0 < \lambda \leqslant 1$, this is easily shown by consideration of a triangle A in E_2. For a λ with $1 < \lambda < 2$, an example of a non-terminating sequence in E_2 is constructed as follows (Fig. 1): Let $\angle pqO = 90°$

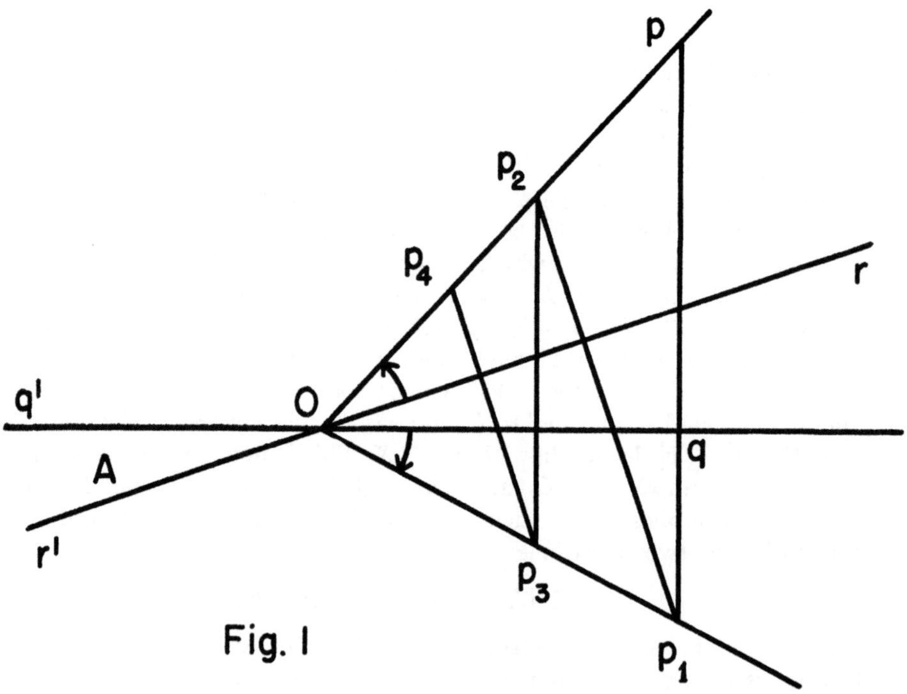

Fig. I

and p, q, p_1 be such that $pp_1/pq = \lambda$, hence $pq > qp_1$. Draw the ray Or such that $\angle pOr = \angle qOp_1$ and produce Or into a full line $r'Or$. Let A be the intersection of the closed half-plane below $q'q$ and the closed half-plane above $r'r$.

Starting with p and iterating the process $p_1 = F_\lambda(p)$, we obtain an *infinite* sequence of points $\{p_\nu\}$ which oscillate between the rays Op and Op_1, converging to O.

(c) Let $\lambda = 2$, $r < n$ and let us suppose that the sequence $\{p_\nu\}$ is infinite. By Theorem 2, Case 2, for $\nu > \nu_0$, all p_ν are on a S_{n-r-1} having L_r as its axis. Since p_ν and $p_{\nu+1}$ are both on S_{n-r-1}, the hyperplane with respect to which p_ν and $p_{\nu+1}$ are symmetric to each other must contain the axis L_r of S_{n-r-1}. We may state this result as

COROLLARY 1. *Let $r < n$ and let the reflexion process* ($\lambda = 2$) *lead to an infinite sequence $\{p_\nu\}$. Then there exists an integer ν_0 such that all the hyperplanes*

$$\pi_i\colon \sum_{j=1}^{n} a_{ij}x_j + b_i = 0 \tag{1.11}$$

which are actually used in the reflexion process for $\nu > \nu_0$ contain the entire polytope A and therefore also the r-flat L_r which contains A. Any such hyperplane, or combination of such independent hyperplanes, may therefore be used to reduce the problem to one of a dimension less than n.

(d) When the inequalities (1.2) are all homogeneous, we wish to find a point of A distinct from its vertex o. The relaxation method may well lead to the trivial solution o, if $0 < \lambda < 2$. Thus if $1 < \lambda < 2$ and if A is the "cone" in E_2 of Fig. 1, we have the infinite sequence $\{p_\nu\}$ converging to o. By Theorem 1, Case 2, and Theorem 2, Case 2, this can never happen if $\lambda = 2$.

In II and III we prove the Theorems 1 and 2 respectively. In IV we discuss the behavior of the reflexion process for a special kind of infinite family of half-spaces, namely all half-spaces of support of a bounded and closed convex set in E_n. A study of this problem, suggested by our previous discussion, seems justified by its own geometric interest.

II. A Proof of Theorem 1

4. On Fejér-monotone sequences of points. Let A be defined by (1.4) and let $q_0, q_1, q_2, \ldots$ be an infinite sequence of points outside A with the following properties:

$$q_i \neq q_{i+1}, \tag{2.1}$$

$$|q_i - a| \geqslant |q_{i+1} - a|, \text{ for all } a \in A,\ i = 0, 1, \ldots. \tag{2.2}$$

The sequence $\{q_\nu\}$ is approaching the set A point-wise and we summarize this situation by saying that the sequence $\{q_\nu\}$ is *Fejér-monotone* with respect to A. Concerning such sequences we prove

LEMMA 1. *Let the sequence $\{q_\nu\}$ be Fejér-monotone with respect to the polytope A, assumed to be of dimension r.*

CASE 1. *If $r = n$ then the sequence $\{q_\nu\}$ converges to a point.*

CASE 2. *If $r < n$ then the sequence $\{q_\nu\}$ either converges to a point or else the set of its limit points lies on a spherical surface S_{n-r-1} whose axis is the r-flat L_r spanned by A.*

Proof. CASE 1. Let $r = n$ and consider the spherical surfaces

$$S^\nu(a) : |x - a| = |q_\nu - a| \qquad (a \in A, \nu = 0, 1, \ldots). \tag{2.3}$$

By (2.2) the surface $S^\nu(a)$ is non-expanding as its center a is kept fixed and ν increases. Therefore the following limits exist

$$\lim_{\nu\to\infty} |q_\nu - a| = R(a) \qquad (a \in A). \tag{2.4}$$

Define the surface

$$S(a) : |x - a| = R(a) \qquad (a \in A), \tag{2.5}$$

and let us consider the set

$$X = \bigcap_{a \in A} S(a). \tag{2.6}$$

(a) *Every limit point l of the sequence $\{q_\nu\}$ is in X.* Indeed by (2.4) we see that $l \in S(a)$, for every $a \in A$, hence $l \in X$, by (2.6). We conclude that X is not void because the bounded sequence $\{q_\nu\}$ has at least one limit point.

(b) *The set X contains exactly one point l.* Indeed, if $l \neq l'$, $l \in X$, $l' \in X$, then let π denote the hyperplane of points equidistant from l and l'. If $a \in A$ then l, l' being both in X, are also both in $S(a)$. Hence $a \in \pi$ and we conclude that $A \subset \pi$ in contradiction to our assumption that $r = n$.

(c) *We conclude that $q_\nu \to l$.* Indeed, by (*a*) and (*b*) we see that $l = X$ is the only limit point of the sequence $\{q_\nu\}$.

CASE 2. Let $r < n$, $A \subset L_r$. If the sequence $\{q_\nu\}$ converges to a point then there is nothing to prove. Let us assume that

(2.7) *the sequence $\{q_\nu\}$ does not converge to a point.*

In any case the bounded sequence $\{q_\nu\}$ has limit points and let p be one of them. Define as before the spheres $S^\nu(a)$, $S(a)$ and the set X by (2.3), (2.4), (2.5) and (2.6). By (2.4) we have that $R(a) = |p - a|$ $(a \in A)$. The set X is therefore identical with the set of points x such that

$$|x - a| = |p - a|, \text{ for every } a \in A.$$

Since A spans L_r we may also define X as the locus of points x such that

$$|x - a| = |p - a|, \text{ for every } a \in L_r. \tag{2.8}$$

As shown in § 2, (2.8) defines a S_{n-r-1}, provided that the locus does not reduce to a point of L_r. Since this locus contains all limit points of $\{q_\nu\}$, our assumption (2.7) excludes this possibility and (2.8) defined a spherical surface S_{n-r-1}. This completes a proof of Lemma 1.

5. Proof of Theorem 1, Case 1. Here $r = n$; assume the sequence $\{p_\nu\}$ to be infinite. As already mentioned in § 1, the sequence $\{p_\nu\}$ is Fejér-monotone with respect to A. By Lemma 1, Case 1, the sequence $\{p_\nu\}$ converges to a point l:

$$\lim p_\nu = l. \tag{2.9}$$

We have to show that $l \in A$. For this purpose we introduce the function

$$d(x) = \max_i \text{dist}\,(x, H_i). \tag{2.10}$$

Observe that $d(x)$ is everywhere continuous and that

$$d(x) \begin{cases} =0, & x \in A, \\ >0, & x \notin A. \end{cases} \tag{2.11}$$

Now (2.9) implies that $|p_{\nu-1} - p_\nu| \to 0$ and therefore also that

$$d(p_\nu) = \frac{1}{\lambda}\,|p_{\nu+1} - p_\nu| \to 0.$$

By the continuity of $d(x)$ we have

$$d(l) = \lim d(p_\nu) = 0$$

and therefore $l \in A$, by (2.11). The point l, being a limit of exterior points p_ν, must be on the boundary of A.

6. Proof of Theorem 1, Case 2. Let $\lambda = 2$, $r = n$, and let us show that the sequence $\{p_\nu\}$ must terminate. Indeed, suppose it were infinite. By Lemma 1, Case 1, we again conclude that

$$\lim p_\nu = l \tag{2.12}$$

and the argument used in the previous paragraph shows that l is on the boundary of A. Let $p_{\nu+1}$ be obtained from p_ν by reflexion in the boundary π_{j_ν} of the half-space H_{j_ν}. By (2.12) it is clear that

$$\lim_{\nu\to\infty} \text{dist}\,(l, \pi_{j_\nu}) = 0. \tag{2.13}$$

The given family of hyperplanes (1.11) being finite, we conclude from (2.13) that

$$\text{dist}\,(l, \pi_{j_\nu}) = 0,$$

provided $\nu \geqslant \nu_0$, hence

$$l \in \pi_{j_\nu}, \qquad \nu \geqslant \nu_0.$$

This conclusion, however, contradicts (2.12). Indeed, every point p_ν $(\nu > \nu_0)$ is obtained from the preceding point $p_{\nu-1}$ by reflexion in a hyperplane through l. All these points must therefore lie on the spherical surface

$$|x - l| = |p_{\nu_0} - l| \;(> 0)$$

and can therefore never converge to l, as (2.12) requires.

III. A Proof of Theorem 2

7. Proof of Theorem 2, Case 1. We assume $r < n$, $0 < \lambda < 2$, and that the relaxation process (1.9) furnishes an infinite sequence $\{p_\nu\}$. We are to show that

p_ν converges to a point l of A. If the sequence $\{p_\nu\}$ converges to a point l then the argument of § 5 shows that $l \in A$ and we are through. Let us now assume that

(3.1) *the sequence* $\{p_\nu\}$ *does not converge*

and show that we shall reach a contradiction. By Lemma 1, Case 2, and (3.1), we conclude that all the limit points of $\{p_\nu\}$ are on a spherical surface S_{n-r-1} having L_r as an axis.

(a) *Our assumptions imply that*

$$\inf |p_{\nu+1} - p_\nu| = c > 0. \tag{3.2}$$

Indeed, consider the function $d(x)$, defined by (2.10), for the points x on the surface S_{n-r-1}. Since $A \cap S_{n-r-1} = 0$, we conclude that

$$d(x) > 0, \text{ if } x \in S_{n-r-1}.$$

Since $d(x)$ is continuous and S_{n-r-1} is compact, we conclude that

$$\gamma = \min d(x) > 0, \qquad x \in S_{n-r-1}.$$

Let us select γ_1 fixed such that $0 < \gamma_1 < \gamma$. Let δ be positive and let N_δ denote the set of points x defined by

$$\text{dist}\,(x, S_{n-r-1}) < \delta.$$

Again by the continuity of $d(x)$ we can select δ so small that

$$d(x) > \gamma_1, \qquad x \in N_\delta. \tag{3.3}$$

By Lemma 1, Case 2, we have $p_\nu \in N_\delta$, provided $\nu > \nu_0$. Now (3.3) implies that

$$|p_{\nu+1} - p_\nu| = \lambda\, d(p_\nu) > \lambda\, \gamma_1,$$

provided $\nu > \nu_0$. This proves (3.2) with $c \geqslant \lambda\, \gamma_1 > 0$.

We may now easily show that the assumption (3.1) leads to a contradiction. Indeed, let the point α on S_{n-r-1} be a limit point of $\{p_\nu\}$. For an appropriate subsequence $\{\nu'\}$ of the sequence of all integers $\{\nu\}$ we have

$$p_{\nu'} \to \alpha \in S_{n-r-1}.$$

For a subsequence $\{\nu''\}$ of $\{\nu'\}$ we may also assume that

$$p_{\nu''} \to \alpha, \text{ and } p_{\nu''+1} \to \beta \in S_{n-r-1}.$$

By (3.2) we conclude that $\alpha \neq \beta$; in fact

$$|\alpha - \beta| \geqslant c > 0.$$

Select on the line through α and β a point η such that $\beta - \alpha = \lambda(\eta - \alpha)$, and notice, because of $0 < \lambda < 2$, that η is nearer to β than to α:

$$|\eta - \alpha| > |\eta - \beta|. \tag{3.4}$$

For the subsequence $\{\nu''\}$, the half-spaces H_j, used in obtaining $p_{\nu''+1}$ from $p_{\nu''}$, must converge to the half-space

$$H : |x - \alpha|^2 - |x - \beta|^2 \geqslant |\eta - \alpha|^2 - |\eta - \beta|^2; \tag{3.5}$$

in fact H_j must already be identical with H, for sufficiently large ν'', because of the finiteness of the number of half-spaces H_i. This, however, leads to a contradiction, for on the one hand $A \subset H_j$ implies that $A \subset H$. Hence $x \in A$ implies $x \in H$ and therefore by (3.5) and (3.4)

$$|x - \alpha|^2 - |x - \beta|^2 \geqslant |\eta - \alpha|^2 - |\eta - \beta|^2 > 0,$$

or

$$|x - \alpha| > |x - \beta|. \tag{3.6}$$

On the other hand, A being on the axis of S_{n-r-1}, we must have in A the equality $|x - \alpha| = |x - \beta|$ in contradiction to (3.6). Thus our assumption (3.1) is untenable and the proof is completed.

8. Proof of Theorem 2, Case 2. We assume $r < n$, $\lambda = 2$, and that the reflexion process produces an infinite sequence $\{p_\nu\}$. This sequence cannot possibly converge, for its limit l would belong to A (§ 5) and would then have to terminate (§ 6). By Lemma 1, Case 2, the only alternative is that the sequence $\{p_\nu\}$ converges to a spherical surface S_{n-r-1} of axis L_r. Then (3.2) or

$$\inf |p_{\nu+1} - p_\nu| > 0 \tag{3.7}$$

again holds. Out of A select $r + 1$ fixed points $a_0, a_1, \ldots, a_r$ spanning L_r and let π_{j_ν} be the reflecting hyperplane used in obtaining $p_{\nu+1}$ as the point symmetric to p_ν. Since all limit points of p_ν are on S_{n-r-1}, it is clear that

$$\lim_{\nu\to\infty} |a_k - p_\nu| = \lim_{\nu\to\infty} |a_k - p_{\nu+1}| = R(a_k) \qquad (k = 0, \ldots, r).$$

This, together with (3.7), shows that

$$\lim_{\nu\to\infty} \operatorname{dist}(a_k, \pi_{j_\nu}) = 0 \qquad (k = 0, \ldots, r).$$

By the finiteness of our supply of reflecting hyperplanes we conclude from the last relations that

$$a_k \in \pi_{j_\nu}, \qquad \nu \geqslant \nu_0;\ k = 0, \ldots, r,$$

or what amounts to the same thing:

$$L_r \subset \pi_{j_\nu}, \qquad \nu \geqslant \nu_0.$$

In other words: there is a number ν_0 such that all reflexions for $\nu \geqslant \nu_0$ are performed with respect to hyperplanes π_{j_ν} which contain the axis L_ν of S_{n-r-1}. This, however, requires that

$$p_\nu \in S_{n-r-1}, \qquad \nu \geqslant \nu_0.$$

Indeed, if $p_{\nu_0} \notin S_{n-r-1}$, then all p_ν $(\nu \geqslant \nu_0)$ would lie on a surface S'_{n-r-1} of axis L_ν, passing through p_{ν_0}, and could then not converge to S_{n-r-1}, as we assumed.

IV. THE REFLEXION PROCESS WITH RESPECT TO A CONVEX DOMAIN

9. Statement of the problem. We have so far discussed the behavior of the reflexion process with respect to a finite family $\{H\}$ of half-spaces in E_n. Do the results obtained extend to infinite classes of half-spaces? We deal here only with the following special case of this problem: Let A be a given closed and bounded convex set in E_n. A closed half-space H belongs to the family F if and only if the boundary of H is a hyperplane of support of A and $A \subset H$. Let $p \notin A$ and let q be the point of A which is nearest to p. Let π_0 be the hyperplane through q which is normal to the segment joining p and q and let H_0 be the closed half-space, bounded by π_0, which does not contain p. Evidently $H_0 \in F$; also

$$\text{dist}\,(p, H_0) = \max_{H \in F} \text{dist}\,(p, H)$$

Indeed, if there were a $H \in F$ such that $\text{dist}\,(p, H) > \text{dist}\,(p, H_0) = |p - q|$, then $q \notin H$, in contradiction to the fact that $q \in A = \cap H$. This shows that the reflexion process with respect to the family $F = \{H\}$ amounts to the construction of the point

$$p_1 = p + 2(q - p) = F(p) \qquad (p \notin A). \tag{3.1}$$

Let us call $p_1 = F(p)$ the *image* of p with respect to A.

If $p_1 \notin A$ we may form $p_2 = F(p_1)$ and continue in like manner obtaining a sequence of points $p = p_0, p_1, p_2, \ldots$ connected by the relation

$$p_{\nu+1} = F(p_\nu) \qquad (\nu = 0, 1, \ldots). \tag{3.2}$$

We have again the old alternative: (1) The process terminates after N steps with $p_N \in A$; (2) The process continues indefinitely, producing an infinite sequence $\{p_\nu\}$.

10. The main result. The behavior of the reflexion process with respect to A is described by the following

THEOREM 3. *Let A be a closed convex and bounded set of E_n of dimension r and let L_r be the r-flat containing A. Suppose $p_0 \notin A$ and let $\{p_\nu\}$ be the sequence obtained by the reflexion process* (3.1), (3.2).

CASE 1. *If $r = n$, then the process always terminates.*

CASE 2. *Let $r < n$. If $p_0 \in L_r$ then the process terminates. If $p_0 \notin L_r$ then the process produces an infinite sequence $\{p_\nu\}$ with the following property. There is a number ν_0 such that for all $\nu > \nu_0$ the points p_ν oscillate between two points which are symmetric with respect to L_r.*

Proof. CASE 1. Let $r = n$ and let us assume, to obtain a contradiction, that the sequence $\{p_\nu\}$ is infinite. We know that $\{p_\nu\}$ is a Fejér-monotone sequence and hence that it converges to a point a by Lemma 1, Case 1. Clearly $a \in A$, and hence a is on the boundary of A, by the argument of § 5. Let P be the projection cone of A at the point a, i.e. the intersection of all H whose boundary hyperplanes pass through a. The cone P is convex and of dimension n, since $P \supset A$. There is therefore a half-space $H \in F$, whose boundary π supports P (and also A) at the point a and whose interior normal ai, at a, is wholly interior to P, except for the point a. Let us think of π as horizontal and its normal ai as pointing vertically downward. The point i being interior to P also a certain small spherical neighborhood S of i is in P. Let us call C the slim circular cone, of vertex a and axis ai, which is circumscribed to S. This convex cone C is wholly in P.

Let us denote by Q the convex cone of vertex a which is generated by the interior normals of all H supporting P (or A) at the point a. These H support also C at a. It follows that the closed convex cone Q (called the polar cone of A at a) has only the point a in common with π, Q being below π. Q was defined as the locus of all interior normals of A at a. Let us denote by N a small neighborhood of a on the boundary of A. Let Q' be a given closed and convex cone satisfying the following conditions: (i) Every ray of Q is interior to Q', (ii) Q' has only the point a in common with π. Clearly, a neighborhood N of a exists such that the interior normals to A at the points of N, if transferred parallel to themselves so as to start at a, will all lie in Q'. In fact otherwise we could find normals at points converging to A whose limit (which obviously being a normal at A) would be outside or on the boundary of Q'.

We now return to the sequence of points $\{p_\nu\}$ which converges to a. Let q_ν be the midpoint of the segment $p_\nu p_{\nu+1}$. It is clear by our construction that $q_\nu \in$ boundary of A and that the vectors $q_\nu p_{\nu+1}$ are interior normals to A. Also $p_\nu \to a$ implies that $q_\nu \to a$, hence $q_\nu \in N$, provided $\nu \geqslant \nu_0$. This leads to a contradiction. Indeed, consider the sequence of points

$$p_{\nu_0}, p_{\nu_0+1}, \ldots .$$

We know from what was said above that the vectors $q_\nu p_{\nu+1}$, if transferred to a, lie in Q'. That means that the vectors

$$p_\nu p_{\nu+1} \qquad (\nu \geqslant \nu_0)$$

have a positive component in the direction of the downward vertical vector ai. Since $p_\nu \to a$, we conclude that all points $p_\nu (\nu \geqslant \nu_0)$ are *above* the horizontal plane π. But this implies that also q_ν are above π. This, however, is absurd since $q_\nu \in A$ and A is below π.

CASE 2. If $r < n$ and $p_0 \in L_r$ then again the process must terminate by the previous case because we have no occasion to leave L_r in the course of our process.

We now assume that $r < n$ while $p_0 \notin L_r$. Let L_{r+1} be the $(r + 1)$-flat containing L_r and p_0. Note that we never leave L_{r+1} which amounts to assuming at the start that $r = n - 1$. Let, therefore, A be $(n - 1)$-dimensional, $A \subset L_{n-1}$ and $p_0 \notin L_{n-1}$. Let p'_0 and p'_1 be the projections of p_0 and p_1, respectively, on L_{n-1}. It should be clear that a point q_0 of A is nearest to p_0 if and only if it is nearest to p'_0. It follows that $p_1 = F(p_0)$ implies that $p'_1 = F(p'_0)$. Consider the sequence of reflexions $\{p_\nu\}$. It is clear that the distance from p_ν to L_{n-1} has the same positive value dist (p_0, L_{n-1}), the points p_ν passing from one side of the plane to the other alternately. However, the sequence $\{p'_\nu\}$ of their projections on L_{n-1} has the property $p'_{\nu+1} = F(p'_\nu)$. By the previous case this sequence "terminates" with a first $p'_N \in A$. From that moment onwards the sequence p_ν must oscillate between two points on the normal to L_{n-1} at the point $p'_N = p'_{N+1} = \ldots$.

References

1. S. Agmon, *The relaxation method for linear inequalities*, Can. J. Math., *6* (1954), 382–392.
2. L. Fejér, *Ueber die Lage der Nullstellen von Polynomen, die aus Minimumforderungen gewisser Arten entspringen*, Math. Annalen, *85* (1922), 41–48.

University of California, Los Angeles *University of Pennsylvania*

Reprinted from
Canad. J. Math. **6** (1954), pp. 393–404

The relaxation method for linear inequalities

The relaxation method for solving systems of linear inequalities was introduced and studied, in the very early fifties, by Motzkin and Schoenberg [**50***] and in a companion paper by Agmon [**A**]. It is a simple approach as it attempts to solve a system of inequalities by solving one inequality at a time, projecting or reflecting on a violated inequality. The method is an extension of the iterative approaches to the solution of systems of linear equations such as Kacmarz's projection method, Southwell's relaxation method or the SOR method.

The loveliest part of Motzkin and Schoenberg's paper is the use of a geometrical proof technique, namely the Fejér-monotonicity of the sequence of iterates (i.e., the fact that successive iterates get closer to **every** point in the feasible set), to prove the finite convergence of the relaxation method if reflexion (i.e., a relaxation parameter equal to 2) is used and the solution set has an interior. They also show the finite convergence of the maximal distance relaxation method, using reflexion, when applied to finding a point in a closed convex set with an interior.

Computer experiments conducted at the National Bureau of Standards in 1952 [**HM**] compared the simplex method, the relaxation method, and the fictitious play method (for matrix games) and concluded that the simplex method was fastest. The relaxation method lay dormant for some years before coming back to prominence under the guise of subgradient optimization, which, when used on the piecewise linear convex function that is defined as the maximum distance to a violated inequality, reduces to the relaxation method.

Held and Karp [**HK**] used subgradient optimization to compute bounds on a Lagrangean relaxation of the travelling salesman problem, thus introducing a very effective branch-and-bound method for that problem. Lagrangean relaxation has been applied with success to a wide range of hard combinatorial optimization problems, and subgradient optimization, here, seems to usually outperform the simplex method.

Very large systems of linear inequalities arising in reconstructing objects from their projections (computerized tomography) have been solved using algebraic reconstruction techniques – another alias for the relaxation method – ; see, for instance, [**H**].

Nondifferentiable optimization, both deterministic and stochastic, was studied extensively in the sixties, mostly in the Soviet literature, and the early solution techniques used Fejér type processes – i.e., the relaxation method or subgradient optimization – ; variable metric extensions of the relaxation method were defined by Shor (see [**S**]), and Yudin and Nemirovsky (see [**NY**]) and became known as the ellipsoid method. Khacian [**K**] showed that in the case of a system of linear inequalities the ellipsoid method finds a solution in polynomial time and space, thus proving that linear programming belongs to P; the ellipsoid method, with shallow cuts, does in fact compute an affine transformation of the space with the property that the relaxation method, with a relaxation parameter equal to 1, is polynomial in this transformed space.

References

[A] S. Agmon, The relaxation method for linear inequalities, *Canadian Journal of Mathematics* **6** (1954) 382–392.

[HM] A. J. Hoffman, M. Mannos, D. Sokolowsky and N. Wiegmann, Computational expe-
rience in solving linear programs, *SIAM Journal* **1** (1953), 17–33.

[HK] M. Held and R. M. Karp, The travelling salesman problem and minimum spannir
trees: Part II, *Mathematical Programming* **6** (1971), 6–25.

[H] G. T. Herman, *Image reconstruction from their projections: The fundamentals
computerized tomography*, Academic Press, New York (1980).

[K] L. G. Khacian, Polynomial algorithms in linear programming, *Zh. vycisl. Mat. ma
Fiz.* **20**, No. 1 (1980), 51–68; translated in *USSR Computational Mathematics ar
Mathematical Physics* **20**, No. 1 (1980), 53–72.

[S] N. Z. Shor, *Minimization methods for non-differentiable functions*, Springer-Verla
Berlin, Heidelberg (1985).

[NY] A. S. Nemirovsky and D. B. Yudin, *Problem complexity and method efficiency
optimization*, John Wiley, Chichester (1983).

Jean-Louis Goff
McGill University, Montre

METRIC GEOMETRY

ANNALS OF MATHEMATICS
Vol. 36, No. 3, July, 1935

REMARKS TO MAURICE FRÉCHET'S ARTICLE "SUR LA DÉFINITION AXIOMATIQUE D'UNE CLASSE D'ESPACE DISTANCIÉS VECTORIELLEMENT APPLICABLE SUR L'ESPACE DE HILBERT[1]

BY I. J. SCHOENBERG

(Received April 16, 1935)

1. Fréchet's developments in the last section of his article suggest an elegant solution of the following problem.

Let

$$a_{ik} = a_{ki} \qquad (i \neq k\,; i, k = 0, 1, \cdots, n)$$

be $\frac{1}{2}n(n+1)$ *given positive quantities. What are the necessary and sufficient conditions that they be the lengths of the edges of a n-simplex* $A_0A_1 \cdots A_n$*? More general, what are the conditions that they be the lengths of the edges of a n-"simplex"*[2] $A_0A_1 \cdots A_n$ *lying in a euclidean space* R_r $(1 \leqq r \leqq n)$ *but not in a* R_{r-1}*?*

This problem is fundamental in K. Menger's metric investigation of euclidean spaces ([6] and [7], particularly his third fundamental theorem in [7], pp. 737–743). It was solved by Menger by means of equations and inequalities involving certain determinants. Theorem 1 below furnishes a complete and independent solution of this problem. Theorem 2 solves the similar problem for spherical spaces previously treated by Menger's methods by L. M. Blumenthal and G. A. Garrett ([1]) and Laura Klanfer ([5]); it may be conveniently applied (Theorems 3 and 3′) to prove and extend a theorem of K. Gödel ([4]). The method of Theorem 1 is finally applied to solve the corresponding problem for spaces with indefinite line element recently considered by A. Wald ([8]) and H. S. M. Coxeter and J. A. Todd ([2]).

Construction of simplexes of given edges in euclidean spaces

2. A complete answer to the questions stated above is given by the following theorem.

THEOREM 1. *A necessary and sufficient condition that the* a_{ik} *be the lengths of the edges of an n-"simplex"* $A_0A_1 \cdots A_n$ *lying in* R_r*, but not in* R_{r-1}*, is that the quadratic form*

---

[1] These Annals, vol. 36 (1935), pp. 705–718.

[2] The quotation marks should indicate that the configuration may lie in a euclidean space of less than n dimensions.

$$(1)\qquad F(x_1, x_2, \cdots, x_n) = \sum_{i=1}^{n} a_{0\,i}^2\, x_i^2 + \sum_{\substack{i,k=1\\(i<k)}}^{n} (a_{0\,i}^2 + a_{0\,k}^2 - a_{i\,k}^2)\, x_i x_k$$

$$= \tfrac{1}{2} \sum_{i,k=1}^{n} (a_{0\,i}^2 + a_{0\,k}^2 - a_{i\,k}^2)\, x_i x_k$$

(with $a_{ik} = 0$ if $i = k$)

be positive, i.e. *always* $\geqq 0$, *and of rank r*.

The condition is necessary. Let $A_0A_1 \cdots A_n$ be an n-"simplex" with $A_iA_k = a_{ik}$. Let $A_0 = 0$ be the origin of a R_n in which A_i has the cartesian coördinates $\alpha_{i1}, \alpha_{i2}, \cdots, \alpha_{in}$. The point (in vector space notation)

$$P = x_1A_1 + x_2A_2 + \cdots + x_nA_n = (\xi_1, \xi_2, \cdots, \xi_n)$$

has the coördinates

$$\xi_\nu = x_1\alpha_{1\nu} + x_2\alpha_{2\nu} + \cdots + x_n\alpha_{n\nu} \qquad (\nu = 1, \cdots, n),$$

whence

$$\overline{OP}^2 = \| P \|^2 = \sum_1^n \xi_\nu^2 = \sum_{\nu=1}^{n} (x_1\alpha_{1\nu} + \cdots + x_n\alpha_{n\nu})^2$$

$$= \sum_{i=1}^{n} x_i^2 \sum_{\nu=1}^{n} \alpha_{i\,\nu}^2 + 2 \sum_{i<k} x_i x_k \sum_{\nu=1}^{n} \alpha_{i\nu}\alpha_{k\nu}.$$

Since

$$\sum_{\nu=1}^{n} \alpha_{i\,\nu}^2 = \overline{OA}_i^2 = a_{0\,i}^2,$$

$$2 \sum_{\nu=1}^{n} \alpha_{i\nu}\alpha_{k\nu} = \sum_{\nu=1}^{n} \alpha_{i\,\nu}^2 + \sum_{\nu=1}^{n} \alpha_{k\,\nu}^2 - \sum_{\nu=1}^{n} (\alpha_{i\nu} - \alpha_{k\nu})^2 = \overline{A_0A}_i^2 + \overline{A_0A}_k^2 - \overline{A_iA}_k^2$$

$$= a_{0\,i}^2 + a_{0\,k}^2 - a_{i\,k}^2,$$

we have

$$(2)\qquad \overline{OP}^2 = \| x_1A_1 + \cdots + x_nA_n \|^2 = F(x_1, x_2, \cdots, x_n).$$

Hence $F(x_1, \cdots, x_n)$ *is positive.* It follows furthermore from our assumptions that $P = 0$, hence $F = 0$, on a linear manifold of $n - r$ dimensions in the variables $x_1, \cdots, x_n$; hence F *is of rank r*.

The condition is sufficient. Let us first assume F to be positive definite, i.e. $r = n$. By means of a certain linear non-singular transformation

$$(3)\qquad (y) = H(x)$$

we get the identity

$$(4)\qquad F(x_1, \cdots, x_n) = y_1^2 + y_2^2 + \cdots + y_n^2.$$

Call A_0 the origin of the cartesian space of the variables $(y_1, \cdots, y_n)$ and

$$A_1, A_2, \cdots, A_n,$$

the n points which in virtue of (3) correspond to

$$(5)\quad (x_1, x_2, \cdots, x_n) = (1, 0, \cdots, 0), (0, 1, 0, \cdots, 0), \cdots, (0, 0, \cdots, 0, 1),$$

respectively. Their y-coördinates are readily found by (3). For their mutual distances we find by (3), (4) and (5),

$$\overline{A_0A_i}^2 = F(0, \cdots, \overset{(i)}{1}, \cdots, 0) = a_{0i}^2,$$

$$\overline{A_iA_k}^2 = F(0, \cdots, \overset{(i)}{1}, \cdots, -\overset{(k)}{1}, \cdots, 0) = a_{0i}^2 + a_{0k}^2 - (a_{0i}^2 + a_{0k}^2 - a_{ik}^2)$$
$$= a_{ik}^2, (i < k),$$

which show that $A_0A_1 \cdots A_n$ is precisely the n-simplex we are looking for. It is indeed an n-simplex because the points (5) are independent and (3) is non-singular.

If $r < n$, then (4) has to be replaced by

$$(6)\qquad F(x_1, \cdots, x_n) = y_1^2 + y_2^2 + \cdots + y_r^2.$$

The above procedure gives an n-simplex $A_0A_1 \cdots A_n$, however the quantities

$$F(1, 0, \cdots, 0) = a_{01}^2, \qquad F(1, -1, 0, \cdots, 0) = a_{12}^2, \cdots$$

are no more the squared lengths of the edges $\overline{A_0A_1}^2, \overline{A_1A_2}^2, \cdots$, but, viewing (6), the squared lengths of their projections on the sub-space $(y_1, \cdots, y_r)$, i.e., on the manifold $y_{r+1} = \cdots = y_n = 0$. Hence the projection $A_0'A_1' \cdots A_n'$ on this manifold of the n-simplex $A_0A_1 \cdots A_n$ is an n-"simplex" of the type we are looking for, i.e. with $A_i'A_k' = a_{ik}$. This n-"simplex" $A_0'A_1' \cdots A_n'$ is by construction contained in a R_r but not in a R_{r-1}, as readily seen.

Remark. If the matrix H of (3) is $H = \| h_{ik} \|$, then the y-coördinates of the vertices A_i and A_i' are

$$A_i = (h_{1i}, h_{2i}, \cdots, h_{ni}), \qquad A_i' = (h_{1i}, h_{2i}, \cdots, h_{ri}, 0, \cdots, 0).$$

The actual construction (i.e. determination of the coördinates of its vertices) of an n-"simplex" of edges a_{ik} is therefore carried out by a reduction of the quadratic form (1) *to its canonical form* (6). This is a problem of the second degree, for the transformation (3) is by no means required to be orthogonal.

As an illustration of this method let us construct a regular n-simplex with $a_{ik} = 1$. By (1) we have

$$F(x_1, \cdots, x_n) = \sum_{i=1}^{n} x_i^2 + \sum_{i<k} x_i x_k.$$

The identity

$$F(x_1, \cdots, x_n) = \sum_{i=1}^{n} \frac{i+1}{2i}\left(x_i + \frac{x_{i+1}}{i+1} + \frac{x_{i+2}}{i+1} + \frac{x_{i+3}}{i+1} + \cdots\right)^2,$$

$$(x_i = 0\,, \text{ if } i > n)\,,$$

shows that F is positive definite, hence the existence of our regular n-simplex is assured. The coördinates of the vertices of one such simplex may be read off from this last identity: one vertex is $A_0 = (0, \cdots, 0)$ while the coördinates of A_ν $(\nu = 1, \cdots, n)$ are

$$\frac{1}{\sqrt{2\cdot 1\cdot 2}}, \frac{1}{\sqrt{2\cdot 2\cdot 3}}, \frac{1}{\sqrt{2\cdot 3\cdot 4}}, \cdots, \frac{1}{\sqrt{2(\nu-1)\nu}}, \sqrt{\frac{\nu+1}{2\nu}}, \overbrace{0, \cdots, 0}^{n-\nu}.$$

Construction of simplexes of given edges in spherical spaces

3. Denote by S_r^ρ the r-dimensional spherical space

$$x_1^2 + x_2^2 + \cdots + x_{r+1}^2 = \rho^2$$

immersed in a R_{r+1}. The problem is as follows.

Given $\binom{n}{2}$ *positive quantities* α_{ik} $(i \neq k;\, i, k = 1, 2, \cdots, n)$ *and a positive* ρ, *to decide whether there exist, on some* S_r^ρ, *n points* $A_1, A_2, \cdots, A_n$, *such that their spherical distances* $\overset{\frown}{A_iA_k} = \alpha_{ik}$.

According to a remark of J. von Neumann this problem may be reduced to the preceding one regarding the construction of simplexes in euclidean spaces.[3] Combining his remark with Theorem 1 we get the following theorem which solves completely the problem stated above.

THEOREM 2. *Let* $\alpha_{ik} = \alpha_{ki}$ $(i \neq k;\, i, k = 1, 2, \cdots, n)$ *be* $\binom{n}{2}$ *given positive quantities. Necessary and sufficient conditions that there be, on some spherical manifold of radius* ρ, *n points* $A_1, A_2, \cdots, A_n$, *of mutual spherical distances equal to the* α_{ik}, i.e. $\overset{\frown}{A_iA_k} = \alpha_{ik}$, *are the inequalities.*

$$(7) \qquad \alpha_{ik} \leqq \pi\rho\,,$$

together with the condition that the quadratic form

$$(8) \qquad \Phi(x_1, x_2, \cdots, x_n) = \sum_{i,k=1}^{n} \cos(\alpha_{ik}/\rho)\, x_i x_k \quad (\alpha_{ik} = 0, \text{ if } i = k)$$

be positive. If r $(\geqq 1)$ *is the rank of* Φ, *then we can find such points in* S_{r-1}^ρ, *but not in* S_{r-2}^ρ *(which is undefined if* $r = 1$).

[3] After Prof. von Neumann's verbal communication I noticed that the same reduction has already been used by Laura Klanfer ([5]) to carry over Menger's results from euclidean spaces to spherical spaces.

The meaning of the inequalities (7) is obvious viewing the fact that no distance on a sphere of radius ρ can exceed $\pi\rho$. Suppose there are required points $A_1, \cdots, A_n$ on some $S_m^\rho (m \geqq 1)$. Call A_0 the sphere's center. Then $A_0 A_1 \cdots A_m$ is an n-"simplex" in R_{m+1}, the lengths of its edges being

$$(9) \qquad A_0A_1 = \rho = a_{0i}, \quad A_iA_k = 2\rho \sin \frac{\alpha_{ik}}{2\rho} = a_{ik} \quad (i, k = 1, \cdots, n; i \neq k).$$

From Theorem 1 we know that the construction of such a "simplex" amounts to the investigation of the quadratic form

$$F = \frac{1}{2} \sum_{i,k=1}^{n} (a_{0i}^2 + a_{0k}^2 - a_{ik}^2)\, x_i x_k = \rho^2 \sum_{i,k=1}^{n} \left(1 - 2 \sin^2 \frac{\alpha_{ik}}{2\rho}\right) x_i x_k$$

$$= \rho^2 \sum_{i,k=1}^{n} \cos (\alpha_{ik}/\rho)\, x_i x_k = \rho^2\, \Phi .$$

Its positivity is necessary and sufficient for the existence of $A_0 A_1 \cdots A_n$ with the properties (9). Its rank r indicates that $A_0 A_1 \cdots A_n$ is contained in R_r but not in R_{r-1}, hence $A_1 A_2 \cdots A_n$ with the desired properties, i.e. $\widehat{A_i A_k} = \alpha_{ik}$, is contained in S_{r-1}^ρ but not in S_{r-2}^ρ.

4. The set of quantities α_{ik} in Theorem 2 could be thought of as the edges of an abstractly defined $(n - 1)$-simplex (in Menger's terminology it is a semi-metric space composed of n points). Theorem 2 answers the question whether or not this abstract simplex can be immersed isometrically, i.e. by congruence, in a spherical space of given radius.

An interesting consequence of Theorem 2 is the following theorem.

Theorem 3. *Let σ_{n-1} be a $(n - 1)$-simplex of a $S_{n-1}^{\rho_0}$; there exists a radius $\rho_1 \leqq \rho_0$ such that σ_{n-1} can be immersed isometrically in $S_{n-2}^{\rho_1}$.*

Thus for $n = 3$ we get the following geometrically obvious statement: Any ordinary spherical triangle of a $S_2^{\rho_0}$ can be placed isometrically on a circumference of suitable radius $\rho_1 \leqq \rho_0$.

We note first that if σ_{n-1} can be immersed in $S_{n-2}^{\rho_0}$, which happens when the rank of

$$(10) \qquad \Phi(x; \rho) = \sum_{i,k=1}^{n} \cos (\alpha_{ik}/\rho)\, x_i x_k$$

is $\leqq n - 1$ for $\rho = \rho_0$, our theorem is proved with $\rho_1 = \rho_0$. Let us now assume $\Phi(x; \rho_0)$ to be of rank n, hence

$$\Phi(x; \rho_0) \text{ positive definite and } \frac{\alpha_{ik}}{\pi} \leqq \rho_0 ,$$

by Theorem 2. Note that $\Phi(x; \rho)$ can not be positive definite for all ρ with $0 < \rho \leqq \rho_0$, for it fails to be so if e.g. $\rho = \alpha_{12}/\pi$ since the first principal minor of order 2 of the discriminant of $\Phi(x; \alpha_{12}/\pi)$ vanishes. Call ρ_1 the greatest lower bound of the values σ with the property that $\Phi(x; \rho)$ is positive definite if $\sigma \leqq \rho \leqq \rho_0$. By a previous remark necessarily

$$\alpha_{ik} \leqq \pi\rho_1. \tag{11}$$

Now $\Phi(x; \rho)$ can not be positive definite if $\rho = \rho_1$ for it would still be so (by continuity) for all values ρ sufficiently close to ρ_1 in contradiction to the definition of ρ_1. But $\Phi(x; \rho_1)$ is necessarily positive, as the limit of positive definite forms $\Phi(x; \rho)$, for $\rho \to \rho_1 + 0$. Hence $\Phi(x; \rho_1)$ is positive and of rank $< n$. Now the proof is completed by (11) and Theorem 2.[4]

5. We shall now extend Theorem 3 to cover the case when $\rho_0 = \infty$, that is when σ_{n-1} is in R_{n-1}. We assume σ_{n-1}, of edges α_{ik}, to be a $(n-1)$-simplex of R_{n-1}, i.e.

$$\tfrac{1}{2} \sum_{i,k=2}^{n} (\alpha_{1i}^2 + \alpha_{1k}^2 - \alpha_{ik}^2)\, x_i x_k \text{ positive definite.} \tag{12}$$

Let us prove that *σ_{n-1} can be immersed isometrically in S_{n-1}^{ρ}, provided ρ is sufficiently large.* This is proved if we can show that

$$\Phi(x; \rho) = \sum_{i,k=1}^{n} \cos(\alpha_{ik}/\rho)\, x_i x_k$$

is positive definite if ρ is sufficiently large. A well known criterion states that a quadratic form is positive definite if and only if all the n principal minors of its discriminant chosen as follows

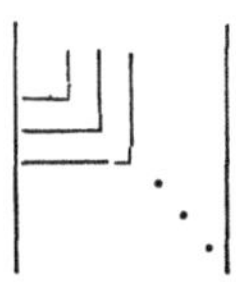

are positive (see Dickson [3], §40). If in the matrix of coefficients

$$\begin{Vmatrix} 1 & \cos\dfrac{\alpha_{1k}}{\rho} \\ \cos\dfrac{\alpha_{i1}}{\rho} & \cos\dfrac{\alpha_{ik}}{\rho} \end{Vmatrix} \qquad (i, k = 2, \cdots, n)$$

of $\Phi(x; \rho)$ we subtract the first line from all the other lines and then the first column from all the other columns we get the symmetric matrix

$$\begin{Vmatrix} 1 & \cos\dfrac{\alpha_{1k}}{\rho} - 1 \\ \cos\dfrac{\alpha_{i1}}{\rho} - 1 & \cos\dfrac{\alpha_{ik}}{\rho} - \cos\dfrac{\alpha_{i1}}{\rho} - \cos\dfrac{\alpha_{1k}}{\rho} + 1 \end{Vmatrix} \tag{13}$$

[4] Note that $\rho = \rho_1$ is the first value $< \rho_0$ which is a root of the transcendental equation $\det \| \cos(\alpha_{ik}/\rho) \| = 0$. It would be interesting to decide whether $\rho = \rho_1$ is necessarily a simple root of this equation.

which, as a result of the above criterion, will be the matrix of a positive definite form if and only if $\Phi(x; \rho)$ is positive definite itself. Noting that (13) can be written as follows

$$\left\| \begin{matrix} 1 & -\dfrac{\alpha_{1k}^2}{2\rho^2} + O\left(\dfrac{1}{\rho^4}\right) \\ -\dfrac{\alpha_{i1}^2}{2\rho^2} + O\left(\dfrac{1}{\rho^4}\right) & \dfrac{1}{2\rho^2}(\alpha_{1i}^2 + \alpha_{1k}^2 - \alpha_{ik}^2) + O\left(\dfrac{1}{\rho^4}\right) \end{matrix} \right\|, \quad (\rho \to \infty),$$

we see that the νth $(\nu > 1)$ principal minor of (13) is = to $\rho^{-2(\nu-1)}$ times the $(\nu - 1)^{\text{st}}$ principal minor of the discriminant of (12), plus a remainder $O(\rho^{-2\nu})$. By (12) all these minors are positive if ρ is sufficiently large, hence $\Phi(x; \rho)$ is positive definite and σ_{n-1} can be immersed in S_{n-1}^{ρ}. For any such $\rho = \rho_0$, Theorem 3 proves the existence of $S_{n-2}^{\rho_1}$, with $\rho_1 < \rho_0$, in which σ_{n-1} can be immersed. We have thus proved the following

THEOREM 3′ (of Gödel). *If σ_n is a n-simplex of R_n, then there always exists a S_{n-1}^{ρ} in which σ_n can be immersed isometrically.*[5]

The case of indefinite spaces

6. Consider the space of real variables $(y_1, \cdots, y_m)$ with the property that the square of the distance PP' of two points is given by the formula

$$\overline{PP'}^2 = \sum_{\nu=1}^{m} \epsilon_\nu (y_\nu - y'_\nu)^2,$$

with $\epsilon_\nu = +1$ for $\nu = 1, \cdots, p$, $\epsilon_\nu = -1$ for $\nu = p + 1, \cdots, p + q$ $(= m)$. We denote this space by $R_{p,q}$; thus $R_m = R_{m,0}$. The linear geometry of $R_{p,q}$ is obviously the same as that of $R_{p+q} = R_m$.

Let now $\frac{1}{2}n(n + 1)$ real numbers $c_{ik}(c_{ii} = 0, c_{ik} = c_{ki}; i, k = 0, \cdots, n)$ be given. Are there $n + 1$ points $A_0, A_1, \cdots, A_n$ in some space $R_{p,q}$ such that $\overline{A_iA_k}^2 = c_{ik}$, and what is the space $R_{p,q}$ of the least number of dimensions in which there are such points? A complete answer is furnished by the following theorem.

THEOREM 1′. *Consider the quadratic form*

$$F(x_1, x_2, \cdots, x_n) = \frac{1}{2} \sum_{i,k=1}^{n} (c_{0i} + c_{0k} - c_{ik})\, x_i x_k. \tag{14}$$

[5] A heuristic proof of this theorem for $n = 3$ is as follows. Think of the edges of σ_3 to be made of flexible strings; place in the interior of σ_3 a small sphere which is gradually inflated. This sphere will reach a certain definite size when it will become tightly packed within the 6 strings (edges) of σ_3. Note that in the rigorous proof above a very large sphere was used which was gradually deflated to its proper size.

Let it be of type (p, q).[6] *The necessary and sufficient conditions that there be* $n + 1$ *points* $A_0, A_1, \cdots, A_n$ *in* $R_{p',q'}$ *with* $\overline{A_iA}_k^2 = c_{ik}$, *are the inequalities*

$$p' \geqq p, \qquad q' \geqq q.$$

Thus $R_{p,q}$ *is the least space in which there are such points.*

The condition is necessary. Let the points $A_0 = 0, A_1, \cdots, A_n$ in $R_{p',q'}$ have the required property and let $R_{p,q}$ be the least linear subspace containing these points. We know that $p \leqq p'$, $q \leqq q'$, $p + q \leqq n$. Let $p + q = m$ and let $A_i = (\alpha_{i1}, \cdots, \alpha_{im})$ be the coördinates of A_i in $R_{p,q}$ with respect to an orthogonal coördinate system. For the point

$$P = x_1A_1 + \cdots + x_nA_n = (\xi_1, \cdots, \xi_m)$$

of coördinates $\xi_\nu = x_1\alpha_{1\nu} + \cdots + x_n\alpha_{n\nu}$ we find as in section 2 the identity

$$\overline{OP}^2 = \sum_{\nu=1}^{m} \epsilon_\nu \xi_\nu^2 = \sum_{\nu=1}^{m} \epsilon_\nu(x_1\alpha_{1\nu} + \cdots + x_n\alpha_{n\nu})^2 = F(x_1, \cdots, x_n).$$

Viewing our assumption that the matrix of the $\alpha_{\mu\nu}$ is of rank m and the law of inertia (Dickson, [3], p. 72), we see that $F(x)$ is of type (p, q).

The condition is sufficient. Assume first $p + q = n$. By a non-singular transformation

$$(3') \qquad (y) = H(x)$$

we get the identity

$$F(x_1, \cdots, x_n) = y_1^2 + \cdots + y_p^2 - y_{p+1}^2 - \cdots - y_n^2.$$

Consider in the space $R_{p,q}$ of the variables $(y_1, \cdots, y_n)$ the points whose x-coördinates are given by (5). We find as in section 2 $\overline{A_iA}_k^2 = c_{ik}$ and the theorem is proved, for $R_{p,q}$ can be considered as a subspace of $R_{p',q'}$, if $p' \geqq p$, $q' \geqq q$.

If $p + q = m < n$, then we get

$$F(x_1, \cdots, x_n) = y_1^2 + \cdots + y_p^2 - y_{p+1}^2 - \cdots - y_m^2.$$

To get the desired points we have to project the points $A_0, \cdots, A_n$ on the manifold $y_{m+1} = \cdots = y_n = 0$, which is a $R_{p,q}$.

7. It should be remarked that F defined by (14) is the most general real quadratic form in n variables. We thus have the following

Corollary. *Let*

$$(15) \qquad F = \sum_1^n b_{ik}\, x_ix_k$$

[6] That is of *index* p and rank $p + q$. See Dickson [3], p. 71.

be a non-degenerate real quadratic form of type (p, q). *If by means of*

$$(y) = H(x) \tag{3''}$$

we have

$$F = y_1^2 + \cdots + y_p^2 - y_{p+1}^2 - \cdots - y_n^2, \tag{16}$$

then the columns of the matrix

$$H = \begin{Vmatrix} h_{11} & \cdots & h_{1n} \\ \vdots & & \vdots \\ h_{n1} & \cdots & h_{nn} \end{Vmatrix}$$

are the y-coördinates in $R_{p,q}$ *of n points* $A_1, \cdots, A_n$, *which together with* $A_0 = (0)$ *have the property* $\overline{A_iA_k}^2 = c_{ik}$, *where*

$$c_{0i} = b_{ii}, \qquad c_{ik} = b_{ii} + b_{kk} - 2b_{ik} \qquad (i, k > 0).$$

A geometric interpretation of the reduction of (15) to the canonical form (16) by means of an *orthogonal* linear transformation is well known from the theory of quadrics. The above Corollary furnishes a geometric interpretation of this reduction by any linear non-singular transformation.

Probably the most concise description of the result of Theorems 1 and 1′ is as follows. If the squares of the edges of a simplex $A_0A_1 \cdots A_n$ are given real numbers, $\overline{A_iA_k}^2 = c_{ik}$, then this defines uniquely a (indefinite) space which, if referred to the coördinate unit-vectors $A_0A_1, A_0A_2, \cdots, A_0A_n$, has the line element

$$ds^2 = \tfrac{1}{2} \sum_{i,k=1}^{n} (c_{0i} + c_{0k} - c_{ik})\, dx_i\, dx_k.$$

SWARTHMORE COLLEGE, SWARTHMORE, PA.

REFERENCES

[1] L. M. BLUMENTHAL AND G. A. GARRETT: *Characterization of spherical and pseudo-spherical sets of points*, American J. of Math. vol. 53 (1931), pp. 619–640.

[2] H. S. M. COXETER AND J. A. TODD: *On points with arbitrarily assigned mutual distances*, Proc. of the Cambridge Phil. Soc., vol. 30 (1934), pp. 1–3.

[3] L. E. DICKSON: *Modern algebraic theories*, Sanborn Co., Chicago, 1926.

[4] K. GÖDEL: *Über die metrische Einbettbarkeit der Quadrupel des R_3 in Kugelflächen*, Ergebnisse eines math. Kolloquiums, Leipzig und Berlin, Heft 4 (1933), pp. 16–17.

[5] L. KLANFER: *Metrische Charakterisierung der Kugel* (Vienna Dissertation), Ergebnisse eines math. Koll., Heft 4 (1933), pp. 43–45.

[6] K. MENGER: *Bericht über metrische Geometrie*, Jahresber. der deutschen Math.-Ver., vol. 40 (1931), pp. 201–219.

[7] K. MENGER: *New foundations of euclidean geometry*, American J. of Math., vol. 53 (1931), pp. 721–745.

[8] A. WALD: *Komplexe und indefinite Räume*, Ergebnisse eines math. Koll., Heft 5 (1933), pp. 32–42.

METRIC SPACES AND POSITIVE DEFINITE FUNCTIONS*

BY

I. J. SCHOENBERG

1. **Introduction.** Let E_m denote the m-dimensional euclidean space and generally E_m^p the pseudo-euclidean space of m real variables with the distance function

$$(|x_1 - x_1'|^p + \cdots + |x_m - x_m'|^p)^{1/p}, \qquad p > 0.$$

As $p \to \infty$ we get the space E_m^∞ with the distance function $\max_{i=1,\dots,m} |x_i - x_i'|$. Let, furthermore, l^p stand for the space of real sequences with the series of pth powers of the absolute values convergent. Similarly let L^p denote the space of real measurable functions in the interval (0, 1) which are summable to the pth power, while C shall mean the space of real continuous functions in the same interval. In all these spaces a distance funct is assumed to be defined as usual.† L^2 is equivalent to the real Hilbert space $\mathfrak{H}$. The spaces E_m^p, l^p, and L^p are metric only if $p \geqq 1$, but we shall consider them also for positive values of $p<1$. Finally, if $\mathfrak{S}$ is a (not necessarily metric) space with the distance function $\overline{PP'}$, we shall denote by $\mathfrak{S}(\gamma)$ the new space which arises by changing the distance function from $\overline{PP'}$ to $\overline{PP'}^\gamma$, $(\gamma > 0)$.

A general theorem of Banach and Mazur ([1], p. 187) states that any separable metric space $\mathfrak{S}$ may be imbedded isometrically in the space C. Furthermore, as a special case of a well known theorem of Urysohn, any such space $\mathfrak{S}$ may be imbedded *topologically* in $\mathfrak{H}$. *Isometric* imbeddability of $\mathfrak{S}$ in $\mathfrak{H}$ is, however, a much more restricted property of $\mathfrak{S}$.

The chief purpose of this paper is to point out the intimate relationship between the problem of isometric imbedding and the concept of positive definite functions, if this concept is properly enlarged. As a first approach to this connection we consider here isometric imbedding in Hilbert space only. It turns out that the possibility of imbedding‡ in $\mathfrak{H}$ is very easily expressible in terms of the elementary function e^{-t^2} and the concept of positive definite functions (Theorem 1). The author's previous result ([10]) to the effect that $\mathfrak{H}(\gamma)$, $(0<\gamma<1)$, which is the space arising from $\mathfrak{H}$ by raising its metric to a

* Presented to the Society, December 29, 1937; received by the editors December 14, 1937.

† See, for example, Banach [1], pp. 11–12. The numbers in square brackets refer to the list of references at the end of the paper.

‡ Here and below the word "imbedding" stands for "isometrical imbedding."

fractional power, is imbeddable in $\mathfrak{H}$,* appears again as a simple consequence (Corollary 1). For the class of spaces $\mathfrak{S}_m$ arising from the euclidean space E_m by a general change of metric of the vector type (11) below, the condition of imbeddability in $\mathfrak{H}$ is directly expressible in a simple way in terms of the *usual* concept of positive definite functions as described by Mathias and Bochner (Theorem 2). The solution of this problem for $m=1$ (the problem of "screw lines" in $\mathfrak{H}$, von Neumann and Schoenberg [8]) allows us now to derive purely analytical results in the theory of positive definite functions with which it is equivalent. Two readily defined classes of positive definite functions are completely determined (Theorems 3 and 4). In particular two new proofs are given for the already known fact that the function exp $[-|x|^p]$ is positive definite for values of p in the range $0<p\leqq 2$, and not positive definite if $p>2$† (§5). One proof is geometrical, the other proof covering the case $0<p\leqq 2$ is analytical and may be read independently of the rest of this paper.

All previous results now allow us to conclude that the spaces $E_m^p(\gamma)$, $l^p(\gamma)$, and $L^p(\gamma)$ are imbeddable in $\mathfrak{H}$ for values of γ in the range $0<\gamma\leqq p/2$, where p is restricted to the range $0<p\leqq 2$ (Theorem 5). For $p=2$ we regain our previous result concerning $\mathfrak{H}(\gamma)$. It is interesting to compare and combine this result about L^p with a theoem of Banach and Mazur ([1], p. 203) concerning the linear dimensions of L^2 and L^q. According to this theorem L^2 is imbeddable in L^q if $q>1$. Now as $L^p(\gamma)$, $(0<p\leqq 2, 0<\gamma\leqq p/2)$, is imbeddable in $\mathfrak{H}$ or L^2, it follows that $L^p(\gamma)$, $(0<p\leqq 2, 0<\gamma\leqq p/2)$, is imbeddable in any L^q if $q>1$.

Similar as yet unsolved problems concerning the case of $p>2$ are shown to be equivalent to further knowledge as to the positive definite character of certain special functions of m variables. One of these unsolved problems suggests a probably possible way of extending an interesting theorem of L. M. Blumenthal on metric sets of four points to such sets of any number of points.

The reader primarily interested in the geometrical results of this paper may omit §4 and §5 entirely by taking the known Corollary 3 for granted.

2. **Positive definite functions.** A real continuous function $f(x_1, x_2, \cdots, x_m)$ which is defined for all real values of its variables and is even, that is for which $f(-x_1, \cdots, -x_m)=f(x_1, \cdots, x_m)$, is said to be positive definite (p.d.) if

$$(1)\qquad \sum_{i,k=1}^{n} f(x_1^{(i)} - x_1^{(k)}, \cdots, x_m^{(i)} - x_m^{(k)})\rho_i\rho_k \geqq 0$$

* W. A. Wilson [12], had previously remarked that $E_1(1/2)$ is imbeddable in $\mathfrak{H}$.

† Due in various parts to Pólya, Mathias, and P. Lévy. For references see Bochner [3], pp. 76–77. For still another treatment of the case $0<p\leqq 2$ see Bochner [5].

for arbitrary real ρ_i and any n points $(x^{(i)})$, $(i=1,\cdots,n)$, for $n=2,3,\cdots$. For $n=2$, (1) gives $|f(x_1,\cdots,x_m)|\leq f(0,\cdots,0)$, hence p.d. functions are bounded throughout space and take their maximum value at the origin. We shall use in the sequel only the following most simple properties of this interesting class of functions.

I. The function defined by

$$f(x_1,\cdots,x_m)=\int_{-\infty}^{\infty}\cdots\int_{-\infty}^{\infty}\psi(u_1,\cdots,u_m)\cdot\exp\,[i(x_1u_1+\cdots+x_mu_m)]du_1\cdots du_m, \tag{2}$$

where ψ is a non-negative even function such that its integral over the whole space exists, is p. d.

II. Any finite linear combination of p.d. functions with non-negative coefficients is again p.d. The product of two p.d. functions is again p.d.

III. A continuous function which is the limit of a sequence of p.d. functions is itself p.d.

For completeness we sketch the simple proofs. Property I follows from the fact that the left-hand side of (1) reduces to

$$\int\cdots\int\left|\sum_{k=1}^{n}\rho_k\exp\,[i(x_1^{(k)}u_1+\cdots+x_m^{(k)}u_m)]\right|^2\psi du_1\cdots du_m\geq 0.$$

The additive property is clear from the fact that (1) is a linear inequality in f. The multiplicative property is a direct consequence of a lemma of I. Schur ([11],p.10)* which states that if $\sum_1^n a_{ik}\rho_i\rho_k$, $\sum_1^n b_{ik}\rho_i\rho_k$, are two positive quadratic forms, then $\sum_1^n a_{ik}b_{ik}\rho_i\rho_k$ is also positive. Property III is immediately clear by continuity. We shall later on use the fact that $f(x)=\cos\lambda x$ is a p.d. function.

A closely allied concept is as follows. Let $\mathfrak{S}$ be a space in which a distance function $\overline{PP'}$ is defined subject to the following conditions: (1) $\overline{PP'}=\overline{P'P}\geq 0$ for arbitrary points P, P' in $\mathfrak{S}$, (2) $\overline{PP}=0$. A real continuous even function $g(t)$, which is defined in the range of values of $\pm\overline{PP'}$, $(P, P'$ in $\mathfrak{S})$, is said to be positive definite in $\mathfrak{S}$ if

$$\sum_{i,k=1}^{n} g(\overline{P_iP_k})\rho_i\rho_k\geq 0 \tag{3}$$

for arbitrary real ρ_i and any n points P_i (different or not) of $\mathfrak{S}$, $(n=2,3,\cdots)$.

* For a discussion and consequences of Schur's result see also Pólya and Szegö [9], pp. 106–107, 307–308.

This class of functions $g(t)$ has for a given space $\mathfrak{S}$ properties II and III above for similar reasons. Both definitions agree if $m=1$, while $\mathfrak{S}$ is the one-dimensional euclidean space E_1.

The peculiar relationship between the definitions, which will be clearer later on, is already exhibited by the following simple example needed in the sequel. From the formula

$$e^{-x^2} = 2^{-1}\pi^{-1/2}\int_{-\infty}^{\infty} e^{ixu}e^{-u^2/4}du$$

we get, replacing x by x_j, $(j=1, \cdots, m)$, and multiplying the resulting equations,

$$\begin{aligned} \exp\left[-(x_1^2+\cdots+x_m^2)\right] &= \exp\left[-[(x_1^2+\cdots+x_m^2)^{1/2}]^2\right] \\ &= 2^{-m}\pi^{-m/2}\int_{-\infty}^{\infty}\cdots\int_{-\infty}^{\infty}\exp\left[i(x_1u_1+\cdots+x_mu_m)\right] \qquad (4) \\ &\qquad\cdot \exp\left[-(u_1^2+\cdots+u_m^2)/4\right]du_1\cdots du_m, \end{aligned}$$

which shows at a glance (property I) that the function

$$f = \exp\left[-(x_1^2+\cdots+x_m^2)\right]$$

is p.d. and also that $g(t)=e^{-t^2}$ is p.d. in E_m. As m is arbitrary, this implies that *the function* e^{-t^2} *is positive definite in the real Hilbert space* $\mathfrak{H}$.

3. **Conditions for isometric imbedding in Hilbert space in terms of positive definite functions.** It was pointed out by K. Menger and by the author (for references see [10]) that a necessary and sufficient condition that a separable space $\mathfrak{S}$ be imbeddable in $\mathfrak{H}$ is that for any $n+1$ points of $\mathfrak{S}$, $(n\geqq 2)$, we have

$$\sum_{i,k=1}^{n}(\overline{P_0P_i}^2+\overline{P_0P_k}^2-\overline{P_iP_k}^2)\rho_i\rho_k \geqq 0,$$

for arbitrary real ρ_i.* Let us now put this condition in a slightly more symmetrical form. By summing over the three terms separately, we may write this as

$$2\sum_1^n \rho_k\cdot\sum_1^n \overline{P_0P_k}^2\rho_k - \sum_1^n \overline{P_iP_k}^2\rho_i\rho_k \geqq 0,$$

* This was proved for the case when $\mathfrak{S}$ is a separable *semi-metric* space; that is, when the metric $\overline{PP'}$ satisfies the additional condition (3) $\overline{PP'}>0$ if $P\neq P'$, whereas we postulated only that (1) $\overline{PP'}=\overline{P'P}\geqq 0$, (2) $\overline{PP}=0$. However, our quadratic inequality, for $n=2$, insures the triangle inequality $\overline{PQ}+\overline{QR}\geqq\overline{PR}$ for any three points of $\mathfrak{S}$. If we now *identify* with P all points Q such that $\overline{PQ}=0$ (which is now allowed, since $\overline{PQ}=0$ implies $\overline{RP}=\overline{RQ}$ for any R, on account of the triangle inequality) and do this for all points of $\mathfrak{S}$, we get a new space which is not only semi-metric but even metric.

and if we set $\rho_0 = -\sum_1^n \rho_k$, this last inequality is equivalent to

$$-\rho_0^2\overline{P_0P_0}^2 - 2\sum_0^n \overline{P_0P_k}^2\rho_0\rho_k - \sum_1^n \overline{P_iP_k}^2\rho_i\rho_k \geqq 0,$$

or, finally,

$$\sum_{i,k=0}^n \overline{P_iP_k}^2\rho_i\rho_k \leqq 0. \tag{5}$$

Hence the inequality (5), as a consequence of the relation

$$\sum_{i=0}^n \rho_i = 0, \tag{6}$$

is equivalent to the above stated condition of imbeddability.

Now we are prepared to express this condition in terms of p.d. functions. We have seen in the previous section that the function e^{-t^2} is p.d. in $\mathfrak{H}$. Clearly also $e^{-\lambda^2t^2}$ (λ real) is p.d. in $\mathfrak{H}$ as can be seen from (4) or by direct reasoning. Hence $e^{-\lambda^2t^2}$ is also p.d. in any subset of $\mathfrak{H}$. Now if $\mathfrak{S}$ is to be imbeddable in $\mathfrak{H}$, it is clearly a necessary condition that $e^{-\lambda^2t^2}$ be p.d. in $\mathfrak{S}$.

Let us prove now that this condition, together with the separability of $\mathfrak{S}$, is also sufficient. To prove this we have to show that (5) holds as a consequence of (6) for any $n+1$ points P_i of $\mathfrak{S}$. Now as $e^{-\lambda^2t^2}$ is p.d. in $\mathfrak{S}$, we have

$$\sum_0^n \rho_i\rho_k \exp\left[-\lambda^2\overline{P_iP_k}^2\right] \geqq 0 \tag{7}$$

by (3). We complete the proof in two different ways.

First proof. By expanding the left-hand side of (7) in power series we have, in view of (6),

$$-\lambda^2\left(\sum_0^n \overline{P_iP_k}^2\rho_i\rho_k\right) + \lambda^4\left(\sum_0^n \overline{P_iP_k}^4\rho_i\rho_k\right)/2 - \cdots \geqq 0,$$

which clearly implies (5) for small values of λ.

Second proof. Using the formulas

$$\begin{aligned} t^\alpha &= c(\alpha)\int_0^\infty (1 - e^{-\lambda^2t^2})\lambda^{-1-\alpha}d\lambda, \\ c(\alpha) &= \left[\int_0^\infty (1 - e^{-\lambda^2})\lambda^{-1-\alpha}d\lambda\right]^{-1}, \qquad 0 < \alpha < 2,\ t \geqq 0, \end{aligned} \tag{8}$$

which are immediately proved by substituting in the first integral λt^{-1} for λ, we have

$$\overline{P_iP_k}^\alpha = c(\alpha)\int_0^\infty (1 - \exp[-\lambda^2\overline{P_iP_k}^2])\lambda^{-1-\alpha}d\lambda.$$

If now ρ_i are any real numbers satisfying (6), we get, for $0<\alpha<2$,

$$(9)\qquad \sum_{i,k=0}^{n} \overline{P_iP_k}^\alpha\rho_i\rho_k = -c(\alpha)\int_0^\infty \left\{\sum_0^n \rho_i\rho_k \exp[-\lambda^2\overline{P_iP_k}^2]\right\}\lambda^{-1-\alpha}d\lambda \leqq 0,$$

by (7) and the obvious fact that $c(\alpha)>0$. Now we get again the desired inequality (5) on allowing α in (9) to approach the limit 2. We have thus proved the following theorem:

THEOREM 1. *A necessary and sufficient condition that a separable space $\mathfrak{S}$ with a distance function $\overline{PP'}$ with the properties $\overline{PP'}=\overline{P'P}\geqq 0$, $\overline{PP}=0$, be isometrically imbeddable in $\mathfrak{H}$, is that the family of functions $e^{-\lambda t^2}$, $(\lambda>0)$, be positive definite in $\mathfrak{S}$.*

Notice that the condition of this theorem may be restricted to require that $e^{-\lambda t^2}$ be p.d. in $\mathfrak{S}$ only for a set of positive values of λ admitting the origin $\lambda=0$ as a point of accumulation. The properties II and III (§2) will then imply that $e^{-\lambda t^2}$ is p.d. in $\mathfrak{S}$ for all positive values of λ.

Recalling that we denote by $\mathfrak{S}(\gamma)$, $(\gamma>0)$, the space obtained from $\mathfrak{S}$ by replacing its metric $\overline{PP'}$ by $\overline{PP'}^\gamma$, it is of interest to point out the further fact, implicitly contained in the previous second proof, that if $\mathfrak{S}$ is imbeddable in $\mathfrak{H}$, then so is $\mathfrak{S}(\gamma)$ for any value of γ in the range $0<\gamma<1$. Indeed, let $\alpha=2\gamma$; if $\mathfrak{S}$ is imbeddable in $\mathfrak{H}$, then $e^{-\lambda t^2}$, $(\lambda>0)$, is p. d. in $\mathfrak{S}$; hence (9) holds in virtue of (6), and $\mathfrak{S}(\gamma)$ is therefore imbeddable in $\mathfrak{H}$ on account of the form (5) of the imbeddability condition. Applying this conclusion to $\mathfrak{S}=\mathfrak{H}$ itself, we have the following corollary:

COROLLARY 1. *The space $\mathfrak{H}(\gamma)$, $(0<\gamma<1)$, obtained from Hilbert space $\mathfrak{H}$ by raising its metric to a power γ, is imbeddable in $\mathfrak{H}$.**

* We may even state the following more general theorem: *Let*

$$(8')\qquad F(t) = \int_0^\infty (1 - e^{-\lambda^2 t^2})\lambda^{-2}d\sigma(\lambda),$$

where $\sigma(\lambda)$ is non-decreasing for $\lambda\geqq 0$ and is such that $\int_1^\infty \lambda^{-2}d\sigma(\lambda)$ exists. If we change the metric of $\mathfrak{H}$ from $\overline{PP'}$ to $[F(\overline{PP'})]^{1/2}$, then the new space thus arising is imbeddable in $\mathfrak{H}$. Indeed (6), (7), and (8′) imply

$$\sum_1^n F(\overline{P_iP_k})\rho_i\rho_k = -\int_0^\infty \left(\sum_{i,k=1}^n \rho_i\rho_k \exp[-\lambda^2\overline{P_iP_k}^2]\right)\lambda^{-2}d\sigma(\lambda) \leqq 0,$$

and the theorem follows on account of the form (5) and (6) of the imbeddability condition. We leave open the question whether or not (8′) gives the most general function $F(t)$ with this property.

Added in proof, August, 1938: Formula (8′) gives indeed all functions with the property stated above. See the following paper *Metric spaces and completely monotone functions*, to appear in the Annals of Mathematics.

We now turn our attention to a special class of spaces of the type $\mathfrak{S}$ which are defined as follows. Let $\phi(x_1, \cdots, x_m)$ be a continuous function defined for all real values of its variables with the following properties

$$\phi(x_1, \cdots, x_m) \geqq 0, \qquad \phi(0, \cdots, 0) = 0, \tag{10}$$
$$\phi(-x_1, \cdots, -x_m) = \phi(x_1, \cdots, x_m);$$

and call $\mathfrak{S}_m$ the space of points $P=(x_1, \cdots, x_m)$ with the distance function

$$\overline{PP'} = [\phi(x_1 - x_1', \cdots, x_m - x_m')]^{1/2}. \tag{11}$$

For such a space, which is obviously separable, the condition of Theorem 1 that $e^{-\lambda t^2}$ be p.d. in $\mathfrak{S}_m$ may be expressed in a more familiar form. Indeed, on comparing the formulas (1), (3), and (11), we now obtain from Theorem 1 the following theorem:

THEOREM 2. *The space $\mathfrak{S}_m$ of m real numbers with the metric* (11) *is imbeddable in $\mathfrak{H}$ if and only if the functions*

$$f_\lambda = \exp[-\lambda\phi(x_1, \cdots, x_m)], \qquad \lambda > 0, \tag{12}$$

are positive definite in the sense (1) *of Mathias and Bochner for all positive values of* λ.

If $\phi(x_1, \cdots, x_m)$ is a homogeneous function, the conditions* (12) *may be replaced by the single condition that the function*

$$f_1 = \exp[-\phi(x_1, \cdots, x_m)] \tag{13}$$

be positive definite.

We mention the following corollary:

COROLLARY 2. *If $\phi(x_1, \cdots, x_m)$ is homogeneous and such that $e^{-\phi}$ is positive definite, then*

$$e^{-\phi^\gamma}, \qquad 0 < \gamma < 1,$$

is also positive definite.

For if $e^{-\phi}$ is p.d., then by Theorem 2, $\mathfrak{S}_m$ is imbeddable in $\mathfrak{H}$ and therefor $\mathfrak{S}_m(\gamma)$ is also (Corollary 1). Hence $\exp[-\phi^\gamma]$ is seen to be p.d. by applying Theorem 2 the other way around.

4. **Determination of certain classes of positive definite functions.** In this section we shall assume $m=1$, $\phi(x)$ being therefore a continuous non-negative even function vanishing at the origin. In this case we know precisely when $\mathfrak{S}_1$,

* We say that ϕ is homogeneous of degree κ if $\phi(tx_1, \cdots, tx_m) = t^\kappa\phi(x_1, \cdots, x_m)$ holds identically in the x_i and for $t>0$. A continuous homogeneous function ϕ with the properties (10) must, unless it vanishes identically, have a positive degree κ.

which is the space of real numbers with the metric $[\phi(x-x')]^{1/2}$ is imbeddable in $\mathfrak{H}$ (von Neumann and Schoenberg [8]). The most general function $\phi(x)$ for which this is the case is of the form

$$\phi(x) = \int_0^\infty \frac{\sin^2 xu}{u^2}\, d\sigma(u), \tag{14}$$

where $\sigma(u)$ is non-decreasing for $u \geqq 0$ such that

$$\int_1^\infty \frac{d\sigma(u)}{u^2} \text{ exists.} \tag{15}$$

Hence by Theorem 2 we infer that if

$$[f(x)]^\lambda = e^{-\lambda\phi(x)}$$

is p.d. for $\lambda > 0$, then $\phi(x)$ is of the form (14) and conversely. If now $f(x)$ is any positive function whose positive powers $[f(x)]^\lambda$ are all p.d., then considering

$$f(x)/f(0) = e^{-\phi(x)},$$

where $\phi(x)$ is necessarily non-negative, even, and vanishing at the origin, we conclude as before that $\phi(x)$ is of the form (14). We have thus proved the following theorem:

THEOREM 3. *The most general positive function $f(x)$ whose positive powers $[f(x)]^\lambda$, $(\lambda > 0)$, are all positive definite is of the form*

$$f(x) = \exp\left\{c - \int_0^\infty \frac{\sin^2 xu}{u^2}\, d\sigma(u)\right\}, \tag{16}$$

where $\sigma(u)$ is a non-decreasing function subject to the restriction (15), *while c is any real constant.*

A few remarks are called for regarding the condition of this theorem that $[f(x)]^\lambda$ be p.d. for $\lambda > 0$. In the first place, as remarked after the statement of Theorem 1, the range of λ in this condition may be restricted to a sequence of positive numbers tending to zero.

A second and more important remark is that Theorem 3 becomes false if we assume only that the positive function $f(x)$ is p.d., or in other words: formula (16) always represents a positive and p.d. function; it does not, however, represent all such functions but only those whose fractional powers are also p. d. To prove this statement it suffices to exhibit a p.d. function $f(x) > 0$ such that $[f(x)]^\lambda$, $(0 < \lambda < 1)$, are not all p.d. Such a function is

$$f_\epsilon(x) = \epsilon^2 + \cos^2 x,$$

if ϵ is sufficiently small, for it is easy to see that

$$[f_\epsilon(x)]^{1/2} = (\epsilon^2 + \cos^2 x)^{1/2}$$

is not p.d. for such values of ϵ. Indeed, if it were p.d. for arbitrarily small ϵ, it would follow that the limiting function

$$\lim_{\epsilon\to 0} [f_\epsilon(x)]^{1/2} = |\cos x|$$

were also p.d. (property III). But this is not the case as is seen from the fact that the cosine series

$$|\cos x| = \frac{2}{\pi} + \frac{4}{\pi}\sum_{n=1}^{\infty}(-1)^{n-1}\frac{\cos 2nx}{4n^2-1}$$

has some negative coefficients, or even more directly from the fact that the form

$$\sum_{i,k=1}^{4} |\cos (x_i - x_k)| \rho_i\rho_k,$$

where $x_1=0$, $x_2=\pi/4$, $x_3=\pi/2$, $x_4=3\pi/4$, is not positive. Indeed, its determinant is readily found to be -1, hence negative. The essence of the matter is that Schur's theorem, "If $\sum a_{ik}\rho_i\rho_k$ is positive, then so is $\sum a_{ik}^n\rho_i\rho_k$ for $n=1, 2, 3, \cdots$," can not be so extended that we may conclude that $\sum |a_{ik}|^\lambda\rho_i\rho_k$, $(\lambda>0)$, is positive.

The third and last remark is that Theorem 3 is now equivalent to the theorem that (14) and (15) give the most general $\phi(x)$ such that the space of real numbers with the metric $[\phi(x-x')]^{1/2}$ be imbeddable in $\mathfrak{H}$ (see [8]). Hence a direct proof of Theorem 3 would furnish a new proof of that theorem.

While the problem of determining all positive and positive definite functions is yet unsolved, there is another subclass of this class of functions which can now be readily determined. It will in fact be a subclass of the class determined by Theorem 3.

Let $\psi(x)$ be a p.d. function, c a real constant. Clearly

$$(17)\qquad f(x) = \exp[c + \psi(x)] = e^c\left(1 + \frac{1}{1!}\psi(x) + \frac{1}{2!}\psi^2(x) + \cdots\right)$$

is also p.d. in virtue of the properties II and III. It has moreover the following additional two properties: (α) It is bounded away from zero, since $f(x) \geqq \exp[c-\psi(0)]>0$. ($\beta$) All its positive powers $[f(x)]^\lambda$, $(\lambda>0)$, are also p.d. Let us show that the converse is true, that any function $f(x)$ having the properties (α) and (β) is of the form (17) where $\psi(x)$ is p.d. Now an $f(x)$ with the properties (α) and (β) clearly belongs to the class described by Theorem 3, and all we have to do is to decide which functions of the form (16) are bounded away from zero. Now this is the case if and only if the exponent (14), namely

$$\phi(x) = \int_0^\infty \frac{\sin^2 xu}{u^2} d\sigma(u)$$

is bounded in $-\infty < x < \infty$. The characteristic conditions for this have likewise been determined by von Neumann and the author in [8] and they are as follows: $\phi(x)$ is bounded if and only if $\sigma(0) = \sigma(+0)$ and $\int_{+0}^\infty u^{-2} d\sigma(u)$ exists. But if these conditions are fulfilled, then the non-decreasing function

$$\tau(u) = \int_{+0}^u \frac{d\sigma(u)}{u^2}$$

is bounded for $u > 0$, and (14) can be written as

$$(18) \qquad \begin{aligned} \phi(x) &= \int_{+0}^\infty \frac{\sin^2 xu}{u^2} d\sigma(u) = \int_{+0}^\infty \sin^2 xu \, d\tau(u) \\ &= \frac{1}{2} \int_{+0}^\infty (1 - \cos 2xu) d\tau(u) = \psi(0) - \psi(x), \end{aligned}$$

where

$$\psi(x) = \frac{1}{2} \int_{+0}^\infty \cos 2ux \, d\tau(u)$$

is p.d. Hence, by (14), (18), and (16),

$$f(x) = \exp [c - \phi(x)] = \exp [c - \psi(0) + \psi(x)]$$

and is indeed of the desired form (17). We have thus proved the following theorem:

THEOREM 4. *The most general positive function $f(x)$ which is bounded away from zero and whose positive powers $[f(x)]^\lambda$, $(\lambda > 0)$, are positive definite is of the form*

$$f(x) = \exp [c + \psi(x)],$$

where $\psi(x)$ is positive definite and c is a real constant.

EQUIVALENT STATEMENT: *If $f(x)$ is a positive function, then* $\log f(x)$ *differs by a constant from a positive definite function if and only if $f(x)$ is bounded away from zero and its powers $[f(x)]^\lambda$, $(\lambda > 0)$, are all positive definite.*

5. **On the positive definite character of** $\exp [-|x|^p]$. We shall need in the next section the following well known result:

COROLLARY 3. *The function*

$$\exp\,[-|x|^p] \tag{19}$$

is positive definite if $0<p\leqq 2$ *and not positive definite if* $p>2$.

First proof. We know that $\exp\,[-x^2]$ is p.d. from the formula just preceding (4) and property I. Hence $\exp\,[-|x|^{2\gamma}]$, $(0<\gamma<1)$, is p.d. by Corollary 2, as $\phi(x)=|x|^2$ is homogeneous, and the first part of the statement is proved. Suppose now that (19) were p.d. for a value $p>2$. By Theorem 2 this would imply that the space $E_1(p/2)$, which is the real axis with the metric $|x-x'|^{p/2}$, were imbeddable in $\mathfrak{H}$. But this is clearly impossible as $E_1(p/2)$ is not even a metric space since the distances among the points $x=0, 1, 2$ are $\overline{01}=1$, $\overline{12}=1$, $\overline{02}=2^{p/2}>2$, and they violate the triangle inequality.

Second direct proof for the case $0<p\leqq 2$. First we shall prove directly on the basis of properties II and III (§2) that the function $f(x)$ defined by (16) and (15) is p.d. By property III $(T\to\infty)$ it suffices to show that

$$\exp\left[-\int_0^T \frac{\sin^2 xu}{u^2}\,d\sigma(u)\right]$$

is p.d. By the definition of the Stieltjes integral, this is a limit of functions of the form

$$\exp\left[-\sum_{\nu=1}^{n} 2A_\nu \sin^2 xu_\nu\right], \qquad A_\nu>0,$$

and it suffices to show that each of the factors is p.d. Now

$$\exp\,[-2A\sin^2 xu]=\exp\,[-A]\,\exp\,[A\cos 2xu],$$

and this is p.d. because $\cos 2xu$ is p.d. and the exponential series has positive coefficients only. This point being disposed of, it suffices to show that $|x|^p$, $(0<p\leqq 2)$, is of the form (14).* This is apparent on account of the formula

$$|x|^p=\int_0^\infty \sin^2 xu\cdot u^{-1-p}du\Big/\int_0^\infty \sin^2 u\cdot u^{-1-p}du,$$

valid for $0<p<2$, which is as easily established as the similar representation (8). An obvious step-function for $\sigma(u)$ in (14), with only one jump at the origin, settles the case $p=2$.

6. **On the imbedding of the spaces E_m^p, l^p, L^p, $(0<p\leqq 2)$, in $\mathfrak{H}$ by a change of metric.** We learned in §2 that e^{-t^2} is p.d. in $\mathfrak{H}$ or L^2. We shall now show

* It is not by accident that the function (19) is of the form (16) if $0<p\leqq 2$. For *if* (19) is p.d., its positive powers must also be p.d., and the function is necessarily of the form (16), by Theorem 3. Notice that it is not bounded away from zero and hence is not of the form (17).

that $e^{-|t|^p}$ *is p.d. in the spaces* E_m^p, l^p, *and* L^p *for values of* p *in the range* $0<p\leqq 2$.

The function $\exp[-|x|^p]$ was found to be p.d. if $0<p\leqq 2$. Hence the functions $\exp[-|x_i|^p]$, $(i=1, 2, \cdots, m)$, are also p.d. when regarded as functions of the m variables x_i, and we may infer, by property II (§2), that their product

$$f = \exp[-(|x_1|^p + \cdots + |x_m|^p)], \qquad 0 < p \leqq 2,$$

is p.d. But this is equivalent to the statement that $\exp[-|t|^p]$ is p.d. in E_m^p. As the similar statement for l^p is proved in the same way as for L^p, only requiring less care, we shall limit ourselves to the consideration of L^p. Let

$$P_i = x_i(t),$$

where

$$\int_0^1 |x_i(t)|^p dt, \qquad i = 1, 2, \cdots, n,$$

exists, be n points of L^p. For

$$\overline{P_iP_k} = \left(\int_0^1 |x_i(t) - x_k(t)|^p dt\right)^{1/p}$$

we have to show that

$$\sum_{i,k=1}^{n} \rho_i\rho_k \exp[-\overline{P_iP_k}^p] \geqq 0 \tag{20}$$

for real ρ_i. There is no loss of generality in assuming that the functions $x_i(t)$ are continuous, as continuous functions are everywhere dense in L^p. For

$$x_{\nu i} = m^{-1/p}\cdot x_i(\nu/m), \quad \nu = 1, 2, \cdots, m;\ i = 1, 2, \cdots, n,$$

we have

$$\left(\sum_{\nu=1}^{m} |x_{\nu i} - x_{\nu k}|^p\right)^{1/p} = \left(\frac{1}{m}\sum_{\nu=1}^{m} |x_i(\nu/m) - x_k(\nu/m)|^p\right)^{1/p} \tag{21}$$

$$\rightarrow \overline{P_iP_k}, \qquad m \rightarrow \infty,$$

which proves the inequality (20), for (20) is already known to hold for all values of m if the $\overline{P_iP_k}$ of (20) are replaced by the quantities on the left-hand side of (21).

The fact that $\exp[-|t|^p]$, $(0<p\leqq 2)$, is p.d. in E_m^p, l^p, L^p implies that $\exp[-|t|^2]$, and therefore also $\exp[-\lambda|t|^2]$, $(\lambda>0)$, is p.d. in $E_m^p(p/2)$, $l^p(p/2)$, and $L^p(p/2)$. We have thus proved the following theorem:

THEOREM 5. *The spaces E_m^p, l^p, and L^p, $(0<p\leqq 2)$, become imbeddable in $\mathfrak{H}$ if we raise their metrics to the power $p/2$ or less. In other words, $E_m^p(\gamma)$, $l^p(\gamma)$, and $L^p(\gamma)$ are imbeddable in $\mathfrak{H}$ for values of γ in the range $0<\gamma\leqq p/2$.**

Note that for $p=2$ this result coincides with Corollary 1.

The fact that E_m^p, $(0<p<1)$, which is not a metric space, may be made metric by raising its metric to a suitable power was already known in a different terminology. Indeed, the following readily established "substitute" for Minkowski's inequality ([7], p. 32)

$$\sum_1^m |x_i+y_i|^p \leqq \sum_1^m |x_i|^p + \sum_1^m |y_i|^p, \qquad 0<p<1,$$

means precisely that $E_m^p(p)$, and therefore $E_m^p(\gamma)$, $(0<\gamma\leqq p)$, is a metric space. Now Theorem 5 states that if we restrict this range of γ to $0<\gamma\leqq p/2$, then not only can any *three* points of $E_m^p(\gamma)$ be imbedded isometrically in a euclidean space, but the same is true for any number of such points, since the whole space is imbeddable in Hilbert space.

7. **Some unsolved problems.** We shall devote this last section to a few unsolved analytical problems and to a brief discussion of their geometrical implications.

PROBLEM 1. *Let p be a real number exceeding 2, and let $m=2, 3, \cdots$. Do there exist positive exponents κ such that the function*

$$\exp\left[-(|x_1|^p+\cdots+|x_m|^p)^\kappa\right] \tag{22}$$

is positive definite?

If there are such positive exponents, then Corollary 2 implies the existence of a positive number κ_m with the properties: (22) is p.d. if $0<\kappa\leqq\kappa_m$ and is not p.d. if $\kappa>\kappa_m$. Furthermore Corollary 3 implies that $p\kappa_m\leqq 2$ or $\kappa_m\leqq 2/p$. Theorem 2 furnishes the following geometrical equivalent to Problem 1. The function (22) is p.d. if and only if $E_m^p(p\kappa/2)$, and therefore also $l^p(p\kappa/2)$ and $L^p(p\kappa/2)$, are imbeddable in $\mathfrak{H}$.

Particularly interesting is the next problem concerning the limiting case $p\to\infty$.

---

* The following is a different phrasing of the fact that $L^p(p/2)$ is imbeddable in L^2: *If $0<p<2$, then there is a functional $y(t)=U[x(t)]$, defined for all functions $x(t)$ in L^p, with values $y(t)$ in L^2 such that the relation*

$$\int_0^1 |x_1(t)-x_2(t)|^p dt = \int_0^1 |y_1(t)-y_2(t)|^2 dt$$

holds for two arbitrary $x_1(t)$, $x_2(t)$ of L^p. This functional $y=U[x]$ is necessarily continuous and univalent. A study of its further properties might prove to be of interest in the theory of L^p.

PROBLEM 2. *Do there exist positive exponents γ such that the function*

$$\exp\left[-\left[\max\left(|x_1|,\cdots,|x_m|\right)\right]^{2\gamma}\right]=\min\left(e^{-|x_1|^{2\gamma}},\cdots,e^{-|x_m|^{2\gamma}}\right) \tag{23}$$

is positive definite?

Again, if such exponents exist, there would exist a positive number γ_m with the properties: (23) is p.d. if $0<\gamma\leqq\gamma_m$ and is not p.d. if $\gamma>\gamma_m$. Moreover the function (23) is p.d. if and only if $E_m^\infty(\gamma)$ is imbeddable in $\mathfrak{H}$.

The particular interest of this second problem is due to the following lemma which is due essentially to Fréchet ([6], pp. 161–162). Let us call a *finite metric set* and denote by σ_m a metric space composed of exactly $m+1$ distinct points.

LEMMA. *Any finite metric set σ_m of $m+1$ points may be imbedded isometrically in E_m^∞.*

Proof. Let $P_0, P_1, \cdots, P_m$ denote the points of σ_m and $\overline{P_iP_k}$ their distances. Consider in E_m^∞ the following $m+1$ points:

$$Q_i=(x_{1i},x_{2i},\cdots,x_{mi})\equiv(\overline{P_1P_i},\overline{P_2P_i},\cdots,\overline{P_mP_i}),\qquad i=0,1,\cdots,m.$$

For their distance $\overline{Q_iQ_k}$ in E_m^∞ we have

$$\overline{Q_iQ_k}=\max_{j=1,\cdots,m}|x_{ji}-x_{jk}|=\max_{j=1,\cdots,m}|\overline{P_jP_i}-\overline{P_jP_k}|=\overline{P_iP_k},$$

for certainly $|\overline{P_jP_i}-\overline{P_jP_k}|\leqq\overline{P_iP_k}$, $(j=1,\cdots,m)$, on account of the triangle inequality in σ_m, while the equality sign holds if j is equal to whichever of the numbers i or k happens to be different from zero and hence within the range of j. If both $i=k=0$, the result $\overline{Q_0Q_0}=0$ was clear from the start.

On the basis of this lemma it is natural to classify finite metric sets according to their "dimension" as follows: *A set σ_n is said to be of dimension m (always less than or equal to n, on account of the lemma) if it is imbeddable in E_m^∞, but not in E_{m-1}^∞. If the last requirement is removed, the dimension of σ_n does not exceed m.*

If Problem 2 were solved in the affirmative, we would get the following statement: If σ_n is any finite metric set of dimension less than or equal to m, then $\sigma_n(\gamma)$ is imbeddable in E_n if $0<\gamma\leqq\gamma_m$, and γ_m is the best constant.

This would generalize in a certain sense the following theorem due to L. M. Blumenthal ([2], p. 402): *If σ_3 is a finite metric set composed of four points, then $\sigma_3(\gamma)$ is imbeddable in E_3 if $0<\gamma\leqq 1/2$, and $1/2$ is the best constant.*

In conclusion let us point out the following perhaps not trivial remark concerning Problem 2: *γ_2 exists and is greater than or equal to* 1/2. Indeed, in view of the formula

$$e^{-|t|} = \frac{1}{\pi}\int_{-\infty}^{\infty} \frac{e^{itx}}{1+x^2}\,dx,$$

we get by the transformation of variables $u=(\xi+\eta)/2$, $v=(\xi-\eta)/2$ in the double integral

$$\int_{-\infty}^{\infty}\int_{-\infty}^{\infty} \exp\,[i(xu+yv)]\,\frac{dudv}{[1+(u+v)^2][1+(u-v)^2]}$$

$$= \frac{1}{2}\int_{-\infty}^{\infty} \exp\,[i\xi(x+y)/2]\,\frac{d\xi}{1+\xi^2}\cdot\int_{-\infty}^{\infty} \exp\,[i\eta(x-y)/2]\,\frac{d\eta}{1+\eta^2}$$

$$= \frac{\pi^2}{2}\exp\,[-\tfrac{1}{2}|\,x+y\,|-\tfrac{1}{2}|\,x-y\,|\,] = \frac{\pi^2}{2}\exp\,[-\max\,(|\,x\,|,|\,y\,|)],$$

due to the relation $2\max\,(|x|,\,|y|)=|x+y|+|x-y|$. This formula shows that the function (23) is p.d. if $m=2$, $\gamma=1/2$. Hence this much is proved:

If σ_n is any finite metric set of dimension not exceeding 2, then $\sigma_n(\gamma)$, $(0<\gamma\leqq 1/2)$, is imbeddable in E_n.

If γ_3 exists, then it is certainly less than or equal to 1/2, by Blumenthal's theorem, and is probably even less than 1/2. A certain special set σ_4 shows that if γ_4 exists, it must be less than 0.45.

BIBLIOGRAPHY

1. S. Banach, *Théorie des Opérations Linéaires*, Warsaw, 1932.

2. L. M. Blumenthal, *New theorems and methods in determinant theory*, Duke Mathematical Journal, vol. 2 (1936), pp. 396–404.

3. S. Bochner, *Vorlesungen über Fouriersche Integrale*, Leipzig, 1932.

4. ———, *Monotone Funktionen, Stieltjessche Integrale und harmonische Analyse*, Mathematische Annalen, vol. 108 (1933), pp. 378–410.

5. ———, *Stable laws of probability and completely monotone functions*, Duke Mathematical Journal, vol. 3 (1937), pp. 726–728.

6. M. Fréchet, *Les dimensions d'un ensemble abstrait*, Mathematische Annalen, vol. 68 (1910), pp. 145–168.

7. G. H. Hardy, J. E. Littlewood, and G. Pólya, *Inequalities*, Cambridge, 1934.

8. J. von Neumann and I. J. Schoenberg, *Fourier integrals and metric geometry*, to appear in the Transactions of the American Mathematical Society.

9. G. Pólya and G. Szegö, *Aufgaben und Lehrsätze aus der Analysis*, vol. 2, Berlin, 1925.

10. I. J. Schoenberg, *On certain metric spaces arising from euclidean spaces by a change of metric and their imbedding in Hilbert space*, Annals of Mathematics, (2), vol. 38 (1937), pp. 787–793.

11. I. Schur, *Bemerkungen zur Theorie der beschränkten Bilinearformen mit unendlich vielen Veränderlichen*, Journal für die reine und angewandte Mathematik, vol. 140 (1911), pp. 1–28.

12. W. A. Wilson, *On certain types of continuous transformations of metric spaces*, American Journal of Mathematics, vol. 57 (1935), pp. 62–68.

COLBY COLLEGE,
WATERVILLE, ME.

Reprinted from *Trans. Amer. Math. Soc.* **44** (1938), pp. 522–536

Annals of Mathematics
Vol. 39, No. 4, October, 1938

METRIC SPACES AND COMPLETELY MONOTONE FUNCTIONS[1]

By I. J. Schoenberg

Edmund Landau in memoriam

(Received May 31, 1938)

Introduction

A set $\mathfrak{S}$ of elements, or points, $P, P', \cdots$, is said to be a semi-metric space if it is provided with a distance function PP' with the following two properties: 1. $PP' = P'P \geqq 0$, 2. $PP' = 0$ if and only if $P = P'$. Let now $F(t)$ be a real continuous function defined for $t \geqq 0$, $F(t) > 0$ if $t > 0$, $F(0) = 0$. According to L. M. Blumenthal we call the new semi-metric space which arises if we replace the metric PP', of $\mathfrak{S}$, by $F(PP')$ the *metric transform* of $\mathfrak{S}$ by $F(t)$ and denote it by the suggestive symbol $F(\mathfrak{S})$ ([2], p. 8).[2] This concept was also investigated by W. A. Wilson ([21], p. 64). Certain metric transforms have long been in use, for it is well known that if $\mathfrak{S}$ is metric, also the space with the new distance function $d(P, P') = PP'/(1 + PP')$ is metric. Moreover, these two metrics are topologically equivalent and the new space has a finite diameter. The new space is clearly the metric transform $F(\mathfrak{S})$ by the function

$$F(t) = \frac{t}{1+t}. \tag{0.1}$$

A general type of question suggested by the concept of metric transforms is whether functions $F(t)$ may be found such that $\mathfrak{S}$, if metrically transformed by $F(t)$, becomes more nearly comparable to certain simple spaces; or whether non-trivial functions $F(t)$ exist which will preserve certain simple characteristics of the original space $\mathfrak{S}$. To illustrate these points we shall state a few recent results.

If $\mathfrak{S}$ is a metric space it is readily seen that the metric transform $(\mathfrak{S})^\kappa$ (of $\mathfrak{S}$ by the function $F(t) = t^\kappa$) is also metric, provided the exponent κ is chosen in the range $0 < \kappa \leqq 1$. Blumenthal has shown that if this exponent is further restricted within the range $0 < \kappa \leqq \frac{1}{2}$, then $(\mathfrak{S})^\kappa$ has the 4-point property, i.e., any *four* points of $(\mathfrak{S})^\kappa$ may be imbedded[3] in a euclidean space. Hence $(\mathfrak{S})^\kappa$ $(0 < \kappa \leqq \frac{1}{2})$ is more "nearly euclidean" as compared with $\mathfrak{S}$, for $\mathfrak{S}$ enjoys only the 3-point property which is equivalent to the triangle inequality (see [2]).

As an example of a metric transformation which preserves certain properties

[1] Presented to the American Mathematical Society, April 16, 1938.

[2] The numbers in square brackets refer to the list of references at the end of this paper.

[3] Here and throughout this paper the word *imbedding* is meant in the sense of *isometric imbedding.*

of a space we mention the result of Wilson ([21], p. 64) to the effect that the metric transform $(E_1)^{\frac{1}{2}}$, of E_1[4] by the function $t^{\frac{1}{2}}$, which is not any more a euclidean space, is still isometrically imbeddable in $\mathfrak{H}$. This result was recently extended in two different directions. 1. It was shown by the author ([15]) that $(\mathfrak{H})^\kappa$ $(0 < \kappa \leqq 1)$ is imbeddable in $\mathfrak{H}$.[5] More significant is the following second result. 2. John von Neumann and the author ([12]) have determined *all* those functions $F(t)$ with the property that the metric transform $F(E_1)$ be imbeddable in $\mathfrak{H}$, or that it be imbeddable in E_n. This problem is equivalent with the problem of determining all screw lines in Hilbert space and in euclidean spaces.

These developments suggest the following general problem: *To determine all functions $F(t)$ such that the metric transform $F(E_m)$ be isometrically imbeddable in E_n, where the dimensions m and n are given in advance* $(1 \leqq m \leqq \infty, 1 \leqq n \leqq \infty)$.

We shall refer to this as the problem $\{E_m ; E_n\}$. The problems of the screw lines in E_n and $\mathfrak{H}$ are identical with the problems $\{E_1 ; E_n\}$ and $\{E_1 ; \mathfrak{H}\}$ in the present notation. The existence of *closed* screw lines in $E_n (n \geqq 2)$ and in $\mathfrak{H}$ may suggest that interesting solutions of a problem $\{E_m ; E_n\}$ are omitted by restricting the transforming function $F(t)$ to be > 0 if $t > 0$. However, if $F(t)$ does vanish for positive values of t, then $F(E_m)$ is by no means a semi-metric space. However, the problem of isometric imbedding of $F(E_m)$ in E_n retains also in this case its old meaning in an obvious way. In the present paper we are primarily interested in the case when $m \geqq 2$ and a simple lemma (Lemma 6, §5) will show that if we exclude the trivial solution $F(t) \equiv 0$, a problem $\{E_m ; E_n\}$, where m exceeds one, does admit, if any, *only* solutions $F(t)$ which are > 0 for $t > 0$.

The case when $m > n$ is immediately ruled out as trivial, for a moment's reflection shows that $F(t) \equiv 0$ is the only solution of $\{E_m ; E_n\}$. In the Appendix, §7, we solve the problem $\{E_m ; E_n\}$, where $2 \leqq m \leqq n < \infty$, with the rather negative result that the obvious solutions $F(t) = ct$ $(c \geqq 0)$, which correspond to similitude transformations of E_m, are the only possible ones. The problems $\{E_m ; \mathfrak{H}\}$, where $2 \leqq m < \infty$, will be partially solved in §8 of the Appendix where all "rectifiable" solutions of these problems are determined by a method suggested by recent investigations on homogeneous random processes in the calculus of probabilities. The chief objective of this paper is a complete solution of the problem $\{\mathfrak{H}; \mathfrak{H}\}$ (Part II, §5). The solutions $F(t)$ of this problem, as described by Theorems 6 and 6′ of §5, turn out to be essentially integrals of completely monotone functions. In particular it is found that (0.1) is a solution of this problem. Thus $F(\mathfrak{H})$, where $F(t) = t/(1 + t)$, is not only a bounded homeomorph of $\mathfrak{H}$, but is actually congruent to a subset of $\mathfrak{H}$.

A convenient way of attacking analytically these geometrical problems is afforded by combining an imbedding theorem of Menger ([9]) with a connection

[4] We denote by E_m the m-dim. euclidean space, while E_∞ or $\mathfrak{H}$ shall mean the real Hilbert space.

[5] This is a special case (for $p = 2$) of the following result: $(L_p)^\kappa (0 < \kappa < p/2; 0 < p \leqq 2)$ is imbeddable in $\mathfrak{H}$.

between euclidean simplices and quadratic forms, which goes back to Gauss, and which is so familiar in the geometry of numbers. The author had previously emphasized the usefulness of this combination in problems of isometric imbedding in Hilbert space ([14], [15], [16]). It allows us to attack our problems with such powerful tools as the Fourier-Stieltjes and Laplace-Stieltjes integrals.

Part I of this paper, on which the subsequent work is based, incidentally answers a question which has puzzled the author long before he ever thought of metric transforms of spaces, namely the connection between the class $\mathfrak{P}$ of Fourier-Stieltjes integrals (positive definite functions)

$$(0.2) \qquad g(t) = \int_{-\infty}^{\infty} e^{itu}\, d\alpha(u), \qquad (-\infty < t < \infty),$$

and the class $\mathfrak{M}$ of Laplace-Stieltjes integrals (completely monotone functions)

$$(0.3) \qquad f(t) = \int_{0}^{\infty} e^{-tu}\, d\beta(n), \qquad (0 \leqq t < \infty),$$

where $\alpha(u)$ and $\beta(u)$ are bounded non-decreasing functions. In spite of the entirely different analytical character of these two classes,[6] a certain kinship was to be expected for the following two reasons: 1. In both cases the defining kernel is the exponential function. 2. The less formal reason of the similarity of the closure properties of both classes, for both classes are *convex*, i.e., $a_1 f_1 + a_2 f_2$ ($a_1 \geqq 0$, $a_2 \geqq 0$) belongs to the class if f_1 and f_2 belong to it, *multiplicative*, i.e., also $f_1 \cdot f_2$ belongs to the class, and finally *closed* with respect to ordinary convergence to a continuous limit function. The answer, as given by Theorem 3 below, is based on a result of S. Bochner on characteristic functions of distribution functions in E_m([4], p. 407) and may best be stated in terms of such functions. A subclass of the class of characteristic functions $f(x_1, \cdots, x_m)$ in E_m (defined by (1.4) below) are those which are of radial symmetry. These are completely described by Theorem 1 below. While there is not as yet at present an explicit determination of the characteristic functions in Hilbert space $\mathfrak{H}$ by an appropriate Stieltjes integration process over $\mathfrak{H}$, these functions may easily enough be defined as positive definite vector functions in $\mathfrak{H}$. Those positive definite vector functions in $\mathfrak{H}$ which are of *radial symmetry* are now explicitly determined and turn out to be identical with the functions $f(t^2)$, where $f(t) \in \mathfrak{M}$. Incidentally, Theorem 3 exhibits a new property (3.3) of completely monotone functions which is sufficient to characterize this class of functions.

The sixth section of Part II is devoted to further applications to the theory of completely monotone functions. It is shown that the solution of the problem $\{\mathfrak{H}; \mathfrak{H}\}$ (Theorem 6) is equivalent with a complete description of those trans-

[6] Indeed $f(t)$, defined by (0.3), is analytic in the half-plane $\Re t > 0$, while Weierstrass' nowhere differentiable function $g(t) = \sum_0^\infty a^n \cos b^n t$, ($0 < a < 1$, $ab \geqq 1$), is readily representable as a Fourier integral (0.2) with an appropriate step function $\alpha(t)$. Thus $\mathfrak{P}$ contains "arbitrary" functions.

formations of the variable $t \mid \phi(t)$ which turn completely monotone functions $f(t)$ into functions $f(\phi(t))$ of the same class (Theorem 8). We venture to call them *inner transformations* of the class $\mathfrak{M}$. The fact that an essential part of this class of transformations does have this property was pointed out quite recently by S. Bochner ([5], p. 498).

I. Positive Definite Functions and Completely Monotone Functions

1. Positive definite functions in euclidean space E_m

1.1. Let $\mathfrak{S}$ be an abstract set of elements $P, P', Q, \cdots$. A real or complex-valued function $F(P, Q)$ of two arbitrary points of $\mathfrak{S}$ is called positive definite, if it enjoys the following two properties:

1. Hermitean symmetry

$$F(P, Q) = \overline{F(Q, P)}. \tag{1.1}$$

2. For any n points $P_1, \cdots, P_n$ of $\mathfrak{S}(n = 2, 3, \cdots)$ we have

$$\sum_{j,k=1}^{n} F(P_j, P_k)\rho_j\bar{\rho}_k \geqq 0 \tag{1.2}$$

for arbitrary ρ_j.[7]

If $\mathfrak{S}$ is also endowed with a metric PP', we would also require our function $F(P, Q)$ to be continuous in both points with respect to this metric. By imposing successive restrictions on $\mathfrak{S}$, as well as $F(P, Q)$, we shall now examine certain subclasses of Moore's general class of positive definite functions.

Let us assume first that $\mathfrak{S}$ is a linear vector space with the norm (metric) $| P - P' | = PP'$. In this case we limit ourselves to positive definite functions $F(P, Q)$ which are functions of the vector $P - Q$ only:

$$F(P, Q) = f(P - Q). \tag{1.3}$$

Our continuous vector function $f(P - Q)$ is now subject to the conditions

$$f(P - Q) = \overline{f(Q - P)}, \tag{1.1'}$$

$$\sum_{j,k=1}^{n} f(P_j - P_k)\rho_j\bar{\rho}_k \geqq 0. \tag{1.2'}$$

If our underlying space is the m-dimensional euclidean space E_m, these functions $f(P - Q)$ are precisely the positive definite functions of m variables as defined by M. Mathias ($m = 1$) and S. Bochner (m arbitrary) ([4], p. 406). Let us refer E_m to rectangular coördinates and write $P = x = (x_1, \cdots, x_m)$, $f(P - 0) = f(x_1, \cdots, x_m) = f(x)$, $xy = x_1y_1 + \cdots + x_my_m$. Bochner established the identity of this class of positive definite functions with the class of characteristic functions of distribution functions in E_m:

$$f(x) = \int_{E_m} e^{ixy}\, d\phi(y), \tag{1.4}$$

[7] This general concept is due to E. H. Moore. See [11], pp. 3–4, 173, 181–190, 209–220. Moore calls $F(P, Q)$ a positive Hermitean matrix.

where $\phi(y)$ is the monotone point function of a non-negative, bounded, totally additive set function $\phi(S)$ defined for all Borel sets of E_m.

A subclass of the class of vector functions $f(P - Q)$ associated with $\mathfrak{S}$ and enjoying the properties (1.1′) and (1.2′), is made up of those functions which are functions of the length $| P - Q |$ of the vector only:

$$f(P - Q) = g(| P - Q |) = g(PQ), \tag{1.5}$$

where $g(t)$ is a continuous function defined for $t \geqq 0$. They are necessarily real, for (1.5) implies the ordinary symmetry $f(P - Q) = f(Q - P)$, which together with the hermitean symmetry (1.1′) obviously implies the reality of $f(P - Q)$ and $g(t)$. These real functions $g(t)$ will be called positive definite "in $\mathfrak{S}$" throughout this paper. This final restriction implies on the other hand a gain in generality, for this restricted definition of positive definite functions may at once be extended to any semi-metric space $\mathfrak{S}$.

DEFINITION 1. *Let $\mathfrak{S}$ be a semi-metric space with the metric PP'. A real continuous function $g(t)$ defined in the range of values of $t = PP'(P, P' \epsilon \mathfrak{S})$, is said to be positive definite in $\mathfrak{S}$, if for any n points $P_1, \cdots, P_n$ of $\mathfrak{S}(n = 2, 3, \cdots)$ we have*

$$\sum_{j,k=1}^{n} g(P_j P_k)\rho_j \rho_k \geqq 0, \tag{1.2″}$$

for arbitrary real ρ_j. We shall denote this class of functions by the symbol $\mathfrak{P}(\mathfrak{S})$.[8]

The class $\mathfrak{P}(\mathfrak{S})$ is never empty, for $g(t) \epsilon \mathfrak{P}(\mathfrak{S})$, if $g(t) \equiv 1$. From (1.2″) we conclude that $| g(t) | \leqq g(0)$. It was pointed out before ([16], §2) that the class $\mathfrak{P}(\mathfrak{S})$ is convex, multiplicative and closed with respect to convergence to a continuous limit function.

1.2. Let us determine the functions $g(t)$ of the positive definite class $\mathfrak{P}(E_m)$. From (1.5) and Bochner's theorem we learn that our $g(t)$ are in a one-to-one correspondence with those characteristic functions (1.4) which are functions of $| x |$ only, i.e., of radial symmetry, and that the relationship valid both ways is

$$f(x) = g(| x |). \tag{1.5′}$$

Let $d\omega(\xi)$ denote the area element of the spherical shell $| \xi | = 1$ in E_m, and $\omega_m = \int d\omega(\xi)$ be its total area. The mean value of $e^{ix\xi}$ over $| \xi | = 1$ is obviously invariant with respect to rotations in E_m about the origin and hence is a function of $| x |$ only. We write for it

$$M_\xi\{e^{ix\xi}\} = \frac{1}{\omega_m} \int_{|\xi|=1} e^{ix\xi}\, d\omega(\xi) = \Omega_m(| x |). \tag{1.6}$$

By m-dimensional polar coördinates we readily find

$$\Omega_m(r) = \int_0^\pi e^{ir\cos\theta} \sin^{m-2}\theta\, d\theta \Big/ \int_0^\pi \sin^{m-2}\theta\, d\theta, \qquad (m \geqq 2), \tag{1.7}$$

[8] We see that $g(t) \epsilon \mathfrak{P}(\mathfrak{S})$ if and only if $F(P, Q) = g(PQ)$ is positive definite in the sense of Moore.

whence the power series expansion and expression in terms of Bessel functions

$$(1.8)\qquad \begin{aligned}\Omega_m(r) &= 1 - \frac{r^2}{2m} + \frac{r^4}{2\cdot 4\cdot m(m+2)} - \frac{r^6}{2\cdot 4\cdot 6m(m+2)(m+4)} + \cdots \\ &= \Gamma\left(\frac{m}{2}\right)\left(\frac{2}{r}\right)^{\frac{1}{2}(m-2)} J_{\frac{1}{2}(m-2)}(r), \qquad (m \geqq 1),\end{aligned}$$

showing that $\Omega_m(r)$ is an integral transcendental function.[9] We may now readily derive the following theorem.

THEOREM 1. *The class $\mathfrak{P}(E_m)$ of functions which are positive definite in E_m is identical with the class of functions of the form*

$$(1.9)\qquad g(t) = \int_0^\infty \Omega_m(tu)\, d\alpha(u),$$

where $\alpha(u)$ is non-decreasing and bounded for $u \geqq 0$, and $\Omega_m(r)$ is the integral function defined above.

Indeed, let $f(x)$ of (1.4) be of radial symmetry: $f(x) = g(|x|)$. Since $f(x) = f(|x|\xi)$, for all unit vectors ξ, we have

$$(1.10)\qquad \begin{aligned} g(|x|) = f(x) &= \frac{1}{\omega_m}\int_{|\xi|=1} f(|x|\xi)\, d\omega(\xi) \\ &= \frac{1}{\omega_m}\int_{|\xi|=1}\left(\int_{E_m} e^{i|x|\cdot\xi y}\, d\phi(y)\right) d\omega(\xi) \\ &= \int_{E_m}\left(\frac{1}{\omega_m}\int_{|\xi|=1} e^{i|x|\cdot y\xi}\, d\omega(\xi)\right) d\phi(y) \\ &= \int_{E_m} \Omega_m(|x|\cdot|y|)\, d\phi(y) = \int_0^\infty \Omega_m(|x|u)\, d\alpha(u),\end{aligned}$$ [10]

where we put

$$\alpha(u) = \int_{|y|\leqq u} d\phi(y) = \phi(|y| \leqq u).$$

Thus (1.10) shows that functions of $\mathfrak{P}(E_m)$ are of the form (1.9). The converse is obvious, for $\Omega_m(|x|u)$ is a characteristic function on account of (1.6).

2. Positive definite functions in Hilbert space $\mathfrak{H}$

2.1. Definition 1 implies a certain monotoneity property of the class $\mathfrak{P}(\mathfrak{S})$ in its dependence on the space $\mathfrak{S}$, for if $\mathfrak{S}$ is a subset of a space $\mathfrak{S}'$, or congruent to a subset of $\mathfrak{S}'$, then $\mathfrak{P}(\mathfrak{S}') \subset \mathfrak{P}(\mathfrak{S})$. Now as $E_1 \subset E_2 \subset \cdots \subset E_m \subset \cdots \subset \mathfrak{H}$, we see that

$$\mathfrak{P}(E_1) \supset \mathfrak{P}(E_2) \supset \cdots \supset \mathfrak{P}(E_m) \supset \cdots \supset \mathfrak{P}(\mathfrak{H}),$$

[9] The $\Omega_m(r)$ may be called Poisson functions as they were discovered by Poisson in connection with problems on heat conduction prior to Bessel's investigation of his functions. See G. N. Watson [18], p. 24.

[10] The interchange of integrations in (1.10) is justified by a general argument found in [4], p. 393.

with $\mathfrak{P}(\mathfrak{H})$ actually identical with the intersection (logical product) of this non-increasing sequence of classes $\{\mathfrak{P}(E_m)\}$.

It was pointed out elsewhere ([16], §2) that $g(t) = \exp\{-t^2u^2\}$ (u a real constant) is positive definite in E_m, for all m, hence also positive definite in $\mathfrak{H}$. This fact is immediately apparent from the elementary formula

$$(2.1)\quad \begin{aligned} &e^{-(x_1^2+\cdots+x_m^2)u^2} \\ &= (2u)^{-m}\pi^{-m/2}\int_{-\infty}^{\infty}\cdots\int_{-\infty}^{\infty} e^{i(x_1y_1+\cdots+x_my_m)}e^{-(y_1^2+\cdots+y_m^2)/4u^2}\,dy_1\cdots dy_m, \end{aligned}$$

showing that $f(x) = \exp\{-|x|^2u^2\}$ is a characteristic function in E_m of radial symmetry about the origin. That we get all functions of the class $\mathfrak{P}(\mathfrak{H})$ as linear combinations of these particular functions, with non-negative coefficients, is stated by the following theorem.

THEOREM 2. *The class $\mathfrak{P}(\mathfrak{H})$ of functions which are positive definite in $\mathfrak{H}$ is identical with the class of functions of the form*

$$(2.2)\qquad g(t) = \int_0^\infty e^{-t^2u^2}\,d\alpha(u),$$

where $\alpha(u)$ is non-decreasing and bounded for $u \geqq 0$.[11]

Indeed, since $\mathfrak{P}(\mathfrak{H})$ is convex and $\exp\{-t^2u^2\} \in \mathfrak{P}(\mathfrak{H})$, we see that $g(t)$, of (2.2), belongs to $\mathfrak{P}(\mathfrak{H})$. There remains to prove the converse. Now let $g(t)$ be an element of $\mathfrak{P}(\mathfrak{H})$. As $g(t) \in \mathfrak{P}(E_m)$, we know by Theorem 1 that it is of the form

$$(2.3)\qquad g(t) = \int_0^\infty \Omega_m(tu)\,d\alpha_m(u), \qquad (m = 1, 2, 3, \cdots),$$

[11] Theorem 2 and Theorem 4 below (i.e. Theorem 1 of [16]) allow us to state the following result: *A separable semi-metric space $\mathfrak{S}$ is imbeddable in $\mathfrak{H}$ if and only if $\mathfrak{P}(\mathfrak{H}) \subset \mathfrak{P}(\mathfrak{S})$.* If we replace $\mathfrak{H}$ by E_m, then the theorem is no longer true without further restrictions on $\mathfrak{S}$. To show this, consider E_3, and let $\mathfrak{S}$ be a regular simplex σ_4 of E_4 of side 1, i.e., $\mathfrak{S}$ be the set of five points $P_0, \cdots, P_4$ with $P_iP_k = 1 (i \neq k)$. The statement $\mathfrak{P}(E_3) \subset \mathfrak{P}(\mathfrak{S})$ means (by Theorem 1 below for $m = 3$) that the family of functions $\Omega_3(tu)$ is positive definite in $\mathfrak{S} = \sigma_4$. As the $\Omega_3(tu)$ are known to be positive definite in E_3, our last statement amounts to the inequality

$$(*)\qquad \Delta(u) = \det \| \Omega_3(u\cdot P_iP_k) \|_{0,4} \geqq 0, \text{ for real } u.$$

Now $P_iP_k = \delta_{ik}$ implies $\Delta(u) = (1 + \Omega_3(u))^4(1 + 4\Omega_3(u))$. Hence (*) holds, since the inequality

$$(**)\qquad \Omega_3(u) = \frac{\sin u}{u} > -\tfrac{1}{4}$$

happens to be valid for all real u. Thus $\mathfrak{P}(E_3) \subset \mathfrak{P}(\sigma_4)$, without, of course, σ_4 being imbeddable in E_3. In order that $\mathfrak{P}(E_m) \subset \mathfrak{P}(\mathfrak{S})$ should imply that $\mathfrak{S}$ is imbeddable in E_m, it will probably be sufficient to assume that $\mathfrak{S}$ is made up of at least $m + 3$ points, an assumption suggested by certain results of Menger, [10].

$\{\alpha_m(u)\}$ being a certain sequence of uniformly bounded monotone functions. From the power series expansion (1.8) we derive the limiting relation

$$\lim_{m\to\infty} \Omega_m(r\sqrt{(2m)}) = e^{-r^2}. \tag{2.4}$$[12]

As on the other hand (2.3) may be written in the form

$$g(t) = \int_0^\infty \Omega_m(tu\sqrt{(2m)})\, d\alpha_m(u\sqrt{(2m)}), \qquad (m = 1, 2, 3, \cdots), \tag{2.5}$$

we shall expect to be able to pass from (2.5) to the representation (2.2), of $g(t)$, by means of (2.4). Now it is easy to show that (2.4) holds uniformly within any finite interval of the r-axis (as well as in any finite domain of the complex r-plane). This, however, is insufficient to insure the applicability of Helly's well known convergence theorem for Stieltjes integrals, as our interval of integration is infinite. We shall have to show that (2.4) holds uniformly *for all real values of* r.[13]

2.2. The following proof of the uniformity of convergence in (2.4) is elementary.[14] For the sake of clarity we present it in the form of three lemmas.

LEMMA 1. *Let*

$$F(x) = \int_0^1 \cos(x\tau(u))\, du, \tag{2.6}$$

where $\tau(u)$ *is continuous for* $0 \leqq u \leqq 1$, $\tau(0) = 0$, *and* $\tau'(u)$ *non-decreasing for* $0 \leqq u < 1$, $\tau'(0) > k > 0$. *We have*

$$|F(x)| < \frac{1}{kx} \qquad \textit{for } x > 0. \tag{2.7}$$

---

[12] The author is indebted to J. von Neumann for pointing out to him this relation.

[13] To stress the decisive rôle of the uniformity of convergence in (2.4) for all real r, we point out the following fact. The expansion (1.8) shows that

$$\lim_{m\to\infty} \Omega_m(r) = 1, \tag{2.4'}$$

uniformly in any finite r-interval. From this and (2.3) we may be tempted to expect, as $m \to \infty$, that

$$g(t) = \int_0^\infty 1\cdot d\alpha(u),$$

and therefore conclude that $\mathfrak{P}(\mathfrak{H})$ is made up of constant functions only. This however, is not true, as e.g. $\exp\{-t^2\} \in \mathfrak{P}(\mathfrak{H})$. The point is that (2.4′) does not hold uniformly for all real r.

[14] None of the available asymptotic estimates of Bessel functions was found to apply in the present situation.

Indeed, as $\tau'(u) > k$ holds throughout [0, 1], the second mean value theorem gives

$$F(1) = \int_0^1 \cos(\tau(u))\,du = \int_0^1 \frac{1}{\tau'(u)} \cos(\tau(u))\tau'(u)\,du$$
$$= \frac{1}{\tau'(0)} \int_0^\xi \cos(\tau(u))\tau'(u)\,du = \frac{1}{\tau'(0)}(\sin\tau(\xi) - \sin\tau(0)) = \frac{\sin\tau(\xi)}{\tau'(0)},$$

which implies

$$|F(1)| = \left|\int_0^1 \cos(\tau(u))\,du\right| < \frac{1}{k}.$$

Applying this inequality to $x\tau(u)$ instead of $\tau(u)$ (for a fixed positive value of x), and therefore with kx, instead of k, we immediately get (2.7).

LEMMA 2. *We have*

$$|\Omega_m(x\sqrt{(2m)})| < \frac{1}{x} \quad for \quad x > 0, \qquad m = 3, 4, 5, \cdots, \tag{2.8}$$

Indeed, passing to the variable $t = \cos\theta$ in

$$\Omega_m(x\sqrt{(2m)}) = \int_0^{\pi/2} \cos(x\sqrt{(2m)}\cos\theta)\sin^{m-2}\theta\,d\theta \Big/ \int_0^{\pi/2} \sin^{m-2}\theta\,d\theta,$$

we get

$$\Omega_m(x\sqrt{(2m)}) = \int_0^1 \cos(x\sqrt{(2m)}t)(1 - t^2)^{(m-3)/2}\,dt \Big/ \int_0^1 (1 - t^2)^{(m-3)/2}\,dt,$$
$$(m \geqq 3).$$

By means of the function

$$u = \int_0^t (1 - t^2)^{(m-3)/2}\,dt \Big/ \int_0^1 (1 - t^2)^{(m-3)/2}\,dt, \qquad (0 \leqq t \leqq 1), \tag{2.9}$$

whose inverse we denote by $t = t(u)(0 \leqq u \leqq 1)$, this goes over into

$$\Omega_m(x\sqrt{(2m)}) = \int_0^1 \cos(x\sqrt{(2m)}t(u))\,du.$$

With

$$\tau(u) = \sqrt{(2m)}t(u), \tag{2.10}$$

we finally have

$$\Omega_m(x\sqrt{(2m)}) = \int_0^1 \cos(x\tau(u))\,du. \tag{2.11}$$

The change of variable and the inversion $t = t(u)$, of (2.9) are justified by the fact that the function (2.9) is continuous and strictly increasing for $0 \leqq t \leqq 1$, with $u(0) = 0$, $u(1) = 1$. Moreover

$$\tau'(u) = \sqrt{(2m)}t'(u) = \sqrt{(2m)}(1-t^2)^{-(m-3)/2} \cdot \int_0^1 (1-t^2)^{(m-3)/2}\,dt, \qquad (0 \leqq t < 1),$$

is an increasing function and

$$(2.12)\qquad \begin{aligned}\tau'(0) &= \sqrt{(2m)} \int_0^1 (1-t^2)^{(m-3)/2}\,dt\\ &= \sqrt{(2m)} \int_0^{\pi/2} \sin^{m-2}\theta\,d\theta \geqq \sqrt{(2m)}\,\frac{(m-3)!!}{(m-2)!!} > 1.\end{aligned}$$ [15]

Now (2.11), (2.12) and Lemma 1 imply the required inequality (2.8).

LEMMA 3. *The limiting relation*

$$(2.13)\qquad \lim_{m\to\infty} \Omega_m(x\sqrt{(2m)}) = e^{-x^2}$$

holds uniformly for $-\infty < x < \infty$.

Indeed, the difference

$$|\,\Omega_m(x\sqrt{(2m)}) - e^{-x^2}\,|$$

will be $< \epsilon$ for real x with $|\,x\,| > A(\epsilon)$, and $m \geqq 3$, since it falls below $x^{-1} + e^{-x^2}$ by Lemma 2. Now for sufficiently large m it will likewise be $< \epsilon$ for $|\,x\,| \leqq A(\epsilon)$, since (2.13) holds uniformly in every finite interval as already mentioned.

2.3. We can now readily complete a proof of Theorem 2. Returning to the relation (2.5), Helly's compactness theorem insures the existence of a subsequence $\{\alpha_{m_\nu}(u\sqrt{(2m_\nu)})\}$ converging to the monotone function $\alpha(u)(\alpha(0) = 0)$ in all points of continuity of $\alpha(u)$ ([8], p. 286). Let t assume a fixed positive value. By Lemma 3 we know that $\Omega_{m_\nu}(tu\sqrt{(2m_\nu)})$ tends to $\exp\{-t^2u^2\}$ uniformly for all real values of u, hence (2.5) implies

$$g(t) = \int_0^\infty e^{-t^2u^2}\,d\alpha_{m_\nu}(u\sqrt{(2m_\nu)}) + \sigma,$$

---

[15] Here we write $k!! = k(k-2)(k-4)\cdots$. The last inequality (2.12) can be verified by Stirling's formula. I owe the following interesting direct proof to F. Bohnenblust. Let $m = 2n+2$ be even. Then

$$\begin{aligned}\frac{(2n-1)!!}{(2n)!!}&\sqrt{(4n+4)}\\ &= \frac{(2n-1)(2n-3)\cdots 3\cdot 1}{\sqrt{(2n)}(\sqrt{(2n)}\sqrt{(2n-2)})(\sqrt{(2n-2)}\sqrt{(2n-4)})\cdots(\sqrt{4}\sqrt{2})\sqrt{2}}\sqrt{(4n+4)}\\ &\geqq \frac{(2n-1)(2n-3)\cdots 3\cdot 1}{\sqrt{(2n)}(2n-1)(2n-3)\cdots 3\cdot\sqrt{2}}\sqrt{(4n+4)} = \sqrt{\left(\frac{n+1}{n}\right)} > 1,\end{aligned}$$

since $\sqrt{(2n(2n-2))} < 2n-1$. Similar argument in case m is odd.

with $|\sigma| < \epsilon$ for sufficiently large values of ν. But this implies that

$$g(t) = \int_0^l e^{-t^2u^2}\, d\alpha_{m_\nu}(u\sqrt{(2m_\nu)}) + \rho,$$

with $|\rho| < 2\epsilon$, provided both l and ν are sufficiently large. If $u = l$ is a continuity point of $\alpha(u)$, letting $\nu \to \infty$, we get ([8], pp. 288–289)

$$g(t) = \int_0^l e^{-t^2u^2}\, d\alpha(u) + \rho_1. \qquad |\rho_1| \leqq 2\epsilon.$$

Now letting $l \to \infty$ and then $\epsilon \to 0$ we get the desired representation (2.2). This formula, established for $t > 0$, holds also if $t = 0$ in view of the continuity of its two sides.

3. Completely monotone functions

3.1. A real function $f(t)$ is said to be *completely monotone for* $t \geqq 0$, if

(3.1) $$(-1)^n f^{(n)}(t) \geqq 0 \quad \text{for} \quad 0 < t < \infty, \qquad (n = 0, 1, 2, \cdots),$$

and

(3.1′) $$f(0) = f(+0),$$

the last condition expressing the continuity of $f(t)$ at the origin. A fundamental theorem[16] states the identity of this class with the class of functions representable as a Laplace-Stieltjes integral

(3.2) $$f(t) = \int_0^\infty e^{-tu}\, d\beta(u),$$

where $\beta(u)$ is non-decreasing and bounded for $u \geqq 0$. *We shall denote this class of functions throughout this paper by the symbol* $\mathfrak{M}$.

If $g(t)$ is positive definite in $\mathfrak{H}$, then we may write in view of Theorem 2

$$g(t) = \int_0^\infty e^{-t^2u^2}\, d\alpha(u) = \int_0^\infty e^{-t^2u}\, d\alpha(\sqrt{u}) = \int_0^\infty e^{-t^2u}\, d\beta(u) = f(t^2).$$

Thus Theorem 2 and the Laplace integral representation (3.2) immediately prove the following theorem.

THEOREM 3. *A function $f(t)$ is completely monotone for $t \geqq 0$, if and only if $f(t^2)$ is positive definite in Hilbert space $\mathfrak{H}$. In symbols: $f(t) \in \mathfrak{M}$ if and only if $f(t^2) \in \mathfrak{P}(\mathfrak{H})$.*

The elements $g(t)$ of $\mathfrak{P}(\mathfrak{H})$ and $f(t)$ of $\mathfrak{M}$ are therefore in a one-to-one correspondence expressed by the relation $g(t) = f(t^2)$, which is valid both ways.

Definition 1 of positive definite functions and Theorem 3 allow us to state that the conditions (3.1) and (3.1′) for complete monotoneity of a function $f(t)$ are entirely equivalent with the following two requirements:

1. *$f(t)$ be real and continuous for $t \geqq 0$.*

[16] See [7], [1], [19], and [13].

2. *If $P_1, P_2, \cdots, P_n$ are any n points of a euclidean space* ($n = 1, 2, 3, \cdots$), *we should have*

$$\begin{vmatrix} f(P_1P_1^2) & f(P_1P_2^2) & \cdots & f(P_1P_n^2) \\ f(P_2P_1^2) & f(P_2P_2^2) & \cdots & f(P_2P_n^2) \\ \vdots & \vdots & & \vdots \\ f(P_nP_1^2) & f(P_nP_2^2) & \cdots & f(P_nP_n^2) \end{vmatrix} \geqq 0. \tag{3.3}$$

[17]

A direct derivation of (3.3) as a consequence of (3.1) and (3.1′) would furnish a new proof of the theorem that (3.1) and (3.1′) imply the representation (3.2).

The connection between the class $\mathfrak{P}$ of characteristic functions (0.2) and the class $\mathfrak{M}$ is now clear. We have the infinite sequence of classes

$$\mathfrak{P} \supset \mathfrak{P}(E_1) \supset \mathfrak{P}(E_2) \supset \cdots \supset \mathfrak{P}(E_m) \supset \cdots \supset \mathfrak{P}(\mathfrak{H}). \tag{3.4}$$

The class $\mathfrak{P}(E_1)$ is made up of those elements of $\mathfrak{P}$ which are of radial symmetry in E_1, i.e., are *even,* or merely *real.* Their form is obviously

$$g(t) = \int_0^\infty \cos tu \, d\alpha(u),$$

which is in accord with Theorem 1, since $\Omega_1(r) = \cos r$. A function $g(t)$ of $\mathfrak{P}(E_1)$ belongs to $\mathfrak{P}(E_m)$ if and only if

(3.5) $\quad g(\sqrt{(x_1^2 + \cdots + x_m^2)})$ is a characteristic function of E_m.

The completely monotone $f(t^2)(f(t) \in \mathfrak{M})$ are identical with those elements $g(t)$ of $\mathfrak{P}(E_1)$ which enjoy the property (3.5) for *all* values of m.

3.2. Further light is thrown on the gradual narrowing down of the classes (3.4) by the requirement of positive definiteness in E_m, for increasing m, by the following lemma.

LEMMA 4. *The functions*

$$g(t) = \int_0^\infty \Omega_m(tu) \, d\alpha(u) \tag{3.6}$$

of the class $\mathfrak{P}(E_m)$ are $\left[\frac{m-1}{2}\right]$-times differentiable.

This fact shows clearly the smoothing effect on $g(t)$ of the requirement of positive definiteness in higher euclidean spaces. It implies in particular that the elements of $\mathfrak{P}(\mathfrak{H})$, and those of $\mathfrak{M}$, are ∞-often differentiable. Theorem 2, moreover, shows that the elements $g(t)$ of $\mathfrak{P}(\mathfrak{H})$ are even analytic in the sector $|\arg t| < \pi/4$ of the complex t-plane.

In order to prove Lemma 4 it suffices to consider

$$g^{(\nu)}(t) = \int_0^\infty \frac{\partial^\nu}{\partial t^\nu} \Omega_m(tu) \, d\alpha(u) \tag{3.7}$$

[17] See Theorem 3′ below which excludes the equality sign in (3.3), except in trivial cases.

and to see for what values of $\nu = 1, 2, 3, \cdots$ this integral converges for all values of t. The integrand is

$$\frac{\partial^\nu}{\partial t^\nu}\Omega_m(tu) = \Omega_m^{(\nu)}(tu)\cdot u^\nu. \tag{3.8}$$

From $\Omega_m(r) = c_m J_k(r) r^{-k}$, where $k = (m-2)/2$ and c_m is a positive constant, we get by the rule of Leibnitz

$$c_m^{-1}\Omega_m^{(\nu)}(r) = J_k^{(\nu)}(r)r^{-k} - \binom{\nu}{1} k J_k^{(\nu-1)}(r)r^{-k-1}$$
$$+ \cdots \pm k(k+1)\cdots(k+\nu-1)J_k(r)r^{-k-\nu}.$$

As $J_k^{(\nu)}(r) = O(r^{-\frac{1}{2}})$ we have

$$\Omega_m^{(\nu)}(r) = O(r^{-k-\frac{1}{2}}) = O(r^{-\frac{1}{2}(m-1)}),$$

and hence by (3.8)

$$\frac{\partial^\nu}{\partial t^\nu}\Omega_m(tu) = O(u^{\nu-\frac{1}{2}(m-1)}).$$

Thus (3.7) converges absolutely as long as $\nu \leqq (m-1)/2$, proving our lemma.

In view of Lemma 4 there are strong reasons to believe that the continuity assumption on $g(t)$ in Definition 1, section 1.1, may be dropped as soon as $\mathfrak{S}$ is an euclidean space E_m of dimension m exceeding one. More precisely: The inequality (1.2″) should imply the continuity of $g(t)$ everywhere, with the exception of the origin $t = 0$. The last discontinuity must obviously remain, for if $g(t)$ satisfies the inequalities (1.2″), it will still satisfy them if the value of $g(0)$ is increased.

3.3. We close this section with the following theorem.

THEOREM 3′. *If $P_1, P_2, \cdots, P_n$ are any n distinct points of E_{n-1} and $f(t) \in \mathfrak{M}$, then*

$$\det \| f(P_jP_k^2) \|_{1,n} > 0, \tag{3.9}$$

unless $f(t)$ reduces to a non-negative constant.

Indeed, let us show first that the quadratic form $\sum_{j,k=1}^{n} \exp\{-P_jP_k^2\}\rho_j\rho_k$ is positive definite. Transforming the integral (2.1) ($u = 1$) to polar coördinates and using again vector notation in $E_m (m = n-1)$, with $P_j = x^j = (x_1^j, \cdots, x_m^j)$, we get

$$e^{-|x|^2} = 2^{1-m}[\Gamma(m/2)]^{-1}\int_0^\infty M_\xi\{e^{i(x\xi)r}\}e^{-r^2/4}r^{m-1}\,dr.$$

Replacing the vector $x = x - 0$ by $P_j - P_k = x^j - x^k$, we get

$$e^{-P_jP_k^2} = c\cdot\int_0^\infty M_\xi\{e^{i(x^j\xi)r}\cdot e^{-i(x^k\xi)r}\}e^{-r^2/4}r^{m-1}\,dr,$$

Now

$$\sum_{j,k=1}^{n} e^{-P_jP_k^2}\rho_j\rho_k = c\cdot\int_0^\infty M_\xi\left\{\left|\sum_{j=1}^{n} e^{i(x_j\xi)r}\cdot\rho_j\right|^2\right\}e^{-r^2/4}r^{m-1}\,dr,$$

which proves the positive definite character of this form by an elementary argument previously used (see [15], p. 790). Hence

$$\sum_{j,k=1}^{n} e^{-u\cdot P_jP_k^2}\rho_j\rho_k > 0, \qquad (u > 0,\ \rho_1^2 + \cdots + \rho_n^2 > 0). \tag{3.10}$$

Now (3.2) and (3.10) show that

$$\sum_1^n f(P_jP_k^2)\rho_j\rho_k = \int_0^\infty\left(\sum_1^n e^{-uP_jP_k^2}\rho_j\rho_k\right)d\beta(u)$$

is positive, if $\rho_1^2 + \cdots + \rho_n^2 > 0$, unless $\beta(u)$ reduces to a step function with a single saltus at the origin while $\sum\rho_j\rho_k = (\rho_1 + \cdots + \rho_n)^2 = 0$. But then $f(t)$ is a constant as was to be proved.

4. Integrals of completely monotone functions

4.1. We have denoted by $\mathfrak{M}$ the class of functions $f(t)$ completely monotone for $t \geqq 0$. If we drop the second condition (3.1′), which requires continuity at the origin, we get a larger class of functions $\psi(t)$ defined for $t > 0$ and satisfying the inequalities

$$(-1)^n\psi^{(n)}(t) \geqq 0 \quad \text{for} \quad 0 < t < \infty, \qquad (n = 0, 1, 2, \cdots). \tag{4.1}$$

These functions are also representable as Laplace integrals

$$\psi(t) = \int_0^\infty e^{-tu}\,d\gamma(u) \quad \text{for} \quad 0 < t < \infty, \tag{4.2}$$

where $\gamma(u)$ is non-decreasing. $\psi(0) = \psi(+0)$ exists if and only if $\gamma(u)$ is bounded, in which case $\psi(0) = \psi(+0) = \gamma(\infty)$. Integrating (4.2) between ϵ and $t(0 < \epsilon < t)$ we get

$$\int_\epsilon^t \psi(t)\,dt = \int_0^\infty \frac{e^{-u\epsilon} - e^{-ut}}{u}\,d\gamma(u). \tag{4.3}$$

Letting here $\epsilon \to 0$ we get

$$\phi(t) = \int_{+0}^t \psi(t)\,dt = \int_0^\infty \frac{1 - e^{-ut}}{u}\,d\gamma(u), \qquad (0 < t < \infty), \tag{4.4}$$

where both improper integrals converge or diverge simultaneously. We have convergence for all $t > 0$, if and only if

$$\int_1^\infty \frac{d\gamma(u)}{u} \tag{4.5}$$

exists.

DEFINITION 2. *We denote by* T *the class of functions* $\phi(t)$ *defined by* (4.4), *and* $\phi(0) = 0$, *as integrals of those functions* $\psi(t)$ *completely monotone for* $t > 0$ *for which the improper integral* $\int_{+0}\psi(t)dt$ *converges. In terms of the Laplace representation* (4.2), *this last mentioned condition is equivalent with the assumption* (4.5).

This class T is fundamental in our subsequent work. In this section we only prove a certain closure property of this class expressed by the following lemma.

LEMMA 5. *The class* T *is closed with respect to ordinary convergence to a continuous limit function, that is, if* $\{\phi_n(t)\}$ $(n = 1, 2, 3, \cdots)$ *is a sequence of elements of* T *and* $\phi_0(t)$ *is continuous for* $t \geqq 0$, *then*

$$\lim_{n\to\infty} \phi_n(t) = \phi_0(t), \qquad (0 \leqq t < \infty), \tag{4.6}$$

implies that $\phi_0(t) \in T$.

Indeed, by assumption we have

$$\phi_n(t) = \int_{+0}^{t} \psi_n(t)\, dt, \qquad (n = 1, 2, 3, \cdots), \tag{4.7}$$

where $\psi_n(t)$ are completely monotone for $t > 0$ and all improper integrals converge. Now the $\psi_n(t)$ are analytic in the half-plane $\Re t > 0$ and therefore also their integrals $\phi_n(t)$ are of this nature. Let $t = \sigma + i\tau$. We want to show that the functions of the sequence $\{\phi_n(t)\}$ are uniformly bounded in the rectangular domain

$$D\colon \sigma_1 \leqq \sigma \leqq \sigma_2, \qquad |\tau| \leqq \tau_1, \qquad (0 < \sigma_1 < \sigma_2).$$

Indeed

$$\phi_n(t) = \phi_n(\sigma + i\tau) = \int_0^\infty (1 - e^{-\sigma u - iu\tau})u^{-1}\, d\gamma_n(u)$$

$$= \int_0^\infty (1 - e^{-\sigma u})u^{-1}\, d\gamma_n(u) + \int_0^\infty e^{-\sigma u}(1 - e^{-iu\tau})u^{-1}\, d\gamma_n(u).$$

Hence

$$|\phi_n(t)| \leqq \phi_n(\sigma) + |\tau| \int_0^\infty e^{-\sigma u} \left|\sin\frac{u\tau}{2} \Big/ \frac{u\tau}{2}\right| d\gamma_n(u) \leqq \phi_n(\sigma) + \tau_1 \int_0^\infty e^{-\sigma u}\, d\gamma_n(u)$$

$$\leqq \phi_n(\sigma_2) + \tau_1 \int_0^\infty e^{-\sigma_1 u}\, d\gamma_n(u) \leqq \phi_n(\sigma_2) + \tau_1 \int_0^\infty (1 - e^{-\sigma_1 u})(\sigma_1 u)^{-1}\, d\gamma_n(u),$$

since $e^{-x} < (1 - e^{-x})x^{-1} (x > 0)$. We therefore have

$$|\phi_n(t)| \leqq \phi_n(\sigma_2) + \tau_1\sigma_1^{-1}\phi_n(\sigma_1) \qquad (t \text{ in } D)$$

This proves, in view of (4.6), that the $\phi_n(t)$ are uniformly bounded in D. By Vitali's convergence theorem we conclude that (4.6) holds uniformly in any bounded domain inside the half-plane $\Re t > 0$ and that $\phi_0(t)$ is analytic. (See Titchmarsh, [17], p. 168.)

Let $\psi_0(t) = \phi_0'(t)$; now (4.6) implies

$$\lim_{n\to\infty} (-1)^k \psi_n^{(k)}(t) = (-1)^k \psi_0^{(k)}(t), \qquad (k = 0, 1, 2, \cdots),$$

if $\Re t > 0$, hence in particular for all positive values of t. Thus $(-1)^k\psi_n^{(k)}(t) \geqq 0$ implies $(-1)^k\psi_0^{(k)}(t) \geqq 0$ $(t > 0;\ k = 0, 1, 2, \cdots)$, hence $\psi_0(t)$ is completely monotone for $t > 0$. The proof is now immediately completed, for

$$\phi_0(t) - \phi_0(\epsilon) = \int_\epsilon^t \psi_0(t)\,dt, \quad (0 < \epsilon < t), \quad \text{and} \quad \phi_0(+0) = \phi_0(0) = 0,$$

imply, as $\epsilon \to 0$, the existence of $\int_{+0}\psi_0(t)dt$ and the relation

$$\phi_0(t) = \int_{+0}^t \psi_0(t)\,dt, \qquad (0 < t < \infty).$$

Hence $\phi_0(t) \in T$.

We conclude this section with two remarks concerning Lemma 5. The function theoretic argument shows that if 1. $\phi_n(t)$ are uniformly bounded in any finite interval $0 \leqq t \leqq \sigma_2$, 2. the limit (4.6) exists in a set of positive values of t with a positive limit point, then (4.6) holds for all t with positive real part, and $\phi_0(t)$ is analytic for $\Re t > 0$. In concluding, however, that $\phi_0(t) \in T$, it is essential to assume that $\phi_0(+0) = 0$, i.e., with $\phi_0(0) = 0$, $\phi_0(t)$ should be continuous at the origin. This fact was used in the last step of our proof. That $\phi_0(+0) = 0$ is not implied by (4.6), is seen by the following sequence: $\phi_n(t) = 1 - e^{-nt}$, $(\psi_n(t) = ne^{-nt})$, for which (4.6) holds for $t \geqq 0$ with $\phi_0(t) = 1 (t > 0)$, $\phi_0(0) = 0$. Our second remark is that a somewhat more involved but elementary proof of Lemma 5 (free of any appeal to analytic function theory) is available.

II. Determination of All Metric Transforms of Hilbert Space Which Are Isometrically Imbeddable in Hilbert Space. Applications to Completely Monotone Functions

5. Isometric imbedding in $\mathfrak{H}$ and positive definite functions

5.1. Before we enter into a discussion of our specific topic, a few general remarks concerning metric transforms are necessary. Let $\mathfrak{S}$ be a semi-metric space with the metric PP'. In a study of its metric transforms $F(\mathfrak{S})$ by a transforming function $F(t)$, the following distinction is of interest. We shall say that $F(t)$ is a *proper* transforming function if $F(t) > 0$ for all positive values of t; if, however, $F(t)$ has positive zeros, we call $F(t)$ an *improper* transforming function. The reason for this distinction is as follows. If $F(t)$ is proper, then $F(\mathfrak{S})$ is obviously semi-metric. This, however, is not always the case if $F(t)$ is improper, for the only properties of the metric $F(PP')$ of $F(\mathfrak{S})$ are 1. $F(PP') = F(P'P) \geqq 0$, 2. $F(PP) = 0$.[18] Operating with $\mathfrak{S}$, rather than $F(\mathfrak{S})$, we are

[18] Of course *some* transforms $F(\mathfrak{S})$ may still be semi-metric. Thus if $F(t) = |\sin t|$ $(t \geqq 0)$, we see that $F(\mathfrak{S})$ is semi-metric if the diameter of $\mathfrak{S}$ is $< \pi$.

thus led to consider spaces $\mathfrak{S}$ whose distance function PP' enjoys only the following weaker properties

$$(5.1) \qquad 1.\ PP' = P'P \geqq 0, \qquad 2.\ PP = 0.$$

We shall refer to such spaces as *quasi-metric* spaces.

A quasi-metric space $\mathfrak{S}$ can not, in general, be turned into a semi-metric space by the identification of points P, Q for which $PQ = 0$ (similar to the identification of measurable functions equal almost everywhere), for this identification does not enjoy the necessary transitive property. Indeed, if $PQ = 0$ and $QR = 0$, there is no reason why $PR = 0$ should hold. Moreover, even if this were always true, then $PQ = 0$ need not imply $PS = QS$, where S is a third point of $\mathfrak{S}$. It becomes now apparent that this identification is possible and turns $\mathfrak{S}$ into a semi-metric space $\mathfrak{R}$, if and only if $\mathfrak{S}$ is isometrically imbeddable in a semi-metric space $\mathfrak{R}$.

For the sake of generality we include also *improper* transforming functions $F(t)$ in our discussion, even $F(t) \equiv 0$ not being excluded. Accordingly we are bound to deal with *quasi-metric* spaces and their imbedding in $\mathfrak{H}$.

5.2. Our imbedding problem $\{E_m ; \mathfrak{H}\}$ (see Introduction) is of the following general type: *Given a separable semi-metric space $\mathfrak{S}$, determine all functions $F(t)$, proper or improper, such that the metric transform $F(\mathfrak{S})$ be imbeddable in $\mathfrak{H}$. We shall denote by $\Pi(\mathfrak{S})$ the class of functions $F(t)$ which are solutions of this problem $\{\mathfrak{S}; \mathfrak{H}\}$.* In particular $\Pi(E_m)$ and $\Pi(\mathfrak{H})$ represent the totality of solutions of the problems $\{E_m ; \mathfrak{H}\}$ and $\{\mathfrak{H} ; \mathfrak{H}\}$, respectively.

A characteristic property of a function $F(t)$ in $\Pi(\mathfrak{S})$ is as follows. If $P_0, P_1, \cdots, P_n$ are any $n + 1$ points of $\mathfrak{S}$, then the distances $F(P_iP_k)$ are the edges of a euclidean simplex, i.e., there exist $n + 1$ points $Q_0, \cdots, Q_n$ in E_n, such that $F(P_iP_k) =$ the euclidean distance between Q_i and Q_k. As $F(t)$ thus turns every finite set of points of $\mathfrak{S}$ into a euclidean simplex, $F(t)$ may be called a *simplicial function* of $\mathfrak{S}$[19], and the whole class $\mathfrak{P}(\mathfrak{S})$ may be referred to as the *simplicial class* of $\mathfrak{S}$. Note that $\Pi(\mathfrak{S})$ is never empty, for $F(t) \in \Pi(\mathfrak{S})$, if $F(t) \equiv 0$.

5.3. A connecting link between problems of isometric imbedding in $\mathfrak{H}$ and the theory of positive definite functions is furnished by the following theorem.

THEOREM 4. *A necessary and sufficient condition that a separable quasi-metric space $\mathfrak{S}$ be isometrically imbeddable in $\mathfrak{H}$ is that the family of functions* $\exp\{-\lambda t^2\}$ *(where λ is a positive parameter) be positive definite in $\mathfrak{S}$.*[20]

[19] The term "simplicial function" was suggested to the author by L. M. Blumenthal. Blumenthal had previously called $F(t)$ a "tetrahedral function" of $\mathfrak{S}$, if $F(\mathfrak{S})$ enjoys the 4-point property. His theorem mentioned in the Introduction may be stated as follows: $F(t) = t^\kappa (0 < \kappa \leqq \frac{1}{2})$ is a tetrahedral function of any *metric* space $\mathfrak{S}$. See [3].

[20] Definition 1, §1.1, of functions $g(t)$ which are positive definite in a semi-metric space $\mathfrak{S}$, applies unchanged in case $\mathfrak{S}$ is only quasi-metric.

A proof of this theorem is found in [16], §3. Let now $\mathfrak{S}$ be semi-metric and let $\mathfrak{P}(\mathfrak{S})$ be its positive definite class and $\Pi(\mathfrak{S})$ its simplicial class. An immediate consequence of Theorem 4 is the following characterization of the class $\Pi(\mathfrak{S})$ in terms of the class $\mathfrak{P}(\mathfrak{S})$.

THEOREM 5. *A function $F(t)(F(0) = 0)$ belongs to the simplicial class $\Pi(\mathfrak{S})$ of a semi-metric space $\mathfrak{S}$, if and only if the family of functions $\exp\{-\lambda F^2(t)\}$ $(\lambda > 0)$ belongs to the positive definite class $\mathfrak{P}(\mathfrak{S})$.*

Indeed, $F(t) \in \Pi(\mathfrak{S})$ is equivalent with the statement that $F(\mathfrak{S})$ is imbeddable in $\mathfrak{H}$. By Theorem 4 this is the case if and only if $\exp\{-\lambda t^2\}$ is positive definite in $F(\mathfrak{S})$, for all $\lambda > 0$, or, which is the same thing, if $\exp\{-\lambda F^2(t)\}$ is positive definite in $\mathfrak{S}$. This proves our theorem.

This general theorem will be used in this paper only in case $\mathfrak{S}$ is either a euclidean space or Hilbert space. For convenience we restate it in each of these cases as special corollaries.

COROLLARY 1. *$F(t) \in \Pi(E_m)$ if and only if $\exp\{-\lambda F^2(t)\} \in \mathfrak{P}(E_m)$ for all $\lambda > 0$.*

COROLLARY 2. *$F(t) \in \Pi(\mathfrak{H})$ if and only if $\exp\{-\lambda F^2(t)\} \in \mathfrak{P}(\mathfrak{H})$ for all $\lambda > 0$.*

Corollary 1 was already discussed, in case $m = 1$, in [16], §4, in connection with the class $\Pi(E_1)$ of "screw functions" of $\mathfrak{H}$; here it will be applied in §8 of the Appendix.

5.4. We shall presently use Corollary 2 in establishing the following theorem which determines the class $\Pi(\mathfrak{H})$, i.e., the totality of solutions of the problem $\{\mathfrak{H} ; \mathfrak{H}\}$.

THEOREM 6. *The simplicial class $\Pi(\mathfrak{H})$, that is, the totality of continuous functions $F(t)$ such that the metric transform $F(\mathfrak{H})$ be isometrically imbeddable in $\mathfrak{H}$, is identical with the class of functions of the form*

$$F(t) = \left\{\int_0^\infty \frac{1 - e^{-t^2 u}}{u}\, d\gamma(u)\right\}^{1/2}, \qquad (t \geqq 0), \tag{5.2}$$

where $\gamma(u)$ is non-decreasing for $u \geqq 0$ and such that

$$\int_1^\infty \frac{d\gamma(u)}{u} \text{ exists.} \tag{5.3}$$

The proof that (5.2) furnishes functions $F(t)$ belonging to $\Pi(\mathfrak{H})$ is very simple (see [16], 2nd footnote of §3). For let $P_0, \cdots, P_n$ be any $n + 1$ points of $\mathfrak{H}$. In order to prove that $F(t) \in \Pi(\mathfrak{H})$, i.e., $F(\mathfrak{H})$ is imbeddable in $\mathfrak{H}$, it suffices to show that $F(P_iP_k)$ are distances of $n + 1$ points of E_n. As $\exp\{-ut^2\}(u \geqq 0)$ belongs to $\mathfrak{P}(E_m)$, we have

$$\sum_0^n e^{-uP_iP_k^2}\rho_i\rho_k \geqq 0 \qquad \text{for} \qquad u \geqq 0. \tag{5.4}$$

Hence if

$$\sum_0^n \rho_i = 0, \tag{5.5}$$

we conclude from (5.2), (5.4) and (5.5) that

$$\sum_0^n F^2(P_iP_k)\rho_i\rho_k = -\int_0^\infty \left(\sum_0^n e^{-uP_iP_k^2}\rho_i\rho_k\right) u^{-1}\, d\gamma(u) \leqq 0. \tag{5.6}$$

The fact that $\sum F^2(P_iP_k)\rho_i\rho_k \leqq 0$ for any set of real ρ_i subject to the relation (5.5), is equivalent to the statement that $F(P_iP_k)$ are distances of $n + 1$ points of E_n.[21] This completes the first part of the proof.

Let us now prove the converse to the effect that an element $F(t)$ of $\Pi(\mathfrak{H})$ is necessarily of the form (5.2). From Corollary 2 we learn that $\exp\{-\lambda F^2(t)\}$ is positive definite in $\mathfrak{H}$ for all positive values of λ. By Theorem 2 it is therefore of the form

$$e^{-\lambda F^2(t)} = \int_0^\infty e^{-t^2u}\, d\alpha(u, \lambda), \qquad (\lambda > 0), \tag{5.7}$$

where $\alpha(u, \lambda)$ is a family of non-decreasing functions of $u \geqq 0$, defined for all $\lambda > 0$, with

$$\alpha(0, \lambda) = 0, \qquad \alpha(\infty, \lambda) = 1. \tag{5.8}$$

For convenience we introduce the function

$$\phi(t) = F^2(\sqrt{t}) \tag{5.9}$$

in terms of which (5.7) becomes

$$e^{-\lambda\phi(t)} = \int_0^\infty e^{-tu}\, d\alpha(u, \lambda), \qquad (\lambda > 0). \tag{5.10}$$

In view of (5.2) and (5.9) our objective is to show that $\phi(t)$ is of the form

$$\phi(t) = \int_0^\infty \frac{1 - e^{-tu}}{u}\, d\gamma(u), \tag{5.11}$$

with $\gamma(u)$ subject to the restriction (5.3). In different words: That $\phi(t)$ belongs to the class T of §4. This fact is now readily established. Indeed (5.10) and the relation $1 = \int_0^\infty d\alpha(u, \lambda)$ imply

$$\frac{1 - e^{-\lambda\phi(t)}}{\lambda} = \int_0^\infty \frac{1 - e^{-tu}}{\lambda}\, d\alpha(u, \lambda) = \int_0^\infty \frac{1 - e^{-tu}}{u} \cdot \frac{u}{\lambda}\, d\alpha(u, \lambda),$$

whence

$$\frac{1 - e^{-\lambda\phi(t)}}{\lambda} = \int_0^\infty \frac{1 - e^{-tu}}{u}\, d\beta(u, \lambda), \qquad (t \geqq 0, \lambda > 0), \tag{5.12}$$

[21] Here we use the following theorems: 1. *The non-negative quantities* $a_{ik}(a_{ik} = a_{ki}$; $i, k = 0, 1, \cdots n$; $a_{ii} = 0)$ *are distances of* $n + 1$ *points of* E_n, *if and only if* (5.5) *implies* $\sum_0^n a_{ik}^2\rho_i\rho_k \leqq 0$. (See [16], §3). 2. *A separable quasi-metric space is imbeddable in* $\mathfrak{H}$, *if and only if any* $n + 1$ *of its points* ($n = 2, 3, \cdots$) *are congruent with* $n + 1$ *points of* $\mathfrak{H}$. (See Menger, [9].)

where the new family of monotone functions $\beta(u, \lambda)$ is defined by

$$\beta(u, \lambda) = \frac{1}{\lambda} \int_0^u v \, d\alpha(v, \lambda), \qquad (\lambda > 0). \tag{5.13}$$

From (5.12) we see that

$$(1 - e^{-\lambda\phi(t)})\lambda^{-1}$$

belongs to the class T for all $\lambda > 0$. Letting $\lambda \to 0$ we find this function converging to the continuous limit function $\phi(t)$. On the basis of the closure property expressed by Lemma 5 (§4) we conclude that $\phi(t)$ belongs to T. This completes the proof of Theorem 6.

What has just been established may also be stated as follows: *There is a one-to-one correspondence between the classes T and $\Pi(\mathfrak{H})$ expressed by the relation*

$$\phi(t) = F^2(\sqrt{t}) \quad \text{or} \quad F(t) = \sqrt{\phi(t^2)}, \qquad (\phi \in T, F \in \Pi(\mathfrak{H})). \tag{5.14}$$

It is likewise interesting to connect $F(t)$ directly with the completely monotone functions. From (5.14) and (4.4) we get

$$\frac{d}{dt}\phi(t) = \frac{d}{dt}F^2(\sqrt{t}) = \psi(t). \qquad (t > 0).$$

We may therefore state the following theorem.

THEOREM 6′. *A necessary and sufficient condition that the non-negative continuous function $F(t)$, vanishing at the origin, have the property that $F(\mathfrak{H})$ be isometrically imbeddable in $\mathfrak{H}$, is that*

$$\frac{d}{dt}F^2(\sqrt{t}) = \psi(t) \tag{5.15}$$

be completely monotone for $t > 0$.

5.5. Let us apply the criterion expressed by Theorem 6′ to a few examples. Let $F_1(t) = t^\kappa (\kappa > 0)$. By (5.15) we get $\psi(t) = \kappa t^{\kappa-1} = \kappa/t^{1-\kappa}$ and this function is completely monotone if $1 - \kappa \geqq 0$, or $\kappa \leqq 1$. Hence

$$F_1(t) = t^\kappa, \qquad (0 < \kappa \leqq 1). \tag{5.16}$$

belongs to $\Pi(\mathfrak{H})$. Let now

$$F_2(t) = \frac{t}{1+t}, \tag{5.17}$$

which was mentioned in the Introduction. In this case $\psi(t) = (1 + \sqrt{t})^{-3}$ is likewise completely monotone. This fact may be verified directly by successive differentiations or more elegantly as follows: $(1 + t)^{-1}$ is obviously in $\mathfrak{M}$ and therefore also its cube $(1 + t)^{-3}$. Finally it will follow from the results of §6 (Theorem 8) that if $f(t)$ is in $\mathfrak{M}$, then so is $f(\sqrt{t})$. Our last two examples are the functions

(5.18) $$F_3(t) = (1 - e^{-t^2})^{\frac{1}{2}}, \qquad F_4(t) = (\log (1 + t^2))^{\frac{1}{2}}.$$

We find that $\psi_3(t) = e^{-t}$, $\psi_4(t) = (1 + t)^{-1}$ are both in $\mathfrak{M}$, hence F_3, F_4 are elements of $\Pi(\mathfrak{H})$.

A glance at either (5.2) or (5.15) will show that, if we exclude from consideration the trivial function $F(t) \equiv 0$, all elements of $\Pi(\mathfrak{H})$ are *proper* transforming functions, that is, $F(t) > 0$ for $t > 0$. Moreover, if $F(t)$ is not of the form $F(t) = ct$ ($c \geqq 0$), we actually have

(5.19) $$F^2(\sqrt{t}) > 0, \qquad \frac{d}{dt} F^2(\sqrt{t}) > 0,$$
$$\frac{d^2}{dt^2} F^2(\sqrt{t}) < 0, \qquad \frac{d^3}{dt^3} F^2(\sqrt{t}) > 0, \cdots \qquad (t > 0).$$

Hence: *The continuity of* $F(t)$ *for* $t \geqq 0$, $F(0) = 0$, *and the inequalities* (5.19) *characterize completely all elements of* $\Pi(\mathfrak{H})$, *with the exception of the elements* $F(t) = ct$ ($c \geqq 0$).

Ruling out the case $F(t) \equiv 0$, we see that $\mathfrak{H}$ and $F(\mathfrak{H})$ are homeomorphs of each other, since both metrics PP' and $F(PP')$ are topologically equivalent.

The class $\Pi(\mathfrak{H})$ contains bounded as well as unbounded functions. Thus $F_2(t)$ and $F_3(t)$, defined above, are bounded. In terms of the function $\gamma(u)$ of (5.2) it is readily seen that $F(t)$ is bounded, if and only if

(5.20) $$\gamma(0) = \gamma(+0) \quad \text{and} \quad \int_{+0}^{1} \frac{d\gamma(u)}{u} \quad \text{exists.}$$

5.6. In [15] it was shown that the function (5.16) ($0 < \kappa < 1$) enjoys the following further property: If $P_0, \cdots, P_n$ are any $n + 1$ *distinct* points of $F_1(\mathfrak{H})$, the distances $F_1(P_iP_k) = (P_iP_k)^\kappa$ are the edges of a *non-degenerate* simplex in E_n. This property extends as follows to the whole class $\Pi(\mathfrak{H})$. *If* $F(t)$ *is any element of* $\Pi(\mathfrak{H})$, *which is not of the form* $F(t) = ct$, *then any* $n + 1$ *distinct points of* $F(\mathfrak{H})$ *are congruent with a non-degenerate n-dim. euclidean simplex.*

Indeed, it suffices to show that if $P_0, \cdots, P_n$ are distinct points of $\mathfrak{H}$, then

$$\sum_0^n \rho_i = 0, \qquad \sum_0^n \rho_i^2 > 0, \qquad \text{imply} \qquad \sum_0^n F^2(P_iP_k)\rho_i\rho_k < 0.$$

Now (5.2) gives

(5.21) $$F^2(t) = c^2t^2 + \int_{+0}^{\infty} \frac{1 - e^{-t^2u}}{u} \, d\gamma(u), \qquad (c^2 = \gamma(+0) - \gamma(0)),$$

and therefore

(5.22) $$\sum_0^n F^2(P_iP_k)\rho_i\rho_k = c^2 \sum_0^n (P_iP_k)^2\rho_i\rho_k - \int_{+0}^{\infty} \left(\sum_0^n e^{-uP_iP_k^2}\rho_i\rho_k \right) u^{-1} \, d\gamma(u).$$

But $\sum (P_iP_k)^2\rho_i\rho_k \leqq 0$ and the integrand in (5.22) is positive by (3.10). Therefore the vanishing of the quadratic form (5.22) implies that $\gamma(u)$ is constant for $u > 0$, hence $F(t) = ct$.

5.7. It was pointed out in 5.5 that $\Pi(\mathfrak{H})$ contains only one improper transforming function, namely $F(t) \equiv 0$. That the same conclusion may be reached directly, independently of Theorem 6, and for all classes $\Pi(E_m)(2 \leqq m \leqq \infty)$, is shown by the following elementary lemma.

LEMMA 6. *Let $F(t)$ be continuous for $t \geqq 0$, $F(t) \geqq 0$, $F(0) = 0$, $F(t) \not\equiv 0$. If $F(E_m)(2 \leqq m \leqq \infty)$ is imbeddable in a semi-metric space $\mathfrak{R}$, then $F(t) > 0$ for $t > 0$.*

Indeed, let $P \rightarrow P'(P \in F(E_m), P' \in \mathfrak{R})$ indicate the mapping of $F(E_m)$ in $\mathfrak{R}$. Denoting by $[P', Q']$ the distance in $\mathfrak{R}$, the isometricity of this imbedding is expressed by the identical relation

$$[P', Q'] = F(PQ), \qquad (P, Q \in E_m). \tag{5.23}$$

Assume that for $t = \tau > 0$ we have $F(\tau) = 0$. Take in E_m an ordinary circle Γ of radius τ, center O, and let A and B be any two points on Γ. Now, by (5.23) we have

$$[O', A'] = F(OA) = F(\tau) = 0, \qquad [O', B'] = F(OB) = F(\tau) = 0,$$

hence ($\mathfrak{R}$ being semi-metric) $O' = A'$, $O' = B'$, and therefore $A' = B'$ or $[A', B'] = 0$. But this implies

$$F(AB) = [A', B'] = 0.$$

As A, B were arbitrary points of Γ, we conclude that $F(t) = 0$ in the interval $0 \leqq t \leqq 2\tau$. Repeating the argument with 2τ, instead of τ, we conclude that $F(t)$ vanishes in the interval $(0, 4\tau)$, and so forth. Hence $F(t) \equiv 0$, in contradiction to our assumption that $F(t) \not\equiv 0$.

Notice that the proof as well as the result fail if $m = 1$. This is already shown by the simplest *closed* euclidean "screw line", namely the circle; for $F(E_1)$, where $F(t) = |\sin t|$ $(0 \leqq t < \infty)$, is imbeddable in E_2. Indeed, taking E_1 to be the real θ-axis with the euclidean metric $|\theta_1 - \theta_2|$, the imbedding of $F(E_1)$ in E_2 is performed by mapping the point $P = \theta$ onto the point P': $x = \frac{1}{2} \cos 2\theta$, $y = \frac{1}{2} \sin 2\theta$, of E_2. For we have identically

$$F^2(P_1P_2) = |\sin|\theta_1 - \theta_2||^2 = \sin^2(\theta_1 - \theta_2) = \tfrac{1}{4}(\cos 2\theta_1 - \cos 2\theta_2)^2 + \tfrac{1}{4}(\sin 2\theta_1 - \sin 2\theta_2)^2 = [P_1', P_2']^2, \text{ hence } [P_1', P_2'] = F(P_1P_2).$$

6. The inner transformations of completely monotone functions

6.1. The simplicial class $\Pi(\mathfrak{H})$ enjoys a further analytical property expressed by the following theorem.

THEOREM 7. *The class of non-negative continuous functions $F(t)$, $(F(0) = 0)$, with the property that if $g(t)$ is an element of $\mathfrak{P}(\mathfrak{H})$, also $g(F(t))$ should be in $\mathfrak{P}(\mathfrak{H})$, coincides with the simplicial class $\Pi(\mathfrak{H})$.*

Indeed, let $F(t) \in \Pi(\mathfrak{H})$ and $g(t) \in \mathfrak{P}(\mathfrak{H})$. In order to prove that $g(F(t)) \in \mathfrak{P}(\mathfrak{H})$, we have to show that if $P_1, \cdots, P_n$ are any points of $\mathfrak{H}$, then

$$\sum_1^n g(F(P_iP_k))\rho_i\rho_k \geqq 0.$$

But this is evident, since there are points Q_i of $\mathfrak{H}$, with $F(P_iP_k) = Q_iQ_k$ (since $F(t) \in \Pi(\mathfrak{H})$), and $\sum g(Q_iQ_k)\rho_i\rho_k \geqq 0$ (since $g(t) \in \mathfrak{P}(\mathfrak{H})$).

To prove the converse, let $F(t)$ be such that $g(t) \in \mathfrak{P}(\mathfrak{H})$ implies $g(F(t)) \in \mathfrak{P}(\mathfrak{H})$. As in particular $g(t) = \exp\{-\lambda t^2\} \in \mathfrak{P}(\mathfrak{H})$, we have

$$e^{-\lambda F^2(t)} \in \mathfrak{P}(\mathfrak{H}), \qquad \text{for all positive } \lambda.$$

By Corollary 2 we now conclude that $F(t) \in \Pi(\mathfrak{H})$, which proves the theorem.

Let us now state Theorem 7 in terms of completely monotone functions by means of Theorem 3. Let $\phi(t)$ be a continuous non-negative function vanishing at the origin. Let us determine all such functions with the property that if $f(t) \in \mathfrak{M}$, also $f(\phi(t))$ should be an element of $\mathfrak{M}$. But if $\phi(t)$ is of this nature and $g(t) \in \mathfrak{P}(\mathfrak{H})$, we may conclude successively that the following relations hold: $g(\sqrt{t}) \in \mathfrak{M}$, $g(\sqrt{\phi(t)}) \in \mathfrak{M}$, $g(\sqrt{\phi(t^2)}) \in \mathfrak{P}(\mathfrak{H})$. Therefore $\sqrt{\phi(t^2)} = F(t) \in \Pi(\mathfrak{H})$, by Theorem 7. In view of (5.14) we thus see that $\phi(t)$ belongs to the class T of §4. This proves the following theorem.

THEOREM 8. *The class of non-negative continuous functions $\phi(t)$, ($\phi(0) = 0$), with the property that if $f(t)$ is completely monotone for $t \geqq 0$, also $f(\phi(t))$ should be completely monotone for $t \geqq 0$, is identical with the class T of integrals of completely monotone functions*:

$$\phi(t) = \int_{+0}^{t} \psi(t)\, dt, \qquad (t > 0). \tag{6.1}$$

In view of this theorem we may call these functions $\phi(t)$ *inner transformations* of the class $\mathfrak{M}$ of completely monotone functions. In a recent paper ([5], p. 498) S. Bochner has already pointed out this property of the functions (6.1) with the restriction that $\psi(t)$ itself belong to the class $\mathfrak{M}$. This restriction, which amounts to assuming the existence of $\psi(+0)$ rules out such simple elements of T as $\phi(t) = t^\kappa (0 < \kappa < 1)$ for $\psi(t) = \phi'(t) = \kappa/t^{1-\kappa}$ is not bounded near the origin, although $\int_{+0}\psi(t)dt$ obviously exists. However, the principal new contribution of Theorem 8 is not the *enlargement* of Bochner's class of inner transformations, but the statement that the enlarged class T contains *all* such transformations.

6.2. The class T enjoys a further closure property which we state as a corollary.

COROLLARY 3. *If $\phi_1(t)$ and $\phi_2(t)$ are any two elements of T, i.e., of the form* (6.1) *or* (4.4), *also $\phi_1(\phi_2(t))$ belongs to T.*

For if $f(t)$ is completely monotone for $t \geqq 0$, then so is $f(\phi_1(t))$ and therefore also $f(\phi_1(\phi_2(t)))$. Hence $\phi_1(\phi_2(t))$ must belong to T by Theorem 8.

This closure property of T appears here as a consequence of virtually all previous results of this paper. It seems of sufficient interest to deserve a direct proof to which we now proceed.[22]

We prove first: *If* $\phi(t) \in T$, *then*

$$1 - e^{-\phi(t)} \in T. \tag{6.2}$$

It suffices to show that

$$\psi(t) = \frac{d}{dt}(1 - e^{-\phi(t)}) = e^{-\phi(t)}\phi'(t)$$

is completely monotone for $t > 0$, that is, verifies the inequalities (4.1). Now

$$\psi'(t) = e^{-\phi}(\phi'' - \phi'^2), \qquad \psi''(t) = e^{-\phi}(\phi''' - 3\phi'\phi'' + \phi'^3),$$

and generally

$$\psi^{(n-1)}(t) = \sum (-1)^{0\cdot k_1 + 1\cdot k_2 + \cdots + (n-1)k_n + n-1} \cdot c_{(k)} e^{-\phi}(\phi')^{k_1}(\phi'')^{k_2} \cdots (\phi^{(n)})^{k_n}, \quad (c_{(k)} > 0), \tag{6.3}$$

where all terms of this sum have constant weight $k_1 + 2k_2 + \cdots + nk_n = n$. This last statement as well as the fact that the sign before each term has the expression indicated, is readily proved (by induction) by differentiating once both sides of (6.3). Now $\phi'(t)$ being completely monotone for $t > 0$, we have $(-1)^{m-1}\phi^{(m)}(t) \geqq 0$, which, in view of (6.3), clearly implies $(-1)^{n-1}\psi^{(n-1)}(t) \geqq 0$.

We can now complete a proof of Corollary 3. For let

$$\phi_1(t) = \int_0^\infty \frac{1 - e^{-tu}}{u}\, d\gamma(u) \tag{6.4}$$

be a second arbitrary element of T. This integral is the limit, for any fixed $t \geqq 0$, of an appropriate sequence of functions of the form

$$\phi_n^*(t) = \sum_{\nu=1}^{n} \frac{1 - e^{-tu_\nu}}{u_\nu}(\gamma(u_\nu) - \gamma(u_{\nu-1})) \qquad (0 = u_0 < u_1 < \cdots < u_n)$$

where the u_ν depend on n as well. Hence

$$\phi_n^*(\phi(t)) \to \phi_1(\phi(t)) \qquad \text{as} \qquad n \to \infty.$$

As $\phi_n^*(\phi(t)) \in T$, by (6.2), we conclude that also the limit functions $\phi_1(\phi(t))$ is in T, on the basis of the closure property of Lemma 5, §4.

6.3. We conclude this section with two further theorems on the class $\mathfrak{M}$ of functions $f(t)$ which are completely monotone for $t \geqq 0$. We find it convenient to normalize our functions by requiring that $f(0) = 1$. Let $\mathfrak{M}_0$ denote the subclass of elements of $\mathfrak{M}$ thus normalized.

---

[22] This direct proof was suggested to the author by von Neumann.

THEOREM 9. *The class of elements $f(t)$ of $\mathfrak{M}_0$ with the property that $(f(t))^\lambda$ also belongs to $\mathfrak{M}_0$, for all $\lambda > 0$, is identical with the class of functions of the form*

$$f(t) = e^{-\phi(t)} \tag{6.5}$$

where $\phi(t) \in T$.

THEOREM 10. *If to the previous theorem's assumptions on $f(t)$ we add the further assumption that $f(t)$ be bounded away from zero, $f(t) \geqq \epsilon_0 > 0$, then this new class of functions is identical with the class of functions of the form*

$$f(t) = e^{f_1(t) - f_1(0)}, \tag{6.6}$$

where $f_1(t) \in \mathfrak{M}$.

These theorems are immediate consequences of Corollary 2 and Theorems 3 and 6. Indeed, in order to prove Theorem 9, define $\phi(t)$ by (6.5). Now $(f(t))^\lambda = \exp\{-\lambda\phi(t)\} \in \mathfrak{M}_0$, i.e., $\exp\{-\lambda\phi(t^2)\} \in \mathfrak{P}(\mathfrak{H})$ (Theorem 3), if and only if $\sqrt{\phi(t^2)} \in \Pi(\mathfrak{H})$ (Corollary 2), hence $\phi(t) \in T$ (Theorem 6 and (5.14)). To prove Theorem 10, notice that $f(t)$ is bounded away from zero if and only if $\phi(t)$, of (6.5), is bounded. Now

$$\phi(t) = \int_0^\infty \frac{1 - e^{-tu}}{u}\, d\gamma(u)$$

is bounded, if and only if $\gamma(+0) = \gamma(0)$ and $\int_{+0}^\infty u^{-1}\, d\gamma(u)$ exists. But then

$$\phi(t) = \int_{+0}^\infty \frac{1 - e^{-tu}}{u}\, d\gamma(u) = \int_{+0}^\infty u^{-1}\, d\gamma(u) - \int_{+0}^\infty e^{-tu}u^{-1}\, d\gamma(u) = f_1(0) - f_1(t),$$

where $f_1(t) \in \mathfrak{M}$. Now (6.5) goes over into (6.6).

Incidentally we have proved the following result: *The elements $\phi(t)$ of T which are bounded are of the form*

$$\phi(t) = f(0) - f(t), \qquad (f(t) \in \mathfrak{M}), \tag{6.7}$$

and conversely.

Theorem 9 may also be stated in the following suggestive form: *The elements $f(t)$ of $\mathfrak{M}_0$ which belong to $\mathfrak{M}_0$ together with all their positive powers $(f(t))^\lambda (\lambda > 0)$ are identical with those elements of $\mathfrak{M}_0$ which arise out of the particular function*

$$f(t) = e^{-t} \tag{6.8}$$

by the general inner transformation $t \mid \phi(t)$, where $\phi(t) \in T$.

APPENDIX

7. On metric transforms of euclidean spaces which are isometrically imbeddable in euclidean spaces

7.1. The solution of the problem $\{E_m ; E_n\}$, $(2 \leqq m \leqq n < \infty)$, is given by the following theorem.

THEOREM 11. *If the metric transform $F(E_m)$, of $E_m (m \geqq 2)$ by $F(t)$, is isometrically imbeddable in the euclidean space $E_n (m \leqq n < \infty)$, then necessarily $F(t) = ct$ $(c > 0)$, unless $F(t) \equiv 0$.*

In the proof of this theorem we shall use the solution of the problem $\{E_1 ; E_n\}$ $(n < \infty)$ (von Neumann-Schoenberg, [12]) which is as follows: *$F(E_1)$ is imbeddable in E_n if and only if $F^2(t)$ is of the form*

$$F^2(t) = c^2t^2 + \sum_{\nu=1}^{r} A_\nu^2 \sin^2 (k_\nu t), \tag{7.1}$$

where $2r + 1 \leqq n$ if $c > 0$, or $2r \leqq n$ if $c = 0$, and c, A_ν, k_ν are non-negative constants.

Let E_1 be the real x-axis with the metric $| x_1 - x_2 |$. The statement that $F(E_1)$ is imbeddable in E_n means that there is a continuous curve $P' = \phi(x)$ $(P' \in E_n , -\infty < x < \infty)$ such that we have identically

$$F^2(| x_1 - x_2 |) = (P_1'P_2')^2 = | \phi(x_1) - \phi(x_2) |^2, \qquad (-\infty < x_1 , x_2 < \infty), \tag{7.2}$$

in standard vector notation. The curve $P' = \phi(x)$, of E_n , is what was called a *screw line* of E_n , while the corresponding $F(t)$ is a *screw function* of E_n . Thus whenever a continuous function $F(t)$ satisfies an identity of the form $F^2(| t - t'|) = | \phi(t) - \phi(t') |^2$, $(\phi \in E_n)$, for all real t, t', we conclude that $F^2(t)$ is of the form (7.1).

We can now readily prove Theorem 11. Ruling out the trivial case $F(t) \equiv 0$, by Lemma 6, §5.7, we may assume that $F(t) > 0$ for $t > 0$. Consider now a straight line E_1 in E_n . The assumption of Theorem 11 clearly implies that $F(E_1)$ is imbeddable in E_n , hence $F^2(t)$ is of the form (7.1). Now consider in E_n an ordinary circle Γ, of radius $\frac{1}{2}$, referred to the angular parameter θ. Its image in E_n is a simple closed curve $P' = \phi(\theta)$ (of period 2π, $P' \in E_n$). The isometricity of the imbedding of $F(\Gamma)$ in E_n gives

$$F^2(P_1P_2) = F^2\left(\left|2 \sin \frac{\theta_1 - \theta_2}{2}\right|\right) = (P_1'P_2')^2 = | \phi(\theta_1) - \phi(\theta_2) |^2,$$

for any real θ_1, θ_2. This shows that also $F\left(\left|2 \sin \frac{\theta}{2}\right|\right)$ is a screw function of E_n. By (7.1) we therefore have

$$F^2\left(\left|2 \sin \frac{\theta}{2}\right|\right) = \sum_{\mu=1}^{s} B_\mu^2 \sin^2 (h_\mu \theta), \qquad (2s \leqq n), \tag{7.3}$$

this time without a term in θ^2, since the left side of (7.3) is bounded. From (7.1) and (7.3) we now derive the identity

$$\sum_{\mu=1}^{s} B_\mu^2 \sin^2 (h_\mu \theta) = 4c^2 \sin^2 \frac{\theta}{2} + \sum_{\nu=1}^{r} A_\nu^2 \sin^2 \left(2k_\nu \sin \frac{\theta}{2}\right),$$

from which we readily conclude that for each value of $\nu (\nu = 1, \cdots , r)$ we must have either $A_\nu = 0$, or else $k_\nu = 0$. Now (7.1) reduces to $F^2(t) = c^2t^2$ and the theorem is established.

7.2. In conclusion we remark that Theorem 11 may be easily proved without reference to the result (7.1) about screw functions of E_n, *if we limit ourselves to the case when* $m = n < \infty$. Omitting details we shall merely sketch the proof. The case $m = n = 1$ is easily disposed of in a manner similar to the treatment of Cauchy's functional relation $F(x + y) = F(x) + F(y)$. Now assume the theorem proved up to the dimension $m - 1$, and let $F(E_m)$ be imbeddable in E_m'. Take two distinct points A and B in E_m and let E_{m-1} be their "perpendicular bisector" in E_m, i.e., the locus of points equidistant from A and B. Now if the imbedding of $F(E_m)$ in E_m' maps A, B into A', B', it is clear that it will map E_{m-1} into the perpendicular bisector E_{m-1}' of A', B'. Therefore $F(t) = ct(c > 0)$.

8. Determination of the rectifiable solutions of the problem $\{E_m ; \mathfrak{H}\}$, $(1 \leqq m < \infty)$

8.1. The class of solutions of the problem $\{E_m ; \mathfrak{H}\}$ may be described in the terminology of §5.2 as the simplicial class $\Pi(E_m)$. As $E_m \subset E_{m+1}$ clearly implies $\Pi(E_m) \supset \Pi(E_{m+1})$, we have before us the following non-increasing sequence of classes

$$\Pi(E_1) \supset \Pi(E_2) \supset \cdots \supset \Pi(E_m) \supset \cdots \supset \Pi(\mathfrak{H}). \tag{8.1}$$

In view of the inequality (5.6) it now becomes apparent that all functions $F(t)$ given by the formula

$$F^2(t) = \int_0^\infty \frac{1 - \Omega_m(tu)}{u^2}\, d\gamma(u), \tag{8.2}$$

where $\gamma(u)$ is non-decreasing for $u \geqq 0$ and such that

$$\int_1^\infty \frac{d\gamma(u)}{u^2} \qquad \text{exists,} \tag{8.3}$$

belong to the simplicial class $\Pi(E_m)$. For $\sum_0^n \rho_i = 0$ and (8.2) imply

$$\sum_0^n F^2(P_iP_k)\rho_i\rho_k = -\int_0^\infty \left\{\sum_0^n \Omega_m(P_iP_k \cdot u)\rho_i\rho_k\right\} u^{-2}\, d\gamma(u) \leqq 0, \qquad (P_i \in E_m),$$

since $\Omega_m(tu)$ is positive definite in E_m for $u \geqq 0$.

Conversely, let $F(t)$ be an element of $\Pi(E_m)$, that is, such that the metric transform $F(E_m)$ be imbeddable in $\mathfrak{H}$. A closure property for integrals of the form (8.2), analogous to the closure property of Lemma 5 for the class T, would allow us to conclude by the method of §5.4 that $F^2(t)$ is of the form (8.2). This required closure property has not been established as yet. We shall show very simply, however, that a significant subclass of the class $\Pi(E_m)$ is necessarily given by (8.2) for *bounded* functions $\gamma(u)$. Let us explain first what we mean by a *rectifiable* simplicial function $F(t)$.

We need the following lemma (see Wilson, [21], p. 65, Property II)

LEMMA 7. *If* $F(t) \in \Pi(E_m)$, $F(t) \not\equiv 0$, *then* $F'(0)$ *exists and*

$$0 < F'(0) \leqq \infty. \tag{8.4}$$

We know by Lemma 6, §5.7, that $F(t) > 0$ for $t > 0$, provided $m \geqq 2$. In any case ($m \geqq 1$) it is readily seen that $F(t)$ can not possess arbitrarily small positive zeros, for this would again imply that $F(t) \equiv 0$. This secures a positive quantity τ such that $F(t) > 0$ if $0 < t \leqq \tau$. Now take for E_1 the real x-axis and consider in $F(E_1)$ the simple arc Γ: $0 \leqq x \leqq \tau$. Denote its length in the ordinary sense by s ($s \leqq \infty$), that is, the least upper bound of the length of inscribed polygonal lines. Note that $s \geqq F(\tau) > 0$, hence $s > 0$. Let now δ be a small positive quantity and choose the integer n such that $n\delta \leqq \tau < (n + 1)\delta$. The length of the inscribed polygonal line of vertices $x = 0, \delta, 2\delta, \cdots, n\delta, \tau$, is $nF(\delta) + F(\tau - n\delta)$. As $\delta \to 0$, this length converges to the arc-length s; moreover $F(\tau - n\delta) \to 0$, for $\tau - n\delta \to 0$. Hence $nF(\delta) \to s$. From this and the fact that $n\delta \to \tau$, we get

$$\lim_{\delta \to 0} \frac{F(\delta)}{\delta} = \frac{s}{\tau} \qquad (> 0 \text{ and } \leqq \infty), \tag{8.5}$$

which proves our lemma.

The arc Γ of $F(E_1)$ is rectifiable ($s < \infty$) if and only if $F'(0)$ is finite and (8.5) gives $s = \tau F'(0)$. In fact *any* rectifiable arc Γ of E_m of length τ is also rectifiable in $F(E_m)$ and its length in $F(E_m)$ is $s = \tau F'(0)$. For this reason we call $F(t)$ *rectifiable* if $F'(0) < \infty$. Let $\Pi'(E_m)$ denote the subclass of rectifiable elements of $\Pi(E_m)$.

8.2. The rectifiable simplicial functions are determined by the following theorem.

THEOREM 12. *The class $\Pi'(E_m)$ of rectifiable simplicial functions $F(t)$ of $E_m (m \geqq 1)$ is identical with the class of functions such that $F^2(t)$ is of the form (8.2) with bounded $\gamma(u)$.*

Indeed, if $\gamma(\infty)$ is finite, then (8.2) and (1.8) give

$$\lim_{t \to 0} \left(\frac{F(t)}{t}\right)^2 = \lim_{t \to 0} \int_0^\infty \frac{1 - \Omega_m(tu)}{t^2 u^2}\, d\gamma(u) = \frac{1}{2m}\, \gamma(\infty) = \sigma^2, \tag{8.6}$$

where we write $2m\sigma^2 = \gamma(\infty)$. Hence $F(t) \in \Pi'(E_m)$.

Conversely, let $F(t) \in \Pi'(E_m)$. By Corollary 1, §5.3, we have $\exp\{-\lambda F^2(t)\} \in \mathfrak{P}(E_m)$ for all $\lambda > 0$, hence by Theorem 1

$$e^{-\lambda F^2(t)} = \int_0^\infty \Omega_m(tu)\, d\alpha(u, \lambda), \qquad (t, \lambda > 0),$$

where $\alpha(u, \lambda)$ is a family of monotone functions with $\alpha(0, \lambda) = 0$, $\alpha(\infty, \lambda) = 1$. As in §5.4 we may now write

$$\frac{1 - e^{-\lambda F^2(t)}}{\lambda} = \int_0^\infty \frac{1 - \Omega_m(tu)}{\lambda}\, d\alpha(u, \lambda) = \int_0^\infty \frac{1 - \Omega_m(tu)}{u^2}\, d\beta(u, \lambda), \tag{8.7}$$

where

$$\beta(u, \lambda) = \frac{1}{\lambda} \int_0^u u^2\, d\alpha(u, \lambda). \tag{8.8}$$

On the other hand the finiteness of $\sigma = F'(0)$ implies $F(t) < (\sigma + \epsilon)t$, for t sufficiently small, and therefore

$$t^2(\sigma + \epsilon)^2 > F^2(t) \geqq \frac{1 - e^{-\lambda F^2(t)}}{\lambda}.$$

By (8.7) we now have

$$(\sigma + \epsilon)^2 > \int_0^\infty \frac{1 - \Omega_m(tu)}{u^2 t^2} \, d\beta(u, \lambda),$$

for t sufficiently small. As $t \to 0$ we now get $\beta(\infty, \lambda) \leqq 2m(\sigma + \epsilon)^2$, showing that the family of monotone functions (8.8) is bounded uniformly for all $\lambda > 0$. For a suitable sequence $\lambda_n \to 0$ we shall therefore have $\beta(u, \lambda_n) \to \gamma(u)$ in all continuity points of $\gamma(u)$. Now, if t is kept fast in (8.7) while $\lambda = \lambda_n \to 0$, we get in the limit the desired representation (8.2).[23]

Let $F(t)(\not\equiv 0)$ be a rectifiable element of $\Pi(E_m)$ and let us assume $m \geqq 2$. The space $F(E_m)$ is a homeomorph of E_m and its geodesics are the straight lines of E_m. The geodesic distance, or intrinsic distance, $l(P, Q)$ between two points of $F(E_m)$ (in the terminology and notation of Wilson, [20], p. 420) is $l(P, Q) = F'(0) \cdot PQ$. Thus by passing in $F(E_m)$ from its metric $F(PQ)$ to the intrinsic metric $l(P, Q)$ which it generates, we merely get a similar image of the original space E_m, the ratio of similitude being $F'(0)$. Notice furthermore that an arc of a geodesic line in $F(E_m)$ (i.e., a straight segment of the underlying E_m) never loses its minimizing property of an absolute minimum no matter how far it is extended in both directions. In the isometric image Σ of $F(E_m)$ in $\mathfrak{H}$, the geodesics of $F(E_m)$ go over in screw lines of $\mathfrak{H}$, all of which are congruent among themselves. In fact any two congruent figures in E_m are mapped in congruent figures of Σ. To any rigid motion of E_m in itself corresponds a rigid motion of Σ in itself.

8.3. We close the paper with the following analogue of Theorem 6′.

THEOREM 12′. *A necessary and sufficient condition that the non-negative continuous function $F(t)$, vanishing at the origin, be an element of $\Pi'(E_m)$, is that $F^2(\sqrt{t})$ be continuously differentiable for $t \geqq 0$ and, on setting*

$$\psi(t) = \frac{d}{dt} F^2(\sqrt{t}), \tag{8.9}$$

the function $\psi(t^2)$ be positive definite in E_{m+2}.

Indeed, denoting by c_m, c_m', c_m'' positive constants and setting $k = (m - 2)/2$, we have by (1.8) (see [18], p. 45)

$$\Omega_m'(r) = c_m(d/dt)(J_k(r)r^{-k}) = -c_m J_{k+1}(r)r^{-k},$$

hence

$$-\Omega_m'(r)r^{-1} = c_m' \Omega_{m+2}(r).$$

[23] The basic idea of this proof was suggested by the proof of Theorem 27, p. 91, in H. Cramér's monograph [6]. References are there found to the work of Kolmogoroff, De Finetti, and P. Lévy.

An element $F(t)$ of $\Pi'(E_m)$ is given by (8.2) with a bounded $\gamma(u)$. Hence

$$F^2(\sqrt{t}) = \int_0^\infty \frac{1 - \Omega_m(\sqrt{(tu)})}{u}\, d\gamma(\sqrt{u}).$$

By differentiation we get

$$\psi(t) = \frac{d}{dt} F^2(\sqrt{t}) = -\int_0^\infty \frac{\Omega_m'(\sqrt{(tu)})}{2\sqrt{(tu)}}\, d\gamma(\sqrt{u}) = c_m'' \int_0^\infty \Omega_{m+2}(\sqrt{(tu)})\, d\gamma(\sqrt{u}),$$

or

$$\psi(t^2) = c_m'' \int_0^\infty \Omega_{m+2}(tu)\, d\gamma(u).$$

Hence $\psi(t^2) \in \mathfrak{P}(E_{m+2})$, by Theorem 1. The converse is proved by retracing our steps.

We discuss only one particular example, namely

(8.10) $$F_5(t) = (1 - \Omega_3(t))^{\frac{1}{2}} = (1 - (\sin t/t))^{\frac{1}{2}}.$$

By (8.2) we have $F_5(t) \in \Pi'(E_3)$. As above we get

$$\psi(t) = \frac{d}{dt} F_5^2(\sqrt{t}) = c_3''\, \Omega_5(\sqrt{t})$$

which is *not* completely monotone for $t > 0$, since $\Omega_5(r)$ has infinitely many positive zeros. Hence $F_5(t)$ is not an element of $\Pi(\mathfrak{H})$, by Theorem 6′. Thus $F_5(E_3)$ is imbeddable in $\mathfrak{H}$, but $F_5(\mathfrak{H})$ is not imbeddable in $\mathfrak{H}$. A further argument, which we omit, would show that already $F_5(E_4)$ is not congruent to a subset of $\mathfrak{H}$.

Colby College,
Waterville, Maine.

References

[1] S. Bernstein, *Sur les fonctions absolument monotones*, Acta Mathematica, vol. 52 (1929), pp. 1–66.

[2] L. M. Blumenthal, *Remarks concerning the euclidean four-point property*, Ergebnisse eines math. Koll., issue 7, Vienna 1936, pp. 8–10.

[3] ——— *Distance Geometries*, University of Missouri Studies, vol. 13 (1938), 142 pp.

[4] S. Bochner, *Monotone Funktionen, Stieltjes Integrale und harmonische Analyse*, Math. Annalen, vol. 108 (1933), pp. 378–410.

[5] ——— *Completely monotone functions of the Laplace operator for torus and sphere*, Duke Math. Journal, vol. 3 (1937), pp. 488–502.

[6] H. Cramér, *Random variables and probability distributions*, Cambridge Tract no. 36, Cambridge, 1937.

[7] F. Hausdorff, *Summationsmethoden und Momentfolgen* II, Math. Zeitschrift, vol. 9 (1921), pp. 280–299.

[8] E. Helly, *Über lineare Funktionaloperationen*, Sitzungsber. der Wiener Akad., vol. 121, IIa (1912), pp. 265–297.

[9] K. Menger, *Die Metrik des Hilbertschen Raumes*, Anzeiger der Akad. der Wiss. in Wien, Math.-Nat. Klasse, vol. 65 (1928), pp. 159–160.

[10] K. MENGER, *New foundations of euclidean geometry*, Amer. J. of Math., vol. 53 (1931), pp. 721–745.
[11] E. H. MOORE AND R. W. BARNARD, *General Analysis*, Part I, Memoirs of the American Phil. Soc., vol. 1, Philadelphia, 1935.
[12] JOHN VON NEUMANN AND I. J. SCHOENBERG, *Fourier integrals and metric geometry*, to appear in the Transactions of the Amer. Math. Soc.
[13] I. J. SCHOENBERG, *On finite-rowed systems of linear inequalities in infinitely many variables*, Trans. of the Amer. Math. Soc., vol. 34 (1933), pp. 594–619.
[14] ——— *Remarks to Maurice Fréchet's article . . .*, these Annals, vol. 36 (1935), pp. 724–732.
[15] ——— *On certain metric spaces arising from euclidean spaces by a change of metric and their imbedding in Hilbert space*, these Annals, vol. 38 (1937), pp. 787–793.
[16] ——— *Metric spaces and positive definite functions*, to appear in the Trans. of the Amer. Math. Soc.
[17] E. C. TITCHMARSH, *The theory of functions*, Oxford, 1932.
[18] G. N. WATSON, *Theory of Bessel functions*, Cambridge, 1922.
[19] D. V. WIDDER, *Necessary and sufficient conditions for the representation of a function as a Laplace integral*, Trans. of the Amer. Math. Soc., vol. 33 (1932), pp. 851–892.
[20] W. A. WILSON, *On rectifiability in metric spaces*, Bulletin of the Amer. Math. Soc., vol. 38 (1932), pp. 419–426.
[21] ——— *On certain types of continuous transformations of metric spaces*, Amer. J. of Math., vol. 57 (1935), pp. 62–68.

FOURIER INTEGRALS AND METRIC GEOMETRY

BY

J. VON NEUMANN AND I. J. SCHOENBERG

Introduction

1. Let S be a metric space, the distance $d(f, g)$ between two elements of S satisfying the usual postulates. Let f_t $(-\infty < t < \infty)$ be a function whose values lie in S and which is metrically continuous, that is, $d(f_{t+h}, f_t) \to 0$ if $h \to 0$. If this function has the property that

$$d(f_t, f_s) = F(t - s) \tag{0.1}$$

is a function of the difference $t-s$ only, we call the curve Γ in S defined by f_t $(-\infty < t < \infty)$ a *screw line* of S and $F(t) = d(f_t, f_0)$ a *screw function* of S. The reason for this terminology is as follows. If τ is a real parameter, the two curves $\Gamma_0: f_t$ $(-\infty < t < \infty)$ and $\Gamma_\tau: f_{t+\tau}$ $(-\infty < t < \infty)$ which are identical as point sets, are isometrically mapped on each other by the correspondence $f_t \leftrightarrow f_{t+\tau}$, for

$$d(f_t, f_s) = F(t - s) = d(f_{t+\tau}, f_{s+\tau}),$$

in view of (0.1). These congruent mappings of Γ into itself form a one-parameter group.

The following properties of a screw function $F(t)$ of S are obvious: $F(t)$ is a continuous non-negative even function and $F(0) = 0$; if $F(t)$ is a screw function, then all functions $F(kt)$ (k real) are screw functions.

A different point of view which puts the emphasis on the screw function $F(t)$ rather than on the screw line Γ is as follows. Consider the real axis $-\infty < t < \infty$ as a euclidean space E_1 and change its metric from $|t-s|$ to $F(t-s)$. We thus get a new space which, following Blumenthal, we shall call the metric transform of E_1 by $F(t)$ and denote by $F(E_1)$. For what functions $F(t)$ may this metric transform $F(E_1)$ be isometrically imbedded in S? Clearly $F(E_1)$ enjoys this property if and only if $F(t)$ is a screw function of S. For if the mapping of the point t, of $F(E_1)$, into the point f_t, of S, performs the imbedding of $F(E_1)$ into S, then (0.1) expresses the isometricity of this imbedding.

Presented to the Society, September 1, 1936, and December 29, 1938; received by the editors September 23, 1940. The first communication to the Society [Bulletin of the American Mathematical Society, abstract 42-9-353] covered the contents of Part I and Part II of the present paper; the contents of Part III were the subject of the second communication of 1938 [abstract 47-1-47]. Thus the contents of Parts I and II precede in time the articles [4] and [5], listed in the bibliography at the end of this paper, which were partly suggested by this earlier work frequently referred to in these articles. Part III, however, carries on the work presented in [4] and [5] and is essentially based on some of that work.

2. As examples we mention

$$F_3(t) = (t^2 + \sin^2 t)^{1/2},$$

which is a screw function of the euclidean space E_3, in view of the identity

$$F_3^2(t-s) = (t-s)^2 + \sin^2(t-s)$$
$$= (t-s)^2 + \tfrac{1}{4}(\cos 2t - \cos 2s)^2 + \tfrac{1}{4}(\sin 2t - \sin 2s)^2.$$

Similarly

$$(0.2) \qquad F_{2N}(t) = \left(\sum_{\nu=1}^{N} A_\nu \sin^2 u_\nu t\right)^{1/2}, \qquad A_\nu > 0,\ 0 < u_1 < u_2 < \cdots < u_N,$$

is a screw function of E_{2N}, as seen from

$$(0.2') \qquad F_{2N}^2(t-s) = \sum_{\nu=1}^{N} \{\tfrac{1}{4}A_\nu(\cos 2u_\nu t - \cos 2u_\nu s)^2 + \tfrac{1}{4}A_\nu(\sin 2u_\nu t - \sin 2u_\nu s)^2\}.$$

Finally

$$(0.3) \qquad F_{2N-1}(t) = \left(Ct^2 + \sum_{\nu=1}^{N-1} A_\nu \sin^2 u_\nu t\right)^{1/2}, \qquad C > 0,\ A_\nu > 0,\ 0 < u_1 < \cdots < u_{N-1},$$

is a screw function of E_{2N-1}, as shown by

$$(0.3') \qquad F_{2N-1}^2(t-s) = C(t-s)^2 + \sum_{\nu=1}^{N-1} \{\tfrac{1}{4}A_\nu(\cos 2u_\nu t - \cos 2u_\nu s)^2 + \tfrac{1}{4}A_\nu(\sin 2u_\nu t - \sin 2u_\nu s)^2\}.$$

We shall see that (0.2) and (0.3) are the most general screw functions of E_{2N} and E_{2N-1}, respectively, which are not also screw functions of a euclidean space of lower dimensions.

The starting point of the present investigation was W. A. Wilson's recent remark that if

$$(0.4) \qquad F(t) = |t|^{1/2}$$

then the metric transform $F(E_1)$ is imbeddable in the Hilbert space $\mathfrak{H}$, hence

$$F^2(t-s) = |t-s| = \|f_t - f_s\|^2,$$

for a suitable function f_t $(-\infty < t < \infty)$ with values in $\mathfrak{H}$ [8, p. 64]. According to our present point of view (0.4) is a screw function of $\mathfrak{H}$. We shall see that this $F(t)$ is not a screw function of any euclidean space.

3. The principal purpose of this paper is to determine all screw functions of Hilbert space. It consists of three parts. In Part I we state our fundamental

result (Theorem 1) and show by means of elementary results of Menger and Schoenberg that all functions $F(t)$ there described are screw functions of $\mathfrak{H}$. Furthermore, those screw functions of $\mathfrak{H}$ are characterized which correspond to screw lines Γ with one of the following properties: Γ is euclidean, bounded, rectifiable or closed.

The converse statement to the effect that Theorem 1 yields *all* screw functions of $\mathfrak{H}$ is established in two essentially different ways in Part II and Part III respectively. In Part II this is proved by a direct investigation of the group of isometric mappings of $\mathfrak{H}$ into itself which is induced by the group of isometric mappings of a screw line into itself. Free use is made of the theory of Hermitian operators in Hilbert space. In Part III $\mathfrak{H}$ intervenes only by its metric in accordance with the ideas of Menger on the metric characterization of metric spaces. The method used is an elaboration of the metrical approach of Part I which is made more effective by an appeal to the theory of positive definite functions, i.e., the characteristic functions of the theory of probabilities. This connection was pointed out in two recent papers by one of us [4, 5]. In [5, p. 837], it was shown that the question as to when the metric transform $F(E_m)$ is imbeddable in $\mathfrak{H}$ depended on certain limit (closure) theorems. Here we establish these theorems in a form similar to P. Lévy's limit theorem concerning characteristic functions.

In conclusion we want to say that the operator method of Part II, dealing with the imbedding of $F(E_1)$ in $\mathfrak{H}$, may also be extended to cover the general case of the imbedding of $F(E_m)$ in $\mathfrak{H}$. Although this extension has been fully worked out, for reasons of conciseness we treat in Part II only the case of screw lines ($m=1$).

Part I. The fundamental theorem on screw functions in Hilbert space and elementary consequences

1.1. Let $\mathfrak{H}$ be a real Hilbert space. For every $t>-\infty$, and $<+\infty$, let a point f_t of $\mathfrak{H}$ be given, such that

(i) f_t is a metrically continuous function of t,

(ii) the distance of f_t and f_s depends on $t-s$ only.

According to the general definition of our Introduction, the curve $\Gamma: f_t$ is a *screw line* of $\mathfrak{H}$. Condition (ii) means

$$\|f_t - f_s\| = F(t - s). \tag{1.1}$$

Then (i) merely states that

$$F(t) \to 0 \quad \text{if} \quad t \to 0, \tag{1.2}$$

and conversely, (1.2) implies the continuity of f_t and therefore of $F(t)$ everywhere. $F(t)$ is called a *screw function* of $\mathfrak{H}$.

Let $f_{t_0}, f_{t_1}, \cdots, f_{t_n}$ ($t_0=0$) be $n+1$ points of Γ. By (1.1) their mutual distances are $\|f_{t_i}-f_{t_k}\|=F(t_i-t_k)$. As these points may be thought of as lying in a n-dimensional linear subspace of $\mathfrak{H}$, i.e., a E_n, we have by a known theorem [3, Theorem 1]

$$\sum_{i,k=1}^{n} \{F^2(t_i) + F^2(t_k) - F^2(t_i - t_k)\} \rho_i \rho_k \geqq 0, \tag{1.3}$$

for arbitrary real ρ_i. Conversely, let $F(t)$ be a continuous even non-negative function, with $F(0)=0$, enjoying the property (1.3) for arbitrary real t_i, ρ_i $(n=2, 3, \cdots)$. Geometrically this means: The points $t_0=0, t_1, \cdots, t_n$, of the space $F(E_1)$, which is obtained from E_1 by changing its metric from $|t-s|$ to $F(t-s)$, may be imbedded in E_n. By a theorem of Menger [2] this is sufficient to insure the possibility of imbedding $F(E_1)$ in $\mathfrak{H}$. We can therefore state

LEMMA 1. *A continuous even non-negative function $F(t)$ vanishing at the origin is a screw function of $\mathfrak{H}$ if and only if it satisfies the inequality* (1.3) *for arbitrary real t_i, ρ_i and for $n=2, 3, 4, \cdots$.*

The interest of this analytical characterization of screw function lies in its obvious consequence that the squares $F^2(t)$ of screw functions form a convex class of functions: If $F_1^2(t)$ and $F_2^2(t)$ are squares of screw functions then also $F_1^2(t)+F_2^2(t)$ is the square of a screw function. In view of the euclidean screw functions (0.2), (0.3), exhibited in the Introduction, this convexity property suggests the following explicit expression of the screw functions of $\mathfrak{H}$.

THEOREM 1. (Fundamental theorem.) *The class of screw functions $F(t)$ of Hilbert space is identical with the class of functions whose squares are of the form*

$$F^2(t) = \int_0^\infty \frac{\sin^2 tu}{u^2} d\gamma(u), \tag{1.4}$$

where $\gamma(u)$ is non-decreasing for $u \geqq 0$ and such that

$$\int_1^\infty u^{-2} d\gamma(u) \text{ exists.} \tag{1.5}$$

A proof that all functions $F(t)$ furnished by (1.4) are screw functions is exceedingly simple. For it suffices to show that $F^2(t)$ satisfies the inequality (1.3) (Lemma 1). Indeed, in view of the identity

$$\sin^2 t_i u + \sin^2 t_k u - \sin^2 (t_i - t_k)u = 2 \sin^2 t_i u \sin^2 t_k u + \tfrac{1}{2} \sin 2t_i u \sin 2t_k u,$$

we have

$$\begin{aligned} &\sum_{i,k=1}^{n} \{F^2(t_i) + F^2(t_k) - F^2(t_i - t_k)\} \rho_i \rho_k \\ &\quad = \int_0^\infty \left\{ 2\left(\sum_1^n \rho_i \sin^2 t_i u \right)^2 + \frac{1}{2}\left(\sum_1^n \rho_i \sin 2t_i u \right)^2 \right\} u^{-2} d\gamma(u) \geqq 0. \end{aligned} \tag{1.6}$$

The converse statement that the square of a screw function is necessarily of the form (1.4) is proved in two different ways in Part II and Part III of this paper.

1.2. Let us now see which of the screw functions of $\mathfrak{H}$ are euclidean screw functions, that is, belong to some euclidean space. It is convenient for this purpose to write (1.4) in the form

$$(1.7) \qquad F^2(t) = Ct^2 + \int_{+0}^{\infty} \frac{\sin^2 tu}{u^2} d\gamma(u), \qquad C = \gamma(+0) - \gamma(0).$$

THEOREM 2. *A screw function $F(t)$ defined by* (1.4) *is euclidean if and only if $\gamma(u)$ has a finite number N of points of increase. $F(t)$ belongs to E_m and to no lower space $E_{m'}$ $(m' < m)$ if and only if*

$$(1.8) \qquad C = 0, \qquad N = m/2 \qquad \textit{for } m \textit{ even},$$

$$(1.9) \qquad C > 0, \qquad N = (m+1)/2 \ \textit{for } m \textit{ odd}.$$

Indeed, let N be finite. If $C=0$, then $F(t)$ is of the form (0.2), and $u_1, \cdots, u_N$ are precisely the points of increase of $\gamma(u)$. We know by (0.2′) that $F(t)$ is a screw function of E_{2N}. Moreover its screw line

$$(\tfrac{1}{2}A_\nu^{1/2} \cos 2u_\nu t, \tfrac{1}{2}A_\nu^{1/2} \sin 2u_\nu t), \qquad \nu = 1, 2, \cdots, N; \ -\infty < t < \infty,$$

lies in E_{2N} but in no linear subspace of E_{2N}. Similarly, if $C>0$, it is shown by (0.3), (0.3′), that $F(t)$ belongs to E_m, $m=2N-1$.

Let now $\gamma(u)$ have infinitely many points of increase. We want to show that $F(t)$ is not euclidean, that is, the metric transform $F(E_1)$ cannot be imbedded in a euclidean space. For if n is an arbitrary positive integer, we may find n points of increase $u=u_\nu$ $(\nu=1, \cdots, n)$ of $\gamma(u)$, such that $0<u_1<u_2<\cdots<u_n$. We shall now locate in $F(E_1)$ $n+1$ points $t=0, t_1, \cdots, t_n$, which cannot be imbedded in E_{n-1}. Such points will enjoy this property if the quadratic form (1.6) is positive definite [3, Theorem 1]. The linear independence of the functions $\sin 2tu_1, \cdots, \sin 2tu_n$, implies the existence of values t_i, such that

$$(1.10) \qquad \det \|\sin (2t_i u_k)\|_{1,n} \neq 0.$$

The form (1.6) is now positive definite; for otherwise there would exist values ρ_i, with

$$(1.11) \qquad \sum \rho_i^2 > 0,$$

which make the form (1.6) vanish. But the vanishing of the integral of (1.6) implies the vanishing of its integrand at all points of increase of $\gamma(u)$, hence in particular

$$\sum_{i=1}^{n} \rho_i \sin (2t_i u_k) = 0, \qquad k = 1, \cdots, n. \tag{1.12}$$

The incompatibility of the three relations (1.10), (1.11), (1.12) completes this indirect proof; for as n may be taken arbitrarily large, $F(t)$ is not euclidean.

Wilson's screw function (0.4) is of the form (1.4) since

$$F^2(t) = |t| = \frac{2}{\pi}\int_0^\infty \frac{\sin^2 tu}{u^2}\,du. \tag{1.13}$$

Since $\gamma(u) = 2u/\pi$ has infinitely many points of increase, it is not a euclidean screw function. A somewhat more general example is

$$F(t) = |t|^\kappa, \qquad 0 < \kappa < 1. \tag{1.14}$$

Indeed

$$F^2(t) = |t|^{2\kappa} = \int_0^\infty \sin^2 tu \cdot u^{-1-2\kappa}du \Big/ \int_0^\infty \sin^2 u \cdot u^{-1-2\kappa}du, \qquad 0 < \kappa < 1,$$

is of the form (1.4).

1.3. Let us find conditions which insure the boundedness of a screw function $F(t)$. Later we shall see that this occurs when the corresponding screw line lies on a sphere of $\mathfrak{H}$. For the present we prove

THEOREM 3. *The screw function $F(t)$ given by*

$$F^2(t) = Ct^2 + \int_{+0}^\infty \frac{\sin^2 tu}{u^2}\,d\gamma(u), \qquad C \geqq 0, \tag{1.15}$$

is bounded if and only if

$$C = 0 \quad \text{and} \quad \int_{+0}^\infty u^{-2}d\gamma(u) \quad \text{exists}. \tag{1.16}$$

More precisely, if $C = 0$, we have

$$\frac{1}{2}\int_{+0}^\infty \frac{d\gamma(u)}{u^2} \leqq \limsup_{t\to\infty} F^2(t) \leqq \int_{+0}^\infty \frac{d\gamma(u)}{u^2}, \tag{1.17}$$

where all three members may also be infinite.

As $C = 0$ is obviously necessary in order that $F^2(t)$ be bounded, it suffices to prove the inequalities (1.17). Let

$$F^2_{\epsilon,a}(t) = \int_\epsilon^a \frac{\sin^2 tu}{u^2}\,d\gamma(u), \qquad 0 < \epsilon < a < +\infty.$$

We have

$$\frac{1}{T}\int_0^T F_{\epsilon,a}^2(t)dt = \int_\epsilon^a \left(\frac{1}{T}\int_0^T \sin^2 tu\, dt\right) u^{-2} d\gamma(u)$$
$$= \int_\epsilon^a \left(\frac{1}{2} - \frac{\sin 2Tu}{4Tu}\right) u^{-2} d\gamma(u),$$

hence

$$\lim_{T\to\infty} \frac{1}{T}\int_0^T F_{\epsilon,a}^2(t)dt = \frac{1}{2}\int_\epsilon^a u^{-2} d\gamma(u).$$

Whence, since $F^2(t) \geqq F_{\epsilon,a}^2(t)$, we derive

$$\limsup_{t\to\infty} F^2(t) \geqq \limsup_{t\to\infty} F_{\epsilon,a}^2(t) \geqq \frac{1}{2}\int_\epsilon^a u^{-2} d\gamma(u).$$

By comparing the extreme terms only and allowing $\epsilon \to 0$ and $a \to \infty$, we get the first inequality (1.17). The second inequality (1.17) follows from

$$F^2(t) = \int_{+0}^\infty \frac{\sin^2 tu}{u^2} d\gamma(u) \leqq \int_{+0}^\infty \frac{1}{u^2} d\gamma(u).$$

The euclidean screw functions (0.2) are always bounded; (0.3) are never bounded.

1.4. We shall now investigate when a screw line Γ of a screw function $F(t)$ is rectifiable. With this purpose in view we show first that

$$\lim_{t\to 0}\left(\frac{F(t)}{t}\right)^2 = \int_0^\infty d\gamma(u), \tag{1.18}$$

where both sides may also be infinite.

Indeed, to each $\epsilon > 0$ there corresponds a $\delta > 0$, such that $\sin x^2/x^2 > 1 - \epsilon$ if $x \geqq 0$ and $\leqq \delta$, hence $\sin^2 tu/(tu)^2 > 1 - \epsilon$ if $0 \leqq u \leqq \delta t^{-1}$, $t > 0$. But then

$$\left(\frac{F(t)}{t}\right)^2 = \int_0^\infty \frac{\sin^2 tu}{t^2u^2} d\gamma(u) \geqq \int_0^{\delta t^{-1}} \frac{\sin^2 tu}{t^2u^2} d\gamma(u) \geqq (1 - \epsilon)\int_0^{\delta t^{-1}} d\gamma(u),$$

and therefore

$$\liminf_{t\to 0}\left(\frac{F(t)}{t}\right)^2 \geqq (1 - \epsilon)\int_0^\infty d\gamma(u).$$

On allowing $\epsilon \to 0$, we have

$$\liminf_{t\to 0}\left(\frac{F(t)}{t}\right)^2 \geqq \int_0^\infty d\gamma(u).$$

This, and the obvious inequality

$$\left(\frac{F(t)}{t}\right)^2 = \int_0^\infty \frac{\sin^2 tu}{t^2u^2}\, d\gamma(u) \leqq \int_0^\infty d\gamma(u)$$

imply (1.18).

Let now $t_0=0<t_1< \cdots <t_{n-1}<t_n=t$ be a subdivision of the fixed interval $(0, t)$ and $\delta=\max(t_i-t_{i-1})$. (1.1) and (1.18) now imply

$$\begin{aligned}\lim_{\delta\to 0} \sum_{i=1}^{n} \|f_{t_i} - f_{t_{i-1}}\| &= \lim_{\delta\to 0} \sum_{1}^{n} F(t_i - t_{i-1}) \\ &= \lim_{\delta\to 0} \sum_{1}^{n} \frac{F(t_i - t_{i-1})}{t_i - t_{i-1}}\cdot(t_i - t_{i-1}) = t\cdot\left(\int_0^\infty d\gamma(u)\right)^{1/2}.\end{aligned}$$

This proves the following theorem.

THEOREM 4. *A screw line f_t $(-\infty<t<\infty)$ corresponding to the screw function $F(t)$ defined by* (1.4) *is rectifiable if and only if $\gamma(u)$ is bounded. The length s of the arc $(0, t)$ of the screw line is then connected with the parameter t by the relation*

$$s = t(\gamma(\infty) - \gamma(0))^{1/2} = t\cdot\lim_{\tau\to 0}\frac{F(\tau)}{\tau}\cdot \tag{1.19}$$

In particular, if $\gamma(\infty)-\gamma(0)=1$, t is identical with the length of arc along the screw line of $F(t)$.

The euclidean screw lines corresponding to (0.2), (0.3) are always rectifiable, the relations between s and t being respectively

$$s = t\left(\sum_{1}^{N} A_\nu u_\nu^2\right)^{1/2}, \qquad s = t\left(C + \sum_{1}^{N-1} A_\nu u_\nu^2\right)^{1/2}.$$

The screw lines of (1.14) are non-rectifiable, since $F(t)/t\to\infty$ as $t\to 0$.

1.5. When is a screw line Γ, corresponding to the screw function $F(t)$, a closed curve? Let f_{t_0} be a double point of Γ, that is, $f_{t_0}=f_{t_0+\tau}$ for some $\tau>0$. Then $0=\|f_{t_0+\tau}-f_{t_0}\|=F(\tau)$, hence $\|f_{t+\tau}-f_t\|=F(\tau)=0$ or $f_{t+\tau}=f_t$ for all real t. This means that f_t and therefore also $F(t)=\|f_t-f_0\|$ has the period τ and Γ is a closed curve. Conversely, all this is implied if $F(t)$ is periodic. Let this be the case and let τ be the least positive period of $F(t)$, which exists if $F(t)\not\equiv 0$. From (1.7) we get

$$F^2(\tau) = C\tau^2 + \int_{+0}^\infty \frac{\sin^2 \tau u}{u^2}\, d\gamma(u) = 0.$$

This implies that $C=0$ and that $\gamma(u)$ is constant in all the intervals

$(k-1)\pi/\tau < u < k\pi/\tau$ $(k=1, 2, 3, \cdots)$ where $\sin^2 \tau u > 0$. But then $F^2(t)$ reduces to the series

$$(1.20) \qquad F^2(t) = \sum_{k=1}^{\infty} c_k \sin^2\left(\frac{k\pi}{\tau} t\right), \qquad \tau > 0,\ c_k \geqq 0,\ \sum_1^{\infty} c_k \text{ convergent.}$$

THEOREM 5. *A screw line of $\mathfrak{H}$ is a closed curve if and only if its screw function is periodic and its square is of the form* (1.20).

The screw lines of (1.20) are rectifiable if and only if $\sum_1^\infty k^2 c_k$ converges.

A euclidean screw line is closed only if it is even-dimensional, that is, only if the form (0.2) and its frequencies u_ν $(\nu=1, \cdots, N)$ have rational ratios(1).

The Fourier series developments of the Bernoulli polynomials $B_2(t)$, $B_4(t)$ furnish the following simple examples of periodic screw functions of period one:

$$(1.21) \qquad F_1^2(t) = t(1-t) = \frac{1}{6} - B_2(t) = \sum_1^\infty \frac{1-\cos 2\pi n t}{n^2\pi^2} = \frac{2}{\pi^2}\sum_1^\infty \frac{\sin^2 \pi n t}{n^2}, \qquad 0 \leqq t \leqq 1,$$

$$(1.21') \qquad F_2^2(t) = t^2(1-t)^2 = \frac{1}{30} + B_4(t) = 3\sum_1^\infty \frac{1-\cos 2\pi n t}{n^4\pi^4} = \frac{6}{\pi^4}\sum_1^\infty \frac{\sin^2 \pi n t}{n^4}, \qquad 0 \leqq t \leqq 1.$$

The screw lines of $F_1(t)$ are non-rectifiable; those of $F_2(t)$ are rectifiable with t as length of arc, since $\lim_{t\to 0} F_2(t)/t = 1$.

1.6. We conclude this survey of these most elementary and characteristic features of screw lines by proving the following theorem.

THEOREM 6. *A screw line Γ of $\mathfrak{H}$ is bounded if and only if it can be placed on a sphere of $\mathfrak{H}$.*

The sufficiency of the condition is clear. Conversely, let the screw line and therefore also its screw function $F(t)$ be bounded. By Theorem 3 we have

$$F^2(t) = \int_{+0}^{\infty} \frac{\sin^2 tu}{u^2}\, d\gamma(u).$$

Setting

$$\alpha(0) = 0, \qquad \alpha(u) = \int_{+0}^{u} u^{-2} d\gamma(u), \qquad u > 0,$$

(1) The projections of such screw lines on any of the coordinate planes x_iOx_k are closed Lissajou curves. Conversely, any Lissajou curve whether open or closed may be regarded as a plane projection of a screw line of E_4.

we get

$$(1.22) \qquad F^2(t) = \int_0^\infty \sin^2 tu \, d\alpha(u),$$

where $\alpha(u)$ is non-decreasing and bounded. Define $r>0$ by

$$(1.23) \qquad 4r^2 = \alpha(\infty).$$

We shall prove that Γ can be placed on a sphere of $\mathfrak{H}$ of radius r. Consider the metric transform $F(E_1)$, i.e., the real axis $-\infty<t<\infty$ with the distance $d(t, t')=F(t-t')$; add to this space a new point A and complete the definition of distance throughout $F(E_1)+A$ by setting

$$(1.24) \qquad d(A, t) = r.$$

Not only $F(E_1)$ can be inbedded in $\mathfrak{H}$, as shown by (1.6), but also $F(E_1)+A$ may so be imbedded. It suffices to show that the points $A, t_1, \cdots, t_n$ may be placed in E_n. Indeed, we have in view of (1.22), (1.23), (1.24)

$$\begin{aligned}
\sum_{i,k=1}^{n} (d^2(A, t_i) &+ d^2(A, t_k) - d^2(t_i, t_k))\rho_i\rho_k \\
&= \sum_1^n (r^2 + r^2 - F^2(t_i - t_k))\rho_i\rho_k \\
&= \sum_1^n \left\{2r^2 - \frac{1}{2}\int_0^\infty [1 - \cos 2(t_i - t_k)u]d\alpha(u)\right\}\rho_i\rho_k \\
&= \frac{1}{2}\int_0^\infty \left\{\sum_1^n \cos 2(t_i - t_k)u\cdot\rho_i\rho_k\right\}d\alpha(u) \\
&= \frac{1}{2}\int_0^\infty \left\{\left(\sum_1^n \rho_i \cos 2t_iu\right)^2 + \left(\sum_1^n \rho_i \sin 2t_iu\right)^2\right\}d\alpha(u) \geqq 0.
\end{aligned}$$

Hence $F(E_1)+A$ may be imbedded in $\mathfrak{H}$ by Menger's theorem. Now (1.24) shows that $F(E_1)$ is mapped into a screw line Γ lying on the sphere of radius r whose center is the image of A.

Part II. First proof of the fundamental theorem by the theory of operators

2.1. Let f_t $(-\infty<t<\infty)$ be a screw line of $\mathfrak{H}$, and $F(t)$ the corresponding screw function. For greater generality we assume that $\mathfrak{H}$ is a real complete unitary space (not necessarily separable). As in §1.1 we have

$$(2.1) \qquad \|f_t - f_s\| = F(t - s),$$

$$(2.2) \qquad F(t) \to 0 \quad \text{if} \quad t \to 0.$$

Using the reality of $\mathfrak{H}$ we have

$$
\begin{aligned}
(f_t - f_u, f_s - f_u) &= \tfrac{1}{2}\{\|f_t - f_u\|^2 + \|f_s - f_u\|^2 - \|(f_t - f_u) - (f_s - f_u)\|^2\} \\
&= \tfrac{1}{2}\{\|f_t - f_u\|^2 + \|f_s - f_u\|^2 - \|f_t - f_s\|^2\} \\
&= \tfrac{1}{2}\{F^2(t-u) + F^2(s-u) - F^2(t-s)\}
\end{aligned}
$$

Hence setting

$$F^2(p) + F^2(q) - F^2(p-q) = 2G(p, q),$$

we have

$$(f_t - f_u, f_s - f_u) = G(t-u, s-u). \tag{2.3}$$

Let $\mathfrak{H}_1$ be the closed linear subset of $\mathfrak{H}$, which is spanned by all

$$f_t - f_0, \qquad t \text{ rational.}$$

Write all rational numbers as a sequence $t_1, t_2, \cdots$, and then orthogonalize the sequence

$$f_{t_1} - f_0, f_{t_2} - f_0, \cdots$$

by the Gram–E. Schmidt procedure, thus obtaining the (finite or enumerably infinite) normalized orthogonal set

$$\phi_1, \phi_2, \cdots.$$

Thus

$$\phi_i = \sum_{j=1}^{i} \alpha_{ij}(f_{t_j} - f_0), \qquad \text{all } \alpha_{ij} \text{ are real numbers.} \tag{2.4}$$

The $f_t - f_0$, t rational, are linear aggregates of the ϕ_i's, hence (owing to f_t's continuity in t) all $f_t - f_0$, $-\infty < t < \infty$, are limit points of such linear aggregates. Therefore all $f_t - f_0$ belong to the closed, linear set which is spanned by $\phi_1, \phi_2, \cdots$, and which therefore coincides with $\mathfrak{H}_1$. Hence

$$f_t - f_0 = \sum_i \beta_i(t)\phi_i \tag{2.5}$$

where $\beta_i(t)$ are real continuous functions of t.

Combining (2.4) and (2.5) we get

$$f_t - f_0 = \sum_i \beta_i(t)\left\{\sum_{j=1}^{i} \alpha_{ij}(f_{t_j} - f_0)\right\}, \tag{2.6}$$

that is

$$\left\|(f_t - f_0) - \sum_i \beta_i(t)\left\{\sum_{j=1}^{i} \alpha_{ij}(f_{t_j} - f_0)\right\}\right\|^2 = 0. \tag{2.7}$$

Consider the equation

$$(2.6') \qquad f_{t+s} - f_s = \sum_i \beta_i(t)\left\{\sum_{j=1}^{i} \alpha_{ij}(f_{t_j+s} - f_s)\right\},$$

and its equivalent

$$(2.7') \qquad \left\| (f_{t+s} - f_s) - \sum_i \beta_i(t)\left\{\sum_{j=1}^{i} \alpha_{ij}(f_{t_j+s} - f_s)\right\}\right\|^2 = 0.$$

(2.7′) can be written as a relation of the

$$(f_{t'+s} - f_s, f_{t''+s} - f_s) \qquad \text{for } t', t'' = t, t_1, t_2, \cdots,$$

and the

$$\beta_i(t), \qquad \alpha_{ij}.$$

By (2.3) this means a relation of the

$$G(t', t'') \qquad \text{for } t', t'' = t, t_1, t_2, \cdots,$$

and the

$$\beta_i(t), \qquad \alpha_{ij}.$$

Hence (2.7′), and consequently (2.6′), are independent of s. But for $s=0$ (2.6′), (2.7′) coincide with (2.6), (2.7), and hence are true. Therefore they hold for all s.

Let

$$(2.4') \qquad \phi_i(s) = \sum_{j=1}^{i} \alpha_{ij}(f_{t_j+s} - f_s).$$

Then (2.6′) gives

$$(2.5') \qquad f_{t+s} - f_s = \sum_i \beta_i(t)\phi_i(s).$$

The relation

$$(2.8) \qquad (\phi_i(s), \phi_k(s)) = \delta_{ik}\begin{cases} = 1 & \text{if } i = k, \\ = 0 & \text{if } i \neq k, \end{cases}$$

means

$$\left(\sum_{j=1}^{i} \alpha_{ij}(f_{t_j+s} - f_s), \sum_{j=1}^{k} \alpha_{kj}(f_{t_j+s} - f_s)\right) = \delta_{ik},$$

which is a relation of the

$$(f_{t'+s} - f_s, f_{t''+s} - f_s), \qquad \text{for } t', t'' = t_1, t_2, \cdots,$$

and the

$$\alpha_{ij}, \qquad \alpha_{kj}.$$

By (2.3) this means a relation of

$$G(t', t''), \qquad \text{for } t', t'' = t_1, t_2, \cdots,$$

and the

$$\alpha_{ij}, \qquad \alpha_{kj}.$$

Hence (2.8) is independent of s. It expresses that $\phi_1(s), \phi_2(s), \cdots$ is a normalized orthogonal set. Since $\phi_i(0) = \phi_i$, this is true for $s=0$. This proves that

(2.9) $\phi_1(s), \phi_2(s), \cdots$ is a normalized orthogonal set for every s.

2.2. Consider a fixed s. The $\phi_1(s), \phi_2(s), \cdots$ span the same closed, linear set as the $f_{t_1+s} - f_s, f_{t_2+s} - f_s, \cdots$, that is, as the $f_{t+s} - f_s$, t rational. Owing to the continuity of f_u in u, this is the same set as spanned by the $f_{t+s} - f_s$, $-\infty < t < +\infty$, or, if we write t for $t+s$, by the $f_t - f_s$, $-\infty < t < +\infty$.

This set contains in particular $f_0 - f_s$, hence every $f_t - f_0 = (f_t - f_s) - (f_0 - f_s)$. Hence it contains $\mathfrak{H}_1$. But $\mathfrak{H}_1$ contains all $f_t - f_0$ (by (2.5), along with the ϕ_i), hence $f_s - f_0$ and the $f_t - f_s = (f_t - f_0) - (f_s - f_0)$. Hence it contains our set, too. In other words:

(2.10) $\phi_1(s), \phi_2(s), \cdots$ span the closed linear set $\mathfrak{H}_1$ for every s.

By (2.9), (2.10) the equations

$$U(s)\phi_i(0) = \phi_i(s), \qquad \text{for } i = 1, 2, \cdots, \tag{2.11}$$

define a unitary transformation $U(s)$ in $\mathfrak{H}_1$. (2.5′) gives immediately

$$U(s)(f_t - f_0) = f_{t+s} - f_s, \tag{2.12}$$

and by (2.4′) this relation (2.12) again implies (2.11).

If we write (2.12) for $t = u$ and v, and subtract, we get

$$U(s)(f_u - f_v) = f_{u+s} - f_{v+s}. \tag{2.13}$$

Since (2.12) is a special case of (2.13) ($u = t$, $v = 0$) we see that (2.13) is also characteristic for $U(s)$. Application of (2.13) for $s = s_1, s_2, s_1 + s_2$ shows that

$$U(s_2)U(s_1) = U(s_1 + s_2). \tag{2.14}$$

As f_{t_j+s}, f_s are continuous functions of s, so is $\phi_i(s)$ by (2.4′). Now (2.11) shows that

(2.15) $U(s)$ is a (strongly) continuous function of s.

2.3. Extend now $\mathfrak{H}$ to a complex unitary space $\overline{\mathfrak{H}}$, and correspondingly $\mathfrak{H}_1$ to $\overline{\mathfrak{H}}_1$. Then $U(s)$ extends to a unitary operator in $\overline{\mathfrak{H}}_1$. Since the extended meaning of $\|f-g\|$ and $U(s)$ remains unchanged in $\mathfrak{H}_1$, it is clear that (2.1), (2.12) and (2.15) remain true.

As $\overline{\mathfrak{H}}_1$ is spanned by the enumerable set $f_{t_1}-f_0, f_{t_2}-f_0, \cdots$, it is separable, hence a finite-dimensional unitary or a complex Hilbert space. Now we may apply, considering (2.14), (2.15), the theorem of M. H. Stone ([6, 7]; this theorem is independent of the separability of the underlying unitary space). This theorem insures the existence of a self-adjoint operator $\overline{A}$ in $\overline{\mathfrak{H}}_1$, such that

$$U(s) = \exp(is\overline{A}). \tag{2.16}$$

Now equation (2.12) may be written as

$$f_{t+s} = U(s)(f_t - f_0) + f_s,$$

or, using (2.16),

$$f_{t+s} = \exp(is\overline{A})(f_t - f_0) + f_s. \tag{2.17}$$

2.4. Let $\overline{E}(x)$ be the resolution of unity which belongs to $\overline{A}$ in $\overline{\mathfrak{H}}_1$. Define the projections $\overline{F}_{n,\epsilon}$, $n=0, \pm 1, \pm 2, \cdots, \epsilon = \pm 1$, and $\overline{F}^0$ as follows:

$$\begin{aligned} \overline{F}_{n,+1} &= \overline{E}(2^n) - \overline{E}(2^{n-1}), \\ \overline{F}_{n,-1} &= \overline{E}(-2^{n-1}) - \overline{E}(-2^n), \\ \overline{F}^0 &= \overline{E}(0) - \overline{E}(-0) \qquad \left(E(-0) = \lim_{x<0,\, x\to 0} \overline{E}(x)\right). \end{aligned} \tag{2.18}$$

The projection operators $\overline{F}_{n,\epsilon}$, $\overline{F}^0$ are clearly orthogonal and their sum is 1.

Let

$$\begin{aligned} &\overline{\mathfrak{M}}_{n,\epsilon} \text{ be the closed linear set with the projection } \overline{F}_{n,\epsilon}, \\ &\qquad\qquad (n = 0, \pm 1, \pm 2, \cdots, \epsilon = \pm 1), \\ &\overline{\mathfrak{M}}^0 \text{ be the closed linear set with the projection } \overline{F}^0. \end{aligned} \tag{2.19}$$

Again the $\overline{\mathfrak{M}}_{n,\epsilon}$, $\overline{\mathfrak{M}}^0$ are mutually orthogonal and they span together the closed linear set $\overline{\mathfrak{H}}_1$. Clearly every $\overline{\mathfrak{M}}_{n,\epsilon}$, $\overline{\mathfrak{M}}^0$ reduces $\overline{A}$, and hence, by (2.16), all $U(s)$, too.

Denote the projections of f_t in $\overline{\mathfrak{M}}_{n,\epsilon}$, and $\overline{\mathfrak{M}}^0$, by $\bar{f}_{n,\epsilon/t}$ and $\bar{f}_t^0$, respectively.

2.5. Let us consider first the situation in $\overline{\mathfrak{M}}^0$. Here clearly $\overline{A}=0$, and (2.17) gives

$$\bar{f}_{t+s}^0 = \bar{f}_t^0 - \bar{f}_0^0 + \bar{f}_s^0, \tag{2.20}$$

or

$$(\bar{f}_{t+s}^0 - \bar{f}_0^0) = (\bar{f}_t^0 - \bar{f}_0^0) + (\bar{f}_s^0 - \bar{f}_0^0).$$

This, and the t-continuity of $\bar{f}_t$, give immediately

$$\bar{f}_t^0 - \bar{f}_0^0 = t(\bar{f}_1^0 - \bar{f}_0^0),$$

that is,

$$\bar{f}_t^0 = t\bar{\omega}^0 + \bar{f}_0^0, \qquad \bar{\omega}^0 \text{ being fixed.} \tag{2.21}$$

2.6. Let us consider next the situation in $\overline{\mathfrak{M}}_{n,\epsilon}$. Here the spectrum $S_{n,\epsilon}$ of $\bar{A}$ lies clearly between 2^{n-1} and 2^n, and between -2^n and -2^{n-1} respectively. Hence $x \in S_{n,\epsilon}$ implies $2^{n-1} \leqq |x| \leqq 2^n$.

Assume

$$0 < |t| \leqq \pi/2^n. \tag{2.22}$$

Then $x \in S_{n,\epsilon}$ implies $0 < 2^{n-1}|t| \leqq |tx| \leqq \pi$. Hence the function

$$f_t(x) = (\exp(itx) - 1)^{-1}$$

is everywhere defined, continuous, and bounded in $x \in S_{n,\epsilon}$. Hence we may form $f_t(\bar{A})$. As $(\exp(itx) - 1) f_t(x) = f_t(x) (\exp(itx) - 1) = 1$ for all $x \in S_{n,\epsilon}$, we have $(\exp(it\bar{A}) - 1) f_t(\bar{A}) = f_t(\bar{A})(\exp(it\bar{A}) - 1) = 1$. Hence $f_t(\bar{A})$ is the inverse of $\exp(it\bar{A}) - 1$.

We have thus proved that

$$(\exp(it\bar{A}) - 1)^{-1} \text{ exists and is bounded.} \tag{2.23}$$

Consider now (2.17) for t, s and for s, t. As the left side is the same in both cases, the right sides are equal, too, giving

$$\exp(is\bar{A})(\bar{f}_{n\epsilon/t} - \bar{f}_{n\epsilon/0}) + \bar{f}_{n\epsilon/s} = \exp(it\bar{A})(\bar{f}_{n\epsilon/s} - \bar{f}_{n\epsilon/0}) + \bar{f}_{n\epsilon/t},$$

that is

$$\{\exp(is\bar{A}) - 1\}(\bar{f}_{n\epsilon/t} - \bar{f}_{n\epsilon/0}) = \{\exp(it\bar{A}) - 1\}(\bar{f}_{n\epsilon/s} - \bar{f}_{n\epsilon/0}).$$

Assume (2.22) for both t and s. Since $\exp(it\bar{A}) - 1$ and $\exp(is\bar{A}) - 1$ commute, and by (2.23) for t and s, we may write this as

$$\{\exp(it\bar{A}) - 1\}^{-1}(\bar{f}_{n\epsilon/t} - \bar{f}_{n\epsilon/0}) = \{\exp(is\bar{A}) - 1\}^{-1}(\bar{f}_{n\epsilon/s} - \bar{f}_{n\epsilon/0}).$$

That is,

$$\{\exp(it\bar{A}) - 1\}^{-1}(\bar{f}_{n\epsilon/t} - \bar{f}_{n\epsilon/0}) = \bar{v}_{n\epsilon},$$

where $\bar{v}_{n\epsilon}$ is fixed. But this may be written as

$$\bar{f}_{n\epsilon/t} = \{\exp(it\bar{A}) - 1\}\bar{v}_{n\epsilon} + \bar{f}_{n\epsilon/0}, \qquad \bar{v}_{n\epsilon} \text{ constant.} \tag{2.24}$$

This holds provided t satisfies (2.22). Since it is obviously true for $t = 0$, it holds whenever $|t| \leqq \pi/2^n$. But if (2.24) holds for both t and s then (2.17) extends it to $t+s$. Hence (2.24) holds for all t.

We may now formulate (2.21), (2.24) as follows:

(2.25) The projections of $f_t - f_0$ in $\overline{\mathfrak{M}}^0$ and $\overline{\mathfrak{M}}_{n,\epsilon}$ are $t\bar{\omega}^0$ and $\{\exp(it\bar{A}) - 1\}\bar{v}_{n\epsilon}$ respectively, the $\bar{\omega}^0$, $\bar{v}_{n\epsilon}$ being fixed elements of $\overline{\mathfrak{M}}^0$ and $\overline{\mathfrak{M}}_{n,\epsilon}$ respectively.

2.7. We know that $\bar{\mathfrak{H}}_1$ is the closed linear set spanned by the mutually orthogonal sets $\bar{\mathfrak{M}}^0$ and $\bar{\mathfrak{M}}_{n,\epsilon}$ ($n=0, \pm 1, \pm 2, \cdots, \epsilon = \pm 1$). As $f_t - f_0$ is in $\mathfrak{H}_1$, we see that (2.25) implies

$$f_t - f_0 = t\bar{\omega}^0 + \sum_{n=-\infty}^{\infty} \sum_{\epsilon=\pm 1} \{\exp(it\bar{A}) - 1\}\bar{v}_{n\epsilon}, \tag{2.26}$$

the convergence of the right side being certain for all t.

Since the addends of the right side of (2.26) are mutually orthogonal, we conclude that

$$\|f_t - f_0\|^2 = t^2\|\bar{\omega}^0\|^2 + \sum_{n=-\infty}^{\infty} \sum_{\epsilon=\pm 1} \|\{\exp(it\bar{A}) - 1\}\bar{v}_{n\epsilon}\|^2.$$

But for each term of this sum we have

$$\begin{aligned} \|\{\exp(it\bar{A}) - 1\}\bar{v}_{n\epsilon}\|^2 &= (\{\exp(it\bar{A}) - 1\}\bar{v}_{n\epsilon}, \{\exp(it\bar{A}) - 1\}\bar{v}_{n\epsilon}) \\ &= (\{\exp(it\bar{A}) - 1\}^*\{\exp(it\bar{A}) - 1\}\bar{v}_{n\epsilon}, \bar{v}_{n\epsilon}). \end{aligned}$$

By well known properties of functions of Hermitian operators, and since

$$\begin{aligned} \overline{(\exp(itx) - 1)}(\exp(itx) - 1) &= (\exp(-itx) - 1)(\exp(itx) - 1) \\ &= 2 - \exp(itx) - \exp(-itx) = 2(1 - \cos tx), \end{aligned}$$

the above expression is equal to

$$(2(1 - \cos(t\bar{A}))\bar{v}_{n\epsilon}, \bar{v}_{n\epsilon}),$$

that is, to

$$\int 2(1 - \cos tx)d(\|\bar{E}(x)\bar{v}_{n\epsilon}\|^2).$$

It suffices to extend this Stieltjes integral from 2^{n-1} to 2^n, if $\epsilon = +1$; and from -2^n to -2^{n-1} if $\epsilon = -1$. Denote this interval by $I_{n,\epsilon}$.

We have thus obtained

$$F^2(t) = \|f_t - f_0\|^2 = t^2\|\bar{\omega}^0\|^2 + \sum_{n=-\infty}^{\infty} \sum_{\epsilon=\pm 1} \int_{I_{n,\epsilon}} 2(1 - \cos tx)d(\|\bar{E}(x)\bar{v}_{n\epsilon}\|^2). \tag{2.27}$$

Integrating this relation from 0 to 1 we obtain

$$\int_0^1 F^2(t)dt = \tfrac{1}{3}\|\bar{\omega}^0\|^2 + \sum_{n=-\infty}^{\infty} \sum_{\epsilon=\pm 1} \int_{I_{n,\epsilon}} 2\left(1 - \frac{\sin x}{x}\right) d(\|\bar{E}(x)\bar{v}_{n\epsilon}\|^2). \tag{2.28}$$

Since

$$2\left(1 - \frac{\sin x}{x}\right)\begin{cases} \geqq \frac{1}{4}x^2, & \text{if } |x| \leqq 1, \\ \geqq \frac{1}{4}, & \text{if } |x| \geqq 1, \end{cases}$$

we derive from (2.28) the finiteness of the following expression:

$$(2.28') \qquad \sum_{n=-\infty}^{0} \sum_{\epsilon=\pm 1} \int_{I_{n,\epsilon}} \tfrac{1}{4}x^2 d\|\bar{E}(x)\bar{v}_{n\epsilon}\|^2 + \sum_{n=1}^{\infty} \sum_{\epsilon=\pm 1} \int_{I_{n,\epsilon}} \tfrac{1}{4} d\|\bar{E}(x)\bar{v}_{n\epsilon}\|^2.$$

2.8. We may now readily complete a proof of Theorem 1 by showing that the last side of (2.27) can be put in the form (1.4). Indeed, define two non-decreasing functions $\beta_-(x)$ $(-\infty < x < 0)$ and $\beta_+(x)$ $(0 < x < +\infty)$ as follows: Set

$$\beta_-(x) = \|\bar{E}(x)\bar{v}_{0,-1}\|^2 \quad \text{in} \quad I_{0,-1}\,(-1 \leqq x \leqq -\tfrac{1}{2}),$$

and extend the definition successively to $I_{1,-1}$, $I_{2,-1}$, $\cdots$, and also to $I_{-1,-1}$, $I_{-2,-1}$, $\cdots$, in such a way that

(i) $\beta_-(x) = \|\bar{E}(x)\bar{v}_{n,-1}\|^2 + \text{const.}$, in $I_{n,-1}$ $(-2^n \leqq x \leqq -2^{n-1})$,

(ii) the constants are successively determined by the requirement that we have agreement of values of $\beta_-(x)$, at every point $x = -2^n$, resulting from this function's definition (i) in the adjoining intervals $I_{n,-1}$ and $I_{n+1,-1}$.

Set similarly

$$\beta_+(x) = \|\bar{E}(x)\bar{v}_{0,1}\|^2 \quad \text{in} \quad I_{0,1} \quad (\tfrac{1}{2} \leqq x \leqq 1),$$

and define successively

$$\beta_+(x) = \|\bar{E}(x)\bar{v}_{n,1}\|^2 = \text{const.}, \quad \text{in} \quad I_{n,1} \quad (2^{n-1} \leqq x \leqq 2^n),$$

determining the constants successively by the same requirement as above. The finite expression (2.28′) now becomes

$$(2.28'') \qquad \int_{-1}^{-0} \tfrac{1}{4}x^2 d\beta_-(x) + \int_{+0}^{1} \tfrac{1}{4}x^2 d\beta_+(x) + \int_{-\infty}^{-1} \tfrac{1}{4} d\beta_-(x) + \int_{1}^{\infty} \tfrac{1}{4} d\beta_+(x),$$

and (2.27) may be written as

$$F^2(t) = t^2\|\bar{\omega}^0\|^2 + \int_{-\infty}^{-0} 2(1 - \cos tx)d\beta_-(x) + \int_{+0}^{\infty} 2(1 - \cos tx)d\beta_+(x).$$

Setting

$$\alpha(x) = 2\beta_+(x) - 2\beta_-(-x), \qquad 0 < x < \infty,$$

this becomes

$$(2.27') \qquad F^2(t) = t^2\|\bar{\omega}^0\|^2 + \int_{+0}^{\infty} (1 - \cos tx)d\alpha(x)$$

where $\alpha(x)$ is non-decreasing such that

$$\alpha(\infty) = 2\beta_+(\infty) - 2\beta_-(-\infty),$$

$$\int_{+0}^{1} x^2 d\alpha(x) = 2\int_{+0}^{1} x^2 d\beta_+(x) + 2\int_{-1}^{-0} x^2 d\beta_-(x)$$

are finite, on account of the finiteness of (2.28″). If we introduce a new monotone function by setting

$$\gamma(0) = 0, \qquad \gamma(x) = \int_{+0}^{x} 2x^2 d\alpha(2x), \qquad x > 0,$$

we have

$$\int_{+0}^{\infty} (1 - \cos tx) d\alpha(x) = \int_{+0}^{\infty} \frac{1 - \cos 2tx}{2x^2} 2x^2 d\alpha(2x) = \int_{+0}^{\infty} \frac{\sin^2 tx}{x^2} d\gamma(x).$$

By changing, finally, the value of $\gamma(0)$ from 0 to $-\|\bar{\omega}^0\|^2$, we see that (2.27′) now assumes the form (1.4). Also (1.5) is satisfied, since $\alpha(x)$ is bounded.

Part III. Second proof and extension of fundamental theorem by the theory of positive definite functions

3.1. We know that the class of screw functions of Hilbert space is identical with the class of continuous, even, non-negative functions $F(t)$ ($F(0) = 0$) such that the metric transform $F(E_1)$ is isometrically imbeddable in $\mathfrak{H}$ (Introduction, §1). Let us denote this class of functions by the symbol $\Pi(E_1)$. We now extend the problem of determining the class $\Pi(E_1)$, as follows: *Determine the class $\Pi(E_m)$ of functions $F(t)$ such that the metric transform $F(E_m)$ be isometrically imbeddable in $\mathfrak{H}$ $(1 \leqq m \leqq \infty, E_\infty = \mathfrak{H})$.* (See [5, Introduction and §5.2].)

The definition of $\Pi(E_m)$ implies the relations

$$\Pi(E_1) \supset \Pi(E_2) \supset \cdots \supset \Pi(E_m) \supset \cdots \supset \Pi(\mathfrak{H}),$$

showing that we have to deal with an infinite sequence of non-increasing subsets of $\Pi(E_1)$. The class $\Pi(\mathfrak{H})$ was found to be identical with the class of functions $F(t)$ whose squares are of the form

$$F^2(t) = \int_0^{\infty} \frac{1 - e^{-t^2 u}}{u} d\gamma(u),$$

where $\gamma(u)$ is non-decreasing for $u \geqq 0$ and such that $\int_1^\infty u^{-1} d\gamma(u)$ exists [5, §5.4, Theorem 6]. We shall now determine the classes $\Pi(E_m)$, $(1 \leqq m < \infty)$.

The statement of our result requires the integral function

$$(3.1) \quad \begin{aligned} \Omega_m(t) &= 1 - \frac{t^2}{2 \cdot m} + \frac{t^4}{2 \cdot 4 \cdot m(m+2)} - \frac{t^6}{2 \cdot 4 \cdot 6 \cdot m(m+2)(m+4)} + \cdots \\ &= \Gamma\left(\frac{m}{2}\right)\left(\frac{2}{t}\right)^{(m-2)/2} J_{(m-2)/2}(t). \end{aligned}$$

In particular

$$\Omega_1(t) = \cos t, \qquad \Omega_2(t) = J_0(t), \qquad \Omega_3(t) = \sin t/t.$$

THEOREM 7. *The class* $\Pi(E_m)$ *of non-negative continuous functions* $F(t)$ $(t \geqq 0,\ F(0)=0)$ *such that the metric transform* $F(E_m)$ *be isometrically imbeddable in* $\mathfrak{H}$, *is identical with the class of functions whose squares are of the form*

$$(3.2)\qquad F^2(t) = \int_0^\infty \frac{1-\Omega_m(tu)}{u^2}\,d\gamma(u),$$

where $\gamma(u)$ *is non-decreasing for* $u \geqq 0$, $\gamma(0)=0$, *and such that*

$$(3.3)\qquad \int_1^\infty \frac{d\gamma(u)}{u^2} \quad \text{exists.}$$

If $m=1$ we obtain our Theorem 1 on screw functions, since $1-\Omega_1(tu) = 1-\cos tu = 2\sin^2(tu/2)$.

The reasoning which leads to the characterization of the class $\Pi(E_1)$ given by Lemma 1, §1.1, extends, of course, to the class $\Pi(E_m)$ and gives the following statement:

A continuous non-negative function $F(t)$ $(t \geqq 0,\ F(0)=0)$ *belongs to* $\Pi(E_m)$ *if and only if it satisfies the inequality*

$$(3.4)\qquad \sum_{i,k=1}^{n} \{F^2(P_0P_i) + F^2(P_0P_k) - F^2(P_iP_k)\}\rho_i\rho_k \geqq 0,$$

for any points $P_0, \cdots, P_n$ *of* E_m *and arbitrary real* ρ_i $(n=2, 3, \cdots)$, *the quantities* P_iP_k *denoting euclidean distances in* E_m.

We refer to [5, §8.1], for a simple proof, based on the statement above, that any $F(t)$ given by (3.2) belongs to $\Pi(E_m)$.

3.2. The proof of the converse, i.e., that formula (3.2) furnishes *all* elements of $\Pi(E_m)$, requires the following limit theorem:

LEMMA 2. *Let*

$$(3.5)\qquad f_n(t) = \int_0^\infty \frac{1-\Omega_m(tu)}{u^2}\,d\gamma_n(u), \qquad n = 1, 2, 3, \cdots,$$

be a sequence of functions with non-decreasing $\gamma_n(u)$, $(\gamma_n(0)=0)$, *such that all integrals* $\int_1^\infty u^{-2}d\gamma_n(u)$ *exist. Let* $f_n(t)$ *converge, as* $n\to\infty$, *uniformly in any finite interval, to a function* $f(t)$. *We indicate this type of convergence by the relation*

$$(3.6)\qquad f_n(t) \rightarrowtail f(t).$$

Then $f(t)$ *is also of the form*

$$(3.7)\qquad f(t) = \int_0^\infty \frac{1-\Omega_m(tu)}{u^2}\,d\gamma(u), \quad \gamma(u)\uparrow,\ \gamma(0)=0, \quad \int_1^\infty u^{-2}d\gamma(u) \text{ exists,}$$

and

$$(3.8) \qquad \gamma_n(u) \to \gamma(u)\text{(}^2\text{)},$$

$$(3.9) \qquad \int_a^\infty \frac{d\gamma_n(u)}{u^2} \to \int_a^\infty \frac{d\gamma(u)}{u^2} \quad \textit{in all continuity points } u = a > 0 \textit{ of } \gamma(u).$$

Conversely, if the relations (3.8) *and* (3.9) *hold and* $f_n(t)$, $f(t)$ *are defined by* (3.5) *and* (3.7), *then* (3.6) *holds.*

Postponing a proof of this limit theorem we shall now use it to establish Theorem 7.

Let $F(t) \in \Pi(E_m)$, that is, $F(E_m)$ be imbeddable in $\mathfrak{H}$. By [5, §5.3, Corollary 1] we conclude that the functions $\exp\{-\lambda F^2(t)\}$, $(\lambda > 0)$, belong to the class $\mathfrak{P}(E_m)$ of functions which are positive definite in E_m. By [5, §1.2, Theorem 1], we are allowed to set

$$\exp\{-\lambda F^2(t)\} = \int_0^\infty \Omega_m(tu)d\alpha(u, \lambda), \qquad \lambda > 0,$$

$\alpha(u, \lambda)$ being a family of non-decreasing functions, $\alpha(0, \lambda) = 0$, $\alpha(\infty, \lambda) = 1$. Now

$$\frac{1 - \exp\{-\lambda F^2(t)\}}{\lambda} = \frac{1}{\lambda}\int_0^\infty [1 - \Omega_m(tu)]d\alpha(u, \lambda),$$

and if we set

$$\beta(u, \lambda) = \frac{1}{\lambda}\int_0^u u^2 d\alpha(u, \lambda),$$

we obtain

$$(3.10) \qquad f(t, \lambda) = \frac{1 - \exp\{-\lambda F^2(t)\}}{\lambda} = \int_0^\infty \frac{1 - \Omega_m(tu)}{u^2} d\beta(u, \lambda).$$

On the other hand,

$$\lim_{\lambda \to +0} f(t, \lambda) = F^2(t)$$

holds *uniformly in any finite interval*. For, if $0 \leqq t \leqq T$, we have $F^2(t) < K$ (since $F(t)$ is continuous) and therefore

$$\begin{aligned} |\exp\{-\lambda F^2(t)\} - 1 + \lambda F^2(t)| &= |\lambda^2 F^4(t)/2! - \lambda^3 F^6(t)/3! + \cdots| \\ &\leqq \lambda^2 K^2 e^{\lambda K}, \end{aligned}$$

or

(2) Here and throughout this paper a limiting relation $\gamma_n(u) \to \gamma(u)$ involving monotone functions is assumed to hold for all values of u which are continuity points of the limiting function $\gamma(u)$.

$$\left| \frac{1 - \exp\{-\lambda F^2(t)\}}{\lambda} - F^2(t) \right| \leqq \lambda K^2 e^{\lambda K},$$

a bound which is independent of t and tends to zero with λ. Setting $f_n(t) = f(t, 1/n)$ we therefore have $f_n(t) \rightarrowtail F^2(t)$ and Lemma 2 is applicable, on account of (3.10), showing that $F^2(t)$ is of the form (3.2).

3.3. In this proof of Theorem 7 only a part of Lemma 2 was used, namely that (3.5) and (3.6) imply (3.7). We have stated and shall prove the more inclusive limit theorem for two reasons. First, our limit theorem implies that $\gamma(u)$ of (3.2) is uniquely defined by $F(t)$. Second, a proof of the complete statement is hardly more complicated than a proof of the partial statement actually used above.

Lemma 2 will be a direct consequence of the following analogous limit theorem involving a simpler kernel under the integral sign.

LEMMA 3. *Let*

$$\phi_n(t) = \int_0^\infty \frac{1 - e^{-ut}}{u} d\beta_n(u), \qquad n = 1, 2, 3, \cdots, \tag{3.11}$$

be a sequence of functions with non-decreasing $\beta_n(u)$ $(\beta_n(0) = 0)$, *such that all integrals* $\int_1^\infty u^{-1} d\beta_n(u)$ *exist. Let* $\phi_n(t)$ *converge, as* $n \to \infty$, *to a function* $\phi(t)$, *uniformly in any finite interval* $0 \leqq t \leqq t_0$, *i.e.,*

$$\phi_n(t) \rightarrowtail \phi(t). \tag{3.12}$$

Then $\phi(t)$ *is also of the form*

$$\phi(t) = \int_0^\infty \frac{1 - e^{-ut}}{u} d\beta(u), \quad \beta(u) \uparrow, \ \beta(0) = 0, \ \int_1^\infty u^{-1} d\beta(u) \ \textit{exists}, \tag{3.13}$$

and

$$\beta_n(u) \to \beta(u), \tag{3.14}$$

$$\int_a^\infty \frac{d\beta_n(u)}{u} \to \int_a^\infty \frac{d\beta(u)}{u} \quad \textit{in all continuity points } u = a > 0 \textit{ of } \beta(u). \tag{3.15}$$

Conversely, if the relations (3.14) *and* (3.15) *hold and* $\phi_n(t)$, $\phi(t)$ *are defined by* (3.11) *and* (3.13), *then* (3.12) *holds*(3).

The $\phi_n(t)$ belong to the class T defined in [5, §4]. It was shown there (Lemma 5) that (3.12) implies $\phi(t) \in T$, hence (3.13). It was furthermore

(3) That the condition (3.15) cannot be dispensed with is shown by the example of $\beta_n(u) = 0$ if $0 \leqq u < n$, $\beta_n(u) = n$ if $u \geqq n$, $\beta(u) \equiv 0$, hence $\phi_n(t) = 1 - e^{-nt}$, $\phi(t) \equiv 0$. Now (3.14) holds but not (3.12). Note that the condition (3.15) is not satisfied.

shown by means of the Vitali-Porter convergence theorem that $\phi_n(t)$ and

$$\phi_n'(t) = \int_0^\infty e^{-ut} d\beta_n(u)$$

converge to $\phi(t)$ and $\phi'(t)$, respectively, uniformly in any finite closed domain inside the half-plane $\Re t > 0$ of the complex variable $t = \sigma + i\tau$. In particular we therefore have

$$(3.16) \qquad \phi_n'(1 + i\tau) = \int_0^\infty e^{-u} e^{-iu\tau} d\beta_n(u) \to \phi'(1 + i\tau) = \int_0^\infty e^{-u} e^{-iu\tau} d\beta(u),$$

uniformly in any finite τ-interval. In terms of the new monotone and bounded functions

$$(3.17) \qquad \alpha_n(u) = \int_0^u e^{-u} d\beta_n(u), \qquad \alpha(u) = \int_0^u e^{-u} d\beta(u),$$

the relation (3.16) becomes

$$\int_0^\infty e^{-iu\tau} d\alpha_n(u) \to \int_0^\infty e^{-iu\tau} d\alpha(u).$$

Now we may apply a limit theorem of Paul Lévy and conclude that

$$(3.18) \qquad \alpha_n(u) \to \alpha(u),$$

[1, p. 197]. But (3.17) may be inverted:

$$\beta_n(u) = \int_0^u e^u d\alpha_n(u), \qquad \beta(u) = \int_0^u e^u d\alpha(u).$$

Now (3.18) implies the desired relation (3.14).

There remains the proving of (3.15). On one hand (3.14) implies

$$\int_0^a \frac{1 - e^{-tu}}{u} d\beta_n(u) \to \int_0^a \frac{1 - e^{-tu}}{u} d\beta(u), \qquad t \geqq 0,$$

and therefore (3.11), (3.13) and (3.12) imply

$$(3.19) \qquad \int_a^\infty \frac{1 - e^{-tu}}{u} d\beta_n(u) \to \int_a^\infty \frac{1 - e^{-tu}}{u} d\beta(u), \qquad t \geqq 0.$$

On the other hand, the obvious inequalities

$$(1 - e^{-ta}) \int_a^\infty \frac{d\beta_n(u)}{u} \leqq \int_a^\infty \frac{1 - e^{-tu}}{u} d\beta_n(u) \leqq \int_a^\infty \frac{d\beta_n(u)}{u}, \quad t \geqq 0,$$

imply

$$\int_a^\infty \frac{1-e^{-tu}}{u}\,d\beta_n(u) \leqq \int_a^\infty \frac{d\beta_n(u)}{u} \leqq \frac{1}{1-e^{-ta}}\int_a^\infty \frac{1-e^{-tu}}{u}\,d\beta_n(u),\ t>0.$$

As $n\to\infty$ we get by (3.19)

$$\int_a^\infty \frac{1-e^{-tu}}{u}\,d\beta(u) \leqq \liminf \int_a^\infty \frac{d\beta_n(u)}{u} \leqq \limsup \int_a^\infty \frac{d\beta_n(u)}{u}$$

$$\leqq \frac{1}{1-e^{-ta}}\cdot\int_a^\infty \frac{1-e^{-tu}}{u}\,d\beta(u).$$

As $t\to\infty$ this gives

$$\int_a^\infty \frac{d\beta(u)}{u} \leqq \liminf \int_a^\infty \frac{d\beta_n(u)}{u} \leqq \limsup \int_a^\infty \frac{d\beta_n(u)}{u} \leqq \int_a^\infty \frac{d\beta(u)}{u},$$

which proves (3.15).

In order to prove the converse of Lemma 3, we have to show that (3.11), (3.13), (3.14), and (3.15) imply (3.12). For a given ϵ, we can choose a positive a, continuity point of $\beta(u)$, such that

$$\int_a^\infty u^{-1}d\beta_n(u) < \epsilon, \qquad \int_a^\infty u^{-1}d\beta(u) < \epsilon,$$

provided n is sufficiently large. But then

$$\int_0^a \frac{1-e^{-tu}}{u}\,d\beta_n(u), \qquad \int_0^a \frac{1-e^{-tu}}{u}\,d\beta(u),$$

differ from $\phi_n(t)$ and $\phi(t)$ respectively by less than ϵ. Hence

$$|\phi_n(t)-\phi(t)| < 2\epsilon + \left|\int_0^a \frac{1-e^{-tu}}{u}\,d\beta_n(u) - \int_0^a \frac{1-e^{-tu}}{u}\,d\beta(u)\right|,$$

provided n is sufficiently large. On the other hand, (3.14) implies

$$\int_0^a \frac{1-e^{-tu}}{u}\,d\beta_n(u) \to \int_0^a \frac{1-e^{-tu}}{u}\,d\beta(u),$$

that is,

$$\left|\int_0^a \frac{1-e^{-tu}}{u}\,d\beta_n(u) - \int_0^a \frac{1-e^{-tu}}{u}\,d\beta(u)\right| < \epsilon,$$

in an arbitrarily given interval $0\leqq t\leqq t_0$, provided n is large enough. Hence $|\phi_n(t)-\phi(t)| < 3\epsilon$, in the interval $0\leqq t\leqq t_0$, provided n is large enough. This is precisely identical to (3.12).

By a similar argument we can prove the converse part of Lemma 2. We may, therefore, consider the converse part of Lemma 2 as already established and shall use it in proving the direct part of that lemma.

3.4. In order to prepare a proof of Lemma 2 we need the following statement:

(3.20) The assumptions (3.5) and (3.6), of Lemma 2, imply the existence of two constants K and H, independent of n and t, such that $f_n(t) < Kt^2 + H$, $t \geqq 0$, $n = 1, 2, 3, \cdots$.

Assume first that $m \geqq 2$. Since

$$-1 < \Omega_m(t) < 1, \quad (t > 0); \qquad \lim_{t\to\infty} \Omega_m(t) = 0,$$

we see that (3.5) and (3.6) imply (for $t = 1$) that the two expressions

$$\int_0^1 \frac{1 - \Omega_m(u)}{u^2} d\gamma_n(u), \qquad \int_1^\infty [1 - \Omega_m(u)] \frac{d\gamma_n(u)}{u^2}$$

are both bounded, as $n \to \infty$. Since the integrands $[1 - \Omega_m(u)]/u^2$, and $1 - \Omega_m(u)$, have positive lower bounds in their respective intervals of integration (0, 1) and (1, ∞), we see that

$$\int_0^1 d\gamma_n(u) = \gamma_n(1), \qquad \int_1^\infty \frac{d\gamma_n(u)}{u^2} \tag{3.21}$$

also are bounded, as $n \to \infty$. On the other hand, it is clear from the power series expansion (3.1), that there is a constant c, such that

$$1 - \Omega_m(t) \leqq ct^2,$$

provided $t \geqq 0$. Hence, by (3.21),

$$\begin{aligned} f_n(t) &= \int_0^1 \frac{1 - \Omega_m(tu)}{u^2} d\gamma_n(u) + \int_1^\infty \frac{1 - \Omega_m(tu)}{u^2} d\gamma_n(u) \\ &\leqq \int_0^1 \frac{ct^2u^2}{u^2} d\gamma_n(u) + \int_1^\infty \frac{2}{u^2} d\gamma_n(u) \\ &= t^2 c\gamma_n(1) + 2\int_1^\infty u^{-2} d\gamma_n(u) < Kt^2 + H. \end{aligned}$$

If $m = 1$, hence $\Omega_m(tu) = \cos tu$, we first integrate both sides of (3.6), between 0 and 1, obtaining the relation

$$\int_0^1 f_n(t)dt = \int_0^\infty \frac{1 - u^{-1}\sin u}{u^2} d\gamma_n(u) = \int_0^\infty \frac{1 - \Omega_3(u)}{u^2} d\gamma_n(u) \to \int_0^1 f(t)dt.$$

From this we derive as above that (3.21) are also bounded in the present case. The argument is then concluded as above. Thus (3.20) is established.

3.5. We may now readily prove Lemma 2. From its assumptions (3.5), (3.6), combined with (3.20), we get

$$f_n(t) \to f(t), \qquad 0 \leqq f_n(t) < Kt^2 + H, \qquad 0 \leqq f(t) < Kt^2 + H,\ t \geqq 0.$$

This implies, as readily seen, that

$$\int_0^\infty e^{-t^2/4} f_n(t\mu) t^{m-1} dt \to \int_0^\infty e^{-t^2/4} f(t\mu) t^{m-1} dt, \tag{3.22}$$

uniformly in μ in any finite interval $0 \leqq \mu \leqq M$. Hence also

$$\int_0^\infty e^{-t^2/4} f_n(t\lambda^{1/2}) t^{m-1} dt \to \int_0^\infty e^{-t^2/4} f(t\lambda^{1/2}) t^{m-1} dt,$$

uniformly in any finite interval $0 \leqq \lambda \leqq L$. Taking $t\lambda^{-1/2}$, instead of t, as a new variable of integration, we see that

$$\lambda^{-m/2} \int_0^\infty e^{-t^2/4\lambda} f_n(t) t^{m-1} dt \to \lambda^{-m/2} \int_0^\infty e^{-t^2/4\lambda} f(t) t^{m-1} dt, \tag{3.23}$$

uniformly in λ in any finite interval $0 < \lambda \leqq L$. We shall now make use of the relation [5, p. 823]

$$e^{-\lambda u^2} = c_m \lambda^{-m/2} \int_0^\infty \Omega_m(tu) e^{-t^2/4\lambda} t^{m-1} dt, \quad \lambda > 0,\ c_m = 2^{1-m} \cdot [\Gamma(m/2)]^{-1}, \tag{3.24}$$

and its particular case $(u = 0)$

$$1 = c_m \lambda^{-m/2} \int_0^\infty e^{-t^2/4\lambda} t^{m-1} dt.$$

If we multiply the relation

$$f_n(t) = \int_0^\infty \frac{1 - \Omega_m(tu)}{u^2} d\gamma_n(u)$$

by $c_m \lambda^{-m/2} e^{-t^2/4\lambda} t^{m-1} dt$, we get by integration

$$\phi_n(\lambda) = c_m \lambda^{-m/2} \int_0^\infty e^{-t^2/4\lambda} f_n(t) t^{m-1} dt = \int_0^\infty \frac{1 - e^{-\lambda u^2}}{u^2} d\gamma_n(u)$$
$$= \int_0^\infty \frac{1 - e^{-\lambda u}}{u} d\gamma_n(u^{1/2}).$$

By (3.23) $\phi_n(\lambda)$ converges, as $n \to \infty$, uniformly in any interval $0 < \lambda \leqq L$. As the last expression is defined and continuous for $\lambda = 0$, we see that the uniform convergence extends to any closed interval $0 \leqq \lambda \leqq L$. But then, by Lemma 3, we conclude that there is a monotone function $\gamma(u^{1/2})$, $(u \geqq 0)$, such that

$$\gamma_n(u^{1/2}) \to \gamma(u^{1/2}), \qquad \int_{a^2}^{\infty} \frac{d\gamma_n(u^{1/2})}{u} \to \int_{a^2}^{\infty} \frac{d\gamma(u^{1/2})}{u},$$

that is,

$$\gamma_n(u) \to \gamma(u), \qquad \int_{a^2}^{\infty} \frac{d\gamma_n(u)}{u^2} \to \int_{a^2}^{\infty} \frac{d\gamma(u)}{u^2}. \tag{3.25}$$

By the converse of Lemma 2, we may now conclude that

$$f_n(t) = \int_0^{\infty} \frac{1 - \Omega_m(tu)}{u^2}\, d\gamma_n(u) \to \int_0^{\infty} \frac{1 - \Omega_m(tu)}{\nu^2}\, d\gamma(u).$$

On comparing this relation with the assumption (3.6), we derive the validity of the desired conclusion (3.7).

Bibliography

1. P. Lévy, *Calcul des Probabilités*, Paris, 1925.

2. K. Menger, *Die Metrik des Hilbertschen Raumes*, Anzeiger der Akademie der Wissenschaften in Wien, Mathematisch-Naturwissenschaftliche Klasse, vol. 65 (1928), pp. 159–160.

3. I. J. Schoenberg, *Remarks to Maurice Fréchet's article* · · · , Annals of Mathematics, (2), vol. 36 (1935), pp. 724–732.

4. ———, *Metric spaces and positive definite functions*, these Transactions, vol. 44 (1938), pp. 522–536.

5. ———, *Metric spaces and completely monotone functions*, Annals of Mathematics, (2), vol. 39 (1938), pp. 811–841.

6. M. H. Stone, *Linear transformations in Hilbert space* III, Proceedings of the National Academy of Sciences, vol. 16 (1930), pp. 172–175.

7. ———, *On one-parameter unitary groups in Hilbert space*, Annals of Mathematics, (2), vol. 33 (1932), pp. 643–648.

8. W. A. Wilson, *On certain types of continuous transformations of metric spaces*, American Journal of Mathematics, vol. 57 (1935), pp. 62–68.

Institute for Advanced Study,
Princeton, N. J.,
Colby College,
Waterville, Me.

Reprinted from
Trans. Amer. Math. Soc. **50** (1941), pp. 226–251

Reprinted from DUKE MATHEMATICAL JOURNAL
Vol. 9, No. 1, March, 1942

POSITIVE DEFINITE FUNCTIONS ON SPHERES

BY I. J. SCHOENBERG

1. **Introduction.** Let S_2 denote the ordinary spherical shell of radius one and center o and let $p_1, p_2, \cdots, p_n$ be n arbitrary points of S_2. Let p_ip_k denote the spherical distance between the points p_i, p_k. For n real variables $x_1, x_2, \cdots, x_n$ we have the following inequality

$$(1.1) \qquad \left| \sum \overrightarrow{op_i} \cdot x_i \right|^2 = \sum_{i,k=1}^{n} \cos (p_i p_k) x_i x_k \geqq 0,$$

which is equivalent to the determinant inequality

$$\det | \cos (p_i p_k) |_{1,n} \geqq 0$$

for arbitrary points p_i and arbitrary n. This property of the function $g(t) = \cos t$ in relation to the space S_2 is expressed by saying that $\cos t$ is positive definite in S_2. The general definition is as follows. Let M be a metric space with the distance function pq. A real continuous function $g(t)$ $(0 \leqq t \leqq$ diameter of $M)$ is said to be positive definite (p. d.) in M if we have

$$(1.2) \qquad \sum_{i,k=1}^{n} g(p_i p_k) x_i x_k \geqq 0,$$

for any n points $p_1, \cdots, p_n$ of M, arbitrary real x_i, and all $n = 2, 3, \cdots$.

We denote this class of functions by the symbol $\mathfrak{P}(M)$. It enjoys the following useful closure properties:[1]

I. If $g_1(t) \in \mathfrak{P}(M)$, $g_2(t) \in \mathfrak{P}(M)$, also $c_1g_1(t) + c_2g_2(t) \in \mathfrak{P}(M)$, provided $c_1 \geqq 0$, $c_2 \geqq 0$.

II. The same assumptions imply also that $g_1(t)g_2(t) \in \mathfrak{P}(M)$.

III. If $g_n(t) \in \mathfrak{P}(M)$, $g_n(t) \to g(t)$ as $n \to \infty$, and $g(t)$ is continuous, then also $g(t) \in \mathfrak{P}(M)$.

In the present note we are concerned with the classes $\mathfrak{P}(S_m)$ and $\mathfrak{P}(S_\infty)$ corresponding to the unit spheres in the Euclidean space E_{m+1} and the Hilbert space H respectively.

Returning to S_2, we have noticed that $\cos t \in \mathfrak{P}(S_2)$. It will be shown below that also $P_n(\cos t) \in \mathfrak{P}(S_2)$, where P_n is a Legendre polynomial. By the above mentioned closure properties it is now apparent that also

$$(1.3) \qquad g(t) = \sum_{n=0}^{\infty} a_n P_n(\cos t) \in \mathfrak{P}(S_2),$$

provided $a_n \geqq 0$ $(n = 0, 1, \cdots)$ and $\sum a_n$ converges. This formula will be shown to furnish the most general element of $\mathfrak{P}(S_2)$.

Received August 25, 1941; presented to the American Mathematical Society, September 11, 1940.

[1] See [6], §2. Numbers in brackets refer to the bibliography at the end of the paper.

Let $P_n^{(\lambda)}(\cos t)$ be the ultraspherical polynomials defined by the expansion

$$(1.4)\qquad (1 - 2r\cos t + r^2)^{-\lambda} = \sum_{n=0}^{\infty} r^n P_n^{(\lambda)}(\cos t), \qquad (\lambda > 0).$$

For $\lambda = 0$ we set

$$P_n^{(0)}(\cos t) = \cos nt = T_n(\cos t).$$

Our result (1.3) extends to S_m $(m = 1, 2, \cdots)$ as follows. The most general element of $\mathfrak{P}(S_m)$ is

$$(1.5)\qquad g(t) = \sum_{n=0}^{\infty} a_n P_n^{(\lambda)}(\cos t), \qquad (\lambda = \tfrac{1}{2}(m-1)),$$

provided all $a_n \geqq 0$ and that we have convergence for $t = 0$. This will follow readily from classical properties of ultraspherical polynomials (§3).

An obvious property of the class $\mathfrak{P}(M)$ is as follows. If $M \subset N$, then $\mathfrak{P}(M) \supset \mathfrak{P}(N)$. As we may assume

$$S_1 \subset S_2 \subset \cdots \subset S_m \subset \cdots \subset S_\infty,$$

it follows that

$$\mathfrak{P}(S_1) \supset \mathfrak{P}(S_2) \supset \cdots \supset \mathfrak{P}(S_m) \supset \cdots \supset \mathfrak{P}(S_\infty).$$

In fact, $\mathfrak{P}(S_\infty)$ is identical with the intersection of all classes $\mathfrak{P}(S_m)$ $(m = 1, 2, \cdots)$. Since $\cos t \in \mathfrak{P}(S_m)$ for all m, we have $\cos t \in \mathfrak{P}(S_\infty)$ and, by the closure property II, also $(\cos t)^n \in \mathfrak{P}(S_\infty)$. The closure properties I and III show that

$$(1.6)\qquad g(t) = \sum_{n=0}^{\infty} a_n(\cos t)^n \in \mathfrak{P}(S_\infty),$$

provided $a_n \geqq 0$ and $\sum a_n$ converges. It will be shown that the functions $g(t)$ of the form (1.6) exhaust the class $\mathfrak{P}(S_\infty)$(§4).[2]

[2] This result establishes the *converse* of Problem 37 of Pólya and Szegö, [5], p. 107. This converse may be stated as follows. *Let $F(x)$, $(-1 \leqq x \leqq 1)$, be a continuous function enjoying the following property. If the quadratic form $\sum^{n} a_{ik}x_i x_k$, $(-1 \leqq a_{ii} \leqq 1; i = 1, \cdots, n; n = 2, 3, \cdots)$, is positive, the form $\sum_{1}^{n} F(a_{ik})x_i x_k$ should also be positive. Then $F(x)$ is necessarily of the form*

$$F(x) = \sum_{0}^{\infty} a_\nu x^\nu, \qquad (a_n \geqq 0, -1 \leqq x \leqq 1).$$

This result is implied by the fact that every element of $\mathfrak{P}(S_\infty)$ is of the form (1.6). Indeed, it suffices to assume $a_{11} = a_{22} = \cdots = a_{nn} = 1$ and to notice (i) that then $a_{ik} = \cos(p_i p_k)$, where $p_1, \cdots, p_n$ are n appropriate points of S_{n-1}, (ii) that the assumptions of our proposition imply that $F(\cos t)$ is positive definite in S_∞.

A second application of our results is as follows. The classical expansion of $(\cos t)^n$ in terms of Legendre polynomials has *positive* coefficients. The reason for this is now obvious in view of the relations $(\cos t)^n \in \mathfrak{P}(S_\infty) \subset \mathfrak{P}(S_2)$ and formula (1.3). We may predict in the same way that the expansion of $P_n^{(\lambda)}(\cos t)$ in terms of the polynomials $P_m^{(\mu)}(\cos t)$, $(m = 0, 1, \cdots; 0 \leqq \mu < \lambda)$, will likewise have positive coefficients, provided 2μ, 2λ are integers.

The fifth and last section of this paper is devoted to problems of isometric imbedding in Hilbert space. We determine all metric transforms of the sphere T_∞ which can be imbedded isometrically in H.

2. Some classical results concerning ultraspherical developments. We return to the spherical shell S_m, of E_{m+1}, defined by the equation $x^2 + x_1^2 + \cdots + x_m^2 = 1$. If p is an arbitrary point of S_m and $0 \leqq r < 1$, let p_r be that point on the radius $\overline{op}$ such that $op_r = r$. Furthermore, let $F(p)$ be a real and continuous point function defined in S_m and let[3]

$$(2.1) \qquad F(p_r) = \frac{1}{\omega_m} \int_{S_m} \frac{1 - r^2}{(1 - 2r \cos pp' + r^2)^{\frac{1}{2}(m+1)}} F(p')\, d\omega_{p'},$$

where $\omega_m = 2\pi^{\frac{1}{2}(m+1)}\Gamma^{-1}[\frac{1}{2}(m+1)]$ is the "area" of S_m and pp' denotes a spherical distance. For $m = 2$, (2.1) reduces to the ordinary Poisson integral which solves the boundary value problem for harmonic functions inside the sphere. A classical result concerning the Poisson integral (2.1) is the limiting relation

$$(2.2) \qquad \lim_{r \to 1} F(p_r) = F(p).$$

Differentiating (1.4) with respect to r and setting $t = pp'$, we get

$$(2.3) \qquad \frac{1 - r^2}{(1 - 2r \cos pp' + r^2)^{\lambda+1}} = \sum_{n=0}^{\infty} \left(\frac{n}{\lambda} + 1\right) r^n P_n^{(\lambda)}(\cos pp'), \quad (0 \leqq r < 1).$$

Let $\lambda = \frac{1}{2}(m - 1)$. Multiplying (2.3) by $F(p')\, d\omega_{p'}/\omega_m$ and integrating both sides over S_m we now get, in view of (2.1), the development[4]

$$(2.4) \qquad F(p_r) = \sum_{n=0}^{\infty} \frac{n + \lambda}{\lambda \omega_m} r^n \int_{S_m} F(p') P_n^{(\lambda)}(\cos pp')\, d\omega_{p'}, \qquad (\lambda = \tfrac{1}{2}(m - 1)).$$

The ultraspherical development of $F(p)$ is now given by

$$(2.5) \qquad F(p) \sim \sum_{n=0}^{\infty} \frac{n + \lambda}{\lambda \omega_m} \int_{S_m} F(p') P_n^{(\lambda)}(\cos pp')\, d\omega_{p'}.$$

Our relation (2.2) expresses the well-known fact that the ultraspherical expansion (2.5) is Abel-summable at every point p of S_m to the sum $F(p)$.

We shall now derive the special form of the expansion (2.5) for the case of a

[3] See [1], p. 198.
[4] See [1], formula (18), p. 207.

zonal function $F(p)$. For this purpose we express the cooordinates of the point $p = (x, x_1, \cdots, x_m)$ in terms of polar coordinates as follows:

$$
\begin{aligned}
x &= \cos\theta,\\
x_1 &= \sin\theta\cos\theta_1,\\
x_2 &= \sin\theta\sin\theta_1\cos\theta_2,\\
&\cdot\\
&\cdot\\
&\cdot\\
x_{m-2} &= \sin\theta\sin\theta_1\cdots\cos\theta_{m-2},\\
x_{m-1} &= \sin\theta\sin\theta_1\cdots\sin\theta_{m-2}\cos\phi, \qquad (0 \leqq \theta_\nu \leqq \pi, 0 \leqq \phi \leqq 2\pi),\\
x_m &= \sin\theta\sin\theta_1\cdots\sin\theta_{m-2}\sin\phi, \qquad (\nu = 0, 1, \cdots, m-2),
\end{aligned}
\tag{2.6}
$$

and consider the special case

$$F(p) = f(\cos\theta).$$

The integral on the right hand side of (2.5) now becomes

$$
\begin{aligned}
I_n = \int \overset{\pi}{\underset{0}{\cdots}} \int \int_0^{2\pi} f(\cos\theta')P_n^{(\lambda)}(\cos pp')\\
\cdot\sin^{m-1}\theta' \sin^{m-2}\theta_1' \cdots \sin\theta_{m-2}'\, d\theta' \cdots d\theta_{m-2}'\, d\phi'.
\end{aligned}
\tag{2.7}
$$

Integrating first with respect to the variables $\theta_1', \cdots, \theta_{m-2}', \phi'$, we get

$$I_n = \int_0^\pi f(\cos\theta')J_n \sin^{m-1}\theta'\, d\theta', \tag{2.8}$$

where we set

$$J_n = \int \overset{\pi}{\underset{0}{\cdots}} \int \int_0^{2\pi} P_n^{(\lambda)}(\cos pp') \sin^{m-2}\theta_1' \cdots \sin\theta_{m-2}'\, d\theta_1' \cdots d\theta_{m-2}'\, d\phi'. \tag{2.9}$$

This integral is now readily reducible to a simple integral as follows. Consider the two points p_1 and p_1' of polar coordinates

$$p_1 = (\tfrac{1}{2}\pi, \theta_1, \cdots, \theta_{m-2}, \phi), \qquad p_1' = (\tfrac{1}{2}\pi, \theta_1', \cdots, \theta_{m-2}', \phi'),$$

both lying in the unit sphere S_{m-1} defined by $\theta = \frac{1}{2}\pi$. Obviously

$$\cos pp' = \cos\theta\cos\theta' + \sin\theta\sin\theta'\cos p_1p_1'. \tag{2.10}$$

Now we notice that J_n as given by (2.9) amounts to an integration of $P_n^{(\lambda)}(\cos pp')$ over S_{m-1}. If we take in S_{m-1} a new system of polar coordinates ($\zeta = p_1p_1'$, $\zeta_1, \cdots, \zeta_{m-3}, \psi$), of pole p_1, we obtain

$$J_n = \int \overset{\pi}{\underset{0}{\cdots}} \int \int_0^{2\pi} P_n^{(\lambda)}(\cos pp') \sin^{m-2}\zeta \sin^{m-3}\zeta_1 \cdots \sin\zeta_{m-3}\, d\zeta \cdots d\zeta_{m-3}\, d\psi.$$

Since $\cos pp'$ depends only on ζ, as shown by (2.10), the remaining integrations may be carried out leading to the expression

$$J_n = \omega_{m-2} \int_0^\pi P_n^{(\lambda)}(\cos pp') \sin^{m-2} \zeta \, d\zeta.$$

Now, since[5]

$$\int_0^\pi P_n^{(\lambda)}(\cos pp') \sin^{2\lambda-1} \zeta \, d\zeta = \frac{\Gamma(\lambda)\Gamma(\frac{1}{2})\Gamma(n+1)\Gamma(2\lambda)}{\Gamma(\lambda+\frac{1}{2})\Gamma(n+2\lambda)} P_n^{(\lambda)}(\cos\theta)P_n^{(\lambda)}(\cos\theta'),$$

J_n is explicitly computed. Substituting its value into (2.8) we find that the general expansion (2.5) reduces to[6]

$$\begin{aligned} f(\cos\theta) \sim \sum_{n=0}^{\infty} \frac{(n+\lambda)\Gamma(\lambda)}{\Gamma(\lambda+\frac{1}{2})\Gamma(\frac{1}{2})} \frac{\Gamma(n+1)\Gamma(2\lambda)}{\Gamma(n+2\lambda)} P_n^{(\lambda)}(\cos\theta) \\ \cdot \int_0^\pi P_n^{(\lambda)}(\cos\theta') f(\cos\theta') \sin^{m-1}\theta' \, d\theta'. \end{aligned} \tag{2.11}$$

This expansion is, of course, also Abel-summable. It will be applied in the next section to functions $f(\cos\theta)$ which are positive definite in S_m.

3. **On positive definite functions in S_m.** Let $f(x)$ be real and continuous in the interval $-1 \leqq x \leqq 1$ and such that $f(\cos t)$ is p. d. in S_m. As stated in the introduction this amounts to the inequality

$$\sum_1^n f(\cos p_i p_k) x_i x_k \geqq 0, \qquad (p_i \in S_m,\ x_i \text{ real}). \tag{3.1}$$

According to W. H. Young this quadratic form requirement is equivalent to the integral inequality[7]

$$I(h) = \int_{S_m} \int_{S_m} f(\cos pp') h(p) h(p') \, d\omega_p \, d\omega_{p'} \geqq 0 \tag{3.2}$$

for an arbitrary continuous point function $h(p)$ in S_m.

A consequence of this re-statement is as follows. If $h(p) \equiv 1$, (3.2) becomes

$$I(1) = \int_{S_m} \left\{ \int_{S_m} f(\cos pp') \, d\omega_{p'} \right\} d\omega_p = \omega_m \int_{S_m} f(\cos pp') \, d\omega_{p'}$$

since the last integral is clearly independent of p. Since $I(1) \geqq 0$, we have proved that

$$\int_{S_m} f(\cos pp') \, d\omega_{p'} \geqq 0, \tag{3.3}$$

provided $f(\cos t)$ is p. d. in S_m.

[5] See [4], formula (1), p. 203.

[6] See [4], p. 198. We need below the fact that the expansion (2.11) is a special case of (2.5). Our derivation of (2.11) from (2.5) is well known as an analogue of a classical reduction. However, being unable to find it in the literature, I developed it here in detail.

[7] See [8].

A further result which we need is as follows. *The ultraspherical polynomials*

$$P_n^{(\lambda)}(\cos t), \quad (n = 0, 1, \cdots; \lambda = \tfrac{1}{2}(m-1)),$$

are all p. d. in S_m.

This statement is trivial for $m = 1$ $(\lambda = 0)$ in view of the cosine addition formula. Let $m \geqq 2$. In order to prove that $P_n^{(\lambda)}(\cos t)$ is p. d. in S_m we proceed by induction assuming $P_n^{(\lambda-\frac{1}{2})}(\cos t)$ to be p. d. in S_{m-1}. Suppose $p_i \in S_m$ $(i = 1, \cdots, N)$ and associate with the point p_i a point p_i', on the "equator" S_{m-1} of equation $\theta = \frac{1}{2}\pi$, such that the last $m - 1$ polar coordinates $\theta_1, \cdots, \phi$ of both points p_i and p_i' agree. As remarked before we have

$$\cos p_i p_k = \cos \theta^i \cos \theta^k + \sin \theta^i \sin \theta^k \cos p_i' p_k'.$$

By the addition formula for ultraspherical polynomials we may write[8]

$$P_n^{(\lambda)}(\cos p_i p_k) = \sum_{s=0}^{n} c_{n,\lambda,s} P_n^{\lambda,s}(\cos \theta^i) P_n^{\lambda,s}(\cos \theta^k) P_n^{(\lambda-\frac{1}{2})}(\cos p_i' p_k'),$$

where $P_n^{\lambda,s}$ are the real polynomials associated to $P_n^{(\lambda)}$ and $c_{n,\lambda,s}$ are *positive* coefficients whose values are here of no concern. But then

$$(3.4) \qquad \sum_{i,k=1}^{N} P_n^{(\lambda)}(\cos p_i p_k)\xi_i \xi_k = \sum_{s=0}^{n} c_{n,\lambda,s} \sum_{i,k=1}^{N} P_n^{(\lambda-\frac{1}{2})}(\cos p_i' p_k')\eta_i \eta_k \geqq 0,$$

where $\eta_i = P_n^{\lambda,s}(\cos \theta^i)\xi_i$. The expression (3.4) is indeed non-negative since $P_n^{(\lambda-\frac{1}{2})}(\cos t)$ was assumed to be p. d. in S_{m-1}. This completes our proof.

We may now establish the following theorem.[9]

THEOREM 1. *A necessary and sufficient condition in order that $f(\cos \theta)$ be positive definite in S_m is that the ultraspherical expansion* (2.11) *have non-negative coefficients in which case the series* (2.11) *converges throughout* $0 \leqq \theta \leqq \pi$ *absolutely and uniformly to the sum $f(\cos \theta)$. The most general $f(\cos \theta)$ which is p. d. in S_m is therefore given by the expansion*

$$(3.5) \qquad f(\cos \theta) = \sum_{n=0}^{\infty} a_n P_n^{(\lambda)}(\cos \theta), \qquad (a_n \geqq 0, \lambda = \tfrac{1}{2}(m-1)),$$

provided the series converges for $\theta = 0$.

Indeed, let $f(\cos \theta)$ be p. d. in S_m. The coefficient of $P_n^{(\lambda)}(\cos \theta)$ in (2.11) may be written as

$$\int_0^{\pi} P_n^{(\lambda)}(\cos \theta') f(\cos \theta') \sin^{m-1} \theta' \, d\theta' = \frac{1}{\omega_{m-1}} \int_{S_m} P_n^{(\lambda)}(\cos ap') f(\cos ap') \, d\omega_{p'},$$

where a is the point of S_m of coordinates $x = 1, x_1 = \cdots = x_m = 0$. But then the last integral is visibly positive for the following reason. Since $P_n^{(\lambda)}(\cos t)$ and $f(\cos t)$ are both p. d. in S_m, their product also enjoys this

[8] See [4], formula (6), p. 182.

[9] Recently Bochner has extended this theorem by means of the theory of generalized spherical harmonics of E. Cartan and H. Weyl. See [2], §§III and IV.

property. Now (3.3) shows that all coefficients of (2.11) are non-negative. We may therefore re-write (2.11) in the form

$$f(\cos\theta) \sim \sum_{n=0}^{\infty} a_n P_n^{(\lambda)}(\cos\theta), \qquad (a_n \geqq 0). \tag{3.6}$$

On the other hand, we know this series to be Abel-summable for all θ, hence, in particular, for $\theta = 0$. Thus

$$\sum_{n=0}^{k} a_n \mid P_n^{(\lambda)}(\cos\theta) \mid \leqq \sum_{n=0}^{k} a_n P_n^{(\lambda)}(1) \leqq \lim_{r\to 1} \sum_{0}^{\infty} a_n r^n P_n^{(\lambda)}(1) = f(1).$$

This shows that the series (3.6) is absolutely and uniformly convergent for all θ $(0 \leqq \theta \leqq \pi)$, hence convergent to its Abel-sum which is $f(\cos\theta)$.

The converse to the effect that the convergent series (3.5) defines a p. d. $f(\cos\theta)$ is clear. Indeed, $f(\cos\theta)$ is obviously continuous because the series (3.5) must converge uniformly. As $f(\cos\theta)$ thus appears as the continuous limit of a sequence of p. d. functions, it is p. d. itself.

4. **On positive definite functions in S_∞.** If $f(\cos\theta)$ is p. d. in S_∞ it is also p. d. in S_m. By Theorem 1 we are therefore assured to have an expansion with non-negative coefficients

$$f(\cos\theta) = \sum_{n=0}^{\infty} a_n(\lambda) P_n^{(\lambda)}(\cos\theta), \qquad (a_n(\lambda) \geqq 0,\ 0 \leqq \theta \leqq \pi), \tag{4.1}$$

which is valid for all values of λ of the form $\lambda = \frac{1}{2}(m-1)$, $(m = 1, 2, 3, \cdots)$. Setting

$$p_n^\lambda(\cos\theta) = \frac{P_n^{(\lambda)}(\cos\theta)}{P_n^{(\lambda)}(1)}, \tag{4.2}$$

we have a similar expansion

$$f(\cos\theta) = \sum_{n=0}^{\infty} b_n(\lambda) p_n^\lambda(\cos\theta), \qquad (b_n(\lambda) \geqq 0,\ 0 \leqq \theta \leqq \pi). \tag{4.3}$$

Since (see [4], p. 95)

$$\lim_{\lambda\to\infty} p_n^\lambda(\cos\theta) = \cos^n\theta, \tag{4.4}$$

we should expect to derive from (4.3), by letting $\lambda \to \infty$, the following theorem.

THEOREM 2. *A function $f(\cos\theta)$ which is positive definite in S_∞ is necessarily of the form*

$$f(\cos\theta) = \sum_{n=0}^{\infty} a_n \cos^n\theta, \qquad (a_n \geqq 0). \tag{4.5}$$

Now (4.4) obviously holds uniformly in θ. This fact, however, will not suffice to prove Theorem 2 due to our scant information concerning the coefficients

$b_n(\lambda)$ of (4.3). What we shall actually need is that, for a fixed value of θ, (4.4) holds uniformly for all n. This we state as a separate lemma.

LEMMA 1. *Let θ be such that $0 < \theta < \pi$. If $\epsilon > 0$ is arbitrarily small, we have*

$$(4.6) \qquad | p_n^\lambda(\cos \theta) - \cos^n \theta | < \epsilon \qquad \text{for all } n = 0, 1, 2, \cdots,$$

provided $\lambda > L(\theta, \epsilon)$.

Proof of the lemma.[10] From the known integral representation[11]

$$P_n^{(\lambda)}(\cos \theta) = \int_0^\pi (\cos \theta + i \sin \theta \cos \phi)^n \sin^{2\lambda-1} \phi \, d\phi,$$

we have

$$(4.7) \quad \Delta_n^\lambda = p_n^\lambda(\cos \theta) - \cos^n \theta = \int_0^\pi F_n(\theta, \phi) \sin^{2\lambda-1} \phi \, d\phi \bigg/ \int_0^\pi \sin^{2\lambda-1} \phi \, d\phi,$$

where

$$(4.8) \qquad F_n(\theta, \phi) = (\cos \theta + i \sin \theta \cos \phi)^n - \cos^n \theta.$$

Evidently

$$(4.9) \qquad | F_n(\theta, \phi) | \leqq 2.$$

Now we choose a $\delta = \delta(\theta)$ such that

$$(4.10) \qquad 0 < \delta < \tfrac{1}{2}\pi, \qquad \cos^2 \theta + \sin^2 \theta \sin^2 \delta < 1.$$

From (4.7), (4.8) and (4.9), we have

$$\begin{aligned} | \Delta_n^\lambda | \leqq{} & 4 \int_0^{\frac{1}{2}\pi-\delta} \sin^{2\lambda-1} \phi \, d\phi \bigg/ \int_0^\pi \sin^{2\lambda-1} \phi \, d\phi \\ & + \int_{\frac{1}{2}\pi-\delta}^{\frac{1}{2}\pi+\delta} | F_n(\theta, \phi) | \sin^{2\lambda-1} \phi \, d\phi \bigg/ \int_0^\pi \sin^{2\lambda-1} \phi \, d\phi \\ \leqq{} & 4 \int_0^{\frac{1}{2}\pi-\delta} \sin^{2\lambda-1} \phi \, d\phi \bigg/ \int_0^\pi \sin^{2\lambda-1} \phi \, d\phi \\ & + (\cos^2 \theta + \sin^2 \theta \sin^2 \delta)^{\frac{1}{2}n} + | \cos \theta |^n. \end{aligned}$$

Now let ϵ be given. Let $n_0 = n_0(\theta, \epsilon)$ be such that $n > n_0$ implies

$$(\cos^2 \theta + \sin^2 \theta \sin^2 \delta)^{\frac{1}{2}n} + | \cos \theta |^n < \tfrac{1}{2}\epsilon.$$

[10] This elegant proof of Lemma 1 is due to Professor Szegö. My original proof was more complicated and based on a new estimate which I shall state here since it might possibly be used elsewhere. For $0 < \theta < \frac{1}{2}\pi$, $\lambda \geqq 5$, and $n = 1, 2, 3, \cdots$ the following inequality holds

$$| P_n^{(\lambda)}(\cos \theta) | \leqq P_n^{(\lambda)}(1) \left\{ \left(\frac{1 + \cos^2 \theta}{2} \right)^{\frac{1}{2}n} + \frac{4}{n} \frac{1 + \cos^2 \theta}{\sin (2\theta)} \right\}.$$

[11] See [4], formulas (9) and (11), pp. 193–194.

The existence of such an n_0 is assured by (4.10). Furthermore, suppose $\lambda_0 = \lambda_0(\theta, \epsilon)$ is such that $\lambda > \lambda_0$ implies

$$4 \int_0^{\frac{1}{2}\pi-\delta} \sin^{2\lambda-1} \phi \, d\phi \Big/ \int_0^{\pi} \sin^{2\lambda-1} \phi \, d\phi < \tfrac{1}{2}\epsilon.$$

Hence $| \Delta_n^\lambda | < \epsilon$, provided $\lambda > \lambda_0(\theta, \epsilon)$, $n > n_0(\theta, \epsilon)$. On the other hand, from (4.4) we get that $| \Delta_n^\lambda | < \epsilon$ for $n = 0, 1, \cdots, n_0(\theta, \epsilon)$, provided $\lambda > \lambda_1(\theta, \epsilon)$. Now the lemma is proved if we take as $L(\theta, \epsilon)$ the larger of the two numbers λ_0 and λ_1.

Proof of Theorem 2. Our starting point is the expansion (4.3). Since $p_n^\lambda(1) = 1$ and hence

$$f(1) = \sum_{n=0}^{\infty} b_n(\lambda),$$

we see that the coefficients $b_n(\lambda)$ are uniformly bounded for all n and the range of values of λ. By Cantor's diagonal process we may find a subsequence $\lambda_\nu \to \infty$ such that

$$\lim_{\nu \to \infty} b_n(\lambda_\nu) = a_n \geqq 0, \qquad (n = 0, 1, 2, \cdots). \tag{4.11}$$

Let θ have a fixed value between 0 and π. Let us write the relation (4.3) in the form

$$f(\cos\theta) = \sum_0^\infty b_n(\lambda_\nu) \cos^n \theta + \sum_0^\infty b_n(\lambda_\nu) [p_n^{\lambda_\nu}(\cos\theta) - \cos^n \theta].$$

Since by the lemma we have

$$\left| \sum_0^\infty b_n(\lambda_\nu) [p_n^{\lambda_\nu}(\cos\theta) - \cos^n\theta] \right| < \epsilon \sum b_n(\lambda_\nu) = \epsilon f(1),$$

provided λ_ν is sufficiently large, we may write

$$f(\cos\theta) = \sum_0^\infty b_n(\lambda_\nu) \cos^n\theta + \sigma, \tag{4.12}$$

where $| \sigma | < \epsilon f(1)$ for sufficiently large λ_ν. However, the series

$$\sum_0^\infty b_n(\lambda_\nu) \cos^n \theta \tag{4.13}$$

converges uniformly with respect to the variable λ_ν because it is majorized by the convergent series with constant terms

$$\sum_0^\infty f(1) \, | \cos\theta |^n.$$

But now the limiting relations (4.11) imply that the series (4.13) will tend to $\sum a_n \cos^n \theta$ as $\lambda_\nu \to \infty$. Thus (4.12) may be written as

$$f(\cos\theta) = \sum_0^\infty a_n \cos^n \theta + \sigma',$$

where $|\sigma'| \leqq \epsilon f(1)$. Now letting $\epsilon \to 0$, we conclude that $\sigma' = 0$ and hence that

$$f(\cos\theta) = \sum_0^\infty a_n \cos^n \theta, \qquad (a_n \geqq 0). \tag{4.14}$$

There remains to show that (4.14) is also valid for $\theta = 0$ and $\theta = \pi$. This last point, however, is readily settled. Indeed, (4.14) implies the convergence of the series $\sum a_n$ by letting $\theta \to 0$. Now the continuity of both sides of the relation (4.14) at both ends of the interval $0 \leqq \theta \leqq \pi$ implies its validity throughout this closed interval.

5. **On metric transforms of S_∞ which are imbeddable in Hilbert space.** Suppose $F(t)$, $(0 \leqq t \leqq \pi)$, is continuous, $F(0) = 0$, $F(t) \geqq 0$ if $0 < t \leqq \pi$. Let us remetrize the metric space S_∞ from the original distance function pq to the new distance function $F(pq)$. The new semi-metric space thus obtained is called the metric transform of S_∞ by the function $F(t)$ and denoted by the symbol $F(S_\infty)$. We shall prove the following theorem.[12]

THEOREM 3. *The metric transform $F(S_\infty)$ is isometrically imbeddable in H if and only if*

$$F^2(t) = \sum_{n=1}^\infty a_n (1 - \cos^n t), \qquad (a_n \geqq 0,\ 0 \leqq t \leqq \pi). \tag{5.1}$$

We need the following two lemmas.

LEMMA 2. *The metric transform $F(S_\infty)$ is imbeddable in H if and only if the function* $\exp\{-\lambda F^2(t)\}$ *is positive definite in S_∞ for all $\lambda > 0$.*

This lemma is known to be valid not only for S_∞ but for any separable metric space ([6], Theorem 1, p. 527).

LEMMA 3. *Let K denote the class of functions $\phi(x)$ of the form*

$$\phi(x) = \sum_{n=1}^\infty a_n(1 - x^n), \qquad (a_n \geqq 0,\ -1 \leqq x \leqq 1). \tag{5.2}$$

If $\{\phi_n(x)\}$ is a sequence of functions of this class and

$$\lim_{n\to\infty} \phi_n(x) = \phi_0(x), \qquad (-1 \leqq x \leqq 1), \tag{5.3}$$

where $\phi_0(x)$ is continuous in $-1 \leqq x \leqq 1$, then also $\phi_0(x)$ belongs to the class K.

[12] A similar theorem concerning S_m is as follows. *$F(S_m)$ is imbeddable in H if and only if*

$$F^2(t) = \sum_{n=1}^\infty a_n(1 - p_n^\lambda(\cos t)), \qquad (\lambda = \tfrac{1}{2}(m-1),\ 0 \leqq t \leqq \pi).$$

However, this theorem is implied by a general result of Bochner ([2], Theorem 3) concerning a certain type of compact spaces, of which the simplest example is the finite dimensional sphere S_m. Our method of proving Theorem 3 was used before in the case of Euclidean and Hilbert spaces ([7], §5, [3], Part III).

Proof of Lemma 3. We prove first the following statement. Suppose

$$\psi(x) = \sum_0^\infty c_\nu x^\nu, \qquad (c_\nu \geqq 0, \ -1 < x < 1),$$

such that

$$\int_0^{1-0} \psi(x)\,dx \text{ exists.}$$

Then the class K is identical with the class K_1 of functions of the form

$$\phi(x) = \int_x^{1-0} \psi(x)\,dx, \qquad (-1 \leqq x < 1), \qquad \phi(1) = 0. \tag{5.4}$$

Proof that $K \subset K_1$. From $\phi(x) = \sum_1^\infty a_n(1 - x^n)$ we get $-\phi'(x) = \sum_1^\infty n a_n x^{n-1} = \psi(x)$ for $-1 < x < 1$; hence

$$\phi(x) - \phi(1-\epsilon) = \int_x^{1-\epsilon} \psi(x)\,dx.$$

Letting $\epsilon \to 0$ we get (5.4); hence $\phi(x) \in K_1$.

Proof that $K_1 \subset K$. Suppose $\phi(x) \in K_1$, and hence is of the form (5.4). We have

$$\phi(x) - \phi(1-\epsilon) = \int_x^{1-\epsilon} \left(\sum_0^\infty c_\nu x^\nu\right) dx = \sum_0^\infty \frac{c_\nu}{\nu+1}\,[(1-\epsilon)^{\nu+1} - x^{\nu+1}].$$

Letting $\epsilon \to 0$ we get $\phi(x) = \sum_0^\infty (1 - x^{\nu+1})c_\nu/(\nu+1)$; hence $\phi(x) \in K$.

Returning to the assumptions of Lemma 3, let

$$\phi_n(x) = \sum_{\nu=1}^\infty a_{n\nu}(1 - x^\nu).$$

If $x = re^{i\theta}$ $(0 \leqq r < 1)$, we have

$$\phi_n(x) = \sum_\nu a_{n\nu}(1 - r^\nu e^{i\nu\theta}) = \sum_\nu a_{n\nu}(1 - r^\nu) + \sum_\nu a_{n\nu} r^\nu(1 - e^{i\nu\theta});$$

hence

$$|\,\phi_n(x)\,| \leqq \phi_n(r) + 2\sum_{\nu=1}^\infty a_{n\nu} r^\nu. \tag{5.5}$$

But $r^\nu < A\cdot(1 - r^\nu)$ for all $\nu = 1, 2, \cdots$, for a certain $A = A(r)$. Thus (5.5) implies that

$$|\,\phi_n(x)\,| \leqq \phi_n(r) + 2A(r)\phi_n(r)$$

for $|\,x\,| = r$ and therefore also if $|\,x\,| \leqq r$. From (5.3) we now conclude that the $\phi_n(x)$ are uniformly bounded in every circle $|\,x\,| \leqq r$ $(r < 1)$. By the

Vitali-Porter convergence theorem, (5.3) implies that the $\phi_n(x)$ converge uniformly inside $|x| < 1$ and that $\phi_0(x)$ is therefore analytic and regular in $|x| < 1$. Since $\phi_n(x) \in K_1$, we know that $-\phi_n'(x)$ has all its derivatives non-negative at the origin. From $\phi_n^{(k)} \to \phi_0^{(k)}$ (uniformly inside $|x| < 1$) we conclude that also $-\phi_0'(x)$ has all derivatives non-negative at the origin; hence

$$\psi_0(x) = -\phi_0'(x) = \sum_0^\infty c_\nu x^\nu, \qquad (c_\nu \geqq 0, -1 < x < 1).$$

This also implies

$$\phi_0(x) - \phi_0(1 - \epsilon) = \int_x^{1-\epsilon} \psi_0(x)\, dx.$$

Since $\phi_0(1 - 0) = \phi_0(1) = 0$ we get, by letting $\epsilon \to 0$,

$$\phi_0(x) = \int_x^{1-0} \psi_0(x)\, dx.$$

Hence, $\phi_0(x) \in K_1 = K$ and our proof is completed.

Proof of Theorem 3. In order to prove the direct part of the theorem, let $F(t)$ be defined by (5.1). Now, for $\lambda > 0$,

$$\exp\{-\lambda F^2(t)\} = \exp\left\{-\lambda \sum_0^\infty a_n(1 - \cos^n t)\right\}$$

$$= \exp\left\{-\lambda \sum_0^\infty a_n\right\} \cdot \exp\left\{\lambda \sum_0^\infty a_n \cos^n t\right\}$$

evidently admits an expansion of the form

$$\exp\{-\lambda F^2(t)\} = \sum_0^\infty b_n(\lambda) \cos^n t, \qquad (0 \leqq t \leqq \pi,\quad b_n(\lambda) \geqq 0), \tag{5.6}$$

and is therefore p. d. in S_∞ by Theorem 2.

The converse part is proved as follows. Let $F(S_\infty)$ be imbeddable in H. By Lemma 2 and Theorem 2 the expansion (5.6) is valid for all $\lambda > 0$ and therefore

$$\Phi(t, \lambda) = \frac{1}{\lambda}(1 - \exp\{-\lambda F^2(t)\}) = \sum_0^\infty \lambda^{-1} b_n(\lambda)(1 - \cos^n t).$$

However, $\Phi(t, \lambda) \to F^2(t)$ as $\lambda \to 0$ and $\Phi(t, \lambda)$ is a function of class K if regarded as a function of the variable $x = \cos t$. As the limit function $F^2(t)$ is continuous throughout $0 \leqq t \leqq \pi$, Lemma 3 implies that $F^2(t)$ is of the desired form (5.1).

Remarks. 1. If in (5.1) we set $a_1 = 2$, $a_2 = a_3 = \cdots = 0$, we get $F_1^2(t) = 2(1 - \cos t) = 4 \sin^2(\frac{1}{2}t)$; hence

$$F_1(t) = 2 \sin \tfrac{1}{2}t.$$

By Theorem 3, $F_1(S_\infty)$ is imbeddable in H. This is a trivial result, for $F_1(S_\infty)$ is the sphere S_∞ with the spherical distance pq changed to the Euclidean chord-distance $2 \sin(\frac{1}{2}pq)$, which is already imbedded in H.

By taking only the second term of the expansion (5.1), we get $F_2^2(t) = 1 - \cos^2 t = \sin^2 t$; hence

$$F_2(t) = \sin t.$$

It should be noticed that this remetrization identifies diametrically opposite points on the sphere S_∞. This identification occurs for a $F(t)$ defined by (5.1) if and only if $a_n = 0$ whenever n is odd.

2. Let $F(S_\infty)$ be imbeddable in H. We want to show that $F(S_\infty)$ may also be placed in some appropriate spherical shell of H. Indeed, let us adjoin to the space $F(S_\infty)$ a new point A such that, if $p \in F(S_\infty)$, its distance to A is $d(A, p) = r$. We want to determine the constant r so that the space $F(S_\infty) + A$ can be imbedded in H. This requires (see [3], p. 229) that

$$\sum_{i,k=1}^{N} \{r^2 + r^2 - F^2(p_i p_k)\} x_i x_k \geqq 0$$

for arbitrary points $p_1, \cdots, p_N$ of S_∞. By (5.1) this amounts to

$$\sum_{i,k=1}^{N} \left\{2r^2 - \sum_{n=1}^{\infty} a_n(1 - \cos^n p_i p_k)\right\} x_i x_k \geqq 0.$$

If we set

$$r = \left(\frac{1}{2}\sum_{1}^{\infty} a_n\right)^{\frac{1}{2}}, \tag{5.7}$$

our last inequality reduces to

$$\sum_{n=1}^{\infty} a_n \sum_{i,k=1}^{N} \cos^n (p_i p_k) x_i x_k \geqq 0,$$

which is correct since $\cos^n t$ is p. d. in S_∞. We state this result as a

COROLLARY. *The metric transform $F(S_\infty)$ of Theorem 3, if imbedded isometrically in H, will lie on a spherical shell of H of radius* (5.7).

BIBLIOGRAPHY

1. P. APPELL ET J. KAMPÉ DE FÉRIET, *Fonctions hypergéométriques et hypersphériques. Polynomes d'Hermite*, Paris, 1926.
2. S. BOCHNER, *Hilbert distances and positive definite functions*, Annals of Mathematics, vol. 42(1941), pp. 647–656.
3. JOHN VON NEUMANN AND I. J. SCHOENBERG, *Fourier integrals and metric geometry*, Transactions of the American Mathematical Society, vol. 50(1941), pp. 226–251.
4. N. NIELSEN, *Théorie des fonctions métasphériques*, Paris, 1911.
5. G. PÓLYA UND G. SZEGÖ, *Aufgaben und Lehrsätze aus der Analysis*, vol. 2, Berlin, 1925.
6. I. J. SCHOENBERG, *Metric spaces and positive definite functions*, Transactions of the American Mathematical Society, vol. 44(1938), pp. 522–536.
7. I. J. SCHOENBERG, *Metric spaces and completely monotone functions*, Annals of Mathematics, vol. 39(1938), pp. 811–841.
8. W. H. YOUNG, *A note on a class of symmetric functions and on a theorem required in the theory of integral equations*, Messenger of Mathematics, vol. 40(1910), pp. 37–43.

UNIVERSITY OF PENNSYLVANIA.

Reprinted from the AMERICAN MATHEMATICAL MONTHLY
Vol. LX, No. 9, November, 1953

ON A THEOREM OF KIRZBRAUN AND VALENTINE

I. J. SCHOENBERG, University of Pennsylvania

THEOREM 1. *In the euclidean space* E_n *we are given a set of* m *spheres*

$$S_i: \quad |x - x_i| \leqq r_i, \qquad (i = 1, \cdots, m),$$

which have a point in common:

$$\bigcap_{i=1}^{m} S_i \neq 0.$$

If the spheres

$$S_i': \quad |x - x_i'| \leqq r_i, \qquad (i = 1, \cdots, m),$$

are such that

$$(1) \qquad |x_i' - x_j'| \leqq |x_i - x_j|, \qquad (i, j = 1, \cdots, m),$$

then

$$(2) \qquad \bigcap_{i=1}^{m} S_i' \neq 0.$$

In other words: If a set of m closed spheres in E_n have a common point then again they will have a common point if we displace their centers so as not to be further apart than they were before. This theorem is seen to be equivalent to the following:

THEOREM 2. *Let* x_i, x_i' $(i=1, \cdots, m)$ *be points in* E_n *satisfying* (1). *Let also the point* p *be given. Then there exists a point* p' *such that*

$$(3) \qquad |p' - x_i'| \leqq |p - x_i|, \qquad (i = 1, \cdots, m).$$

Theorem 2 implies Theorem 1, for if $p \in \bigcap S_i$ then, by (3), $|p' - x_i'| \leqq |p - x_i| \leqq r_i$ or $p' \in \bigcap S_i'$, which proves (2). It is equally obvious that Theorem 1 implies Theorem 2. This result is due to Kirzbraun [1]. It was rediscovered by Valentine [3], [4]. Mickle [2] gave a new proof. Valentine uses in his proof a theorem of Knaster, Kuratowski and Mazurkiewicz. Mickle uses certain quadratic forms which were used by the present writer to characterize E_n metrically. The following is a blend of the proofs of Valentine and Mickle which avoids the use of the auxiliary tools just mentioned.

Proof of THEOREM 2: We lose no generality by assuming that $p \neq x_i$ $(i=1, \cdots, m)$, for if $p = x_j$, then $p' = x_j'$ will satisfy (3). Consider the function

$$f(x) = \max_i \frac{|x - x_i'|}{|p - x_i|},$$

which is defined and continuous throughout E_n. Since $f(x) \to +\infty$ as $|x| \to \infty$,

$f(x)$ attains its absolute minimum λ. Let $x=p'$ be such that

(4) $$f(p') = \min_{x\in E_n} f(x) = \lambda,$$

and let us show that p' satisfies (3). If $\lambda=0$ then $p'=x_1'=\cdots=x_m'$ and (3) are trivially verified. Assume, therefore, that $\lambda>0$. Let

(5) $$\frac{|p'-x_i'|}{|p-x_i|}\begin{cases}=\lambda & \text{if } i=1,\cdots,k\\ <\lambda & \text{if } i=k+1,\cdots,m,\end{cases}\qquad (1\leqq k\leqq m),$$

as may always be achieved by a proper numbering of the points. Let $K(x_1',\cdots,x_k')$ denote the convex hull of the points $x_1',\cdots,x_k'$. We claim that

(6) $$p'\in K(x_1',\cdots,x_k').$$

Indeed, otherwise we could diminish all distances $|p'-x_i'|$, $(i=1,\cdots,k)$, by displacing p' so slightly that all m ratios (5) become less than λ. This, however, will contradict the definition (4) of λ. This result implies, in particular, that $k\geqq 2$, for $k=1$ would contradict our assumption that λ is positive.

In view of (5), the inequalities (3) will be established as soon as we can show that

(7) $$\lambda\leqq 1.$$

In order to prove (7) let us assume that

(8) $$\lambda>1$$

and see how we reach a contradiction. In terms of the vectors

$$R_i=x_i-p,\qquad R_i'=x_i'-p',\qquad (i=1,\cdots,k),$$

(5) and (8) imply the inequalities

(9) $$R_i'^2>R_i^2,\qquad (i=1,\cdots,k),$$

where R_i^2 denotes the scalar product of R_i by itself. Now (1) may be written as

(10) $$(R_i'-R_j')^2\leqq(R_i-R_j)^2,\qquad (i,j=1,\cdots,k).$$

Expanding (10) and subtracting appropriate inequalities (9) we have

(11) $$R_i'\cdot R_j'>R_i\cdot R_j,\qquad (i,j=1,\cdots,k).$$

On the other hand (6) shows that p' is a centroid of the points $x_1',\cdots,x_k'$. This means: There exist $c_i\geqq 0$ with $\sum c_i=1$ such that

$$p'=\sum_{i=1}^{k} c_ix_i'.$$

But then

$$0 = \sum c_i x_i' - p' = \sum c_i x_i' - \sum c_i p'$$

or

$$(12) \qquad \sum_{i=1}^{k} c_i R_i' = 0, \qquad c_i \geqq 0, \ \sum c_i = 1.$$

A contradiction is now easily reached: on multiplying (11) by $c_i c_j$ and summing over i and j we obtain the inequality

$$\left(\sum_1^k c_i R_i' \right)^2 > \left(\sum_1^k c_i R_i \right)^2$$

which is obviously impossible for its left side vanishes by (12) while its right side is non-negative.

References

1. M. D. Kirzbraun, Über die zusammenziehende und Lipschitzsche Transformationen, Fund. Math., vol. 22, 1934, pp. 77–108.

2. E. J. Mickle, On the extension of a transformation, Bull. Amer. Math. Soc., vol. 55, 1949, pp. 160–164.

3. F. A. Valentine, On the extension of a vector function so as to preserve a Lipschitz condition, Bull. Amer. Math. Soc., vol. 49, 1943, pp. 100–108.

4. F. A. Valentine, A Lipschitz preserving extension for a vector function, Amer. J. Math., vol. 67, 1945, pp. 83–93.

Metric geometry

As he notes in 'A brief account of my life and work,' Schoenberg's first paper [**16***] on the subject of isometric embeddings of metric spaces was motivated by [**F**] and the main ideas in [**16***] can be seen to be prevalent in [**23**], [**25*–29***] and [**43***]. Schoenberg's work on metric geometry has gained its widest recognition through applications to other areas of mathematics. We examine some of Schoenberg's ideas and their wide ranging impact.

1. Linear Algebra. The real symmetric $n \times n$ matrix $A = (a_{ij})$ is a **distance matrix** if $a_{ii} = 0$ for all i and $a_{ij} > 0$ for all $i > j$; and is Euclidean if the a_{ij} are the distances between n points in a Euclidean space. Schoenberg's main result in [**16***] is that a distance matrix A is Euclidean if and only if the matrix $F = (I - uv^t)A(I - vu^t)$ is positive semi-definite, where $u = (1, \ldots, 1) \in \mathbb{R}^d$ and $v = n^{-1}u$ or $v = e_i$, a unit vector with zero everywhere except at the ith position. [**Go**] and [**Ma**] generalize Schoenberg's condition, explain its geometrical significance, and relate Schoenberg's results to the **statistical problems of multidimensional scaling and clustering.** One problem there is to find the best Euclidean fit to a given matrix A; that is, to determine an $n \times n$ Euclidean matrix $\hat{A}$ such that, with rank$(\hat{A})$ specified, $\|A - \hat{A}\|$ is minimal for a suitable matrix norm $\|\cdot\|$. Schoenberg's characterization of Euclidean matrices leads in many situations to the explicit determination of the minimizing matrix $\hat{A}$.

2. Probability Theory. In [**26***], Schoenberg proves that if $\psi : \mathbb{R}_+ \to \mathbb{R}$ is such that $\psi \circ \|\cdot\|_n (\|\cdot\|_n$ is the Euclidean norm on $\mathbb{R}^n)$ is the Fourier transform of a probability measure on $\mathbb{R}^n$, then ψ can be expressed in terms of Hankel transforms. Using this result, [**CHS**] develops **stochastic representations** for random vectors whose distributions are **elliptically contoured** and uses them to extend to this larger class results which hold for normal distributions.

In [**26***], Schoenberg also characterizes the functions $\psi : \mathbb{R}_+ \to \mathbb{R}$ such that $\psi(\|x - y\|)$ is positive definite on an infinite dimensional Hilbert space. In probabilistic terms this implies that an infinite sequence of random variables is spherically distributed if and only if it is a scale mixture of independent and identically distributed normal variables. This result was extended to infinite dimensional L^p spaces in [**BCK**] and has important implications for stable random processes. Further extensions of Schoenberg's ideas appear in [**CHS**], [**CKS**] and [**R**]. Another probabilistic interpretation of this result of Schoenberg is that if ψ is continuous and $\psi(0) = 1$ then $\psi(\sqrt{s})$ is the Laplace transform of a probability measure on $\mathbb{R}_+$. [**L**] reformulates this result in the context of **Gel'fand pairs**, and relates the more general question of solving these **problems of Schoenberg-type** with the theory of spherical functions.

3. Stochastic Processes. As each positive definite function (pdf) determines a Gaussian process with mean zero, Schoenberg's work on pdf's on various spaces has significant ramifications in the theory of stochastic processes and random fields. From [**25***], it follows that the kernel $K(x,y) = \|x\|_n + \|y\|_n - \|x - y\|_n$, $x, y \in \mathbb{R}^n$, is a pdf. K is the covariance function of Lévy's Brownian motion of several parameters. Generalizing this example, [**Ga**] introduces the class of **Lévy-Schoenberg kernels** on homogeneous spaces, and studies the Gaussian processes with covariance a Lévy-Schoenberg kernel, thus generalizing multiparameter Brownian motion (see also [**AB**]). Other generalizations appear in [**He**] (where much of the geometrical viewpoint is adopted), [**K**], [**P**] and [**V**].

The results in [**28***] provide spectral representations for stochastic processes with stationary increments and for their incremental covariance function. This is closely related to work by Kolmogorov [**K**] and is generalized in [**M**].

Schoenberg's characterization in [**26***] of pdf's on $\mathbb{R}^n \times \mathbb{R}^n$ which depend only on the distance between their arguments (referred to in §2) provides a representation of the covariance function of a homogeneous and isotropic random field in an n-dimensional Euclidean space (see [**Y**]). Schoenberg's simple representation of isotropic pdf's on a sphere in an infinite dimensional space in [**29***] sparked the study of isotropic Gaussian processes on such a sphere in [**Be**].

4. Harmonic Analysis on Semigroups. Perhaps the most complete generalization of Schoenberg's work on positive definiteness and complete monotonicity appears in [**BCR; Chapter 5**], where it is shown that Schoenberg's ideas permit a natural formulation in the setting of semigroups: The notion of a **Schoenberg triple** is formulated and used to develop for semigroups a generalization of the result in [**25***] which was stated in §3. Another connection with probability appears in [**BCR; Chapter 7**] where negative definite functions on semigroups are used to generalize the fundamental inequalities of Hoeffding [**Ho**].

5. Spectral Synthesis. In [**26***] Schoenberg proved that Hankel transforms have certain differentiability properties. This has inspired the results [**Me**] that for $n \geq 3$, the unit sphere S^{n-1} is not a set of spectral synthesis for the Fourier algebra $L^2(\mathbb{R}^n) * L^2(\mathbb{R}^n)$; and that Schoenberg's observations about the Hankel transforms have analogs for the inverse Jacobi transforms, which are then used to analyze sets of spectral synthesis in the setting of symmetric spaces.

REFERENCES

[**AB**] R. Askey and N. H. Brigham, Gaussian processes on compact symmetric spaces, *Z. Wahrsch. verw. Gebiete* **37** (1976), 127–143.

[**BCR**] C. Berg, J. P. R. Christensen and P. Ressel, Harmonic Analysis on Semigroups, Springer-Verlag, New York (1984).

[**Be**] S. Berman, Isotropic Gaussian processes on the Hilbert sphere, *Ann. Probab.* **8** (1980), 1093–1106.

[**BCK**] J. Bretagnolle, D. Dacunha Castelle, et. J.-L. Krivine, Lois stables et spaces L^p, *Ann. Inst. H. Poincaré, Sect. B* **2** (1966), 231–259.

[**CHS**] S. Cambanis, S. Huang and G. Simons, On the theory of elliptically contoured distributions, *J. Multivariate Anal.* **11** (1981), 368–385.

[**CKS**] S. Cambanis, R. Keener and G. Simons, On α-symmetric multivariate distributions, *J. Multivariate Anal.* **13** (1983), 213–233.

[**F**] M. Fréchet, Sur la definition axiomatique d'une classe d'espaces vectoriels distancies applicables vectoriellement sur l'espace de Hilbert, *Ann. of Math.* **36** (1935), 705–718.

[**Ga**] R. Gangolli, Positive definite kernels on homogeneous spaces and certain stochastic processes related to Lévy's Brownian motion parameters, *Ann. Inst. H. Poincaré, Sect. B* **3** (1967), 121–225.

[**Go**] J. C. Gower, Properties of Euclidean and non-Euclidean matrices, *Linear Algebra Appl.* **67** (1985), 81–97.

[**He**] C. S. Herz, A class of negative definite functions, *Proc. Amer. Math. Soc.* **14** (1963), 670–676.

[Ho] W. Hoeffding, On the distribution of the number of successes in independent trials, *Ann. Math. Statist.* **27** (1956), 713–721.
[Ko] A. N. Kolmogorov, Wiener spirals and some other interesting curves in Hilbert spaces, *Dokl. Akad. Nauk SSSR* **26** (1940), 115–118.
[Ku] J. Kuelbs, Positive definite symmetric functions on linear spaces, *J. Math. Anal. Appl.* **42** (1973), 413–426.
[L] G. Letac, Problemes classiques de probabilite sur un couple de Gel'fand, *Lecture Notes in Math.* **861** (1981), 93–120.
[M] P. Masani, On helixes in Hilbert space, I, *Theoret. Probab. Appl.* **17** (1972), 1–19.
[Ma] R. Mathar, The best Euclidean fit to a given distance matrix in prescribed dimensions, *Linear Algebra Appl.* **67** (1985), 1–6.
[Me] C. Meaney, Spherical functions and spectral synthesis, *Compositio Math.* **54** (1985), 311–329.
[P] D. Pickrell, On generalizing a theorem of Schoenberg, preprint, Yale University (1986).
[R] D. St. P. Richards, Positive definite symmetric functions on finite dimensional spaces. I. Applications of the Radon transform, *J. Multivariate Anal.* **19** (1986), 280–298.
[V] A. M. Vershik, Description of invariant measures for the actions of some infinite dimensional groups, *Soviet Math. Dokl.* **15** (1974), 1398.
[Y] A. M. Yaglom, Stationary Random Functions, Prentice-Hall, Englewood Cliffs, New Jersey (1962).

Stamatis Cambanis
Donald Richards
University of North Carolina
at Chapel Hill

Commentary on a footnote

Mathematicians often find that the part of a paper that interests them most is a lemma rather than the main theorem. However it is unusual when the part of the paper that interests them most is a footnote. This happened to me when I read Schoenberg's paper [**29***] on positive definite functions on spheres. The footnote is the second one, and the last sentence is the one that interested me. Here Schoenberg claimed that the expansion of $P_n^{(\lambda)}(\cos t)$ in terms of $P_m^{(\mu)}(\cos t)$ will have nonnegative coefficients when $0 \leq \mu < \lambda$ and 2λ and 2μ are integers. This was interesting, and should continue to hold for real $\lambda, \mu, 0 \leq \mu < \lambda$. It took me a few years to prove this by finding the coefficients explicitly. Shortly thereafter I read an old paper by Gegenbauer [**Ge**], and this explicit formula was given there. This result of Gegenbauer had been forgotten, and it was not included in any handbooks. Shortly before this Hua [**H**] had rediscovered Gegenbauer's identity and used it to find the L_2 norm of an important complete orthogonal system on one of the Cartan domains, the Lie spheres. Most of the applications of Gegenbauer's formula since then have just used the nonnegativity of the coefficients.

The first of these applications is in [**AW**]. After Wainger and I found this application, and the nonnegativity of the coefficients was crucial, I looked for a theorem that would give the nonnegativity for two fairly general sets of orthogonal polynomials. An easy result was given in [$\mathbf{A}_1$], and more importantly a conjecture was made. This conjecture was proven by Trench [**T**], when he showed that the assumptions in a nice result of Wilson [**W**] are satisfied in my conjecture. Micchelli [**M**] then gave necessary and sufficient conditions for the assumptions in Wilson's theorem to be satisfied.

After seeing how useful Gegenbauer's formula could be, I included it in a course on special functions that was taught in 1967. George Gasper audited this course. A few years later he used the nonnegativity of the coefficients in this expansion and two other better known facts to prove the nonnegativity of a sequence of ${}_3F_2$'s (generalized hypergeometric functions). See [$\mathbf{AG}_1$]. This ${}_3F_2$ arose while he was trying to prove the conjectured nonnegativity of a quadrature scheme of Cotes type. This was a minor application, so we were very surprised by another application of this inequality. L. de Branges reduced the Bieberbach conjecture to proving the positivity of a sequence of integrals of Jacobi polynomials. This sequence of polynomials is a special case of the ${}_3F_2$'s that Gasper had proved are positive. See de Branges's proof [$\mathbf{dB}_1$], and [$\mathbf{dB}_2$], [**Ga**], [$\mathbf{A}_2$] for comments on the discovery of this proof and the surprising connection with special functions.

The work of de Branges was so important that someone would have found a proof of the inequality he needed. While there are other proofs, (see [$\mathbf{AG}_2$] for references to three other proofs by Gasper and Koornwinder [**K**] has another proof), it is likely that Gasper's original proof would be the one found. The only part of Gasper's proof that does not use formulas in handbooks is Gegenbauer's formula. Surprisingly the case $2\mu + 1 = 2\lambda = 1, 2, \ldots$ is sufficient to prove the needed inequality, and here Schoenberg's comment suffices. Gasper and I mentioned this in [$\mathbf{AG}_2$]. Koornwinder picked up on this remark and showed [**K**] that integration over an appropriate group leads to a new representaion of solutions of the de Branges differential-difference equation and of the required nonnegativity. Schoenberg's comment continues to lead to new results. Very few footnotes have that staying power.

REFERENCES

[A_1] R. Askey, Orthogonal expansions with positive coefficients, *Proc. Amer. Math. Soc.* **16** (1965), 1191–1194.

[A_2] R. Askey, My reaction to de Branges's proof of the Bieberbach conjecture, Proceedings of the Symposium on the Occasion of the Proof, *Math. Surveys Monogr.* **21** (1986), 213–215.

[AG_1] R. Askey and G. Gasper, Positive Jacobi polynomial sums. II, *Amer. J. Math.* **98** (1976), 709–737.

[AG_2] R. Askey and G. Gasper, Inequalities for polynomials, Proceedings of the Symposium on the Occasion of the Proof, *Math. Surveys Monogr.* **21** (1986), 7–32.

[AW] R. Askey and S. Wainger, A transplantation theorem for ultraspherical coefficients, *Pacific J. Math.* **16** (1966), 393–405.

[Ba] A. Baernstein, II, et al., eds., The Bieberbach Conjecture, Proceedings of the Symposium on the Occasion of the Proof, *Math. Surveys Monogr.* **21** (1986).

[dB_1] L. de Branges, A proof of the Bieberbach conjecture, *Acta Math.* **154** (1985), 137–152.

[dB_2] L. de Branges, The story of the verification of the Bieberbach conjecture, Proceedings of the Symposium on the Occasion of the Proof, *Math. Surveys Monogr.* **21** (1986), 199–203.

[Ga] W. Gautschi, Reminiscences of my involvement in de Branges's proof of the Bieberbach conjecture, Proceedings of the Symposium on the Occasion of the Proof, *Math. Surveys Monogr.* **21** (1986), 205–211.

[Ge] L. Gegenbauer, Zur Theorie der Functionen $C_n^\nu(x)$, *Denkschriften Akad. Wissen. Wien, Math. Klasse* **48** (1884), 293–316.

[H] L. K. Hua, Harmonic Analysis of Functions of Several Complex Variables in the Complex Domains, Chinese original, Science Press, Peking, 1958, English translation, Amer. Math. Soc., 1963.

[K] T. Koornwinder, A group theoretic interpretation of the last part of de Branges' proof of the Bieberbach conjecture, *Complex Variables* **6** (1986), 309–321.

[M] C. A. Micchelli, A characterization of M. W. Wilson's criterion for nonnegative expansions of orthogonal polynomials, *Proc. Amer. Math. Soc.* **71** (1978), 69–72.

[T] W. F. Trench, Proof of a conjecture of Askey on orthogonal expansions with positive coefficients, *Bull. Amer. Math. Soc.* **81** (1975), 954–956.

[W] M. W. Wilson, Nonnegative expansions of polynomials, *Proc. Amer. Math. Soc.* **24** (1970), 100–102.

Richard Askey
University of Wisconsin, Madison

REAL ANALYSIS

ON THE PEANO CURVE OF LEBESGUE*

I. J. SCHOENBERG

Let $f(t)$ be the *even* continuous function of period two which is defined in the interval $(0, 1)$ as follows: $f(t)=0$ in $(0, 1/3)$, $f(t)=1$ in $(2/3, 1)$, and $f(t)$ is linear in $(1/3, 2/3)$. Our curve is defined by the parametric equations

$$
(1) \qquad \begin{aligned} x(t) &= \frac{1}{2} f(t) + \frac{1}{2^2} f(3^2 t) + \frac{1}{2^3} f(3^4 t) + \cdots, \\ y(t) &= \frac{1}{2} f(3t) + \frac{1}{2^2} f(3^3 t) + \frac{1}{2^3} f(3^5 t) + \cdots, \qquad 0 \leqq t \leqq 1. \end{aligned}
$$

The inequalities $0 \leqq f(t) \leqq 1$ imply $0 \leqq x(t) \leqq 1$, $0 \leqq y(t) \leqq 1$, as well as the uniform convergence of both series (1), and hence imply the continuity of $x(t)$, $y(t)$. All there remains to show is that our curve will pass through an arbitrarily given point

$$
(2) \quad x_0 = \frac{a_0}{2} + \frac{a_2}{2^2} + \frac{a_4}{2^3} + \cdots, \quad y_0 = \frac{a_1}{2} + \frac{a_3}{2^2} + \frac{a_5}{2^3} + \cdots, \quad a_\nu = 0, 1,
$$

of the square $0 \leqq x, y \leqq 1$, whose coordinates are given by their binary expansions. Indeed, let

$$
(3) \qquad t_0 = \frac{2a_0}{3} + \frac{2a_1}{3^2} + \frac{2a_2}{3^3} + \cdots + \frac{2a_{k-1}}{3^k} + \frac{2a_k}{3^{k+1}} + \cdots.
$$

If $a_0=0$, we have $0 \leqq t_0 \leqq 2/3^2 + 2/3^3 + \cdots = 1/3$, hence $f(t_0)=0$; if $a_0=1$, we have $2/3 \leqq t_0 \leqq 2/3+1/3=1$, hence $f(t_0)=1$. In either case $f(t_0)=a_0$. Similarly $3^k t_0 =$ even integer $+ 2a_k/3 + 2a_{k+1}/3^2 + \cdots$ shows that

$$
(4) \qquad f(3^k t_0) = a_k, \qquad k = 0, 1, 2, \cdots.
$$

Now (1), (2), and (4) imply $x(t_0)=x_0$, $y(t_0)=y_0$.†

COLBY COLLEGE

* Presented to the Society, April 16, 1938.

† Lebesgue, *Leçons sur l'Intégration*, Paris, 1928, pp. 44–45, defines the functions $x(t)$, $y(t)$, first on Cantor's ternary set T, of points t_0 of the form (3), by means of the equations (2). Having proved their continuity on T, Lebesgue extends their definition throughout (0, 1) by linear interpolation over each one of the denumerable set of open intervals of which the set complementary to T is built up. It should be remarked that our curve (1) coincides with Lebesgue's curve within Cantor's set T but not on the complement of T.

Reprinted from *Bull. Amer. Math. Soc.* **44** (1938), p. 519

AN ISOPERIMETRIC INEQUALITY FOR CLOSED CURVES CONVEX IN EVEN-DIMENSIONAL EUCLIDEAN SPACES*

BY

I. J. SCHOENBERG

in Philadelphia, Pa.

Introduction

1. The main theorem. A closed convex curve in the plane E_2 is usually defined as the boundary of a compact convex set.[1] Alternatively, if the curve is given in parametric form we could say that the curve is *convex* provided it never crosses a straight line more than twice. The second definition has the advantage of extending in a natural way to closed curves in an even-dimensional space E_{2n} as follows:

Let

$$(1) \qquad C: \quad x_i = x_i(t), \qquad (i=1, \ldots, 2n; \ 0 \le t \le 2\pi),$$

where $x_i(t)$ are continuous functions of period 2π, be a closed curve in E_{2n}. We shall say that C is *convex in* E_{2n} provided that it never crosses a hyperplane more than $2n$ times. If C is convex in E_{2n} and spans the space E_{2n}, i.e. is not contained in a lower-dimensional flat space, then we shall say that C is *convex on* E_{2n}. It will be shown below (Article 5) that curves convex in E_{2n} are rectifiable. As an example of a curve convex on E_{2n} we mention the curve

$$(2) \qquad C_0: \quad x_1 = \cos t, \ x_3 = \tfrac{1}{2} \cos 2t, \ \ldots, \ x_{2n-1} = \frac{1}{n} \cos nt,$$

$$x_2 = \sin t, \ x_4 = \tfrac{1}{2} \sin 2t, \ \ldots, \ x_{2n} = \frac{1}{n} \sin nt, \qquad (0 \le t \le 2\pi).$$

Indeed, C_0 is convex in E_{2n}, for if $l(x_1, \ldots, x_{2n})$ is any linear function and if we substitute the x_i as defined by (2), we find that $l = T_n(t)$ is a real trigonometric

* This work was performed on a National Bureau of Standards contract with the University of California, Los Angeles, and was sponsored (in part) by the Office of Scientific Research, USAF.

[1] See [1], page 3, in the list of references at the end of this paper.

10 – 533807. *Acta mathematica*. 91. Imprimé le 27 octobre 1954.

Reprinted from *Acta Math.* **91** (1954), pp. 143–164

polynomial of order n which is known never to change sign more than $2n$ times within a period.

The main result is the following theorem

I. *Let C be a closed curve convex in E_{2n} and let L denote its length. Let $K = K(C)$ be the convex hull of C and let $V = V(K)$ denote the $2n$-dimensional volume of K. Then the following inequality holds*

$$L^{2n} \geq (2\pi n)^n \, n! \, (2n)! \, V(K), \tag{3}$$

with the equality sign if and only if the curve C agrees up to a rigid motion and a similitude, followed perhaps by a reflexion, with the curve C_0 [2] *defined by* (2).

If $n=1$, (3) reduces to the classical isoperimetric inequality $L^2 \geq 4\pi V$ for curves convex in the plane, V now denoting the area enclosed by C, and where equality holds only if C is a circle. The inequality (3) shows that among all closed curves convex in E_{2n} and of given length L, only those which are similar to C_0 will maximize the volume $V(K(C))$ of their convex hull.

2. Two related theorems. Theorem I will follow from the following theorem in which there is no reference to convex curves:

II. *Let $x_i(t)$, $(i = 1, \ldots, 2n)$, be absolutely continuous functions of period 2π, not all being constant and such that $x_i'(t) \in L_2(0, 2\pi)$. Then the following inequality holds*

$$\left\{\int_0^{2\pi} \left(\sum_1^{2n} x_i'^2\right) dt\right\}^n \geq n^n \int_0^{2\pi} \cdots \int_0^{2\pi} \det \| x_i(t_1), x_i'(t_1), x_i(t_2), x_i'(t_2), \ldots, x_i(t_n), x_i'(t_n) \| \, dt_1 \ldots dt_n, \tag{4}$$

the integrand of the right-hand side being a determinant of order $2n$ whose ith row is written out. Moreover, we have the equality sign if and only if the functions $x_i(t)$ arise from the special set of functions (2) *by a right-handed orthogonal transformation followed by a similitude with positive ratio and a translation.*

This theorem will be established by means of Fourier series, a method first used by A. Hurwitz [3] for the special case when $n = 1$. In this classical case, Fourier expansions reduce the problem to a quadratic inequality easily established by completing squares. It will be seen by means of Fourier expansions that Theorem II is equivalent to the following theorem

---

[2] The curve C_0 is a simple example of a closed screw-line in E_{2n} (see [10]). A curve C_1, related to C_0 by similitudes, and the corresponding $K(C_1)$, play an important role in CARATHÉODORY's paper [2] and also in the so-called trigonometric moment problem.

[3] See [8].

III. *Let*

(5) $$M = \| a_{i1}, b_{i1}, a_{i2}, b_{i2}, \ldots, a_{im}, b_{im}, \ldots \|, \quad (i = 1, \ldots, 2n),$$

be a real matrix of $2n$ *rows and infinitely many columns,* $M \neq 0$, *and such that the sum of the squares of the elements in every row converges. Let*

(6) $$D(j_1, j_2, \ldots, j_n) = \det \| a_{ij_1}, b_{ij_1}, a_{ij_2}, b_{ij_2}, \ldots, a_{ij_n}, b_{ij_n} \|,$$
$$(1 \leq j_1 < j_2 < \cdots < j_n).$$

Setting

(7) $$S = \sum_{\nu=1}^{\infty} \sum_{i=1}^{2n} (a_{i\nu}^2 + b_{i\nu}^2)$$

and

(8) $$\Phi = \sum_{j_1 < \cdots < j_n} \frac{1}{j_1 j_2 \cdots j_n} D(j_1, j_2, \ldots, j_n),$$

then the last series converges absolutely and we have the inequality

(9) $$S^n \geq (2n)^n \, n! \, \Phi,$$

with equality holding if and only if the matrix (5) *has only zero elements, except in its first* $2n$ *columns which form a square matrix with elements positively proportional to a right-orthogonal matrix.*

The section-headings describe the contents of the paper. The discussion starts with an account of indispensable properties of convex polygons and curves.

§ 1. On closed curves convex in E_{2n} and the volumes of their convex hulls

3. On convex polygons. Let

(1.1) $$\Pi = P_0 P_1 \ldots P_k, \quad P_i = (x_{i1}, x_{i2}, \ldots, x_{im})$$

be a polygon with vertices in E_m. We assume that Π spans E_m, which is the case provided the matrix

(1.2) $$X = \| 1, x_{i1}, x_{i2}, \ldots, x_{im} \|, \quad (i = 0, \ldots, k; \ k \geq m),$$

is of rank $m+1$. We introduce the following

Definition 1. *We say that the polygon* Π *is convex on* E_m *provided it spans* E_m *and crosses no hyperplane more than m times. If in this definition we do not require that* Π *should span* E_m, *then we say that* Π *is convex in* E_m.

The convexity of Π in (or on) E_m evidently means the following: If $l(P) = l(x_1, \ldots, x_m)$ is an arbitrary linear function of the coordinates, then in the sequence of numbers $l(P_0), l(P_1), \ldots, l(P_k)$, we should have at most m changes of signs. Now the important difference between "convexity in E_m" and "convexity on E_m" becomes apparent. Indeed, for "convexity on E_m" the following lemma is identical, except in terminology, to a theorem of F. Gantmakher and M. Krein[4]:

Lemma 1. *The polygon* Π *is convex on* E_m *if and only if the matrix* X, *defined by* (1.2) *is of rank* $m+1$ *and all its non-vanishing minors of order* $m+1$ *are of the same sign.*

This lemma seems intuitively clear for the case of the plane E_2, for it says that the convexity of Π on E_2 requires that no two among the triangles $P_\alpha P_\beta P_\gamma$ $(\alpha<\beta<\gamma)$ should have opposite orientations.

We are now confronted with a basic distinction depending on the parity of the dimension number m. If *m is even* we readily see that if Π is convex on E_m, then also the closed polygon

$$\Pi_1 = P_0 P_1 \ldots P_k P_0$$

is convex on E_m. Indeed, if $\pi\,(l(\mathrm{P})=0)$ is a hyperplane and Π crosses π less than m times, then also Π_1 will not cross π more than m times. However, if Π crosses π exactly m times, then P_0 and P_k are never on opposite sides of π (m being even), so that the last side $P_k P_0$ does not cross π. For the closed polygon Π_1 it does not matter which vertex is taken to be the first, as long as the correct cyclic order of the vertices is preserved. The convexity of Π_1 may also be described in a way which ignores ab initio which vertex might be the first: Π_1 should never cross π more than m times, as we go once around the polygon. Notice that the number of such crossings is always even. The same indifference to cyclic permutations is also shared by the criterion of Lemma 1: Cyclic permutations of the rows of the matrix X will not change the common sign of its minors of the *odd* order $m+1$.

The situation is quite different if *m is odd.* In this case it can be shown that the first and the last vertex of a polygon, convex on E_m, can never coincide.

4. On convex curves. The above definitions and results extend readily to continuous arcs and closed curves. In view of our particular aim we restrict the discussion to even dimensions $m=2n$ and closed curves. The basic definition is as follows:

---

[4] See their recent book [4], Theorem 3, page 297. Also proved in [14], Theorem 1. ANNE WHITNEY and the author were unaware of the book by GANTMAKHER and KREIN when [14] was published. However, the priority clearly belongs to the Russian authors. See also footnote 5 for the connection with the work of J. HJELMSLEV.

Definition 2. *The closed curve*

$$(1.3)\qquad C: \quad x_i = x_i(t) \quad (i=1, \ldots, 2n;\ 0 \leq t \leq 2\pi),$$

where the continuous functions $x_i(t)$, defined for all t, have the period 2π, is said to be convex on E_{2n} provided C spans E_{2n} and all closed polygons inscribed in C are convex in E_{2n}. If in this definition we do not require that C should span E_{2n}, then we say that C is convex in E_{2n}.[5]

A criterion is given by

Lemma 2. *The closed curve C, defined by* (1.3), *is convex on E_{2n} if and only if the determinants*

$$(1.4)\qquad \Delta = \det \| 1, x_1(t_i), \ldots, x_{2n}(t_i) \| \quad (t_0 < t_1 < \cdots < t_{2n} < t_0 + 2\pi)$$

do not all vanish and the non-vanishing ones have the same sign.

Proof. Let us inscribe in C a closed polygon $\Pi = P_0 P_1 \ldots P_k P_0$ corresponding to the parameter values

$$(1.5)\qquad t_0 < t_1 < \cdots < t_k < t_0 + 2\pi.$$

The conditions are necessary. Assume C convex on E_{2n}. Let Π span E_{2n}. But then, by Lemma 1, the matrix

$$X = \| 1, x_1(t_i), \ldots, x_{2n}(t_i) \| \qquad (i = 0, 1, \ldots, k)$$

---

[5] Professor W. Fenchel kindly called to my attention the fundamental paper [7] of Hjelmslev. Hjelmslev ([7], pp. 4, 6, 7, 23, 41) calls the polygon Π, of (1.1), *monotone.* provided the matrix (1.2) satisfies the conditions of Lemma 1. By Lemma 1 we thus see that Π is convex on E_m if and only if Π is monotone in the sense of Hjelmslev. Similarly, in view of our Definition 2 and Lemma 2 we see that the curve (1.3) is convex on E_{2n} if and only if it is monotone in the sense of Hjelmslev ([7], p. 60). We should also point out the connection with the concept of an arc of order m in E_m due to C. Juel; see Marchaud [9] and Scherk [13], also for references. Arcs of order m in E_m are convex on E_m but not conversely. A theorem of Hjelmslev ([7], p. 62) should be especially quoted, being closely related to our Lemma 2. Its statement for the closed curve (1.3) in E_{2n} is as follows: *If no subarc of C is in a E_{2n-1}, then C is of order $2n$ if and only if all determinants* (1.4) *are positive or all are negative.* One may even omit in the statement the assumption concerning the subarcs of C. In [15] I have recently said (p. 227) that by this theorem Hjelmslev has essentially anticipated the result of Gantmakher–Krein. This view was perhaps exaggerated.

I also owe to W. Fenchel the following reference: E. Egerváry, *On the smallest convex cover of a simple arc of space-curve,* Publ. math. Debrecen, *1* (1949), 65–70, in which its author solves the problem of the present paper for open arcs in E_3. As yet unsolved is the problem for the particularly interesting case of closed curves in E_3 (see [1], footnote on p. 111).

has all non-vanishing minors of order $2n+1$ of the same sign. Given two non-vanishing determinants Δ_1, Δ_2, of the form (1.4), we may so choose the parameter values (1.5), that Δ_1 and Δ_2 are among the minors of X. But then $\Delta_1 \Delta_2 > 0$.

The conditions are sufficient. This follows directly by Lemma 1 and Definitions 1 and 2.

Remarks. 1. Our old definition of a closed polygon convex on E_{2n} agrees with the new definition if applied to the polygon considered as a closed curve.

2. Lemma 2 may be rephrased in geometric terms as follows: The continuous closed curve $P=P(t)$ in E_{2n} is convex on E_{2n} if and only if it has the following properties: If $P_0, P_1, \ldots, P_{2n}$ are $2n+1$ points on C, in correct cyclic order corresponding to increasing values of t, then the $2n$-dimensional simplices $[P_0, P_1, \ldots, P_{2n}]$ should not be all degenerate and the non-degenerate ones should have the same orientation.

3. It seems natural and useful to classify the curves convex on E_{2n} into *positively convex* and *negatively convex* curves, depending on the common sign of the determinants Δ, or the common orientation of the simplices $[P_0, P_1, \ldots, P_{2n}]$.

4. An adaptation of these definitions and results to open arcs in even or odd-dimensional spaces seems perfectly straight-forward. As an example we mention that one full turn (but no longer arc) of a right-handed circular helix in E_3 is positively convex on E_3.[6]

5. Lemma 2 applies easily to show that the curve C_0, defined by (2), is positively convex on E_{2n}. Indeed, in this case the determinant (1.4), familiar from the problem of trigonometric interpolation, may be evaluated explicitly and we find

$$\Delta = \frac{2^{2n^2}}{n!} \prod_{\mu>\nu} \sin \frac{t_\mu - t_\nu}{2} > 0.$$

5. Convex curves are rectifiable.[7] We need the following

Definition 3. *The bounded real function $f(t)$, $a \le t \le b$, is said to be non-oscillatory provided there is a fixed natural integer N such that for every real c the function $f(t)-c$ changes sign at most N times in the range $[a, b]$.*

With this definition we have

---

[6] In this example which can be widely generalized we are especially close to Pólya's paper [12].

[7] The contents of Article 5 where developed in the course of a conversation with Th. S. Motzkin. Here we are close to Marchaud's work [9], Chapitre I.

Lemma 3. *If $f(t)$ is non-oscillatory in the range $[a, b]$, then $f(t)$ is of bounded variation, in fact*

(1.6) $$\text{Total variation of } f(t) \leq N \ (\sup f - \inf f).$$

Proof. Let $a_0 = t_0 < t_1 < \cdots < t_k = b$ and let $P_i = (t_i, f(t_i))$ be the corresponding points on the graph $x = f(t)$. Consider the polygon $P_0 P_1 \ldots P_k$ and its orthogonal projection $P'_0 P'_1 \ldots P'_k$ into the segment $I = [\inf f \leq x \leq \sup f]$ of the x-axis. Since $P_0 P_1 \ldots P_k$ may cross a horizontal line $x = c$ at most N times we conclude that the k open (or void) segments $P'_0 P'_1, P'_1 P'_2, \ldots, P'_{k-1} P'_k$ may cover a point of I at most N times. But then

$$\sum_{\nu=0}^{k-1} P'_\nu P'_{\nu+1} \leq N \cdot m(I)$$

or

$$\sum_{\nu=0}^{k-1} |f(t_{\nu+1}) - f(t_\nu)| \leq N \ (\sup f - \inf f),$$

whence (1.6) follows on taking the supremum of the left side.

This result clearly implies

Lemma 4. *A closed curve C convex in E_{2n} is necessarily rectifiable.*

Indeed, let C be defined by the equations (1.3), say. Now $f(t) = x_i(t)$ being continuous is also bounded in the range $[0, 2\pi]$. However, this function is also non-oscillatory, with $N = 2n$ in Definition 3, because C may cross the hyperplane $x_i = c$ at most $2n$ times. By Lemma 3 all $x_i(t)$ are of bounded variation, which proves Lemma 4.

6. On the volume of polyhedra spanned by convex polygons. Let Π be a closed polygon convex on E_{2n}. We wish to express the volume of the convex polyhedron $K(\Pi)$ in terms of volumes of simplices. We restrict our discussion to the first significant case when $n = 2$. The simplification in notations thus afforded is considerable, it being obvious at all times how the general case is to be treated.

Lemma 5. *Let $\Pi = P_0 P_1 \ldots P_m P_0$ be a closed polygon in E_4 which is convex on E_4. Let $K = K(\Pi)$ denote its convex hull, $V(K)$ the volume of K. Let O be a point in the interior of K. The volume of K is given by the formula*

(1.7) $$V(K) = \frac{1}{2!} \sum_{i,j=0}^{m} V[O, P_i, P_{i+1}, P_j, P_{j+1}], \qquad (P_{m+1} = P_0),$$

where the summand is the volume of the simplex of vertices $O, P_i, P_{i+1}, P_j, P_{j+1}$, it being zero if the simplex is degenerate.

Proof. (i) By Lemma 1 we know that the matrix

$$X = \| 1, x_{i1}, x_{i2}, x_{i3}, x_{i4} \|, \quad (i=0, 1, \ldots, m), \tag{1.8}$$

is of rank 5 and that all non-vanishing minors of order 5 are of the same sign. There is no restriction in assuming these minors to be positive, i.e. the polygon Π to be positively convex on E_4. However, we shall assume for the moment that

(1.9) *All* 5th order minors of X are positive,

showing later how this restriction can be removed.

To abbreviate our notation let

$$D(x, x_i, x_j, x_k, x_l) = \begin{vmatrix} 1 & x_1 & x_2 & x_3 & x_4 \\ 1 & x_{i1} & x_{i2} & x_{i3} & x_{i4} \\ 1 & x_{j1} & x_{j2} & x_{j3} & x_{j4} \\ 1 & x_{k1} & x_{k2} & x_{k3} & x_{k4} \\ 1 & x_{l1} & x_{l2} & x_{l3} & x_{l4} \end{vmatrix}.$$

The assumption (1.9) implies that every four points P_i, P_j, P_k, P_l $(0 \leq i < j < k < l \leq m)$ determine uniquely a 3-flat $\pi(i, j, k, l)$, of equation $D(x, x_i, x_j, x_k, x_l) = 0$, containing no other point P_ν since $D(x_\nu, x_i, x_j, x_k, x_l) \neq 0$ if $\nu \neq i, j, k, l$. We now assert the following: *All 3-dimensional faces of $K(\Pi)$ are precisely in the 3-flats*

$$\begin{aligned} &\pi(i, i+1, j, j+1), \quad (0 \leq i < i+1 < j < j+1 \leq m), \\ &\pi(0, i, i+1, m), \quad (0 < i < i+1 < m). \end{aligned} \tag{1.10}$$

Indeed, that these 3-flats contain 3-faces of $K(\Pi)$ is seen as follows: The assumption (1.9) implies the inequalities

$$\begin{aligned} &D(x_\nu, x_i, x_{i+1}, x_j, x_{j+1}) > 0 \quad \text{if} \quad \nu \neq i, i+1, j, j+1, \\ &D(x_\nu, x_0, x_i, x_{i+1}, x_m) < 0 \quad \text{if} \quad \nu \neq 0, i, i+1, m, \end{aligned} \tag{1.11}$$

which show that each of the planes (1.10) leave all the other vertices strictly on one side. The 3-flats (1.10) are therefore planes of support of $K(\Pi)$. On the other hand none of the 3-flats through four of the points, other than the 3-flats (1.10), can possibly be a plane of support of $K(\Pi)$, for such a 3-flat is seen to have some of the vertices on one of its sides and others on the other side. We conclude[8] that

---

[8] From this result and its extension to E_{2n} it is easy to derive the following: *The curve C convex on E_{2n} is on the boundary of its convex hull $K(C)$.* This is closely related to a result of W. GUSTIN [5].

the 3-dimensional faces of $K(\Pi)$ are the simplices $[P_i, P_{i+1}, P_j, P_{j+1}]$, where the pair (i, j) runs over the set S of pairs (i, j) defined by the conditions

$$0 \leq i < i+1 < j < j+1 \leq m \quad \text{or} \quad 0 < i < i+1 < j = m.$$

We may summarize the situation by the relation

$$\text{Boundary of } K(\Pi) = \sum_S [P_i, P_{i+1}, P_j, P_{j+1}], \qquad (P_{m+1} = P_0).$$

If O is a point interior to $K = K(\Pi)$ we obtain, by central projection from O, the relation

$$K = \sum_S [O, P_i, P_{i+1}, P_j, P_{j+1}] \tag{1.12}$$

which describes a dissection of K into 4-dimensional simplices. On passing to volumes we obtain

$$V(K) = \sum_S V[O, P_i, P_{i+1}, P_j, P_{j+1}].$$

Let us observe now that all simplices on the right-hand side of (1.12) are positively oriented. Indeed, by (1.11) we see that if $(i, j) \in S$ then all determinants $D(x_\nu, x_i, x_{i+1}, x_j, x_{j+1})$ $(\nu \neq i, i+1, j, j+1)$ are positive. But then also $D(x, x_i, x_{i+1}, x_j, x_{j+1}) > 0$ if the point (x) is in the interior of K, this determinant being a linear combination with positive coefficients of non-negative determinants some of which must be positive. On choosing the point O as the origin of the coordinate system we have

$$V(K) = \sum_S V[O, P_i, P_{i+1}, P_j, P_{j+1}] = \frac{1}{4!} \sum_S D(0, x_i, x_{i+1}, x_j, x_{j+1}).$$

Finally, the last expression may be written as

$$V(K) = \frac{1}{2!\,4!} \sum_{i,j=0}^{m} \det \| x_{i\nu}, x_{i+1,\nu}, x_{j\nu}, x_{j+1,\nu} \|, \tag{1.13}$$

since all new terms entering into the double summation will vanish if they have coincident rows while the duplication of old terms is offset by the new factor 2! in the denominator. Now (1.13) is precisely the relation (1.7) we wished to establish.

(ii) We now wish to remove the assumption (1.9) on which we relied heavily in our previous discussion. We do this by the following simple device. We need the matrix

$$\| q^{(i-j)^2} \|_{i,j=0,\ldots,m}, \qquad (0 < q < 1).$$

It follows from a theorem of Pólya[9] that all minors of this matrix, of all orders, are positive. Moreover, this matrix clearly converges to the unit matrix as $q \to 0$.

[9] See [11], Problem 76, page 49.

Dividing the elements of each row by the sum of all elements in that row we obtain a new matrix which we denote by H_q. Returning to the matrix (1.8) we form the product

$$X^* = H_q X = \| 1, x_{i1}^*, x_{i2}^*, x_{i3}^*, x_{i4}^* \|_{i=0,\ldots,m}$$

and easily infer the following: The new matrix X^*, whose elements depend on q, enjoys the property (1.9) and defines a closed polygon

$$\Pi^* = P_0^* P_1^* \ldots P_m^* P_0^* \text{ in } E_4,$$

whose vertices are as close to those of Π as we wish, provided q is sufficiently small. Let $K^* = K(\Pi^*)$. By our previous discussion we know that $V(K^*)$ may be expressed by the analogue of (1.13). Also $V(K^*) \to V(K)$, as $q \to 0$. By continuity we see that (1.13) again holds even if the condition (1.9) is disregarded.

7. The volume $V(K(C))$ expressed as an integral. We are now turning to the main result of this section:

Theorem 1. *Let C be a closed curve convex on E_{2n} defined by*

(1.14) $$C: \quad x_i = x_i(t), \quad (i = 1, \ldots, 2n;\ 0 \leq t \leq 2\pi).$$

We assume the $x_i(t)$ to be absolutely continuous in $[0, 2\pi]$, a condition which is automatically fulfilled if the parameter t is proportional to the arc-length along C. Then the volume of the convex hull $K(C)$ may be expressed by the following Lebesgue integral

(1.15) $$V(K(C)) = \frac{\varepsilon}{n!\,(2n)!} \int_0^{2\pi} \cdots \int_0^{2\pi} \det \| x_i(t_1), x_i'(t_1), \ldots, x_i(t_n), x_i'(t_n) \| \, dt_1 \ldots dt_n,$$

where $\varepsilon = +1$ or -1 depending on whether C is positively or negatively convex on E_{2n}.[10]

Proof. Again, to simplify our notation we assume that $n = 2$. Let us assume that C is positively convex on E_4. Divide the range $[0, 2\pi]$ in 2^k equal parts by the points

$$t_\nu = 2\pi\nu/2^k, \quad (\nu = 0, 1, \ldots, m = 2^k - 1),$$

and let P_ν be the corresponding point on C. The inscribed polygon $\Pi = P_0 P_1 \ldots P_m P_0$ being positively convex on E_4, provided k is sufficiently large, by (1.13) we have

$$V(K(\Pi)) = \frac{1}{2!\,4!} \sum_{i,j=0}^{m} \det \| x_\nu(t_i), x_\nu(t_{i+1}), x_\nu(t_j), x_\nu(t_{j+1}) \|.$$

[10] We lose no generality by restricting our discussion to curves convex *on* E_{2n}, for if C is contained in a lower-dimensional flat space, then both sides of (1.15) evidently vanish.

Subtracting the first and third column of the determinant from those just ahead we obtain

$$V(K(\Pi)) = \frac{1}{2!\,4!} \sum_{i,j=0}^{m} \det \left\| x_\nu(t_i),\ \int_{t_i}^{t_{i+1}} x'_\nu(\tau_1)\,d\tau_1,\ x_\nu(t_j),\ \int_{t_j}^{t_{j+1}} x'_\nu(\tau_2)\,d\tau_2 \right\|$$

and finally

$$(1.16)\qquad V(K(\Pi)) = \frac{1}{2!\,4!} \sum_{i,j=0}^{m} \int_{t_i}^{t_{i+1}} \int_{t_j}^{t_{j+1}} \det \| x_\nu(t_i),\ x'_\nu(\tau_1),\ x_\nu(t_j),\ x'_\nu(\tau_2) \|\, d\tau_1\, d\tau_2.$$

A passage to the limit in this relation, as $k \to \infty$, presents no difficulties. On the left side $V(K(\Pi)) \to V(K(C))$. Indeed, the inclusion $K(\Pi) \subset K(C)$ implies that

$$(1.17)\qquad V(K(\Pi)) \leq V(K(C)).$$

On the other hand the increasing sequence of sets $K(\Pi)$ converges to a limit which includes the set $K_0(C)$ of interior points of $K(C)$. This remark and (1.17) imply

$$\text{measure of } K_0(C) \leq \lim_{k\to\infty} V(K(\Pi)) \leq V(K(C))$$

and the equality of the extreme terms implies the desired conclusion.

There remains to show that the sum S on the right-hand side of (1.16) converges to a double integral which we may write as

$$J = \sum_{i,j=0}^{m} \int_{t_i}^{t_{i+1}} \int_{t_j}^{t_{j+1}} \det \| x_\nu(\tau_1),\ x'_\nu(\tau_1),\ x_\nu(\tau_2),\ x'_\nu(\tau_2) \|\, d\tau_1\, d\tau_2.$$

In order to show that $S \to J$, we expand each of the two determinants into $4!$ terms and find for their difference the expression

$$\sum \pm x'_{\nu_2}(\tau_1)\, x'_{\nu_4}(\tau_2)\, [x_{\nu_1}(t_i)\, x_{\nu_3}(t_j) - x_{\nu_1}(\tau_1)\, x_{\nu_3}(\tau_2)].$$

Given $\varepsilon > 0$, the equi-uniform continuity of all products $x_r(\tau_1)\, x_s(\tau_2)$ shows that all square brackets will be in absolute value less than ε, each within its respective cell

$$t_i \leq \tau_1 \leq t_{i+1}, \quad t_j \leq \tau_2 \leq t_{j+1},$$

provided that k is sufficiently large. But then

$$|S - J| \leq 2\varepsilon \sum \int_0^{2\pi} \int_0^{2\pi} |x'_{\nu_2}(\tau_1)| \cdot |x'_{\nu_4}(\tau_2)|\, d\tau_1\, d\tau_2 = A \cdot \varepsilon.$$

This concludes a proof of Theorem 1 for $n=2$. The alterations necessary to deal with the general case of E_{2n} are deemed to be obvious in every respect.

8. A proof that Theorem II implies the isoperimetric inequality of Theorem I. Let the curve C, defined by (1), be positively convex on E_{2n}. Let us assume, moreover, that the parameter $t=2\pi s/L$, where s is the arc length along C. For these functions $x_i(t)$ we have the inequality (4) of Theorem II which we take for granted for the moment. Both sides of (4) have a geometric meaning which we wish to derive. By (1.15), the right-hand side of (4) is equal to

$$n^n\, n!\,(2n)!\, V(K(C)).$$

Since

$$\sum_1^{2n} x_i'^2 = \left(\frac{ds}{dt}\right)^2 = \frac{L^2}{4\pi^2}$$

holds almost everywhere, the left-hand side of (4) equals

$$\left(\int_0^{2\pi} \frac{L^2}{4\pi^2}\, dt\right)^n = (2\pi)^{-n} L^{2n}.$$

Thus (4) reduces to the isoperimetric inequality (3) with equality only if C is similar to C_0.

§ 2. Further properties of $V(K(C))$ and reformulation of Theorem II in terms of Fourier series

9. An expression for $V(K(C))$ in terms of areas of 2-dimensional projections of C. Assuming the curve C to be positively convex on E_{2n}, we have found the expression

$$V(K(C)) = \frac{1}{n!\,(2n)!} \int_0^{2\pi} \cdots \int_0^{2\pi} \det \| x_i(t_1), x_i'(t_1), \ldots, x_i(t_n), x_i'(t_n) \| \, dt_1 \ldots dt_n. \tag{2.1}$$

We now expand the determinant, by a repeated application of Laplace's rule, by second-order minors formed from the pairs of columns $(1, 2), (3, 4), \ldots, (2n-1, 2n)$, obtaining a sum

$$\sum \pm \begin{vmatrix} x_{\mu_1}(t_1) & x_{\mu_1}'(t_1) \\ x_{\nu_1}(t_1) & x_{\nu_1}'(t_1) \end{vmatrix} \cdot \begin{vmatrix} x_{\mu_2}(t_2) & x_{\mu_2}'(t_2) \\ x_{\nu_2}(t_2) & x_{\nu_2}'(t_2) \end{vmatrix} \cdots \begin{vmatrix} x_{\mu_n}(t_n) & x_{\mu_n}'(t_n) \\ x_{\nu_n}(t_n) & x_{\nu_n}'(t_n) \end{vmatrix}$$

which after integration becomes

$$2^n \sum \pm A(\mu_1, \nu_1)\, A(\mu_2, \nu_2) \ldots A(\mu_n, \nu_n), \tag{2.2}$$

where $A(i, j)$ represents the area of the projection of C on the plane $x_i 0 x_j$, while

$$\mu_1, \nu_1, \mu_2, \nu_2, \ldots, \mu_n, \nu_n \tag{2.3}$$

is an arrangement of the set of numbers 1, 2, ..., $2n$ into pairs such that $\mu_i < \nu_i$. It is clear that those arrangements (2.3) which differ only in the order of their pairs will contribute identical terms to the sum (2.2), there being $n!$ such. Thus (2.2) reduces to

$$2^n n! \sum \pm A(\mu_1, \nu_1) \ldots A(\mu_n, \nu_n) \tag{2.4}$$

where we sum only over such arrangements (2.3) in which the n pairs are arranged lexicographically: $\mu_1 < \mu_2 < \cdots < \mu_n$. The sign of each term of (2.4) is $(-1)^I$, where I is the number of inversions in the permutation

$$\begin{pmatrix} 1 & 2 & \ldots & 2n-1 & 2n \\ \mu_1 & \nu_1 & \ldots & \mu_n & \nu_n \end{pmatrix}.$$

The number of terms in (2.4) is obtained by first counting pairs indiscriminately and then dividing by $n!$, to throw out their order. We thus find in (2.4)

$$\binom{2n}{2}\binom{2n-2}{2}\cdots\binom{4}{2}\binom{2}{2}\Big/ n! = \frac{(2n)!}{2^n\, n!} = 1\cdot 3\cdot 5 \ldots (2n-1)$$

terms. On replacing the integral in (2.1) by its equivalent expression (2.4) we obtain the following

Theorem 2. *If the closed curve C is positively convex on E_{2n} then*

$$V(K(C)) = \frac{2^n}{(2n)!} \sum (-1)^I A(\mu_1, \nu_1)\, A(\mu_2, \nu_2) \ldots A(\mu_n, \nu_n), \tag{2.5}$$

where $A(i, j)$ denotes the area of the projection of C on the plane $x_i 0 x_j$, while the summation runs over all $1\cdot 3\cdot 5 \ldots (2n-1)$ permutations $\mu_1, \nu_1, \mu_2, \nu_2, \ldots, \mu_n, \nu_n$, of the numbers 1, 2, ..., $2n$, such that $\mu_i < \nu_i$ and $\mu_1 < \mu_2 < \cdots < \mu_n$. Finally the exponent I is the number of inversions in the corresponding permutation.

Examples. 1. For $n = 2$, (2.5) becomes

$$V(K(C)) = \tfrac{1}{6}\{A(1, 2)\, A(3, 4) - A(1, 3)\, A(2, 4) + A(1, 4)\, A(2, 3)\}.$$

2. Let us evaluate $V(K(C_0))$ for the curve C_0 defined by the equations (2) of our Introduction. For this case we find on inspection that

$$A(\mu, \nu) = \int_0^{2\pi} x_\mu(t)\, x_\nu'(t)\, dt = 0, \qquad (\mu < \nu),$$

unless μ is odd and $\nu=\mu+1$, when

$$A(1,2)=\pi,\ A(3,4)=\frac{\pi}{2},\ \ldots,\ A(2n-1,2n)=\frac{\pi}{n}.$$

The sum (2.5) reduces now to one term only

$$V(K(C_0))=\frac{2^n}{(2n)!}A(1,2)\,A(3,4)\ldots A(2n-1,2n)=\frac{2^n\pi^n}{n!\,(2n)!}.$$

Also the length of C_0 is easily found from (2):

$$L=\int_0^{2\pi}\sqrt{\Sigma x_i'^2}\,dt=2\pi\sqrt{n}.$$

These values are seen to verify the isoperimetric relation (3) with the equality sign holding.

10. Reformulation of Theorem II in terms of Fourier series. We return to Theorem II of the Introduction and expand $x_i(t)$ in its Fourier series

$$x_i(t)\sim\sum_{\nu=1}^{\infty}\left(a_{i\nu}\frac{\cos\nu t}{\nu}+b_{i\nu}\frac{\sin\nu t}{\nu}\right), \tag{2.6}$$

where we assume the constant term to vanish without loss of generality. But then

$$x_i'(t)\sim\sum_{\nu=1}^{\infty}(-a_{i\nu}\sin\nu t+b_{i\nu}\cos\nu t), \tag{2.7}$$

and Parseval's relation gives

$$\int_0^{2\pi}x_i'^2\,dt=\pi\sum_{\nu=1}^{\infty}(a_{i\nu}^2+b_{i\nu}^2),$$

whence

$$\int_0^{2\pi}\left(\sum_1^{2n}x_i'^2\right)dt=\pi\sum_{i=1}^{2n}\sum_{\nu=1}^{\infty}(a_{i\nu}^2+b_{i\nu}^2). \tag{2.8}$$

We also wish to express the right-hand side of (4) in terms of the Fourier coefficients $a_{i\nu}$, $b_{i\nu}$. This can be done in two ways leading to formally different but necessarily equivalent expressions.

A. We wish to show that the integral

$$J=\int_0^{2\pi}\cdots\int_0^{2\pi}\det\|x_i(t_1),\ x_i'(t_1),\ \ldots,\ x_i(t_n),\ x_i'(t_n)\|\,dt_1\ldots dt_n \tag{2.9}$$

may be expressed as

$$J=(2\pi)^n n! \sum_{j_1<\cdots<j_n} \frac{1}{j_1 j_2 \cdots j_n} D(j_1, j_2, \ldots, j_n), \tag{2.10}$$

where

$$D(j_1, j_2, \ldots, j_n) = \det \| a_{ij_1}, b_{ij_1}, a_{ij_2}, b_{ij_2}, \ldots, a_{ij_n}, b_{ij_n} \|. \tag{2.11}$$

Formally, there is no difficulty whatsoever. Indeed, if we introduce the expansions (2.6) and (2.7) into (2.9), writing the determinant as a sum of determinants obtained from the individual terms of the Fourier series, on using the orthogonality properties of the trigonometric system we find the expansion (2.10). A proof of the validity of (2.10), which includes a proof of the absolute convergence of the n-fold series (2.10), follows from the following remark: The integral J is essentially the constant term of the n-fold Fourier expansion of the determinant under the integral sign. As we operate throughout within the class L_2, we will *actually* obtain that constant term by introducing formally the Fourier expansion of each element and gathering all terms which contribute to it. This, however, is precisely what was done above.

In terms of the quantities S and Φ, defined by (7) and (8), and using (2.8) and (2.10), the inequality (4) becomes

$$\pi^n S^n \geq (2\pi n)^n n! \, \Phi,$$

which is equivalent to (9). The equivalence of the Theorems II and III is now apparent if we refer to the Riesz-Fischer theorem.

B. An alternative expression for J, or equivalently for

$$\Phi = \frac{1}{(2\pi)^n n!} J,$$

may now be derived by observing the following: In Theorem 2 we have expressed $V(K(C))$ in terms of the $A(i, j)$ by the formula (2.5). It is clear that this result amounts to a similar expression for the integral J without any reference to curves convex on E_{2n}. In fact, by (2.1) and (2.5) we find

$$J = 2^n n! \sum (-1)^n A(\mu_1, \nu_1) \ldots A(\mu_n, \nu_n),$$

while Parseval's relation gives

$$A(i, j) = \int_0^{2\pi} x_i x_j' \, dt = \pi \sum_{k=1}^{\infty} \frac{1}{k} (a_{ik} b_{jk} - a_{jk} b_{ik}).$$

Introducing now the new expressions

(2.12) $$(i, j) = \sum_{k=1}^{\infty} \frac{1}{k} (a_{ik} b_{jk} - a_{jk} b_{ik}),$$

we find, on combining the last four relations, the desired expression

(2.13) $$\Phi = \sum (-1)^I (\mu_1, \nu_1)(\mu_2, \nu_2) \dots (\mu_n, \nu_n),$$

which will be found useful in the next section.

§ 3. A proof of Theorem III

11. A first special case of Theorem III. Let us first establish the following

Lemma 6. *Let*

(3.1) $$D = \det \| c_{ij} \|$$

be a real determinant of order m. Then

(3.2) $$\left(\sum_{i,j=1}^{m} c_{ij}^2 \right)^m \geqq m^m D^2,$$

with the equality sign if and only if the elements of D are proportional to the elements of an orthogonal matrix.

Proof. By Hadamard's inequality[11] we have

(3.3) $$D^2 \leq \left(\sum_j c_{1j}^1 \right) \left(\sum_j c_{2j}^2 \right) \cdots \left(\sum_j c_{mj}^2 \right).$$

By the inequality between the arithmetic and geometric mean we have

(3.4) $$\left(\sum_j c_{1j}^2 \right) \cdots \left(\sum_j c_{mj}^2 \right) \leq \left(\frac{1}{m} \sum_{i,j} c_{ij}^2 \right)^m.$$

Notice that (3.3) and (3.4) imply (3.2). Moreover, equality in (3.2) implies the equality signs in both (3.3) and (3.4). Equality in (3.3) shows that the rows of D are orthogonal to each other and the equality in (3.4) requires that

$$\sum_j c_{1j}^2 = \sum_j c_{2j}^2 = \cdots = \sum_j c_{mj}^2,$$

which concludes a proof of Lemma 6.

The special case of Theorem III which we have in mind is the case when all columns of the matrix M vanish except the first $2n$ columns. The inequality (9) reduces then to

---

[11] See [6].

(3.5) $$\left\{\sum_{i=1}^{2n}\sum_{\nu=1}^{n}(a_{i\nu}^2+b_{i\nu}^2)\right\}^n \geq (2n)^n D(1, 2, \ldots, n).$$

However, this inequality is implied by (3.2) for $m=2n$ and $\|c_{ij}\|=M$. Moreover, the equality sign in (3.5) does not only imply, by Lemma 6, that the elements of M are proportional to those of an orthogonal matrix, but even that they are *positively* proportional to a *right*-orthogonal matrix. This establishes Theorem III in this special case.

12. An algebraic lemma. Let

(3.6) $$\Sigma = \|\sigma_{ij}\|_{i,j=1,\ldots,2n}$$

be a real skew-symmetric matrix, which means that $\Sigma' = -\Sigma$. We first establish the following identity in x:

(3.7) $$\begin{vmatrix} -xI & \Sigma \\ \Sigma' & -xI \end{vmatrix} = (P_n(x^2))^2,$$

where $P_n(z)$ is a real polynomial of degree n.

Proof. By matrix multiplication we find

$$\begin{Vmatrix} I & 0 \\ \Sigma' & xI \end{Vmatrix} \cdot \begin{Vmatrix} -xI & \Sigma \\ \Sigma' & -xI \end{Vmatrix} = \begin{Vmatrix} -xI & \Sigma \\ 0 & \Sigma'\Sigma - x^2 I \end{Vmatrix}$$

and passing to determinants we have

$$x^{2n}\begin{vmatrix} -xI & \Sigma \\ \Sigma' & -xI \end{vmatrix} = (-x)^{2n}\,|\Sigma'\Sigma - x^2 I|,$$

whence the identity in x

(3.8) $$\begin{vmatrix} -xI & \Sigma \\ \Sigma' & -xI \end{vmatrix} = |\Sigma'\Sigma - x^2 I|.$$

The right-hand side of this identity may now be factored. Indeed, by the skew-symmetry of Σ we have

$$(\Sigma' - xI)(\Sigma - xI) = \Sigma'\Sigma - x(\Sigma+\Sigma') + x^2 I = \Sigma\Sigma' + x^2 I,$$

whence

(3.9) $$|\Sigma'\Sigma + x^2 I| = |\Sigma - xI|^2.$$

On the other hand

$$\phi(x) \equiv |\Sigma - xI|$$

11 – 533807. *Acta mathematica*. 91. Imprimé le 27 octobre 1954.

is an even polynomial in x, because

$$\phi(-x)=|\Sigma+xI|=|-\Sigma-xI|=|\Sigma'-xI|=|\Sigma-xI|=\phi(x).$$

Thus

$$|\Sigma-xI|=P_n(-x^2),$$

and by (3.9) we have

$$|\Sigma'\Sigma+x^2 I|=(P_n(-x^2))^2.$$

Replacing x^2 by $-x^2$ we now obtain, via (3.8), the identity (3.7) which we wished to establish.

We may now state our

Lemma 7. *If Σ is a real skew-symmetric matrix of order $2n$, then the symmetric matrix*

$$\Omega=\left\|\begin{matrix}0 & \Sigma\\ \Sigma' & 0\end{matrix}\right\|, \tag{3.10}$$

of order $4n$, has at most n distinct and positive characteristic values.

Indeed, the identity (3.7) gives precise information on the characteristic values of the matrix (3.10). It shows that they are all of multiplicity 2 and in pairs symmetric with respect to the origin, from which the conclusion of Lemma 7 follows.

13. A second special case of Theorem III: The matrix M has only a finite number of non-zero columns. We assume throughout this Article 13 that

$$a_{ij}=b_{ij}=0, \qquad (i=1, \ldots, 2n;\ j=m+1,\ m+2, \ldots),$$

where m is a fixed number $>n$. Let us normalize the problem by requiring that

$$S\equiv\sum_{\nu=1}^{m}\sum_{i=1}^{2n}(a_{i\nu}^2+b_{i\nu}^2)=2n. \tag{3.11}$$

We are to show, then, that

$$\Phi\leq\frac{1}{n!}, \tag{3.12}$$

with equality only under conditions as stated in Theorem III. This will be shown by maximizing the function $\Phi=\Phi(a_{i\nu}, b_{i\nu})$ subject to the relation (3.11), a procedure which requires the first partial derivatives of the function Φ. These derivatives are obtained as follows:

Let us denote by $\sigma_{ij}\,(i<j)$ the sum of those terms of the expression (2.13) which contain (i, j) as one of their n factors, *with that factor* $(\mu_s, \nu_s)=(i, j)$ *removed.* Let i be fixed $(=1, \ldots, 2n)$. On replacing the removed factors, (2.13) may now be written as

$$\Phi = \sum_{j=1}^{i-1} (j, i)\, \sigma_{ji} + \sum_{j=i+1}^{2n} (i, j)\, \sigma_{ij}. \tag{3.13}$$

Indeed, this way of writing Φ represents a classification of the original terms of the sum (2.13) where all terms having (j, i) as a factor $(j<i)$ form the class $(j, i)\,\sigma_{ji}$, and all terms having (i, j) as a factor $(i<j)$ form the class $(i, j)\,\sigma_{ij}$. Thus in the case when $n=2$

$$\Phi = (1,2)\,(3,4) - (1,3)\,(2,4) + (1,4)\,(2,3);$$

if $i=3$, the classification (3.13) amounts to writing

$$\Phi = (1,3)\,\sigma_{13} + (2,3)\,\sigma_{23} + (3,4)\,\sigma_{34}$$

where

$$\sigma_{13} = -(2,4), \quad \sigma_{23} = (1,4), \quad \sigma_{34} = (1,2).$$

We may also write (3.13) as

$$\Phi = \sum_{j=1}^{i-1} (i, j)\,(-\sigma_{ji}) + \sum_{j=i+1}^{2n} (i, j)\, \sigma_{ij}.$$

If we extend the meaning of the symbol σ_{ij} by agreeing that it be skew-symmetric in its subscripts, we may write the last relation as

$$\Phi = \sum_{j=1}^{2n} (i, j)\, \sigma_{ij}, \qquad (i = 1, \ldots, 2n). \tag{3.14}$$

However,

$$(i, j) = \sum_{k=1}^{m} \frac{1}{k}\, (a_{ik}\, b_{jk} - a_{jk}\, b_{ik}), \tag{3.15}$$

and σ_{ij} is independent (for all j) of the variables a_{ik}, b_{ik} $(k=1, \ldots, m)$, by its construction. From the last two relations we obtain the partial derivatives

$$\frac{\partial \Phi}{\partial a_{ik}} = \sum_{j=1}^{2n} \sigma_{ij} \frac{1}{k} b_{jk}, \quad \frac{\partial \Phi}{\partial b_{ik}} = -\sum_{j=1}^{2n} \sigma_{ij} \frac{1}{k} a_{jk}.$$

Let us now assume that the elements of the matrix M are such as to make Φ assume its absolutely maximal value subject to the condition (3.11), or $S^n = (2n)^n$. The Lagrange multiplier rule assures us of the existence of two constants μ and λ, not both vanishing, such that all first partial derivatives of the function

$$G \equiv \mu\, \Phi - \lambda\, S^n$$

vanish.[12] We therefore have the $4nm$ equations

[12] See CARATHÉODORY's book [3] for the formulation of the multiplier rule in the form needed here.

$$(3.16)\quad \begin{aligned} k\frac{\partial G}{\partial a_{ik}} &\equiv \mu \sum_j \sigma_{ij} b_{jk} - 2n\lambda k S^{n-1} a_{ik} = 0, \\ k\frac{\partial G}{\partial b_{ik}} &\equiv -\mu \sum_j \sigma_{ij} a_{jk} - 2n\lambda k S^{n-1} b_{ik} = 0, \end{aligned}$$

$$(i=1, \ldots, 2n;\ k=1, \ldots, m).$$

If we multiply these equations by a_{ik}/k and b_{ik}/k, respectively, and add them all together, we obtain, in view of (3.15) and (3.14):

$$0 = \mu \sum_{i,j} \sigma_{ij}(i,j) - 2n\lambda S^n = 2n\mu\Phi - 2n\lambda S^n$$

or

$$(3.17)\qquad \mu\Phi = \lambda S^n.$$

However, the maximal value Φ is clearly positive (because $m > n$) and so is $S = 2n$. We conclude that both constants μ, λ are non-vanishing and we may therefore assume that $\mu = 1$ and hence that λ *is positive.*

The equations (3.16), with $\mu = 1$, may be interpreted as follows: In terms of the matrices

$$(3.18)\qquad \Sigma = \| \sigma_{ij} \|_{1,2n}, \quad \Omega = \begin{Vmatrix} 0 & \Sigma \\ \Sigma' & 0 \end{Vmatrix},$$

and the column vectors

$$(3.19)\qquad v_k = (a_{1k}, \ldots, a_{2n,k}, b_{1k}, \ldots, b_{2n,k}), \qquad (k=1, \ldots, m),$$

the $4nm$ equations (3.16) may be written as m matrix relations

$$(3.20)\qquad \Omega v_k = (2n)^n k\lambda v_k, \qquad (k=1, \ldots, m).$$

We have therefore reached the following conclusion: *If the elements of M maximize Φ subject to the condition* (3.11), *then there exists a positive constant λ satisfying the relations* (3.20).

These relations show that if $v_k \neq 0$, then $(2n)^n k\lambda$ is a characteristic value of the matrix Ω. By Lemma 7 we know that Ω can have at most n distinct positive characteristic values. We conclude that at most n among the vectors (3.19) are non-vanishing. Let these be among the n vectors $v_{j_1}, \ldots, v_{j_n}$ $(1 \leq j_1 < \cdots < j_n \leq m)$. But then, the expansion (8) reduces to a single term:

$$\Phi = \frac{1}{j_1 \cdots j_n} D(j_1, \ldots, j_n).$$

Since $\Phi > 0$, we conclude that none of the vectors $v_{j_1}, \ldots, v_{j_n}$ vanishes.

Now the solution of our problem is clear: 1. If $j_1 = 1, \ldots, j_n = n$, then our problem has already been solved (Article 11). 2. If $j_n > n$, then by the inequality (3.5) applied to $D(j_1, \ldots, j_n)$ we find

$$n!\,\Phi = \frac{n!}{j_1 \cdots j_n} D(j_1, \ldots, j_n) < D(j_1, \ldots, j_n) \leq \frac{S^n}{(2n)^n} = 1$$

whence $\Phi < 1/n!$, in contradiction to the fact that the maximal value of Φ should be at least $1/n!$, by the result in the special case discussed in Article 11. This completes a proof of Theorem III for a "finite" matrix M, or if we wish, a proof of Theorem II for the case when the $x_i(t)$ are trigonometric polynomials.

14. Completing a proof of Theorem III. Let us now drop the assumption of a finite matrix M. Let again

(3.21) $$S = 2n$$

hold and let us show that

(3.22) $$S^n \geq (2n)^n\, n!\,\Phi$$

always holds, with the equality sign only as stated in Theorem III.

1. *We can never have the reverse inequality*

$$S^n < (2n)^n\, n!\,\Phi.$$

Indeed, if this inequality were correct, let us truncate the matrix M by replacing all columns beyond the $2m$th one by zero columns. For m sufficiently large, the reverse inequality would still hold, in contradiction with the finite case already settled.

2. *We can have the equality sign in* (3.22) *only in the case described by Theorem III.* Indeed, let us assume that

$$S^n = (2n)^n\, n!\,\Phi.$$

We may clearly assume M to have infinitely many non-zero columns. We now proceed as follows: Let m be a fixed number $> 2n$ and let us "unfreeze" the first $2m$ columns of the matrix M in such a way that (3.21) is preserved. To these $4mn$ variables we apply the multiplier rule as above, there being no difference from the previous case except that the finite sums (3.15) now become infinite series. Everything is the same as in Article 13, with the exception of the relation (3.17) which now becomes

$$\mu\,\Phi_m = \lambda\, S^{n-1} S_m,$$

where

$$\Phi_m = \sum_{i,j=1}^{2n} \sigma_{ij} \sum_{k=1}^{m} \frac{1}{k} (a_{ik} b_{jk} - a_{jk} b_{ik}),$$

$$S_m = \sum_{k=1}^{m} \sum_{i=1}^{2n} (a_{ik}^2 + b_{ik}^2).$$

However, since $\Phi_m \to \Phi$ and $S_m \to S$, as $m \to \infty$, it is clear that on choosing m sufficiently large, we have $\Phi_m > 0$, $S_m > 0$, hence again $\mu = 1$, $\lambda > 0$. But then we find, as before, that at most n among the vectors (3.19) may be non-zero. For sufficiently large m, this conclusion contradicts our assumption that M has infinitely many non-zero columns.

References

[1]. T. Bonnesen and W. Fenchel, Theorie der konvexen Körper, *Ergebnisse der Mathematik*, 3, Heft 1 (1934).

[2]. C. Carathéodory, Über den Variabilitätsbereich der Fourierschen Konstanten von positiven harmonischen Funktionen, *Rendiconti di Palermo*, 32 (1911), 193–217.

[3]. ——, *Variationsrechnung und partielle Differentialgleichungen erster Ordnung*, Leipzig, 1935.

[4]. F. Gantmakher and M. Krein, *Oscillatory matrices and kernels and small vibrations of mechanical systems* (in Russian), 2nd Ed., Moscow, 1950.

[5]. W. Gustin, Sets of finite planar order, *Duke Math. Journal*, 14 (1947), 51–66.

[6]. J. Hadamard, Résolution d'une question relative aux déterminants, *Bulletin des sciences mathématiques*, (2), 17 (1893), 240–248.

[7]. J. Hjelmslev, Introduction à la théorie des suites monotones, *Oversigt over det Kgl. Danske Videnskabernes Selskabs Forhandlinger*, 1914, 1–74.

[8]. A. Hurwitz, Sur le problème des isopérimètres, *Comptes Rendus*, 132 (1901), 401–403; also in *Math. Werke* I, 490–491.

[9]. A. Marchaud, Sur les continus d'ordre borné, *Acta Mathematica*, 55 (1930), 67–115.

[10]. J. von Neumann and I. J. Schoenberg, Fourier integrals and metric geometry, *Transactions of the Amer. Math. Society*, 50 (1941), 226–251.

[11]. G. Pólya and G. Szegö, *Aufgaben und Lehrsätze aus der Analysis*, II, Berlin, 1925.

[12]. G. Pólya, On the mean-value theorem corresponding to a given linear homogeneous differential equation, *Transactions of the Amer. Math. Society*, 24 (1922), 312–324.

[13]. Peter Scherk, Über differenzierbare Kurven und Bögen, *Časopis pro pěstování Matematiky a Fysiky*, 66 (1937), 165–191.

[14]. I. J. Schoenberg and Anne Whitney, A theorem on polygons in n dimensions with applications to variation-diminishing and cyclic variation-diminishing linear transformations, *Compositio Mathematica*, 9 (1951), 141–160.

[15]. I. J. Schoenberg, On smoothing operations and their generating functions, *Bulletin of the Amer. Math. Society*, 59 (1953), 199–230.

On the Besicovitch-Perron Solution of the Kakeya Problem

I. J. Schoenberg

1. Introduction

In [2] Perron gave a new proof of Besicovitch's solution of the problem of Kakeya [1] by means of certain polygons having the shape of trees. Perron assembled his polygon, for a given value of m, from certain 2^m triangles by making them agree in pairs at different equidistant horizontal levels which proceed from the top down; the figure corresponding to the next value, $m+1$, required a similar but entirely new construction.

In Sec. 2 we show that Perron's polygon T_m $(m = 0, 1, \cdots)$, which we propose to call the *Perron tree*, can be constructed inductively from the bottom up, the way trees grow. This inductive construction yields readily, by induction on m, the fundamental properties of T_m (Sec. 3). In Sec. 4 we give a proof of Besicovitch's result on the Kakeya problem which is a variant of Perron's proof. Section 5 describes a property of Perron's tree which suggests that it is the "best" solution of the problem raised by Besicovitch's lemma, and a conjecture to that effect is formulated.

2. The Construction of the Perron Tree

We start with a triangle $t_1^0 = A_1B_1C_1$ of height 2 having its base A_1B_1 on the x-axis $y = 0$. The shape of this triangle is irrelevant, since the construction will be affine invariant; however, for graphic reasons we assume in Fig. 1 that $A_1B_1C_1$ is isosceles and right-angled at C_1. The triangle $A_1B_1C_1$ will be called *the tree* T_0. The triangle $s_1^0 = T_0 \cap \{y \geqq 1\} = A'B'C_1$ is called the *sprout of* T_0. We let s_1^0 *sprout* by the following process, to be repeated often: *Extend the lateral sides of the sprout upward beyond its vertex until they reach the next integer level line* [here it is the level line $y = 3$], *and join the two points so obtained with the end-points of the base of the sprout.*

We obtain the simple closed polygon $A_1B_1B'C_1^1C_1C_2^1A'A_1$, which is called *the tree* T_1. The intersection $T_1 \cap \{y \geqq 2\}$ is composed of two triangles called *the sprouts of* T_1 and denoted by s_1^1, s_2^1, from right to left. To obtain the next tree, T_2, we let s_1^1 and s_2^1 sprout by the process described above.

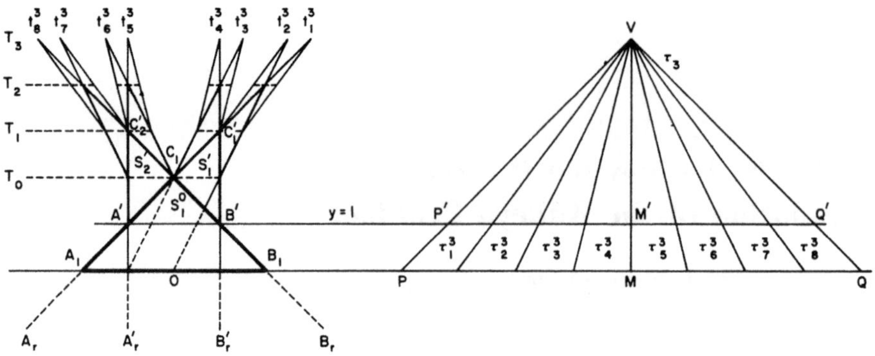

FIG. 1

The intersection $T_2 \cap \{y \geqq 3\}$ gives the four sprouts of T_2, which we denote by $s_1^2, s_2^2, s_3^2, s_4^2$, from right to left. One further step of this process leads to T_3, as shown in Fig. 1. It has eight sprouts $(s_1^3, \cdots, s_8^3)$, which may be grouped into four pairs of *twin* sprouts, where sprouts such as s_1^3 and s_2^3 are called twin sprouts if they arise from the same sprout of the preceding tree (here it is s_1^2).

This process may be continued as far as we wish and will lead, after m steps, to the tree T_m having 2^m sprouts $(s_1^m, s_2^m, \cdots, s_{2^m}^m)$, with bases on the line $y = m + 1$ and vertices on the line $y = m + 2$. The set $T_m - T_{m-1}$ is composed of 2^m triangles, which will be referred to occasionally as the triangles *of new growth of* T_m.

3. Some Properties of the Perron Tree

We shall sometimes write $T_m = T_m(A_1B_1C_1)$ to indicate the dependence of T_m on the initial triangle used in its construction. A number of properties of T_m will now be developed for which we need suitable notations.

Let t_i^m denote the triangle, similar to s_i^m, obtained from s_i^m by extending its lateral sides down to the line $y = 0$, the bases of all the t_i^m being on the line $y = 0$. Let A_i^m denote the (closed) infinite sector (or angle) obtained by extending the lateral sides of s_i^m indefinitely downward. We also extend the sides C_1A_1 and C_1B_1 of $A_1B_1C_1$, indefinitely downward and call the closed infinite region below $A_rA_1B_1B_r$ *the root of* T_m. Referring to Fig. 1, we state the following theorem.

THEOREM 1. (i) *Let $\tau_m = PQV$ be a triangle, similar to $A_1B_1C_1$, of the same height $m + 2$ as T_m, and having its base PQ on the line $y = 0$. If we divide its base PQ into 2^m equal parts and dissect τ_m accordingly into the triangles* $\tau_1^m, \tau_2^m, \cdots, \tau_{2^m}^m$ [numbered from left to right], *then an appropriate horizontal translation will make t_i^m coincide with τ_i^m $(i = 1, \cdots, 2^m)$.*

(ii) *At no stage of the construction of T_m (for $m = 1, 2, \cdots$) is there an overlap between the 2^m triangles of new growth of T_m, so that T_m is bounded by a simple closed polygon.*

(iii) *For every value of* m

(1) $$T_m \cup (\text{root of } T_m) = \bigcup_{i=1}^{2^m} \Lambda_i^m ,$$

so that by cutting off these sets at the ground level $y = 0$ *we obtain the relation*

(2) $$T_m = \bigcup_{i=1}^{2^m} t_i^m .$$

(iv) *If we denote by* $| T_m | = | T_m(A_1B_1C_1) |$ *the area of our tree, then*

(3) $$T_m = \left(1 + \frac{m}{2}\right) \cdot | A_1B_1C_1 |$$

and therefore in our particular case

(4) $$| T_m(A_1B_1C_1) | = 2m + 4 .$$

The truth of these statements is easily verified in Fig. 1 for $m = 0, 1, 2,$ and 3. The proof by induction in m is based on the following observation: The right part of T_3 obtained by cutting along the segments $A'B'$ and $A'C_1$ may also be obtained by applying our construction to the initial triangle $A'B'C_1^1$, obtaining $T_2(A'B'C_1^1)$. The similar left part of T_3 may be described as $T_2(A'B'C_2^1)$. Generally,

$$T_m = T_{m-1}(A'B'C_1^1) \cup T_{m-1}(A'B'C_2^1) \cup (\text{trapezoid } A_1B_1B'A') .$$

Let us now assume the statements (i), (ii), (iii), and (iv) to be true for all values below m, and let us prove them for m (in Fig. 1 $m = 3$ but the reasoning holds for an arbitrary m). Let $\tau_3 = PQV$ be the triangle of height $m + 2 = 5$ similar to $A_1B_1C_1$, hence isosceles and right-angled. We divide it into $2^m = 8$ triangles $\tau_1^3, \cdots, \tau_8^3$ of equal bases. Let

$$t_i^m \cap \{y \geqq 1\} = \tilde{t}_i^m , \qquad \tau_i^m \cap \{y \geqq 1\} = \tilde{\tau}_i^m .$$

Let VM be the median of τ_3 and let the line $y = 1$ intersect VP, VM, VQ in P', M', Q', respectively. However, $P'M'V$ is similar to $A'B'C_1^1$, and by the induction assumption applied to the tree $T_2(A'B'C_1^1)$ (T_2^r for brevity), we know that $\tilde{t}_1^3, \tilde{t}_2^3, \tilde{t}_3^3, \tilde{t}_4^3$ are congruent and parallel to $\tilde{\tau}_1^3, \tilde{\tau}_2^3, \tilde{\tau}_3^3, \tilde{\tau}_4^3$, respectively. The same reasoning applied to $T_2(A'B'C_2^1)$ (T_2^l) shows that $\tilde{t}_5^3$, $\tilde{t}_6^3, \tilde{t}_7^3, \tilde{t}_8^3$ are congruent and parallel to $\tilde{\tau}_5^3, \tilde{\tau}_6^3, \tilde{\tau}_7^3, \tilde{\tau}_8^3$, respectively. But then it is evident that $t_1^3, \cdots, t_8^3$ are congruent and parallel to $\tau_1^3, \cdots, \tau_8^3$, respectively. This completes a proof of (i).

Concerning (ii), we know that by the induction assumption no overlap among triangles of new growth appears in T_2^r or in T_2^l. Then it is evident also that T_3 will be free of such overlap.

To establish (iii) we observe that by assumption,

$$T_{m-1}^r \cup (\text{root of } T_{m-1}^r) = \bigcup_{i=1}^{2^{m-1}} \Lambda_i^m ,$$

$$T_{m-1}^l \cup (\text{root of } T_{m-1}^l) = \bigcup_{i=2^{m-1}+1}^{2^m} \Lambda_i^m .$$

On equating the union of the left sides of these relations with the union of the right sides, we obtain the desired relation (1).

Assertion (iv) remains to be established. Observe that $|A'B'C_1^1| = |A'B'C_2^1| = 2$. By (3) and our induction assumption,

$$|T_{m-1}^r| = |T_{m-1}^l| = \left(1 + \frac{m-1}{2}\right) \cdot 2 = m + 1 .$$

But then (Fig. 1)

$$\begin{aligned}|T_m| &= |T_{m-1}^r| + |T_{m-1}^l| - |A'B'C_1| + |A_1B_1B'A'| \\ &= 2m + 2 - 1 + 3 = 2m + 4 = \left(1 + \frac{m}{2}\right) \cdot |A_1B_1C_1| ,\end{aligned}$$

and this completes a proof of Theorem 1.

4. A Solution of Kakeya's Problem

Besicovitch [1] solved Kakeya's problem by proving the following remarkable theorem. *There exist plane figures of arbitrarily small area in which a segment of unit length may be turned around by* 360 *degrees so as to return to its initial position.*

The main point of the proof is the following.

LEMMA OF BESICOVITCH. *Let ABC be a triangle—for instance, one that is right-angled at C and isosceles. Given a positive* ε, *we can find an integer n having the following property: If we divide the base AB into n equal parts and join the points of subdivision to C so as to dissect ABC into n slim triangles, then these n triangles can be translated independently parallel to AB to produce so much overlap among them that the final figure so obtained has an area less than* ε.

PROOF. Assume the triangle ABC to be of height 1. We take the Perron tree T_m of Sec. 2 and shrink it in the ratio $(m + 2)$ to 1, obtaining a tree T_m^* of height 1, and we place ABC and T_m^* side by side on the line $y = 0$. By Theorem 1 (i) and by (2), we know that T_m^* can also be obtained by dividing ABC into 2^m slim triangles and translating these appropriately along the base-line $y = 0$. By Theorem 1 (iv),

$$|T_m^*| = \frac{|T_m|}{(m+2)^2} = \frac{2m+4}{(m+2)^2} = \frac{2}{m+2} ,$$

and this quantity will be less than ε, provided that m is sufficiently large. For such values of m the number $n = 2^m$ satisfies the requirements of Besicovitch's lemma, which is thereby established.

5. A Conjecture Concerning T_m

We return to the triangle $\tau_m = \bigcup \tau_i^m$ of Theorem 1 (i) (Fig. 1). We know by Theorem 1 (i), (iii), and (iv) that appropriate horizontal translations of the triangles τ_i^m will produce the tree T_m having area $2m + 4$.

CONJECTURE. *Among all figures obtained by independent horizontal translations of the* 2^m *triangles* τ_i^m, *the tree* T_m *has the minimum area* $(2m+4)$.

This I am unable to prove except for the first few values of m. Nevertheless, I wish to record here a result that makes the conjecture at least very plausible. Referring to Fig. 1, let us look at one of the eight triangles t_i^3, for instance t_1^3, and let us see which parts of the boundary of t_1^3 lie on the boundary of T_3. Aside from the base of t_1^3, which we ignore, we see that the right-hand edge of t_1^3 has a segment of vertical component equal to 2 on boundary (T_3), while the left edge of t_1^3 has two segments in common with boundary (T_3), these segments having vertical components of one unit each. If we now displace t_1^3 horizontally by a very small amount δ, then the area of T_3 will increase, up to second-order terms, by $\Delta \mid T_3 \mid = 2\delta - 2\delta = 0$. Similarly, we find that *boundary* $(t_2^3) \cap$ *boundary* (T_3) is composed of three segments, one of vertical component equal to 2 on the left edge and two of unit vertical component each, on the right edge of t_2^3.

THEOREM 2. *The intersection* boundary $(t_{2i-1}^m) \cap$ boundary (T_m) *is composed, ignoring the base-line, of a segment of vertical component equal to* 2 *on the right edge of* t_{2i-1}^m *and of two segments, each of vertical component equal to* 1, *on the left edge of* t_{2i-1}^m. *The same statement applies to* t_{2i}^m, *except that the terms "right" and "left" are to be interchanged.*

A proof by induction in m is so simple that it may be omitted. Theorem 2 supports our conjecture for the following reason. Let us start from the triangle $\tau_m = \bigcup \tau_i^m$ and translate τ_i^m by the amount x_i $(i = 1, \cdots, 2^m)$, denoting by $f(x_1, \cdots, x_{2^m})$ the area of the final figure so obtained. By Theorem 2 and the remark preceding its statement, we see that for the particular configuration T_m we have the relations

$$\frac{\partial f}{\partial x_i} = 0 \qquad (i = 1, \cdots, 2^m) .$$

Thus the so-called *first necessary conditions* for a minimum are satisfied.

In concluding this paper I wish to mention that the infinite tree T_∞, obtained from T_m by letting $m \to \infty$, has interesting properties, which might be discussed on a future occasion.

The University of Pennsylvania

REFERENCES

[1] BESICOVITCH, A. S., On Kakeya's Problem and a Similar One, *Math. Z.*, **27** (1928), 312-20.

[2] PERRON, O., Über einen Satz von Besicovitch, *Math. Z.*, **28** (1928), 383-86.

Reprinted from *Studies in Mathematical Analysis and Related Topics*, Pólya volume, Stanford University Press, Stanford, California, 1962, pp. 359–363

COMPLEX ANALYSIS

ON LINEAL ENTIRE FUNCTIONS OF n COMPLEX VARIABLES[1]

T. S. MOTZKIN AND I. J. SCHOENBERG

1. Statement of problems and main results. A polynomial $P(z_1, \cdots, z_n)$ with real or complex coefficients is called *lineal* if it splits into a product of linear functions of the variables in the complex field of coefficients. Every polynomial $P(z_1)$, of one variable only, is lineal, by the fundamental theorem of algebra; however, for two or more variables lineality is a severe restriction. As an example we mention the cyclic determinant of order n which is a lineal polynomial in the elements of its first row.

We say that a sequence of entire functions $f_k(z_1, \cdots, z_n)$ *converges regularly* to a function $f(z_1, \cdots, z_n)$, and write reg. lim $f_k = f$, meaning thereby that the convergence holds uniformly in every bounded portion of space. Every entire function is a regular limit of polynomials, for instance the sections of its Taylor expansion. But what if we require the approximating polynomials to be lineal? More explicitly: *What entire functions $f(z_1, \cdots, z_n)$ are regular limits of lineal polynomials?* We call such functions *lineal* and denote their class by the symbol $\mathcal{L}^{(n)}$.

By a plane we mean the manifold defined by one linear equation in the variables. It seems plausible that a lineal function will vanish on planes only, by which we mean the following: If f vanishes in a point, then it will vanish on a plane passing through that point. This property does hold and turns out to be characteristic for lineal functions, as stated by the following:

THEOREM 1. *An entire function $f(z_1, \cdots, z_n)$ is lineal if and only if it vanishes on planes only.*

This theorem implies in particular that every entire zero-free function is lineal. As will be shown below, the Weierstrass product representation may be carried over unchanged to functions in $\mathcal{L}^{(n)}$, reducing in the case when $n=1$ to the classical situation. Hadamard's theory of functions of finite order may likely also be so extended.

Noteworthy are the following further concepts. A lineal polynomial

Presented to the Society, December 28, 1951; received by the editors December 11, 1951.

[1] This work was performed on a National Bureau of Standards contract with the University of California, Los Angeles and was sponsored (in part) by the Office of the Air Comptroller, USAF, and by the Office of Naval Research.

$P(z_1, \cdots, z_n)$ is called *really lineal* if it splits into a product of linear functions with only real coefficients. A really lineal polynomial $P(z_1, \cdots, z_n)$ is called *positively lineal* if the coefficients of its linear factors are all greater than or equal to 0. Thus $z_1^2+z_2^2$ is lineal, $z_1^2-z_2^2$ is really lineal, while z_1z_2 is positively lineal. Let us denote by $\mathcal{L}_r^{(n)}$ the class of entire functions $f(z_1, \cdots, z_n)$ which are regular limits of really lineal polynomials and call the elements of $\mathcal{L}_r^{(n)}$ *really lineal functions*. Likewise let $\mathcal{L}_p^{(n)}$ denote the class of *positively lineal functions* which are defined as regular limits of positively lineal polynomials.

While lineal functions of one variable are completely arbitrary entire functions, this is no longer the case for *really* lineal and for *positively* lineal functions of one variable. Indeed, a polynomial $P(z)$ of one variable is really lineal if and only if it has only real zeros and only real coefficients and $P(z)$ is positively lineal if and only if it has only real nonpositive zeros and only non-negative coefficients. Regular limits of polynomials of such special nature have been determined by Laguerre and Pólya (see References [1; 2]) with results which may be described as follows:

THEOREM OF LAGUERRE AND PÓLYA. 1. *$f(z)\in\mathcal{L}_r^{(1)}$ if and only if $f(z)$ admits a Weierstrass product representation of the form*

$$f(z) = Ce^{-\gamma z^2+\delta z}z^m\prod_{\nu=1}^{\infty}(1+\delta_\nu z)e^{-\delta_\nu z}, \tag{1}$$

$$(C,\ \delta,\ \delta_\nu\ \textit{real},\ \gamma \geqq 0,\ \sum \delta_\nu^2 < \infty).$$

2. *$f(z)\in\mathcal{L}_p^{(1)}$ if and only if $f(z)$ admits a representation of the form*

$$f(z) = Ce^{\gamma z}z^m\prod_{\nu=1}^{\infty}(1+\delta_\nu z) \qquad (C,\ \gamma,\ \delta_\nu \geqq 0,\ \sum \delta_\nu < \infty). \tag{2}$$

Using the familiar notation $(a, b)=a_1\bar{b}_1+\cdots+a_n\bar{b}_n$ and $\|a\|^2=(a, a)$ for the scalar product and norm in the unitary space of n complex dimensions, we may now describe the classes $\mathcal{L}_r^{(n)}$ and $\mathcal{L}_p^{(n)}$ by the following theorems.

THEOREM 2. *An entire function is really lineal if and only if it admits a representation of the form*

$$f(z_1, \cdots, z_n) = \exp\left(-\sum_1^n \gamma_{\mu\nu}z_\mu z_\nu + (z, \delta)\right)\prod_{j=1}^{m}(z, c_j) \cdot \prod_{\nu=1}^{\infty}(1+(z, \delta_\nu))e^{-(z,\delta_\nu)}, \tag{3}$$

where all vectors c_j, δ, δ_ν have real components, the series $\sum_{\nu=1}^{\infty} \|\delta_\nu\|^2$ converges, while the quadratic form $\sum \gamma_{\mu\nu} z_\mu z_\nu$ has real coefficients and is positive (definite, or semi-definite).

THEOREM 3. *An entire function is positively lineal if and only if it admits a representation of the form*

$$f(z_1, \cdots, z_n) = e^{(z,\gamma)} \prod_{j=1}^{m} (z, c_j) \prod_{\nu=1}^{\infty} (1 + (z, \delta_\nu)), \tag{4}$$

where all vectors c_j, γ, δ_ν have real, non-negative components, the series $\sum_{\nu=1}^{\infty} \|\delta_\nu\|$ being convergent.

Notice that if $n=1$, Theorems 2 and 3 reduce, as they should, to the statements of the theorem of Laguerre and Pólya.

2. **On functions vanishing on planes only.** Let us denote for convenience by $H^{(n)}$ the class of entire functions $f(z_1, \cdots, z_n)$ vanishing on planes only. An element f of $H^{(n)}$ may be zero-free, in which case it is of the form

$$f(z_1, \cdots, z_n) = e^{G(z_1, \cdots, z_n)} \qquad (G \text{ entire}). \tag{5}$$

If it has zero-planes $L \equiv (z, c) = 0$, or $L \equiv 1-(z, \delta) = 0$, the following facts are readily established: 1. f vanishes on every zero-plane to a certain finite integral order μ, called the multiplicity of the zero-plane, with the property that f/L^μ is entire and nonzero in at least one point of the plane. 2. The totality of the zero-planes of a function in $H^{(n)}$, each counted as often as its multiplicity indicates, form a denumerable set of planes which have no plane of accumulation, i.e., they must recede to infinity. Disregarding for the moment the finitely many zero-planes passing through the origin, we may denote all zero-planes of f by

$$(z, \delta_\nu) = 1 \qquad (\nu = 1, 2, \cdots).$$

Since necessarily $\|\delta_\nu\| \to 0$, we may so number these planes that their distances $\|\delta_\nu\|^{-1}$ from the origin form a nondecreasing sequence, hence $\|\delta_1\| \geq \|\delta_2\| \geq \cdots$.

Now the classical Weierstrass representation goes through without change. Let

$$P_1(u) = 1 - u,$$

$$P_k(u) = (1 - u) \exp\left(u + \frac{1}{2} u^2 + \cdots + \frac{1}{k-1} u^{k-1}\right) \qquad (k = 2, \cdots). \tag{6}$$

A sequence of positive integers k_ν may always be determined such that the series

$$(7) \qquad \sum_{\nu=1}^{\infty} |(z, \delta_\nu)|^{k_\nu} \leqq \sum_{\nu=1}^{\infty} (\|z\| \cdot \|\delta_\nu\|)^{k_\nu}$$

converge for all complex vectors $z=(z_1, \cdots, z_n)$. (For instance the choice $k_\nu=\nu$ will always do.) Denoting by $(z, c_j)=0$ $(j=1, \cdots, m)$ the zero-planes passing through the origin, we see that

$$\prod_{j=1}^{m} (z, c_j) \cdot \prod_{\nu=1}^{\infty} P_{k_\nu}((z, \delta_\nu))$$

is an entire function vanishing on the zero-planes of f to the correct order and nowhere else. If we divide this function into f we obtain a zero-free function of the form (5). Thus

$$(8) \qquad f(z_1, \cdots, z_n) = e^{G(z_1, \cdots z_n)} \prod_{j=1}^{m} (z, c_j) \prod_{\nu=1}^{\infty} P_{k_\nu}((z, \delta_\nu))$$

is a characteristic product representation for the elements of $H^{(n)}$.

3. A proof of Theorem 1. This proof consists of two parts. (i) *$f\in\mathcal{L}^{(n)}$ implies that $f\in H^{(n)}$.* Let $f\in\mathcal{L}^{(n)}$. If f is zero-free there is nothing to prove. If f vanishes in a point z_0 we have to show that f vanishes in a plane passing through z_0. Without loss of generality we may assume that $f(0, \cdots, 0)=0$. Furthermore, without loss of generality we may assume that $f(z_1, 0, \cdots, 0)\not\equiv 0$. Indeed, assuming $f(z_1, \cdots, z_n)\not\equiv 0$, let us examine the Taylor expansion of f at the origin:

$$f = F_k + F_{k+1} + \cdots,$$

where F_k is a homogeneous polynomial in the variables z_ν, not vanishing identically. If F_k has a term in z_1^k, then our assumption $f(z_1, 0, \cdots, 0)\not\equiv 0$ is verified. If F_k has no term in z_1^k, we observe that our classes $\mathcal{L}^{(n)}$ and $H^{(n)}$ are clearly invariant with respect to affine transformations of the variables. An appropriate homogeneous transformation

$$z_\mu = \sum_{\nu=1}^{n} c_{\mu\nu} w_\nu \qquad (\mu = 1, \cdots, n)$$

[$F_k(c_{11}, c_{21}, \cdots, c_{n1})\neq 0$ is the only requirement] will transform F_k into a kth degree form of $w_1, \cdots, w_n$ having a nonvanishing term in w_1^k.

It now follows that $f(z_1, 0, \cdots, 0)$ admits $z_1=0$ as a k-fold zero ($k\geqq 1$). Let $P_\nu(z_1, \cdots, z_n)$ be a sequence of lineal polynomials such that reg. lim $P_\nu=f$. But then also reg. lim $P_\nu(z_1, 0, \cdots, 0) =f(z_1, 0, \cdots, 0)$. By Hurwitz's theorem $P_\nu(z_1, 0, \cdots, 0)$ has at least k zeros in $|z_1|<\rho$, no matter how small ρ is, provided $\nu>N(\rho)$. Since $P_\nu(z_1, \cdots, z_n)$ is lineal, surely one of its linear factors, L_ν say, must vanish in a point which is within a ρ-neighborhood of the origin. Thus $P_\nu(z_1, \cdots, z_n)$ vanishes on a plane $L_\nu=0$ whose distance to the origin converges to zero. Taking a proper subsequence, if necessary, we may assume that the planes $L_\nu=0$ converge to a limit-plane $L=0$. We now claim that $f=0$ on $L=0$. Indeed, let $z=\zeta$ be a point on $L=0$. Then $\zeta=\lim \zeta_\nu$, where ζ_ν is on $L_\nu=0$. But $P_\nu(\zeta_\nu)=0$ and the regular convergence of $P_\nu(z)$ to $f(z)$ implies that $f(\zeta)=\lim P_\nu(\zeta_\nu) =0$.

(ii) *$f\in H^{(n)}$ implies that $f\in\mathcal{L}^{(n)}$.* We know that $H^{(n)}$ is identical with the class of entire functions admitting a representation of the form (8). There remains to show that (8) will always define lineal functions. Now $P_k(z_1)\in\mathcal{L}^{(n)}$, as every entire function of a single variable, and therefore also

$$P_{k_\nu}((z, \delta_\nu)) \in \mathcal{L}^{(n)}.$$

Since $\mathcal{L}^{(n)}$ is obviously closed with respect to multiplication and regular convergence, we conclude that the regularly convergent product of (8) is in $\mathcal{L}^{(n)}$. The proof that f, as defined by (8), is in $\mathcal{L}^{(n)}$ will be complete as soon as we establish the following

LEMMA. *If $G(z_1, \cdots, z_n)$ is entire, then*

$$f(z_1, \cdots, z_n) = e^{G(z_1,\cdots,z_n)}$$

is a lineal function.

PROOF. It suffices to prove the lemma for the special case when G is a polynomial: Indeed, if Q_n is the section of the Taylor expansion of G, then Q_n converges regularly to G, hence e^{Q_n} converges regularly to $f=e^G$. But then the lineality of e^{Q_n} implies the lineality of its regular limit f. Again, by the multiplicative property of $\mathcal{L}^{(n)}$ it suffices to assume that G reduces to a single monomial, hence to prove that

$$\exp\,(az_1^{\nu_1}z_2^{\nu_2}\cdots z_n^{\nu_n}) \in \mathcal{L}^{(n)}. \tag{9}$$

This will readily follow from the fact that a monomial of degree $\nu_1+\cdots+\nu_n=m$ can always be written as a linear combination with constant coefficients of mth powers of linear forms

$$z_1^{\nu_1} z_2^{\nu_2} \cdots z_n^{\nu_n} = c_1 L_1^m + \cdots + c_N L_N^m \qquad (m = \sum \nu_i). \tag{10}$$

In fact the constants c_i as well as the coefficients of the variables z_ν in the forms L_i may be assumed to have rational values.

Taking for granted an identity of the form (10), we indeed conclude that

$$\exp(a z_1^{\nu_1} \cdots z_n^{\nu_n}) = \prod_{i=1}^{N} \exp(a c_i L_i^m)$$

is lineal, because each factor on the right-hand side is lineal, being an entire function of a linear function of our variables.

To establish an identity of the form (10), we consider first the case of two variables and show that $x^p y^q$ can be written as

$$x^p y^q = \sum_{i=0}^{m} A_i (x + \lambda_i y)^m \qquad (m = p + q), \tag{11}$$

with rational λ_i, A_i. In fact, let $\lambda_i = i$; expanding the binomials and comparing coefficients, we obtain for the unknowns A_i a linear system of equations whose determinant is the nonvanishing Vandermondian of the λ's. This identity (11) now generalizes automatically for more than two variables. Indeed, let

$$u^m z^r = \sum_{j=0}^{m+r} B_j (u + \mu_j z)^{m+r}$$

be the identity corresponding to the monomial $u^m z^r$. Multiplying (11) by z^r we find that

$$\begin{aligned} x^p y^q z^r &= \sum_{i=0}^{m} A_i (x + \lambda_i y)^m z^r \\ &= \sum_{i=0}^{p+q} \sum_{j=0}^{p+q+r} A_i B_j (x + \lambda_i y + \mu_j z)^{p+q+r} \end{aligned}$$

is also a linear combination of $(p+q+r)$th powers of linear forms. An obvious induction argument establishes the general validity of (10). This concludes our proof of Theorem 1.

4. A proof of Theorem 2. Let us show first that f, as defined by (3), is always in $\mathcal{L}_r^{(n)}$, i.e., really lineal. Indeed, since $e^{z_1} \in \mathcal{L}_r^{(1)}$, we see that $e^{(z,\delta)} \in \mathcal{L}_r^{(n)}$, provided the vector δ is real. Thus each of the primary factors of (3) is really lineal and therefore so is their product. There remains to show that

$$\exp\left(-\sum_{1}^{n}\gamma_{\mu\nu}z_\mu z_\nu\right)\in\mathcal{L}_{,r}^{(n)},$$

provided the quadratic form in the exponent is positive. This follows from the representation of the quadratic form as a sum of squares of linear forms:

$$\begin{aligned}\exp\left(-\sum\gamma_{\mu\nu}z_\mu z_\nu\right) &= \exp\left(-L_1^2-L_2^2-\cdots-L_s^2\right)\\ &= \prod_{\sigma=1}^{s}\exp\left(-L_\sigma^2\right) \qquad (s\leqq n),\end{aligned}$$

and the remark that $\exp(-z_1^2)\in\mathcal{L}_r^{(1)}$, hence $\exp(-L_\sigma^2)\in\mathcal{L}_r^{(n)}$.

Conversely, let f be really lineal and let us show that f must be of the form (3). Now $f\in\mathcal{L}_r^{(n)}$ implies that $f\in\mathcal{L}^{(n)}$. By Theorem 1 we conclude that f vanishes on planes only and therefore admits a product representation of the form (8). Let $\beta_1, \beta_2, \cdots, \beta_n$ be real constants, not all zero, and let us consider $f(z_1, \cdots, z_n)$ on the ray

$$(12)\qquad z_1=\beta_1 t,\ z_2=\beta_2 t,\ \cdots,\ z_n=\beta_n t.$$

Since a really lineal polynomial, considered as a function of t on such a ray, has only real zeros, it is clear that the function

$$(13)\qquad g(t)=f(\beta_1 t, \beta_2 t, \cdots, \beta_n t)$$

belongs to the Laguerre-Pólya class $\mathcal{L}_r^{(1)}$. If we introduce (12) into the product representation (8) and apply to it the Laguerre-Pólya result that, as a function of t, it must be of the form (1), we obtain the following conclusions:

(i) The nonvanishing zeros of $g(t)$, as given by (8) and (12), are furnished by

$$1-(z,\delta_\nu)\equiv 1-(\beta,\delta_\nu)t=0 \qquad (\nu=1,2,\cdots).$$

Since these zeros must all be real, by (1), we conclude that the scalar product (β, δ_ν) is real, for each choice of the real $\beta_1, \beta_2, \cdots, \beta_n$. This implies that all vectors δ_ν are real. Furthermore, (1) also implies that

$$\sum_{\nu=1}^{\infty}(\beta,\delta_\nu)^2<\infty,$$

again for every choice of the real vector β. If $\delta_\nu=(\delta_{\nu 1}, \delta_{\nu 2}, \cdots, \delta_{\nu n})$, on choosing successively $\beta=(1, 0, \cdots, 0)$, $(0, 1, 0, \cdots, 0)$, $\cdots$, $(0, \cdots, 0, 1)$, we find that the n series $\sum_\nu \delta_{\nu j}^2$ converge, so that

$$\sum_{\nu=1}^{\infty} \|\delta_\nu\|^2 < \infty.$$

But then the series (7) converge if we assume that all $k_\nu = 2$ ($\nu = 1, 2, \cdots$). The canonical product of (8) now becomes of genus unity; writing $-\delta_\nu$, instead of δ_ν, (8) reduces to

$$f(z_1, \cdots, z_n) = e^{G(z_1, \cdots, z_n)} \prod_{j=1}^{m} (z, c_j) \cdot \prod_{\nu=1}^{\infty} (1 + (z, \delta_\nu)) e^{-(z, \delta_\nu)}. \tag{14}$$

(ii) Substituting again (12) into (14) and comparing with (1), we obtain the following identity in t

$$G(\beta_1 t, \cdots, \beta_n t) = -\gamma t^2 + \delta t,$$

for every choice of the β's, where γ and δ are real and depend on the β's only, with $\gamma \geqq 0$. This implies that $G(z_1, \cdots, z_n)$ reduces to a real quadratic polynomial

$$G \equiv -\sum_{\mu,\nu=1}^{n} \gamma_{\mu\nu} z_\mu z_\nu + \sum_{\nu=1}^{n} \epsilon_\nu z_\nu,$$

with the property that

$$\sum \gamma_{\mu\nu} \beta_\mu \beta_\nu = \gamma \geqq 0,$$

for arbitrary real $\beta_1, \cdots, \beta_n$. This entails the positivity of the form.

There remains to show that the homogeneous linear forms (z, c_j) of (3) may all be written with only real coefficients. This we conclude immediately as follows: Observe that we have so far shown that if a really lineal function $f(z_1, \cdots, z_n)$ vanishes on a plane which does not contain the origin, such a plane must be real. The same conclusion, however, may be applied also to planes containing the origin, as seen by a shift of the origin to another real point (the property $f \in \mathcal{L}_r^{(n)}$ is invariant with respect to real translations).

5. A proof of Theorem 3. If $f \in \mathcal{L}_p^{(n)}$, then a fortiori $f \in \mathcal{L}_r^{(n)}$ so that we may already assume f to be of the form (3). We now observe that if $P(z_1, \cdots, z_n)$ is a positively lineal polynomial and

$$z_1 = \beta_1 t,\ z_2 = \beta_2 t, \cdots, z_n = \beta_n t \quad (\beta_1 \geqq 0, \cdots, \beta_n \geqq 0, \textstyle\sum \beta_\nu > 0) \tag{15}$$

is a ray in the positive orthant, then $P(\beta_1 t, \cdots, \beta_n t)$ is a polynomial in t having only real and nonpositive zeros and non-negative coefficients. Thus $g(t)$, defined by (13), is a function of the Laguerre-Pólya class $\mathcal{L}_p^{(1)}$. Confronting the product representation for $g(t)$, as given by (3), with what it ought to be in view of (2), we obtain the following conclusions:

(i) The zeros of $g(t)$, given by $1+(z, \delta_\nu) \equiv 1+(\beta, \delta_\nu)t=0$, are to be negative. This requires that $(\beta, \delta_\nu) \geqq 0$ for every choice of the "positive" vector β. On replacing β successively by the unit vectors along the axes we find that all components of δ_ν are non-negative. Furthermore, (2) also implies that

$$\sum_{\nu=1}^{\infty} (\beta, \delta_\nu) < \infty$$

for every "positive" β. Thus, if $\delta_\nu = (\delta_{\nu 1}, \cdots, \delta_{\nu n})$, all $\sum_{\nu=1}^{\infty} \delta_{\nu j}$ converge, so that also

$$\sum_{\nu=1}^{\infty} \|\delta_\nu\| \leqq \sum_{\nu=1}^{\infty} (\delta_{\nu 1} + \cdots + \delta_{\nu n}) < \infty.$$

The product representation (3) may now be reduced to the form

$$f(z_1, \cdots, z_n) = \exp\left(-\sum \gamma_{\mu\nu} z_\mu z_\nu + (z, \delta)\right) \prod_{j=1}^{m} (z, c_j) \prod_{\nu=1}^{\infty} (1 + (z, \delta_\nu)). \tag{16}$$

(ii) Substituting again (15) into (16) and comparing with (2) we obtain

$$\sum \gamma_{\mu\nu}\beta_\mu\beta_\nu = 0, \qquad (\beta, \delta) = \gamma \geqq 0,$$

for every choice of $\beta_\nu \geqq 0$. The quadratic form therefore vanishes and δ is a "positive" vector.

The final remark to the effect that the vectors c_j, in (4), may all be chosen so as to have non-negative components is seen as follows. It is clearly sufficient to show that none of the vectors c_j can have two components of opposite signs. This, however, is the case, for otherwise we could find positive values $z_\nu = p_\nu$ $(\nu = 1, \cdots, n)$ such that

$$f(p_1, \cdots, p_n) = 0. \tag{17}$$

On the other hand f is the limit of a sequence of positively lineal polynomials P_k. (17) implies that

$$\lim_{k\to\infty} P_k(p_1, \cdots, p_n) = 0. \tag{18}$$

The inequality

$$|P_k(z_1, \cdots, z_n)| \leqq P_k(p_1, \cdots, p_n) \qquad (|z_1| \leqq p_1, \cdots, |z_n| \leqq p_n)$$

shows that (18) implies that $f \equiv 0$.

References

1. E. Laguerre, *Sur les fonctions du genre zéro et du genre un*, Oeuvres de Laguerre, vol. 1, Paris, 1898, pp. 174–177.

2. G. Pólya, *Über Annäherung durch Polynome mit lauter reellen Wurzeln*, Rendiconti di Palermo vol. 36 (1913) pp. 1–17.

University of California, Los Angeles, and
University of Pennsylvania

Reprinted from
Proc. Amer. Math. Soc. **3** (1952), pp. 517–526

Annals of Mathematics
Vol. 62, No. 3, November, 1955
Printed in U.S.A.

ON THE ZEROS OF THE GENERATING FUNCTIONS OF MULTIPLY POSITIVE SEQUENCES AND FUNCTIONS[1]

By I. J. Schoenberg

(Received July 26, 1954)

§1. Introduction

1. Multiply positive sequences. Let $\{a_n\}$, $-\infty < n < \infty$, be a sequence of non-negative numbers such that $\sum a_n$ converges. Such a sequence is said to be *totally positive* provided the matrix

$$(1.1)\qquad A = \| a_{j-i} \| = \left\| \begin{matrix} \cdots & \cdot & \cdot & \cdot & \cdots \\ \cdots & a_0 & a_1 & a_2 & \cdots \\ \cdots & a_{-1} & a_0 & a_1 & \cdots \\ \cdots & a_{-2} & a_{-1} & a_0 & \cdots \\ \cdots & \cdot & \cdot & \cdot & \cdots \end{matrix} \right\|$$

is totally positive, i.e. has no minor which is negative. The problem of the function-theoretic characterization of the corresponding generating function

$$(1.2)\qquad \sum_{-\infty}^{\infty} a_n z^n$$

was recently solved by Edrei ([1]) thereby confirming a conjecture by the writer. For the simplest of all cases when (1.2) reduces to a polynomial we have the following

Theorem 0. *The sequence*

$$(1.3)\qquad \cdots, 0, 0, a_0, a_1, \cdots, a_m, 0, 0, \cdots, \quad (a_0 > 0,\ a_m > 0),$$

is totally positive if and only if the polynomial

$$(1.4)\qquad P(z) = a_0 + a_1 z + \cdots + a_m z^m, \qquad (a_0 > 0,\ a_m > 0),$$

has only real and negative zeros.

A natural generalization of the concept of a totally positive sequence is as follows:[2] *Let k be a given natural integer. We say that the sequence* $\{a_n\}$ ($a_n \geqq 0$, $\sum a_n$ convergent) *is k-times positive, or k-positive, provided the matrix* (1.1) *has no negative minor of order $\leqq k$.*

If we denote by $\mathcal{P}_k$ the class of k-times positive sequences we have a shrinking sequence of classes

$$\mathcal{P}_1 \supset \mathcal{P}_2 \supset \cdots \supset \mathcal{P}_k \supset \cdots$$

[1] The present work was partly sponsored by the Air Research and Development Command, USAF, at the University of Pennsylvania. Presented before the American Mathematical Society on December 28, 1953.

[2] The concept of a k-times positive sequence is essentially due to Fekete [3], page 3. Ostrowski calls a sequence *normal* provided it is 2-times positive. See [7], page 115.

whose intersection $\mathcal{P}_\infty$ is the class of totally positive sequences. The class $\mathcal{P}_1$ is essentially equivalent with the discrete and equidistant distributions of the theory of probabilities. The next class $\mathcal{P}_2$ may also be described directly as follows:

$\{a_n\} \in \mathcal{P}_2$ *if and only if there exist two integers* $\alpha, \beta(-\infty \leqq \alpha < \beta \leqq +\infty)$ *such that*[3] $a_n = 0$ *if* $n \leqq \alpha$ *or* $n \geqq \beta$, *while if* $\alpha < n < \beta$ *then* $a_n > 0$ *and the polygonal line of vertices*

$$x = n, \qquad y = \log a_n, \qquad (\alpha < n < \beta), \tag{1.5}$$

is concave.

See [10], Lemma 1, page 337, for the analogous situation for functions. The present case of sequences is so much simpler that no details are deemed necessary. We may therefore describe a twice-positive sequence $\{a_n\}$ as one which is logarithmically concave on a finite or infinite range of consecutive integers and having vanishing elements outside this range. As examples we mention the sequences:

(i) $a_n = e^{-n^2} \; (-\infty < n < \infty)$,

(ii) $a_n = \begin{cases} 1 \text{ if } 0 \leqq n \leqq m, \\ 0 \text{ if } n < 0 \text{ or } n > m. \end{cases}$

In view of Edrei's theorem concerning totally positive sequences, the problem now arises of discovering analytical properties of the generating function (1.2) of a k-times positive sequence $\{a_n\}$. A contribution in this direction for the very special case when (1.2) reduces to a polynomial (1.4) is furnished by the following

THEOREM 1. *A polynomial* (1.4), *such that the sequence* (1.3) *is k-times positive, has no zeros in the sector*

$$|\arg z| < \frac{k\pi}{m + k - 1}. \tag{1.6}$$

The condition that the sequence (1.3) be k-times positive would seem to require that infinitely many among the minors of (1.1), of orders $\leqq k$, be non-negative. That such is not the case is shown by

THEOREM 2. *The sequence* (1.3) *is k-times positive if and only if the matrix*

$$A_k = \left\| \begin{matrix} a_0 & a_1 & \cdots & a_m & 0 & 0 \\ 0 & a_0 & \cdots & a_{m-1} & a_m & 0 \\ \vdots & \vdots & \ddots & & \ddots & \vdots \\ 0 & 0 & \cdots & a_0 & \cdots & a_m \end{matrix} \right\|, \qquad \textit{having } k \textit{ rows}, \tag{1.7}$$

is totally positive, i.e. has only non-negative minors.

[3] It should be clear that a requirement such as "$a_n = 0$ if $n \leqq \alpha$" is vacuously fulfilled if $\alpha = -\infty$.

The sequence (1.3) *is certainly k-times positive if the following inequalities hold*

$$(1.8)\qquad \begin{vmatrix} a_n & a_{n+1} & \cdots & a_{n+\nu-1} \\ a_{n-1} & a_n & \cdots & a_{n+\nu-2} \\ \vdots & \vdots & & \vdots \\ a_{n-\nu+1} & a_{n-\nu+2} & \cdots & a_n \end{vmatrix} > 0, \qquad \textit{if} \quad 0 \leqq n \leqq m, \quad 1 \leqq \nu \leqq k,$$

where $a_i = 0$ *if* $i < 0$ *or* $i > m$.

The question naturally arises whether the constant on the right-hand side of the inequality (1.6) is the best possible. That this is the case will be shown by the following sequence of polynomials. With

$$(1.9)\qquad \omega = e^{i\theta}, \qquad \theta > 0,$$

we define

$$(1.10)\qquad P_m(z) = \prod_{\nu=0}^{m-1} (z + \omega^{m-1-2\nu}).$$

Thus

$$P_1(z) = z + 1$$

$$P_2(z) = (z + \omega)(z + \omega^{-1})$$

$$P_3(z) = (z + \omega^2)(z + 1)(z + \omega^{-2}) \qquad \text{and so on.}$$

It is fortunate that we may explicitly expand $P_m(z)$ in powers of z. Indeed, in terms of the symbols

$$(1.11)\qquad \begin{aligned} \begin{Bmatrix} m \\ \nu \end{Bmatrix} &= \frac{\sin m\theta \sin (m-1)\theta \cdots \sin (m-\nu+1)\theta}{\sin \theta \sin 2\theta \cdots \sin \nu\theta}, \qquad \text{if} \quad 0 < \nu \leqq m, \\ \begin{Bmatrix} m \\ 0 \end{Bmatrix} &= 1, \end{aligned}$$

which I wish to call *central Gaussian coefficients*, we have

$$(1.12)\qquad P_m(z) = \sum_{\nu=0}^{m} \begin{Bmatrix} m \\ \nu \end{Bmatrix} z^\nu.$$

We may now state

THEOREM 3. *The coefficients*

$$(1.13)\qquad \cdots, 0, 0, \begin{Bmatrix} m \\ 0 \end{Bmatrix}, \begin{Bmatrix} m \\ 1 \end{Bmatrix}, \cdots, \begin{Bmatrix} m \\ m \end{Bmatrix}, 0, 0, \cdots$$

of the polynomials defined by (1.10), (1.9), *are k-times positive if* θ *is in the range*

$$(1.14)\qquad 0 \leqq \theta \leqq \frac{\pi}{m+k-1}.$$

Notice that if θ assumes the right-end value $\theta = \pi/(m + k - 1)$ of the range (1.14), then $P_m(z)$ has two zeros on the boundary rays of the sector (1.6). Indeed, then

$$-\omega^{-m+1} = \exp\left\{\pi i - \frac{\pi i}{m+k-1}(m-1)\right\} = \exp\left\{\frac{k\pi i}{m+k-1}\right\}.$$

If θ would now increase a little beyond the value $\pi/(m+k-1)$, then these two zeros of $P_m(z)$ would penetrate into the sector (1.6) forbidden by Theorem 1. We may therefore state the following two corollaries of the Theorems 3 and 1, respectively.

COROLLARY 1. *The constant $k\pi/(m+k-1)$ on the right side of the inequality* (1.6) *can not be increased.*

COROLLARY 2. *The constant $\pi/(m+k-1)$ on the right side of the inequality* (1.14) *can not be increased.*

REMARK 1. Although covered by the proofs of these Theorems given below, it seems worth while to point out for orientation a few special cases which are more or less evident. Thus, if $k = 1$, Theorem 1 is almost trivial. Indeed, it then states that a polynomial (1.4) with non-negative coefficients can have no zeros in the sector

$$|\arg z| < \frac{\pi}{m}. \tag{1.15}$$

This is evident,[4] for if $z = re^{i\alpha}$, $0 < \alpha < \pi/m$, then $\sin \nu\alpha > 0$ for $\nu = 1, \cdots, m$ and therefore

$$\mathbf{I}P(z) = \sum_1^m a_\nu r^\nu \sin \nu\alpha \geqq a_m r^m \sin m\alpha > 0.$$

Also Theorems 2 and 3 are evident if $k = 1$. Indeed, if $0 \leqq \theta < \pi/m$, then (1.11) immediately shows that $P_m(z)$ has only positive coefficients. If $\theta = \pi/m$ then

$$P_m(z) = 1 + z^m,$$

having two zeros on the boundary of (1.15).

For $k = 2$, Theorem 1 may be stated directly as follows:

If the polygonal line of vertices $(\nu, \log a_\nu)$, $\nu = 0, \cdots, m$, is concave, then the polynomial (1.4) *has no zeros in the sector*

$$|\arg z| < \frac{2\pi}{m+1}. \tag{1.16}$$

This can be proved directly as an interesting exercise. Trivial is the fact that (1.16) is the widest sector with this property (Corollary 1). Indeed, if $\theta = \pi/(m+k-1) = \pi/(m+1)$, then by either (1.10) or by (1.11) and (1.12) we obtain

$$P_m(z) = 1 + z + z^2 + \cdots + z^m = \frac{z^{m+1} - 1}{z - 1},$$

[4] This is also a very special case of a theorem of Obreschkoff: The number of zeros of (1.4) in the sector (1.15) never exceeds the number of variations of sign in the sequence of the coefficients of (1.4). See [6].

whose coefficients are logarithmically concave, while having zeros on both rays bounding the sector (1.16).

REMARK 2. A geometrically suggestive way of viewing Theorem 3 is as follows. Notice that $P_m(z)$ reduces to $(z + 1)^m$ if $\theta = 0$, while (1.13) are then the ordinary binomial coefficients. Thus $P_m(z)$ may be thought of as arising from $(z + 1)^m$ by having the latter's coincident zeros fan out equidistantly to arc-distances 2θ along the unit circle. For $\theta = \pi/m$, $P_m(z) = 1 + z^m$ and the elements of the sequence (1.13) are all zero except the extreme terms $\left\{\begin{matrix} m \\ 0 \end{matrix}\right\}$, $\left\{\begin{matrix} m \\ m \end{matrix}\right\}$ which are always unity. As θ decreases from its initial value π/m to zero, the sequence (1.13) becomes more nearly like the sequence of binomials $\binom{m}{\nu}$, the terms of the decreasing sequence

$$\frac{\pi}{m + k - 1}, \qquad (k = 2, 3, \cdots),$$

marking the exact spots where the sequence (1.13) starts belonging to the class $\mathcal{P}_k$. The binomials $\binom{m}{\nu}$ $(\theta = 0)$ form a totally positive sequence, as also follows from Theorem 0, since all zeros of $P_m(z) = (z + 1)^m$ are now negative.

REMARK 3. Notice that Theorem 1 immediately implies the "only if" conclusion of Theorem 0. Indeed, if the polynomial (1.4) has totally positive coefficients (1.3), then these are also k-times positive for every k. Thus $P(z)$ has no zeros in the sector (1.6) for every k, by Theorem 1. Letting $k \to \infty$ we conclude that (1.4) can have only negative zeros. The "if" conclusion of Theorem 0 follows trivially from the property of $\mathcal{P}_\infty$ to be closed with respect to the operation of convolution of sequences, which property, incidentally, is shared by each of the classes $\mathcal{P}_k$ (See Fekete [3], Satz III, page 6).

2. Multiply positive functions. Let $f(x)$, defined for all real x, be measurable, non-negative and such that

$$0 < \int_{-\infty}^{\infty} f(x)\, dx < \infty. \tag{2.1}$$

Such a function is said to be *totally positive* provided that it enjoys the following property: *If* $x_1 < x_2 < \cdots < x_n$, $y_1 < y_2 < \cdots < y_n$, *then*

$$\det \| f(x_i - y_j) \|_{1,n} \geqq 0. \tag{2.2}$$

This inequality should hold for all $n = 1, 2, \cdots$.

The problem of the function-theoretic characterization of the Laplace transform

$$\int_{-\infty}^{\infty} e^{-sx} f(x)\, dx \tag{2.3}$$

for $f(x)$ with the above property, was solved by the writer, thereby settling an old problem of Pólya and Hamburger (See [10]).

A natural generalization of the concept of a totally positive function is as follows: *Let k be a given natural integer. A function $f(x)$ satisfying* (2.1) *is said to be k-times positive* (we also write $f(x) \in \mathcal{P}_k$ or say that $f(x)$ is an element of $\mathcal{P}_k$) *provided that it enjoys the property* (2.2) *for* $n = 1, n = 2, \cdots, n = k$.

Evidently $\mathcal{P}_1 \supset \mathcal{P}_2 \supset \cdots \supset \mathcal{P}_k \supset \cdots$, the intersection of all these classes being the class P_∞ of the totally positive functions. Clearly $\mathcal{P}_1$ is identical with the class of frequency functions of the theory of probabilities. See [10], page 337, where the class $\mathcal{P}_2$ is shown to allow the following direct description: $f(x) \in \mathcal{P}_2$ *if and only if there exist two numbers* $\alpha, \beta(-\infty \leqq \alpha < \beta \leqq \infty)$ *such that*[5] $f(x) = 0$ *if* $x < \alpha$ or $x > \beta$, *while* $f(x) > 0$ *if* $\alpha < x < \beta$, *and* $\log f(x)$ *is a continuous concave function in the open range* $\alpha < x < \beta$. *Moreover*

$$0 \leqq f(\alpha) \leqq f(\alpha + 0), \qquad 0 \leqq f(\beta) \leqq f(\beta - 0).$$

In view of the writer's results concerning (2.3), the problem arises of investigating the analytical properties of the transform (2.3) of a k-times positive function $f(x)$. A contribution to this problem for the special case when the transform (2.3) is an entire function of exponential type is contained in the following

THEOREM 4. *Let $f(x)$ be a k-times positive function such that $f(x) = 0$ outside the range $-\pi/2 \leqq x \leqq \pi/2$. Its transform*

$$F(s) = \int_{-\pi/2}^{\pi/2} e^{-sx} f(x)\, dx, \tag{2.4}$$

which is an entire function of exponential type, has no zeros in the strip

$$|\,\mathbf{I}s\,| < k. \tag{2.5}$$

Again the question arises whether the constant on the right-hand side of the inequality (2.5) is the best possible. That this is the case will be shown by means of the family of frequency functions

$$f_\alpha(x) = \begin{cases} (\cos x)^\alpha & \text{if} \quad -\pi/2 \leqq x \leqq \pi/2, \\ 0 & \text{elsewhere}, \quad (\alpha > -1). \end{cases} \tag{2.6}$$

Its integral is

$$\int_{-\infty}^{\infty} f_\alpha(x)\, dx = \frac{\pi\Gamma(\alpha + 1)}{2^\alpha(\Gamma(\frac{1}{2}(\alpha + 2)))^2} = \frac{\Gamma(\frac{1}{2}(\alpha + 1))\Gamma(\frac{1}{2})}{\Gamma(\frac{1}{2}(\alpha + 2))},$$

a value first determined by Wallis for integer values of α. For this reason we wish to call (2.6) a *Wallis distribution*. Its transform, due to N. Nielsen, is known to be (see [5], p. 158, [2], p. 12, formula (29))

---

[5] As in footnote 3, all requirements involving α or β drop out if $\alpha = -\infty$ or $\beta = +\infty$, respectively.

$$(2.7)\quad \int_{-\pi/2}^{\pi/2} e^{-sx} (\cos x)^{\alpha}\, dx = \frac{\pi\Gamma(\alpha+1)}{2^{\alpha}\Gamma(\frac{1}{2}(\alpha+2+si))\Gamma(\frac{1}{2}(\alpha+2-si))}, \quad (\alpha > -1),$$

but will also be derived below (Article 10) for the reader's convenience.

We may now state our

THEOREM 5. *The function $f_{\alpha}(x)$, defined by* (2.6), *is k-times positive* ($k \geqq 2$) *for the following range of values of* α:

$$(2.8)\qquad \alpha \geqq k - 2.$$

Notice that the zeros of the entire function (2.7) are

$$s = \pm(\alpha + 2 + n)i, \qquad (n = 0, 1, \cdots).$$

Thus, if α assumes the least value $\alpha = k - 2$ of the range (2.8), then the first two zeros of (2.7) are on the boundary lines of the strip (2.5). Should α now drop by a little below the value $k - 2$, so that $\alpha + 2 < k$, then those two zeros of (2.7) would penetrate within the strip (2.5) which is forbidden by Theorem 4. These remarks prove the following corollaries of Theorems 5 and 4, respectively:

COROLLARY 3. *The constant k on the right side of the inequality* (2.5) *can not be increased.*

COROLLARY 4. *The constant $k - 2$ on the right side of the inequality* (2.8) *can not be decreased.*

Corollary 3 was actually proved for $k \geqq 2$, but the same reasoning applies if $k = 1$: As $\alpha \to -1$, the first zeros $\pm(\alpha + 2)i$, of (2.7), approach $\pm i$.

Remarks similar to those made concerning Theorems 1 and 3 can be made on the subject of Theorems 4 and 5. In the first place Theorem 4 is evident if $k = 1$. Indeed, assuming $s = \sigma + it$ to be a zero of (2.4) within the strip (2.5), hence such that $-1 < t < 1$, we obtain

$$\mathbf{R}F(s) = \int_{-\pi/2}^{\pi/2} e^{-\sigma x} f(x) \cos(tx)\, dx = 0.$$

But this relation is impossible because the cosine factor is positive within the range of integration.

Also the case when $k = 2$ can be settled directly. In view of the direct description of the class $\mathcal{P}_2$ in terms of concave functions, Theorem 4 may be stated directly as follows if $k = 2$: *If $\psi(x)$ is convex in the range $-\pi/2 < x < \pi/2$, then*

$$F(s) = \int_{-\pi/2}^{\pi/2} e^{-sx-\psi(x)}\, dx$$

can not vanish in the strip $|\,\mathbf{I}s\,| < 2$. There are zeros on its boundary if and only if if $\psi(x)$ is a linear function.

Finally notice that Theorem 5 is evident if $k = 2$. Indeed

$$\log f_{\alpha}(x) = \alpha \log \cos x, \qquad (-\pi/2 < x < \pi/2),$$

is a concave function of x, if $\alpha \geqq k - 2 = 0$, which is equivalent to $f_{\alpha}(x) \in \mathcal{P}_2$.

§2. Proofs of Theorems 1 and 2

3. An application of variation-diminishing transformations. A real linear transformation

$$y_i = \sum_{j=1}^{n} a_{ij} x_j , \qquad (i = 1, \cdots, m),$$

is said to be *variation-diminishing* provided it enjoys the following property: If we denote by $v(x)$ the number of variations in sign in the sequence $x_1, \cdots, x_n$ and by $v(y)$ the corresponding number for the sequence $y_1, \cdots, y_m$, then the inequality

$$v(y) \leqq v(x)$$

always holds. It is known that a linear transformation is variation-diminishing whenever its matrix $A = \| a_{ij} \|$ is totally positive, i.e. has no negative minor (See Schoenberg [9]). This result will readily furnish a proof of Theorem 1.

Let the polynomial (1.4) be such that the matrix A_k, defined by (1.7), is totally positive. Let $z = \alpha$ be a zero of (1.4). If $\alpha < 0$, there is nothing to prove. Since α cannot be positive, we may assume

$$\text{(3.1)} \qquad \begin{gathered} a_0 + a_1\alpha + \cdots + a_m\alpha^m = 0, \\ \alpha = re^{i\theta}, \qquad r > 0, \qquad 0 < \theta < \pi, \end{gathered}$$

and we are to show that

$$\text{(3.2)} \qquad \theta \geqq \frac{k\pi}{m + k - 1}.$$

Let n be given positive integer still to be disposed of later. We set up a linear transformation whose matrix M may schematically be described by

$$\text{(3.3)} \qquad M = \begin{Vmatrix} A_k & 0 & \cdots & 0 \\ 0 & A_k & \cdots & 0 \\ \vdots & \vdots & \ddots & \vdots \\ 0 & 0 & & A_k \end{Vmatrix},$$

where the matrix A_k appears n times. The only departure in the structure of M from what the formula (3.3) naturally suggests, is that the last element of each A_k is immediately above and in the same column with the leading element of the next A_k. Thus M has nk rows and $n(m + k) - (n - 1) = n(m + k - 1) + 1$ columns. An explicit expression of our transformation is

$$
(3.4)\qquad
\begin{aligned}
y_1 &= a_0 x_0 + \cdots + a_m x_m \\
&\;\vdots \\
y_k &= \quad a_0 x_{k-1} + \cdots + a_m x_{m+k-1} \\
y_{k+1} &= \qquad a_0 x_{m+k-1} + \cdots + a_m x_{2m+k-1} \\
&\;\vdots \\
y_{2k} &= \qquad\quad a_0 x_{m+2(k-1)} + \cdots + a_m x_{2(m+k-1)} \\
&\;\vdots \\
y_{(n-1)k+1} &= \qquad\qquad a_0 x_{(n-1)(m+k-1)} + \cdots \\
&\;\vdots \\
y_{nk} &= \qquad\qquad\quad a_0 x_{(n-1)m+n(k-1)} + \cdots + a_m x_{n(m+k-1)} .
\end{aligned}
$$

The proof depends essentially on two properties of this transformation:

(i) *The transformation* (3.4) *is variation-diminishing.*

(ii) *The linear forms* $y_1, \cdots, y_{nk}$ *are linearly independent.* By the general theorem mentioned in the previous paragraph, to prove (i) it is sufficient to show that the matrix (3.3) is totally positive. That this is the case is implied by the following general remark: *Let A and B be two finite totally positive matrices. Let us form a new matrix*

$$C = \begin{Vmatrix} A & 0 \\ 0 & B \end{Vmatrix}$$

where, as in (3.3), *A and B have no common row, while the last column of A and first column of B form a single column of C. Then C is also totally positive.*

Indeed, let N be a square submatrix of C. By Laplace's expansion theorem we easily find as only possibilities the following alternatives: Either the minor $| N |$ vanishes because of too many zero elements, or else $| N | = | A' | \cdot |B'|$, where $| A' |$, $| B' |$ are minors of A and B, respectively. Thus $|N| \geqq 0$ in either case and C is totally positive. We remark that C need no longer be totally positive if A and B should have more than one common column.

The second property (ii) is readily revealed by a glance at the transformation (3.4): Observe that the (nk)-order minor of M, formed by all columns of M containing the element a_m, is equal to $(a_m)^{nk}$ and is therefore positive. The rank of M is thus equal to nk.

We now form the successive powers of α and choose for the x's in (3.4) their imaginary parts:

$$(3.5)\qquad x_\nu = \mathbf{I}\alpha^\nu = r^\nu \sin \nu\theta, \qquad (\nu = 0, 1, \cdots).$$

The number v of variations in sign in the sequence $x_0, x_1, \cdots, x_N$ is easily determined. Indeed, as the arguments of α^ν increase successively by the constant amount $\theta(0 < \theta < \pi)$, the number v equals the number of times the point α^ν crosses over from the upper half-plane to the lower half-plane and vice-versa.

Thus clearly

$$v = [N\theta/\pi] + \rho, \qquad \text{where} \quad \rho = 0 \text{ or } -1. \tag{3.6}$$

On the other hand, the first relation (3.1) implies

$$a_0\alpha^{\nu} + a_1\alpha^{\nu+1} + \cdots + a_m\alpha^{\nu+m} = 0$$

and passing to imaginary parts we see that for our choice of values (3.5) all linear forms y_i vanish by (3.4). However, the property (ii) of linear independence insures the following: By changing the values (3.5) of the x_ν's by arbitrarily small amounts (we may even leave $x_0 = 0$ unchanged, which we do) to new values x'_ν, we may achieve that the new values y'_i, of y_i, alternate in sign, i.e.

$$v(y') = nk - 1. \tag{3.7}$$

Surely we also have

$$v(x') \leqq v(x) + 1, \tag{3.8}$$

provided the differences $| x'_i - x_i |$ are sufficiently small. Also

$$v(y') \leqq v(x') \tag{3.9}$$

by the variation-diminishing property (i). Finally

$$v(x) = \left[\frac{n(m + k - 1)\theta}{\pi}\right] + \rho, \qquad (\rho = 0, -1), \tag{3.10}$$

by (3.6). Now (3.7), (3.9), (3.8) and (3.10) give

$$nk - 1 = v(y') \leqq v(x') \leqq v(x) + 1 \leqq \left[\frac{n(m + k - 1)\theta}{\pi}\right] + 1$$

and comparing the extreme terms a fortiori

$$nk - 1 \leqq \frac{n(m + k - 1)}{\pi}\theta + 1.$$

This inequality holds for all values of n. Dividing by n and letting $n \to \infty$ we obtain

$$k \leqq \frac{m + k - 1}{\pi}\theta,$$

that is (3.2) is established and the proof of Theorem 1 is complete.

4. Applications of a Theorem of Fekete. The Theorem of Fekete may be stated as follows: *Let B be a matrix of h rows and l columns ($l > h$) with the following properties:* (i) *All minors of order $h - 1$ from the first $h - 1$ rows of B, are positive,* (ii) *All minors of order h, formed of h consecutive columns of B, are positive. The conclusion is that all minors of order h are positive* (See Fekete [3], pp. 4–6). This Theorem will furnish a proof of Theorem 2.

For convenience we use the following nomenclature. Let B be a matrix which is finite or infinite. A minor of B will be referred to as a *block-minor* provided it is formed of consecutive rows and of consecutive columns of B. The matrix B is said to be *strictly k-times positive* provided that all its minors of orders $\leqq k$ are positive. With these terms we may state the following

LEMMA 1. *A matrix $B = \| b_{ij} \|$ is strictly k-times positive if and only if all its block-minors of orders $\leqq k$ are positive.*

PROOF. The necessity of the condition being clear we may proceed to prove its sufficiency. Let

$$B_h = \begin{vmatrix} b_{i_1j_1} & \cdots & b_{i_1j_h} \\ \vdots & & \vdots \\ b_{i_hj_1} & \cdots & b_{i_hj_h} \end{vmatrix}$$

be a minor of order h. As a measure of the departure of B_h from being a block-minor we introduce the quantity

$$S(B_h) = (i_h - i_1 + 1) + (j_h - j_1 + 1)$$

and call it the span of B_h. Evidently $S(B_h) \geqq 2h$, with equality for and only for block-minors. Our problem is to show that

$$B_h > 0 \quad \text{if} \quad h \leqq k. \tag{4.1}$$

We apply induction by the value of $S(B_h)$. (4.1) is evidently true if $S(B_h) = 2$, since B_h, being of order $h = 1$, is a block-minor, hence positive. Let now $S(B_h) = s$ and let us assume (4.1) to be true for minors of span $< s$. Two cases may arise: B_h is a block-minor or not. In the first case $B_h > 0$ by assumption. If B_h is not a block-minor, without loss of generality we may assume that its columns are not all consecutive columns of B. Let us add to B_h one of the intermediate columns of B, obtaining a $h \times (h + 1)$ submatrix of B which we call B^*. B^* satisfies all assumptions of Fekete's theorem because all minors which appear in these assumptions have spans $< s$, and are therefore positive by the induction assumption. The conclusion of this theorem in particular asserts that $B_h > 0$.

We are now in position to establish

LEMMA 2. *Let $\{a_n\}$, $-\infty < n < \infty$, be a sequence of non-negative numbers, with $\sum a_n$ convergent. Let the matrix*

$$A_k = \| a_{j-i} \|, \tag{4.2}$$

$$(i = 0, 1, \cdots, k - 1; -\infty < j < \infty),$$

of k rows and infinitely many columns be of rank k, i.e. A_k has a non-vanishing minor of order k. Then the sequence $\{a_n\}$ is k-times positive if and only if the matrix A_k is totally positive.

This lemma applied to the special sequence (1.3) evidently implies the first part of Theorem 2. Indeed, the zero columns ($j < 0$ and $j > m + k - 1$) of (4.2) may now be deleted showing that (4.2) and (1.7) are simultaneously totally positive. Moreover, the matrix (1.7) is visibly of rank k.

PROOF. The necessity of the condition being clear, we assume that

$$(4.3) \qquad A_k \text{ is totally positive}$$

and we are to show that the entire matrix

$$A = \| a_{j-i} \|, \qquad (-\infty < i, j < \infty),$$

is k-times positive. For this purpose we use the auxiliary matrix

$$(4.4) \qquad M_\gamma = \| e^{-\gamma(i-j)^2} \|, \qquad (\gamma > 0),$$

which has the following interesting properties: (i) All its minors are positive (See [10], p. 336), (ii) as $\gamma \to +\infty$, M_γ converges to the unit matrix.

Let us form the product matrix

$$(4.5) \qquad AM_\gamma = B_\gamma .$$

We claim that all block-minors of B_γ of orders $\leqq k$ are positive. Indeed, the matrix B^* of a block-minor of B_γ of order $h(\leqq k)$ may be thought of as the product of two matrices A^* and M_γ^* : A^* being formed of h consecutive rows of A, M_γ^* of h consecutive columns of M_γ. The corresponding determinant $| B^* |$, being sum of products of homologous minors of A^* and M_γ^*, is evidently positive; indeed all minors of M_γ^* are positive while all h-order minors of A^* are non-negative with some one such actually positive by our assumption concerning the rank of (4.2). By Lemma 1 B_γ is strictly k-times positive. As $\gamma \to \infty$, $M_\gamma \to I$ and therefore

$$\lim_{\gamma\to\infty} B_\gamma = A,$$

showing (by continuity) that also A as k-times positive. This completes a proof of Lemma 2 and of the first part of Theorem 2.

We now turn to a proof of the second part of Theorem 2. Our assumptions are (1.8) on the basis of which we are again to prove that $A = \| a_{j-i} \|$ is k-positive. Notice first that the one-row matrix

$$A_1 = \| a_0, a_1, \cdots, a_m \|$$

has positive elements by (1.8), for $\nu = 1$. By (1.8), for $\nu = 2$, and Fekete's theorem we conclude that all 2nd order minors of

$$\begin{Vmatrix} a_0 & a_1 & \cdots & a_m \\ 0 & a_0 & \cdots & a_{m-1} \end{Vmatrix}$$

are positive. Evidently all 2nd order minors of

$$A_2 = \begin{Vmatrix} a_0 & a_1 & \cdots & a_m & 0 \\ 0 & a_0 & \cdots & a_{m-1} & a_m \end{Vmatrix}$$

are also positive. But then, again by (1.8), for $\nu = 3$, and Fekete's theorem, the matrix

$$\left\| \begin{matrix} a_0 & a_1 & & \cdots & a_m & 0 \\ 0 & a_0 & & \cdots & & a_m \\ 0 & 0 & a_0 & \cdots & & a_{m-1} \end{matrix} \right\|_\infty$$

and therefore also

$$A_3 = \left\| \begin{matrix} a_0 & a_1 & & \cdots & a_m & 0 & 0 \\ 0 & a_0 & & \cdots & & a_m & 0 \\ 0 & 0 & a_0 & \cdots & & & a_m \end{matrix} \right\|$$

has only positive 3rd order minors.

We may evidently proceed in this fashion to conclude that each of the matrices $A_1, A_2, \cdots, A_k$, the last defined by (1.7), have all their minors, of *maximal* order, positive. We may now prove as follows that $A = \| a_{j-i} \|$ is k-times positive: By what has just been shown, we know that all minors of order ν, in any ν *consecutive* rows of A, are non-negative, some being positive ($\nu = 1, \cdots, k$). If we form the product (4.5) or

$$AM_\gamma = B_\gamma,$$

we conclude, as before, that all block-minors of B_γ, of orders $\leqq k$, are positive But then B_γ is strictly k-times positive by Lemma 1 and by continuity, as $\gamma \to \infty$, A itself is k-times positive. This concludes our proof of Theorem 2.

§3. The Gaussian coefficients and Theorem 3

5. Definitions and elementary properties. Let m be a natural integer and q a variable. The Gaussian coefficients are defined by

(5.1)
$$\begin{bmatrix} m \\ n \end{bmatrix} = \frac{(q^m - 1)(q^{m-1} - 1) \cdots (q^{m-n+1} - 1)}{(q - 1)(q^2 - 1) \cdots (q^n - 1)} \quad \text{if} \quad 0 < n \leqq m,$$
$$\begin{bmatrix} m \\ 0 \end{bmatrix} = 1, \quad \begin{bmatrix} m \\ n \end{bmatrix} = 0 \quad \text{if} \quad n < 0 \quad \text{or} \quad n > m.$$

Notice that the limits

$$\lim_{q \to 1} \begin{bmatrix} m \\ n \end{bmatrix} = \binom{m}{n}$$

are the familiar binomial coefficients. The Gaussian coefficients generalize a number of properties of the binomial coefficients the most noteworthy being the following:

(5.2)
$$\begin{bmatrix} m \\ n \end{bmatrix} = \begin{bmatrix} m \\ m - n \end{bmatrix} \quad \text{(symmetry)},$$

(5.3) $$\begin{bmatrix} m \\ n \end{bmatrix} = \begin{bmatrix} m-1 \\ n-1 \end{bmatrix} + q^n \begin{bmatrix} m-1 \\ n \end{bmatrix} \quad \text{(recurrence relation)},$$

(5.4) $$(1+x)(1+qx)(1+q^2x)\cdots(1+q^{m-1}x) = \sum_{\nu=0}^{m} \begin{bmatrix} m \\ \nu \end{bmatrix} q^{\frac{1}{2}(\nu(\nu-1))} x^\nu,$$

(5.5) $$\frac{1}{(1-x)(1-qx)\cdots(1-q^{m-1}x)} = \sum_{\nu=0}^{\infty} \begin{bmatrix} m-1+\nu \\ m-1 \end{bmatrix} x^\nu, \ (m = 1, 2, \cdots).$$

The last two remarkable expansions[6] are easily established by induction in m on replacing x by qx. By (5.3) or (5.4) we can see that $\begin{bmatrix} m \\ n \end{bmatrix}$ is a polynomial in q.

For our purpose it is convenient to set

(5.6) $$q = \omega^2, \qquad \omega = e^{i\theta}.$$

Now (5.4) becomes

$$(1+x)(1+\omega^2x)\cdots(1+\omega^{2m-2}x) = \sum_0^m \begin{bmatrix} m \\ \nu \end{bmatrix} \omega^{\nu(\nu-1)} x^\nu$$

and setting $x = z\omega^{-m+1}$

(5.7) $$(1+\omega^{-m+1}z)(1+\omega^{-m+3}z)\cdots(1+\omega^{m-1}z) = \sum_0^m \begin{bmatrix} m \\ \nu \end{bmatrix} \omega^{\nu(\nu-m)} z^\nu.$$

We now introduce new symbols

(5.8) $$\begin{Bmatrix} m \\ \nu \end{Bmatrix} = \begin{bmatrix} m \\ \nu \end{bmatrix} \omega^{\nu(\nu-m)}$$

which we call *central Gaussian coefficients*. A simple calculation using the definitions (5.8), (5.1), (5.6) gives

$$\begin{Bmatrix} m \\ n \end{Bmatrix} = \frac{(\omega^m - \omega^{-m})(\omega^{m-1} - \omega^{-m+1})\cdots(\omega^{m-n+1} - \omega^{-m+n-1})}{(\omega - \omega^{-1})(\omega^2 - \omega^{-2})\cdots(\omega^n - \omega^{-n})}$$

and finally the elegant expression

(5.9) $$\begin{Bmatrix} m \\ n \end{Bmatrix} = \frac{\sin m\theta \sin(m-1)\theta \cdots \sin(m-n+1)\theta}{\sin\theta \sin 2\theta \cdots \sin n\theta}.$$

Clearly again

(5.10) $$\begin{Bmatrix} m \\ n \end{Bmatrix} = \begin{Bmatrix} m \\ m-n \end{Bmatrix}, \qquad \begin{Bmatrix} m \\ n \end{Bmatrix} = \binom{m}{n} \text{ if } \theta = 0,$$

while the recurrence relation (5.3) becomes

(5.11) $$\begin{Bmatrix} m \\ n \end{Bmatrix} = \begin{Bmatrix} m-1 \\ n-1 \end{Bmatrix} \omega^{n-m} + \begin{Bmatrix} m-1 \\ n \end{Bmatrix} \omega^n$$

or

(5.11) $$\begin{Bmatrix} m \\ n \end{Bmatrix} = \begin{Bmatrix} m-1 \\ n-1 \end{Bmatrix} \cos 2(n-m)\theta + \begin{Bmatrix} m-1 \\ n \end{Bmatrix} \cos 2n\theta.$$

[6] See [4], Theorems 348, 349 on page 278.

We may now easily obtain expansions for the polynomials

$$P_m(z) = \prod_{\nu=0}^{m-1} (z + \omega^{m-1-2\nu}) \tag{5.12}$$

introduced in §1. From (5.7) and (5.8) we have

$$P_m(z) = \sum_{\nu=0}^{m} \begin{Bmatrix} m \\ \nu \end{Bmatrix} z^\nu. \tag{5.13}$$

Setting $q = \omega^2$ and $x = \omega^{-m+1}z$ in (5.5) we obtain

$$\frac{1}{P_m(-z)} = \sum_{\nu=0}^{\infty} \begin{Bmatrix} m - 1 + \nu \\ m - 1 \end{Bmatrix} z^\nu. \tag{5.14}$$

Thus for $m = 2$

$$\frac{1}{1 - 2z\cos\theta + z^2} = \frac{\sin\theta}{\sin\theta} + \frac{\sin 2\theta}{\sin\theta} z + \frac{\sin 3\theta}{\sin\theta} z^2 + \cdots,$$

which is familiar, while (5.14) for $m = 3$ gives the less known expansion

$$\frac{1}{(1 - z)(1 - 2z\cos 2\theta + z^2)} = \sum_{\nu=0}^{\infty} \frac{\sin(\nu + 1)\theta}{\sin\theta} \frac{\sin(\nu + 2)\theta}{\sin 2\theta} z^\nu.$$

6. A determinant identity for the central Gaussian coefficients. In our previous discussion of $\begin{Bmatrix} m \\ n \end{Bmatrix}$, m was a natural integer. It is now essential to let m be a variable which may assume real or complex values. Thus (5.9) defines $\begin{Bmatrix} m \\ n \end{Bmatrix}$ for $n = 1, 2, \cdots$ while as before we set

$$\begin{Bmatrix} m \\ 0 \end{Bmatrix} = 1, \qquad \begin{Bmatrix} m \\ n \end{Bmatrix} = 0 \quad \text{if} \quad n < 0. \tag{6.1}$$

The key and formal basis of the results of this paper is the following identity

$$\begin{vmatrix} \begin{Bmatrix} m \\ n \end{Bmatrix} & \begin{Bmatrix} m \\ n+1 \end{Bmatrix} & \cdots & \begin{Bmatrix} m \\ n+\nu-1 \end{Bmatrix} \\ \begin{Bmatrix} m \\ n-1 \end{Bmatrix} & \begin{Bmatrix} m \\ n \end{Bmatrix} & \cdots & \begin{Bmatrix} m \\ n+\nu-2 \end{Bmatrix} \\ \vdots & \vdots & & \vdots \\ \begin{Bmatrix} m \\ n-\nu+1 \end{Bmatrix} & \begin{Bmatrix} m \\ n-\nu+2 \end{Bmatrix} & \cdots & \begin{Bmatrix} m \\ n \end{Bmatrix} \end{vmatrix} = \frac{\begin{Bmatrix} m \\ n \end{Bmatrix}\begin{Bmatrix} m+1 \\ n \end{Bmatrix} \cdots \begin{Bmatrix} m+\nu-1 \\ n \end{Bmatrix}}{\begin{Bmatrix} n \\ n \end{Bmatrix}\begin{Bmatrix} n \\ n+1 \end{Bmatrix} \cdots \begin{Bmatrix} n+\nu-1 \\ n \end{Bmatrix}}, \tag{6.2}$$

$(n = 0, 1, \cdots; \nu = 1, 2, \cdots; m$ arbitrary variable).

For $\theta = 0$, (6.2) becomes an identity for the ordinary binomial coefficients which is readily found to be equivalent to a result of Zeipel (See Pascal [8], p. 133). However the simple proof for Zeipel's identity as given by Pascal breaks down in our case because of the more complicated recurrence relation (5.11). Let us look at a few special cases. For $n = 0$ both sides of (6.2) are viz. equal to unity. Also if $\nu = 1$, (6.2) is trivially valid: If $m = 2$, $n = 1$, hence

$$\begin{Bmatrix} 2 \\ 1 \end{Bmatrix} = \frac{\sin 2\theta}{\sin \theta} = 2 \cos \theta,$$

(6.2) becomes

$$\begin{vmatrix} 2\cos\theta & 1 & \cdots & 0 & 0 \\ 1 & 2\cos\theta & \cdots & 0 & 0 \\ \vdots & \vdots & & \vdots & \vdots \\ 0 & 0 & & 2\cos\theta & 1 \\ 0 & 0 & & 1 & 2\cos\theta \end{vmatrix} = \frac{\sin 2\theta \sin 3\theta \cdots \sin(\nu+1)\theta}{\sin\theta \sin 2\theta \cdots \sin \nu\theta}$$

$$= \frac{\sin(\nu+1)\theta}{\sin\theta},$$

which we recognize as a known determinant representation of the Chebychev polynomials of the second kind.

In order to prove (6.2) we apply induction by the order ν of the determinant, assuming the relation true for the order $\nu - 1$. To simplify writing we use the condensed notation

$$\sin k\theta = (k)$$

so that (5.9) becomes

$$\begin{Bmatrix} m \\ n \end{Bmatrix} = \frac{(m)(m+1)\cdots(m-n+1)}{(1)(2)\cdots(n)}.$$

Denote the determinant (6.2) by D_ν and let us factor out from each column its leading element. We obtain

$$(6.3)\quad D_\nu = \prod_{j=0}^{\nu-1} \begin{Bmatrix} m \\ n+j \end{Bmatrix} \cdot \left| \begin{matrix} 1 \\ \dfrac{(n+j)}{(m-n-j+1)} \\ \dfrac{(n+j-1)(n+j)}{(m-n-j+1)(m-n-j+2)} \\ \vdots \\ \dfrac{(n+j-\nu+2)\cdots(n+j)}{(m-n-j+1)\cdots(m-n-j+\nu-1)} \end{matrix} \right|_{j=0,1,\cdots,\nu-1},$$

where the second factor on the right side is a νth order determinant whose j^{th} column ($j = 0, \cdots, \nu - 1$) only is written out. Notice the important point that the relation is valid no matter what the relative sizes of the numbers n and ν might be. Thus any vanishing elements of D_ν, due to the second relation (6.1), are correctly represented by zero elements in the right-hand determinant. To see this point clearly I write out the special case when $\nu = 2$ and $n = 2$:

$$\begin{vmatrix} \begin{Bmatrix} m \\ 2 \end{Bmatrix} & \begin{Bmatrix} m \\ 3 \end{Bmatrix} \\ \begin{Bmatrix} m \\ 1 \end{Bmatrix} & \begin{Bmatrix} m \\ 2 \end{Bmatrix} \end{vmatrix} = \begin{Bmatrix} m \\ 2 \end{Bmatrix}\begin{Bmatrix} m \\ 3 \end{Bmatrix} \begin{vmatrix} 1 & 1 \\ \dfrac{(2)}{(m-1)} & \dfrac{(3)}{(m-2)} \end{vmatrix}.$$

Now we see the reason why m must be thought of as a variable, for otherwise the relation (6.3) would often break down for integer values of m: In our example it makes no sense if $m = 2$.

We now subtract each column of the right-hand determinant (6.3) from its successor, starting from right to left. This knocks out all the unit elements in the first row, except the first one, and depresses the order to $\nu - 1$. The result may be written thus:

$$D_\nu = \prod_{j=0}^{\nu-1} \begin{Bmatrix} m \\ n+j \end{Bmatrix} \cdot$$

$$\text{(6.4)} \quad \begin{vmatrix} \dfrac{(n+j)}{(m-n-j+1)} - \dfrac{(n+j-1)}{(m-n-j+2)} \\ \dfrac{(n+j-1)}{(m-n-j+2)}\left(\dfrac{(n+j)}{(m-n-j+1)} - \dfrac{(n+j-2)}{(m-n-j+3)}\right) \\ \vdots \\ \dfrac{(n+j-\nu+2)\cdots(n+j-1)}{(m-n-j+2)\cdots(m-n-j+\nu-1)} \cdot \\ \left(\dfrac{(n+j)}{(m-n-j+1)} - \dfrac{(n+j-\nu+1)}{(m-n-j+\nu)}\right) \end{vmatrix}_{j=1,\cdots,\nu-1}$$

Here, again, only the jth column ($j = 1, \cdots, \nu - 1$) is written out.

At this point we need a bit of elementary trigonometry. We note the identities

$$\begin{aligned} &\sin(a+h)\sin a - \sin(b+h)\sin b \\ &\qquad = \tfrac{1}{2}(\cos h - \cos(2a+h) - \cos h + \cos(2b+h)) \\ &\qquad = \tfrac{1}{2}(\cos(2b+h) - \cos(2a+h)) \\ &\qquad\qquad = -\sin(a+b+h)\sin(b-a) \end{aligned}$$

or[7]

$$\sin(a+h)\sin a - \sin(b+h)\sin b = \sin(a-b)\sin(a+b+h).$$

If we set $a + h = A$, $b + h = B$ then $a + b + h = \frac{1}{2}(A + B + a + b)$. Substituting these values we conclude that the relation

$$A - B = a - b \tag{6.5}$$

implies

$$\frac{\sin A}{\sin b} - \frac{\sin B}{\sin a} = \frac{\sin(a-b)\sin\frac{1}{2}(A+B+a+b)}{\sin a \sin b}. \tag{6.6}$$

Notice that all the differences of ratios of sines appearing in (6.4) enjoy the property (6.5). By (6.6) we may therfore write

$$D_\nu = \begin{Bmatrix} m \\ n+\nu-1 \end{Bmatrix} \prod_{j=1}^{\nu-1} \begin{Bmatrix} m \\ n+j-1 \end{Bmatrix} \cdot$$

$$\left| \begin{matrix} \dfrac{(1)(m+1)}{(m-n-j+1)(m-n-j+2)} \\ \dfrac{(n+j-1)(2)(m+1)}{(m-n-j+1)(m-n-j+2)(m-n-j+3)} \\ \vdots \\ \dfrac{(n+j-\nu+2)\cdots(n+j-1)(\nu-1)(m+1)}{(m-n-j+1)\cdots(m-n-j+\nu)} \end{matrix} \right|_{j=1,\cdots,\nu-1}$$

or

$$D_\nu = \begin{Bmatrix} m \\ n+\nu-1 \end{Bmatrix} \frac{(1)(2)\cdots(\nu-1)}{\prod_{j=1}^{\nu-1}(m-n-j+1)} \cdot$$

$$\left| \begin{matrix} \begin{Bmatrix} m \\ n+j-1 \end{Bmatrix} \dfrac{(m+1)}{(m-n-j+2)} \\ \begin{Bmatrix} m \\ n+j-1 \end{Bmatrix} \dfrac{(n+j-1)(m+1)}{(m-n-j+2)(m-n-j+3)} \\ \vdots \\ \begin{Bmatrix} m \\ n+j-1 \end{Bmatrix} \dfrac{(n+j-\nu+2)\cdots(n+j-1)(m+1)}{(m-n-j+2)\cdots(m-n-j+\nu)} \end{matrix} \right|_{j=1,\cdots,\nu-1}$$

[7] This identity has the curious property of remaining valid if we drop all four sine symbols: $(a+h)a - (b+h)b = (a-b)(a+b+h)$. This is not an accident: Any identity between sines which is homogeneous in the sines (ours if of the second degree) must obviously remain valid if we drop the sine symbols.

and finally after obvious cancellations in the elements of the determinant

$$(6.7)\qquad D_\nu = \begin{Bmatrix} m \\ n+\nu-1 \end{Bmatrix} \frac{(1)(2)\cdots(\nu-1)}{\prod_{j=1}^{\nu-1}(m-n-j+1)} \cdot \begin{vmatrix} \begin{Bmatrix} m+1 \\ n+j-1 \end{Bmatrix} \\ \begin{Bmatrix} m+1 \\ n+j-2 \end{Bmatrix} \\ \vdots \\ \begin{Bmatrix} m+1 \\ n+j-\nu+1 \end{Bmatrix} \end{vmatrix}_{j=1,\cdots,\nu-1} = F_1 \cdot F_3.$$

The first factor F_1 is easily found to be

$$F_1 = \begin{Bmatrix} m \\ n \end{Bmatrix} \Big/ \begin{Bmatrix} n+\nu-1 \\ n \end{Bmatrix}.$$

To the factor F_2 we may apply (6.2) assumed true for the order $\nu - 1$. Thus (6.7) becomes

$$D_\nu = \frac{\begin{Bmatrix} m \\ n \end{Bmatrix}}{\begin{Bmatrix} n+\nu-1 \\ n \end{Bmatrix}} \cdot \frac{\begin{Bmatrix} m+1 \\ n \end{Bmatrix}\begin{Bmatrix} m+2 \\ n \end{Bmatrix}\cdots\begin{Bmatrix} m+\nu-1 \\ n \end{Bmatrix}}{\begin{Bmatrix} n \\ n \end{Bmatrix}\begin{Bmatrix} n+1 \\ n \end{Bmatrix}\cdots\begin{Bmatrix} n+\nu-2 \\ n \end{Bmatrix}}$$

which agrees with the desired expression described by (6.2). This completes the proof by induction.

7. A proof of Theorem 3. A proof follows almost immediately by combining the second part of Theorem 2 with the determinant identity (6.2). Let us restrict the closed range (1.14) to the open range

$$(7.1)\qquad 0 < \theta < \frac{\pi}{m+k-1}.$$

By the second part of theorem 2 the sequence (1.13) is surely k-positive if the determinant (6.2) is positive for the following values of n and ν: $0 < n < m$, $1 \leqq \nu \leqq k$. This however is evident from the right-hand side of (6.2). Indeed, every one of the 2ν Gauss coefficients appearing on the right side of (6.2) is a quotient of products of sines of the form $\sin \mu\theta$, where $1 \leqq \mu \leqq m+k-1$ and therefore

$$\sin \mu\theta > 0$$

since $0 < \mu\theta < \pi$ by (7.1). By continuity the desired result follows also for the endpoints of the range (7.1).

§4. Proofs of Theorems 4 and 5

8. A proof of Theorem 4. It is clear that it is the length of the range of integration which matters in Theorem 4 rather than its location. We may therefore assume the function $f(x)$ to be k-times positive and vanishing outside the range $(0, \pi)$. With these assumptions we are to show that

$$F(s) = \int_0^\pi e^{-sx} f(x)\, dx \tag{8.1}$$

has no zeros in the strip

$$|\, \mathbf{I}s \,| < k. \tag{8.2}$$

We have already mentioned in Article 2 that Theorem 4 is evidently true if $k = 1$. We may therefore assume that $k \geqq 2$. But then $f(x) \in \mathcal{P}_k \subseteqq \mathcal{P}_2$, showing that $f(x)$ is surely Riemann integrable in the range $[0, \pi]$.

Let us consider the Riemann sum

$$F_m(s) = \frac{\pi}{m} \sum_{\nu=0}^{m} e^{-s(\pi\nu/m)} f\left(\frac{\pi\nu}{m}\right). \tag{8.3}$$

Setting

$$z = e^{-s(\pi/m)} = e^{-(\pi/m)(\sigma+it)} \tag{8.4}$$

we obtain the polynomial

$$F_m(s) = \frac{\pi}{m} \sum_{\nu=0}^{m} f\left(\frac{\pi\nu}{m}\right) z^\nu$$

such that the sequence

$$\cdots, 0, 0, f(0), f\left(\frac{\pi}{m}\right), \cdots, f(\pi), 0, 0, \cdots$$

is k-times positive. By Theorem 1 and (8.4) we know[8] that $F_m(s) \neq 0$ in the strip

$$\left|\frac{t\pi}{m}\right| < \frac{k\pi}{m+k-1}, \quad \text{or} \quad |\, t \,| < k\frac{m}{m+k-1}.$$

Given $\delta > 0$, we conclude that

$$F_m(s) \neq 0 \quad \text{in the strip} \quad |\, t \,| \leqq k - \delta,$$

provided that m is sufficiently large. Since $F_m(s) \to F(s)$, as $m \to \infty$, uniformly in every bounded domain, we conclude by a familiar theorem of A. Hurwitz that

[8] The assumptions $a_0 = f(0) > 0$, $a_m = f(\pi) > 0$, of Theorem 1 are not necessarily fulfilled in our case since these coefficients could well vanish. Evidently, this does not matter, its effect being of lowering the effective degree m in Theorem 1, thereby *increasing* the sector (1.6).

$F(s) \neq 0$ in the strip $|t| < k - \delta$ and hence also in (8.2), δ being arbitrarily small.

9. Wallis distributions are multiply positive. In turning to the proof of our last Theorem 5, we will find it more convenient to show, equivalently, that the function

$$(9.1) \qquad g_\alpha(x) = \begin{cases} (\sin\ x)^\alpha & \text{if} \quad 0 \leqq x \leqq \pi, \\ 0 & \quad\text{elsewhere,} \end{cases}$$

is k-times positive if

$$(9.2) \qquad \alpha \geqq k - 2, \qquad (k \geqq 2).$$

This will be done by showing that, as $m \to \infty$, the discrete distribution (1.13) converges, under appropriate circumstances to the continuous distribution (9.1). These circumstances are described by

LEMMA 3. *Let α be given, $\alpha > -1$, and let us set*

$$(9.3) \qquad \theta = \frac{\pi}{m + \alpha + 1},$$

where m is an arbitrary natural integer. Let x be a given real number such that

$$(9.4) \qquad x \neq 0, \qquad x \neq \pi,$$

and let ν vary with m such that

$$(9.5) \qquad \lim_{m \to \infty} \nu\theta = x.$$

Then

$$(9.6) \qquad \lim_{m\to\infty} \frac{\begin{Bmatrix} m \\ \nu \end{Bmatrix}}{\begin{Bmatrix} m \\ [m/2] \end{Bmatrix}} = g_\alpha(x).$$

If

$$(9.7) \qquad \alpha > 0,$$

then the restrictions (9.4) *may be disregarded, i.e.* (9.5) *implies* (9.6) *for all real values of x.*

PROOF. (9.6) certainly holds if $x < 0$ or $x > \pi$. Indeed, notice that $m\theta \to \pi$. But then if $x > \pi$, say, then (9.5) implies that $\nu > m$, for sufficiently large m, and (9.6) becomes evident by the vanishing of the Gauss coefficients.

Assume now $0 < x < \pi$. In view of the symmetry relation

$$\begin{Bmatrix} m \\ \nu \end{Bmatrix} = \begin{Bmatrix} m \\ m - \nu \end{Bmatrix}$$

and the symmetry of $g_\alpha(x)$ in the point $x = \pi/2$, we may assume x to be in the range

(9.8) $$\frac{\pi}{2} \leqq x < \pi.$$

Let us even assume for the moment that

(9.9) $$\frac{\pi}{2} < x < \pi,$$

and let us look at the left side of (9.6) more closely. By (5.9)

$$\begin{Bmatrix} m \\ \nu \end{Bmatrix} = \frac{\sin m\theta \sin (m-1)\theta \cdots \sin (m-\nu+1)\theta}{\sin \theta \sin 2\theta \cdots \sin \nu\theta}.$$

However, (9.3) implies

$$m\theta + (\alpha+1)\theta = \pi,$$
$$(m-1)\theta + (\alpha+2)\theta = \pi, \cdots$$

which allows us to write

$$\begin{Bmatrix} m \\ \nu \end{Bmatrix} = \frac{\sin (\alpha+1)\theta \sin (\alpha+2)\theta \cdots \sin (\alpha+\nu)\theta}{\sin \theta \sin 2\theta \cdots \sin \nu\theta}$$

and in particular

$$\begin{Bmatrix} m \\ n \end{Bmatrix} = \frac{\sin (\alpha+1)\theta \sin (\alpha+2)\theta \cdots \sin (\alpha+n)\theta}{\sin \theta \sin 2\theta \cdots \sin n\theta}, \qquad (n = [m/2]).$$

Now (9.5) and (9.9) imply that $\nu > n = [m/2]$, for large m, and dividing the last two relations we obtain

$$\begin{Bmatrix} m \\ \nu \end{Bmatrix} \Big/ \begin{Bmatrix} m \\ n \end{Bmatrix} = \frac{\sin (\alpha+n+1)\theta \sin (\alpha+n+2)\theta \cdots \sin (\alpha+\nu)\theta}{\sin (n+1)\theta \sin (n+2)\theta \cdots \sin \nu\theta}.$$

We are to show that (9.3), (9.9) and (9.5) imply that

(9.10) $$\lim_{m\to\infty} \frac{\sin (\alpha+n+1)\theta \cdots \sin (\alpha+\nu)\theta}{\sin (n+1)\theta \cdots \sin \nu\theta} = (\sin x)^\alpha, \qquad (n = [m/2]).$$

On passing to logarithms, we have a problem in elementary calculus. Indeed, by the mean-value theorem we have

(9.11) $$\sum_{j=n+1}^{\nu} (\log \sin (j\theta + \alpha\theta) - \log \sin (j\theta)) = \alpha\theta \sum_{j=n+1}^{\nu} \cot \xi_j,$$

where

(9.12) $$j\theta < \xi_j < j\theta + \alpha\theta, \qquad (j = n+1, \cdots, \nu).$$

Now, as $m \to \infty$, then $(n + 1)\theta = ([m/2] + 1)\theta \to \pi/2$, $\nu\theta \to x$, and it becomes clear that the right side of (9.11) converges to the integral

$$\alpha \int_{\pi/2}^{x} \cot \xi \, d\xi = \alpha \log \sin x = \log (\sin x)^{\alpha}.$$

The fact that the "intermediate" points ξ_j have more freedom than is usual if $\alpha > 1$, in view of (9.12), will obviously not vitiate our argument. A return from logarithms to numbers establishes the relation (9.10).

The case when $x = \pi/2$, which was excluded by (9.9), is settled by an obvious continuity argument. Finally notice that the assumption (9.7) implies the continuity of $g_\alpha(x)$ for all real x. Moreover the sequence (1.13) is then surely twice-positive by Theorem 3 because $\theta = \pi/(m + \alpha + 1) < \pi/(m + 1)$. By its symmetry, the sequence $\left\{ {m \atop \nu} \right\}$ is non-decreasing if $\nu \leqq m/2$ and non-increasing if $\nu \geqq m/2$. These remarks allow to establish easily the last statement of Lemma 3.

A proof of Theorem 5 now runs as follows. Let us assume that (9.2) holds. If $k = 2$ and $\alpha = 0$ then clearly $g_0(x) \in \mathcal{P}_2$. If $\alpha > 0$ then Lemma 3 becomes applicable without exception. However, (9.2) and (9.3) imply that

$$\theta \leqq \frac{\pi}{m + k - 1}.$$

By Theorem 3 the sequence $\left\{ {m \atop \nu} \right\}$, $(-\infty < \nu < \infty)$ is k-times positive. Because of the continuity of $g_\alpha(x)$, the relation (9.6) implies that also $g_\alpha(x)$ is k-times positive, which concludes our proof.

10. The transforms of the Wallis distributions. Our last task is to establish the formula (2.7). Now (2.7) is an integral analogue of the relation (5.13) and could be derived from it by an appropriate passage to the limit. We prefer, however, to establish (2.7) directly as follows. Let $\alpha > 0$ and θ be real. The binomial expansion

$$e^{-s\theta i}(1 + e^{i\theta})^{\alpha} e^{-i\theta\alpha/2} = \sum_{\nu=0}^{\infty} \binom{\alpha}{\nu} e^{(\nu - \frac{1}{2}\alpha - s)i\theta}$$

converges absolutely and uniformly in θ in every finite range in view of the estimate

$$\binom{\alpha}{\nu} = 0(\nu^{-\alpha-1})$$

(see [11], vol. 1, p. 429). By integration we find

$$\text{(10.1)} \quad \begin{aligned} \frac{1}{2\pi} \int_{-\pi}^{\pi} e^{-s\theta i} \left(2 \cos \frac{\theta}{2}\right)^{\alpha} d\theta &= \sum_{\nu} \binom{\alpha}{\nu} \int_{-\pi}^{\pi} e^{(\nu - \frac{1}{2}\alpha - s)i\theta} \, d\theta \\ &= \sum_{\nu} \binom{\alpha}{\nu} \frac{\sin (\nu - \frac{1}{2}\alpha - s)\pi}{\pi(\nu - \frac{1}{2}\alpha - s)} = \frac{\sin (s + \frac{1}{2}\alpha)\pi}{\pi} \sum_{\nu=0}^{\infty} \binom{\alpha}{\nu} \frac{(-1)^{\nu}}{s + \frac{1}{2}\alpha - \nu}. \end{aligned}$$

On the other hand, for $s > 0$, $\alpha > 0$

$$B(s, \alpha + 1) = \frac{\Gamma(\alpha + 1)\Gamma(s)}{\Gamma(\alpha + s + 1)} = \int_0^1 x^{s-1}(1 - x)^\alpha \, dx$$

$$= \sum_{\nu=0}^{\infty} (-1)^\nu \binom{\alpha}{\nu} \int_0^1 x^{\nu+s-1} \, dx = \sum_{\nu=0}^{\infty} (-1)^\nu \binom{\alpha}{\nu} \frac{1}{s + \nu},$$

whence

$$\sum_{\nu=0}^{\infty} (-1)^\nu \binom{\alpha}{\nu} \frac{1}{s - \nu} = -\frac{\Gamma(\alpha + 1)\Gamma(-s)}{\Gamma(\alpha + 1 - s)}.$$

The last expression (10.1) may therefore be written as

$$-\frac{\sin (s + \frac{1}{2}\alpha)\pi}{\pi} \frac{\Gamma(\alpha + 1)\Gamma(-s - \frac{1}{2}\alpha)}{\Gamma(\alpha + 1 - s - \frac{1}{2}\alpha)}$$

$$= -\frac{\Gamma(\alpha + 1)\Gamma(-s - \frac{1}{2}\alpha)}{\Gamma(s + \frac{1}{2}\alpha)\Gamma(1 - s - \frac{1}{2}\alpha)\Gamma(\frac{1}{2}\alpha + 1 - s)}$$

$$= -\frac{\Gamma(\alpha + 1)\Gamma(-s - \frac{1}{2}\alpha)}{\Gamma(s + \frac{1}{2}\alpha)(-s - \frac{1}{2}\alpha)\Gamma(-s - \frac{1}{2}\alpha)\Gamma(\frac{1}{2}\alpha + 1 - s)}$$

$$= \frac{\Gamma(\alpha + 1)}{\Gamma(\frac{1}{2}\alpha + 1 + s)\Gamma(\frac{1}{2}\alpha + 1 - s)}.$$

This establishes the Fourier transformation

$$\frac{1}{2\pi} \int_{-\pi}^{\pi} e^{-s\theta i} (\cos \tfrac{1}{2}\theta)^\alpha \, d\theta = \frac{\Gamma(\alpha + 1)}{2^\alpha \Gamma(\frac{1}{2}(\alpha + 2 + 2s))\Gamma(\frac{1}{2}(\alpha + 2 - 2s))}$$

from which (2.7) follows by trivial changes.

UNIVERSITY OF PENNSYLVANIA

Added in proof: For another application of our Theorem 3, for the special case when $m = 2$, see I. J. SCHOENBERG, *A note on multiply positive sequences and the Descartes rule of signs*, Rend. Circ. Mat. Palermo, Ser. II, 4 (1955), pp. 123–131.

REFERENCES

1. A. EDREI, *On the generating function of a doubly infinite, totally positive sequence*, Trans. Amer. Math. Soc., 74 (1953), pp. 367–383.
2. A. ERDÉLYI, W. MAGNUS, F. OBERHETTINGER and F. TRICOMI, Higher transcendental functions, vol. 1, New York, 1953.
3. M. FEKETE and G. PÓLYA, *Über ein Problem von Laguerre*, Rend. Circ. Mat. Palermo, 34(1912), pp. 1–32.
4. G. H. HARDY and E. M. WRIGHT, An introduction to the theory of numbers, Oxford, 1938.
5. N. NIELSEN, Handbuch der Theorie der Gammafunktion, Leipzig, 1906.
6. N. OBRESCHKOFF, *Über die Wurzeln algebraischer Gleichungen*, Jber. Deutschen Math. Verein., 33(1924), pp. 52–64.

7. A. Ostrowski, *Recherches sur la méthode de Graeffe*, Acta Math., 72 (1940), pp. 99–257.
8. E. Pascal, Die Determinanten, Leipzig, 1900.
9. I. J. Schoenberg, *Über variationsvermindernde lineare Transformationen*, Math. Zeit., 32(1930), pp. 321–328.
10. I. J. Schoenberg, *On Pólya frequency functions* I. *The totally positive functions and their Laplace transforms*, Journal d'Analyse Math., 1(1951), pp. 331–374.
11. Ch.-J. de la Vallée Poussin, Cours d'Analyse Infinitésimale, Louvain and Paris, 1921.

REMARKS ON DE LA VALLÉE POUSSIN MEANS AND CONVEX CONFORMAL MAPS OF THE CIRCLE

G. PÓLYA AND I. J. SCHOENBERG

Introduction. The aims of the present remarks are similar to those pursued by L. Fejér in several papers in the early nineteen thirties and well described by the title of one of his paper: *Gestaltliches über die Partialsummen und ihre Mittelwerte bei der Fourierreihe und der Potenzreihe.* However, the means which we use to realize these aims are different. Fejér discovered the remarkable behavior of certain Cesàro means, especially that of the third Cesàro means for even or odd functions of certain simple basic shapes. In what follows we show that the de la Vallée Poussin means possess such shape-preserving properties to a much higher degree thanks to their variation diminishing character.

Before stating our results, we have to explain a few concepts.

Variation diminishing Transformations on the Circle. If $a_1, a_2, \cdots, a_n$ s a finite sequence of real numbers we shall denote by $v(a)$ or $v(a_\nu)$ the number of variations of sign in the terms of this sequence. By the number $v_c(a)$ of cyclic variations of sign of our sequence we mean the following: If all $a_\nu=0$ we set $v_c(a)=0$. If $a_i \neq 0$ we set

$$v_c(a)=v(a_i, a_{i+1}, \cdots, a_n, a_1, a_2, \cdots, a_{i-1}, a_i) .$$

If we think of the a_ν as arranged clockwise in cyclic order, it becomes obvious that $v_c(a)$ does not depend on the particular non-vanishing term a_i we start with. Notice that $v_c(a)$ is always an even number. Let now $f(t)$ be a real-valued function of period 2π. Let $t_1, t_2, \cdots, t_n$ be such that

$$(1) \qquad t_1<t_2< \cdots <t_n<t_1+2\pi .$$

We may now define the number $v_c(f)$ of *cyclic variations of sign of* $f(t)$ by

$$(2) \qquad v_c(f)=\sup v_c(f(t_\nu)) ,$$

the supremum being taken for all finite sequences $\{t_\nu\}$ subject to (1).

Received October 21, 1957. This paper was prepared partly under the sponsorship of the Office of Naval Research, and partly under the sponsorship of the United States Air Force, Office of Scientific Research, Air Research and Development Command.

Also $v_c(f)$, if finite, is even. Thus $v_c(\sin t)=2$, $v_c(\sin 2t)=4$, $v_c(|\sin t|)=0$.

We now describe what is meant by a *variation diminishing transformation on the circle* (See [4]). Such a transformation is characterized by a non-negative weight-function, or kernel, $\Omega(t)$, of period 2π, of bounded variation and normalized by the conditions

$$(3) \qquad \frac{1}{2\pi}\int_0^{2\pi}\Omega(t)dt=1,\ \Omega(t)=\frac{1}{2}(\Omega(t+0)+\Omega(t-0)) .$$

Let $f(t)$ be an arbitrary periodic function, with period 2π, real-valued and integrable (cf. §1.2); let us form its convolution transform

$$(4) \qquad g(t)=\frac{1}{2\pi}\int_0^{2\pi}\Omega(t-\tau)f(\tau)d\tau .$$

We say that this transformation is *variation diminishing* provided that the inequality

$$(5) \qquad v_c(g)\leqq v_c(f)$$

holds for each f. We mean the same thing if we say that $\Omega(t)$ is a *variation diminishing kernel.*

V-means. One of our aims is to show that the de la Vallée Poussin kernels

$$(6) \qquad \omega_n(t)=\frac{(n\,!)^2}{(2n)\,!}\left(2\cos\frac{t}{2}\right)^{2n},$$

the Fourier expansion of which has the simple form

$$(6') \qquad \omega_n(t)=\frac{1}{\binom{2n}{n}}\sum_{-n}^{n}\binom{2n}{n+\nu}e^{i\nu t}=1+2\sum_{1}^{n}\frac{n\,!}{(n-\nu)\,!}\frac{n\,!}{(n+\nu)\,!}\cos\nu t ,$$

possess the property of being variation diminishing for $n=1, 2, 3, \cdots$. For $\Omega(t)=\omega_n(t)$ the transformation (4) becomes

$$(7) \qquad V_n(t)=\frac{(n\,!)^2}{(2n)\,!}\frac{1}{2\pi}\int_0^{2\pi}\left(2\cos\frac{t-\tau}{2}\right)^{2n}f(\tau)d\tau ,$$

and defines the de la Vallée Poussin means, or simply V-means, of the function $f(t)$. It is easily verified (See [14] and [5, p. 15]) that $V_n(t)$ is a trigonometric polynomial of an order not exceeding n, which is readily expressed in terms of the Fourier coefficients of $f(t)$. Indeed, if

$$(8) \qquad f(t)\sim\sum_{-\infty}^{\infty}c_\nu e^{i\nu t},\ (c_{-\nu}=\bar{c}_\nu) ,$$

we obtain by convoluting (6′) and (8)

$$V_n(t)=\frac{1}{\binom{2n}{n}}\sum_{-n}^{n}\binom{2n}{n+\nu}c_\nu e^{i\nu t}\,. \tag{9}$$

In terms of the real Fourier series ($2c_\nu=a_\nu-ib_\nu$)

$$f(t)\sim\frac{1}{2}a_0+\sum_1^\infty(a_\nu\cos\nu t+b_\nu\sin\nu t) \tag{10}$$

we find

$$V_n(t)=\frac{1}{2}a_0+\frac{1}{\binom{2n}{n}}\sum_1^n\binom{2n}{n+\nu}(a_\nu\cos\nu t+b_\nu\sin\nu t) \tag{11}$$

or

$$V_n(t)=\frac{1}{2}a_0+\sum_1^n\frac{n\,!}{(n+\nu)\,!}\,\frac{n\,!}{(n-\nu)\,!}(a_\nu\cos\nu t+b_\nu\sin\nu t)\,. \tag{12}$$

Main Results. Our principal result is the following

THEOREM 1. *The inequalities*

$$v_c(V_n)\leqq Z_c(V_n)\leqq v_c(f) \tag{13}$$

hold for an arbitrary integrable function $f(t)$. (We let $Z_c(V_n)$ denote the number of real zeros of $V_n(t)$ within a period including multiplicities.)

The first inequality $v_c(V_n)\leqq Z_c(V_n)$, which is obvious, shows that Theorem 1 states considerably more than the variation diminishing property of the kernel $\omega_n(t)$ which amounts to $v_c(V_n)\leqq v_c(f)$. In Part I we give two proofs of Theorem 1, both based on a theorem due to Sylvester [12]. The first proof uses the result of Sylvester's theorem, the second uses the method of one of its proofs.

In Part II we discuss applications of the variation diminishing property of V-means. Theorem 1 gives a useful lower bound for $v_c(f)$ if a certain number of Fourier coefficients of $f(t)$ are known. It is shown how this implies easily some results by Sturm, A. Hurwitz, Pólya and Wiener. In §5 we study the simplest classes of discontinuous periodic functions; the behavior of their V-means is described by Theorems 3 and 4. Fejér's Theorem III [1, p. 86] has an analogue for V-means which is our Theorem 5 below. All this refers to *real* periodic functions. However, the shape-preserving properties of V-means appear to best advantage if applied to *complex-valued* periodic functions.

Let us state here the main result of §6 concerning convex maps of the circle. Let K denote the class of those "schlicht" power series $\sum_1^\infty a_\nu z^\nu$ which map $|z|<1$ onto some convex domain. Let

$$f(z)=\sum_1^\infty c_\nu z^\nu, \ (c_1=1) , \tag{14}$$

$$V_n(z)=\frac{1}{\binom{2n}{n}}\sum_{\nu=1}^{n}\binom{2n}{n+\nu}c_\nu z^\nu \tag{15}$$

be the de la Vallée Poussin mean, or V-mean, of the power series (14). It is known that the partial sums of the series (14) need not belong to K. G. Szegö has shown [13] that if $F(z)\in K$ then all partial sums of (14) are "schlicht" in the circle $|z|<1/4$ and map it onto convex domains, and that 1/4 is here the largest constant. That the V-means belong to K is one part of the following

THEOREM 2. For

$$f(z)\in K \tag{16}$$

it is necessary and sufficient that

$$V_n(z)\in K \text{ for } n=1, 2, \cdots . \tag{17}$$

The sufficiency part does not even assume the regularity of (14) in the unit circle, as for any formal power series (14) the assumption (17) imply that (14) converges and defines an element of K.

Additional Results. Parts I and II are followed by two appendices which contain related materials, but are almost independent of the main text.

Appendix I brings out a certain analogy between approximations to two kinds of functions: periodic functions and functions defined in a finite interval. It will be shown that the shape-preserving properties of the V-means, which approximate functions of period 2π, are analogous to the shape-preserving properties of the so called Bernstein polynomials which approximate functions defined in [0,1]. For the definition of these polynomials see §7 where also their variation diminishing property (Theorem 6) is stated and proved.

Appendix II is devoted to a conjecture on power series which represent a conformal one to one mapping of the unit circle onto a convex domain. The conjecture is that the Hadamard composition, or convolution, of two such power series is again a power series of the same kind (see §9). We do not know whether this conjecture is true or not (it seems to us more likely that it is true) but at any rate, in view of the partial results which we have obtained (§§10 and 11), the problem to prove or to disprove the conjecture seems to us worth while.

PART I. THE DE LA VALLÉE POUSSIN SUMMATION METHOD IS VARIATION DIMINISHING

1.1 A theorem of Sylvester. In the course of his work on Newton's rule of signs J. J. Sylvester discovered a remarkable theorem concerning the real zeros of polynomials of the form

$$\sum_{\nu=1}^{m} c_\nu (x-\xi_\nu)^q$$

(see [12, p. 408], [7] and also [9, vol. 2, Problem 79, p. 50]). In Sylvester's theorem q may assume any positive integral value, a fact which is important for its proof which proceeds by induction in q. We need Sylvester's result only for $q=2n$ and state it as follows.

LEMMA 1. *Let $\xi_1<\xi_2<\cdots<\xi_m$, $(m\geqq 2)$, be given reals and consider the polynomial*

$$P(x)=\sum_{\nu=1}^{m} c_\nu (x-\xi_\nu)^{2n}$$

(with real $c_\nu\neq 0$ for all ν), which we assume not to vanish identically. Then

$$Z(P;-\infty<x<\infty)\leqq v(c_1, c_2, \cdots, c_m, c_1)\ ,$$

where the left side denotes the number of real zeros of $P(x)$ while the right side is the number of variations of sign in the sequence displayed.

The significance for us of Sylvester's results is that it easily yields the following

LEMMA 2. Let

$$-\pi<\tau_1<\tau_2<\cdots<\tau_m<\pi, \qquad m\geqq 2\ , \tag{1.1}$$

be given reals and consider the trigonometric polynomial

$$T(t)=\sum_{\nu=1}^{m} c_\nu \left(\sin\frac{t-\tau_\nu}{2}\right)^{2n}, \tag{1.2}$$

(for real $c_\nu\neq 0$ for all ν), which we assume not to vanish identically. Then

$$Z(T;-\pi<t<\pi)\leqq v(c_1, c_2, \cdots, c_m, c_1)\ .$$

Proof. We introduce the new variable

$$(1.3)\qquad x=\tan\frac{t}{2}\qquad -\pi<t<\pi$$

whose range is $-\infty<x<\infty$. The images of the τ_ν we denote by

$$\xi_\nu=\tan\frac{\tau_\nu}{2}$$

and these give rise to the identities

$$\left(\sin\frac{t-\tau_\nu}{2}\right)^2=\frac{(x-\xi_\nu)^2}{(1+x^2)(1+\xi_\nu^2)},\qquad \nu=1,\cdots,m,$$

Thus (1.2) may be expressed in terms of x by

$$T(t)=\frac{1}{(1+x^2)^n}\sum_{\nu=1}^{m}c_\nu\gamma_\nu(x-\xi_\nu)^{2n},$$

where the γ_ν are positive and so Lemma 2 immediately follows from Sylvester's Lemma 1.

We now recast our result in the following more useful form;

LEMMA 3. *Let* $\tau_1,\tau_2,\cdots,\tau_m$ $(m\geqq 2)$ *be* m *points in counter-clockwise order on the circle such that* τ_m *should not overtake or even reach* τ_1. *We may express these requirements by assuming that*

$$(1.4)\qquad \tau_1<\tau_2<\cdots<\tau_m<\tau_1+2\pi .$$

Let

$$(1.5)\qquad T_n(t)=\sum_{\nu=1}^{m}c_\nu\omega_n(t-\tau_\nu),\qquad T_n(t)\not\equiv 0,$$

where at least two among the c_ν *do not vanish. Then*

$$(1.6)\qquad Z_c(T_n)\leqq v_c(c_\nu).$$

Proof. By omitting vanishing terms in (1.5) we may assume that $c_\nu\neq 0$ for all ν. Moreover, a change of variable by $t=t'-\pi$ will evidently not alter the left hand side of (1.6). This implies that in our statement (1.6) we may replace $T_n(t)$ by the polynomial $T(t)$ defined by (1.2). By a second appropriate transformation $t=t'+c$ we may replace the conditions (1.4) by the more restrictive inequalities (1.1), at the same time making sure that $T(\pi)\neq 0$. But then

$$Z_c(T_n)=Z_c(T)=Z(T;\,-\pi<t<\pi)\leqq v(c_1,c_2,\cdots,c_m,c_1)=v_c(c_\nu)$$

and Lemma 3 is established.

1.2. On the number of variations of a function. The reader may interpret the term "integrable" either according to the definition of Riemann or to that of Lebesgue, or to any other definition that involves the familiar standard properties of the integral. We emphasize the following property: If $f(t)$ and $g(t)$ are integrable and $f(t)\geqq 0$ in the interval I, then

$$\int_I f dt=0$$

implies

$$\int_I fg dt=0 .$$

We consider now a real-valued periodic function $f(t)$ with the period 2π, we assume that it is integrable in the interval $(0,2\pi)$ and that $v_c(f)$, as defined in the Introduction, is finite. We consider t (mod 2π), that is, we consider t as attached to a point on the periphery of the unit circle. If $v_c(f)=2k$, we can, as easily seen, divide the circumference of the unit circle into $2k$ consecutive arcs

$$I_1, I_2, \cdots, I_{2k} \tag{1.7}$$

such that

$$(-1)^{\nu-1}f(t)\geqq 0 \text{ in } I_\nu \tag{1.8}$$

for $\nu=1, 2, \cdots, 2k$; the arcs (1.7) may be open, or closed, or open from one side and closed from the other, some of them may even reduce to a single point. Now, we *normalize* $f(t)$, that is, we change $f(t)$ (if necessary) as follows: we set $f(t)=0$ in all points of any interval (1.7) on which $\int f dt$ vanishes; especially, if an interval listed under (1.7) consists of just one point, we set $f(t)=0$ in that point. This normalization cannot increase (but may decrease) $v_c(f)$ and leaves unchanged the V-means of f (cf. the initial remark of this section). Therefore, it will be sufficient to prove Theorem 1 for normalized functions. If, however, $v_c(f)=2k$ for a normalized $f(t)$, the intervals (1.7), constructed as above, have the property

$$(-1)^{\nu-1}\int_{I_\nu} f(t)dt>0 \text{ for } \nu=1, 2, \cdots, 2k . \tag{1.8'}$$

The foregoing remarks will be useful in the following proof of Theorem 1. Yet we do not need them in establishing the weaker inequality

$$v_c(V_n)\leqq v_c(f) \tag{1.9}$$

for a Riemann-integrable function f.
Indeed, let us consider the integral

$$V_n(t)=\frac{1}{2\pi}\int_0^{2\pi}\omega_n(t-\tau)f(\tau)d\tau$$

and its approximating sums

$$V_{n,m}(t)=\frac{1}{m}\sum_{\nu=0}^{m-1}\omega_n\left(t-\frac{2\pi\nu}{m}\right)f\left(\frac{2\pi\nu}{m}\right).$$

Lemma 3 and definition (2) imply

$$v_c(V_{n,m})\leqq v_c\left(f\left(\frac{2\pi\nu}{m}\right)\right)\leqq v_c(f)$$

or

$$v_c(V_{n,m})\leqq v_c(f)\ .$$

Since $V_{n,m}(t)\to V_n(t)$ for all t, as $m\to\infty$, the last inequality evidently implies (1.9). An "approximation argument" extending (1.9) to a more comprehensive class of functions is easy, but hardly deserves to be presented here.

1.3. A first proof of Theorem 1. The first inequality (13) is immediate and so the essential assertion of Theorem 1 consists in the inequality

$$Z_c(V_n)\leqq v_c(f)\ . \tag{1.10}$$

If $v_c(f)\geqq 2n$ there is nothing to prove; also if $v_c(f)=0$ for then $V_n(t)$ clearly can not vanish. Let us assume, then, that $f(t)$ is "normalized" according to §1.2, and that $0<v_c(f)=2k<2n$, and let us divide the unit circumference into the $2k$ consecutive arcs (1.7) which satisfy the conditions (1.8) and (1.8′). We may then write the Fourier coefficients of $f(t)$ in the form

$$\begin{aligned}
a_0&=\frac{1}{\pi}\int_0^{2\pi}f(t)dt=\sum_{s=1}^{k}\frac{1}{\pi}\int_{I_{2s-1}}|f(t)|dt-\sum_{s=1}^{k}\frac{1}{\pi}\int_{I_{2s}}|f(t)|dt\\
a_\nu&=\frac{1}{\pi}\int_0^{2\pi}f(t)\cos\nu t dt=\sum_1^k\frac{1}{\pi}\int_{I_{2s-1}}|f(t)|\cos\nu t dt\\
&\qquad-\sum_1^k\frac{1}{\pi}\int_{I_{2s}}|f(t)|\cos\nu t dt\ ,\\
b_\nu&=\frac{1}{\pi}\int_0^{2\pi}f(t)\sin\nu t dt=\sum_1^k\frac{1}{\pi}\int_{I_{2s-1}}|f(t)|\sin\nu t dt
\end{aligned}\tag{1.11}$$

$$-\sum_{1}^{k}\frac{1}{\pi}\int_{I_{2s}}|f(t)|\sin\nu t dt\ .$$

$$\nu=1,\cdots,n;$$

Consider in the $2n$-dimensional space E_{2n} the closed curve Γ defined in parametric form by $\cos\nu t$, $\sin\nu t$ ($\nu=1,\cdots,n;\ 0\leqq t\leqq 2\pi$). To the division (1.7) of the circumference into the arcs I_μ, corresponds a division of Γ into arcs

$$\Gamma_1, \Gamma_2, \cdots, \Gamma_{2k}\ , \tag{1.12}$$

where we think of the arc Γ_μ as carrying the *positive* mass

$$\frac{1}{\pi}\int_{I_\mu}|f(t)|dt\ . \tag{1.13}$$

This mass has a centroid the coordinates of which, multiplied by (1.13), are

$$\frac{1}{\pi}\int_{I_\mu}|f(t)|\cos\nu t dt,\ \frac{1}{\pi}\int_{I_\mu}|f(t)\sin\nu t dt \qquad (\nu=1,\cdots,n)\ . \tag{1.14}$$

By a well known theorem of Carathéodory the mass (1.13) of Γ_μ may be concentrated in a finite number of points along Γ_μ so as to produce the same centroid (1.14). This we do for each of the arcs (1.12). Arranging all these points in *cyclic order* along Γ we obtain points $\tau_1, \tau_2, \cdots, \tau_m$ and corresponding coefficients $c_1, c_2, \cdots, c_m$ where $(-1)^{\mu-1}c_j>0$ when τ_j belongs to I_μ. In view of the relations (1.11) we obtain

$$a_0=\sum_{j=1}^{m}c_j,\ a_\nu=\sum_{j=1}^{m}c_j\cos\nu\tau_j,\ b_\nu=\sum_{j=1}^{m}c_j\sin\nu\tau_j\ , \tag{1.15}$$

$$v_c(c_j)=2k=v_c(f)\ . \tag{1.16}$$

We consider now the trigonometric polynomial

$$F(t)=\frac{1}{2}\sum_{j=1}^{m}c_j\omega_n(t-\tau_j) \tag{1.17}$$

and claim that

$$F(t)=V_n(t)\ . \tag{1.18}$$

Indeed, by (6′)

$$F(t)=\frac{1}{2}\sum_{j=1}^{m}c_j\omega_n(t-\tau_j)$$

$$=\sum_j c_j\left\{\frac{1}{2}+\sum_{\nu=1}^{n}\frac{n!}{(n-\nu)!}\frac{n!}{(n+\nu)!}\cos\nu(t-\tau_j)\right\}$$

$$=\sum_j c_j\left\{\frac{1}{2}+\sum_{\nu=1}\frac{n!}{(n-\nu)!}\,\frac{n!}{(n+\nu)!}(\cos\nu t\cos\nu\tau_j+\sin\nu t\sin\nu\tau_j)\right\},$$

and interchanging the order of summations, we obtain by (1.15)

$$F(t)=\frac{1}{2}a_0+\sum_{\nu=1}^{n}\frac{n!}{(n-\nu)!}\,\frac{n!}{(n+\nu)!}(a_\nu\cos\nu t+b_\nu\sin\nu t)$$

which is identical with $V_n(t)$ by (12). Finally, by (1.17), (1.18), (1.16) and Lemma 3

$$Z_c(V_n)=Z_c(F)\leqq v_c(c_j)=v_c(f)\ ,$$

which proves the inequality (1.10).

2. A second proof of Theorem 1. The foregoing proof is based on Sylvester's result which we stated as Lemma 1. We shall now prove Theorem 1 without assuming the knowledge of this result.

We transform (7) by changing the variables. Setting

$$x=\tan\frac{t}{2}\ ,\quad \xi=-\cot\frac{\tau}{2}\ ,$$

we obtain from (7) (by steps similar to those exhibited following (1.3)) that

$$(2.1)\quad (1+x^2)^n V_n(2\operatorname{arc}\tan x)=\int_{-\infty}^{\infty}(x-\xi)^{2n}\frac{(n!)^2 2^{2n}}{(2n)!\,\pi}\,\frac{f(-2\operatorname{arc}\cot\xi)}{(1+\xi^2)^{n+1}}d\xi\ .$$

This relation is contained in the more general

$$(2.2)\qquad P(x)=\int_{-\infty}^{\infty}(x-\xi)^m A(\xi)d\xi$$

where m is a positive integer and the integral $\int_{-\infty}^{\infty}\xi^m A(\xi)d\xi$ is absolutely convergent; $P(x)$ is by the structure of the formula (2.2) a polynomial of degree not higher than m.

We consider the following quantities connected with (2.2):

N the number of real zeros of $P(x)$, counted with multiplicity;

v the number of variations of sign of $A(\xi)$ in the open interval $-\infty<\xi<\infty$;

$\operatorname{sgn}A(\infty)$ is the constant sign, different from 0, that $A(\xi)$ possesses whenever it is different from 0 in a suitably chosen interval $\omega<\xi<\infty$; waessume here that $A(\xi)$ is normalized in the sense of §1.2;

$\operatorname{sgn} A(-\infty)$ is similarly defined;

$$\eta=\frac{1}{2}\left|\operatorname{sgn} A(\infty)-\operatorname{sgn}(-1)^m A(-\infty)\right|$$

so that η is either 0 or 1;

$$V=v+\eta .$$

In fitting (2.1) into the more general pattern (2.2), we can assume without loss of generality (by rotating the circle through an appropriate angle) that $V_n(\pi)\neq 0$, that $f(t)$ is normalized in the sense of §1.2, and that 0 is an interior point of one of the intervals of constant sign considered there, so located that, for some positive ε, $f(t)$ takes some non-vanishing values in both intervals $-\varepsilon<t<0$ and $0<t<\varepsilon$. Under these circumstances, in the particular case (2.1),

$$\begin{gathered} m=2n ,\\ A(-\infty)=A(\infty) ,\\ \eta=0 ,\\ V=v=v_c(f) ,\\ N=Z_c(V_n) , \end{gathered}$$

and so Theorem I is an immediate consequence of the following.

LEMMA 4. $N\leqq V$.

We need several steps to prove Lemma 4.

(a) There are some particular cases in which Lemma 4 is obvious.

If $P(x)$ vanishes identically there is nothing to prove since in this case, by definition, $N=0$.

If $V\geqq m$ there is nothing to prove since, of course, $N\leqq m$.

If $v=0$ and m is even (so that $V=\eta=0$) then $P(x)$ will have for all real x the constant sign of $A(\xi)$ and so $N=0$ as it should be according to Lemma 4.

If $v=0$ and m is odd (so that $V=\eta=1$) then $m-1$ is even and so

$$P'(x)=\int_{-\infty}^{\infty}(x-\xi)^{m-1}mA(\xi)d\xi$$

has a constant sign for all x, by what we have just said. Therefore, $P(x)$ is monotone and $N=1$ which agrees with Lemma 4.

And so we may and shall assume in the sequel that

$$1\leqq v\leqq V\leqq m-1 . \tag{2.3}$$

(b) Let c be a point of change of sign for $A(\xi)$; that is, c is the common endpoint of two contiguous intervals in each of which $A(\xi)$ keeps a constant sign, yet the two signs (cf. §1.2) considered are opposite. The number of such points is v and we have assumed (2.3).

We assert that at least one of the $m-1$ quantities $P'(c), P''(c), \cdots, P^{(m-1)}(c)$ is different from 0. If this assertion were wrong, the integral

$$\int_{-\infty}^{\infty} (\xi-c)^{\mu} A(\xi) d\xi$$

would vanish for $\mu=m-1, \cdots, 2,1$ and, as a linear combination of these integrals,

$$\int_{-\infty}^{\infty} (\xi-c) Q(\xi) A(\xi) d\xi \tag{2.4}$$

would vanish for any polynomial $Q(\xi)$ of degree not exceeding $m-2$. Yet this is certainly false if

$$Q(\xi)=(x-c_1)(x-c_2)\cdots(x-c_{v-1}) \tag{2.5}$$

where $c, c_1, c_2, \cdots, c_{v-1}$ are all the points of change of sign of $A(\xi)$; observe (2.3) in computing the degree of $Q(\xi)$. In fact, with (2.5) the integrand in (2.4) has a constant sign and so the integral (2.4) cannot vanish.

We have seen by the way, that under the condition (2.3) $P(x)$ cannot vanish identically.

(c) Set

$$G(x)=P(x)(x-c)^{-m} \tag{2.6}$$

$$\begin{aligned} P^*(x) &= (x-c)^{m+1} G'(x) \\ &= (x-c)P'(x)-mP(x) \\ &= \int_{-\infty}^{\infty} (x-\xi)^{m-1} A^*(\xi) d\xi \end{aligned} \tag{2.7}$$

where

$$A^*(\xi)=m(\xi-c)A(\xi) \tag{2.8}$$

and let $N^*, m^*, v^*, \eta^*, V^*$ be just so connected with $P^*(x)$ and $A^*(\xi)$ as N, m, v, η and V are with $P(x)$ and $A(\xi)$. Obviously

$$\begin{aligned} m^* &= m-1 \\ v^* &= v-1 \\ \operatorname{sgn} A^*(\infty) &= \operatorname{sgn} A(\infty) \\ \operatorname{sgn} A^*(-\infty) &= -\operatorname{sgn} A(-\infty) \end{aligned} \tag{2.9}$$

and so $\eta^*=\eta$. Combining this with (2.9), we obtain

$$V^* = V - 1 . \tag{2.10}$$

We intend to prove Lemma 4 by mathematical induction with respect to V. In fact, we have already proved Lemma 4 in the particular case $V=0$ under (a). We therefore assume $V \geqq 1$, cf. (2.3), and that Lemma 4 has been proved for the preceding value (2.10), and so we take for granted that

$$N^* \leqq V^* . \tag{2.11}$$

(d) Let k denote the number of those zeros of $P(x)$ that coincide with the point c; obviously $k \geqq 0$, and, by (b),

$$k \leqq m-1 . \tag{2.12}$$

Let k^* denote the number of those zeros of $P^*(x)$ that coincide with c. We set

$$N = k + l , \quad N^* = k^* + l^* . \tag{2.13}$$

The quantities l and l^*, defined by (2.13), enumerate those zeros of $P(x)$ and $P^*(x)$, respectively, that fall into one or the other of the two open intervals $-\infty < x < c$ and $c < x < \infty$.

We note the critical term of the expansion of $P(x)$ around the point $x=c$,

$$P(x) = \frac{P^{(k)}(c)(x-c)^k}{k!} + \cdots, \; P^{(k)}(c) \neq 0 .$$

By (2.6) and (2.12), $G(x)$ has a pole at the point c and (2.7) yields

$$P^*(x) = \frac{(k-m)P^{(k)}(c)(x-c)^k}{k!} + \cdots .$$

We infer that $P^*(x)$ has just as many zeros at the point c as $P(x)$:

$$k^* = k . \tag{2.14}$$

By the way, we have seen that $P^*(x)$ does not vanish identically.

(e) It remains to consider the real zeros different from c; $P(x)$ or, which is the same, $G(x)$ has l such zeros, and $P^*(x)$ or, which is the same, $G'(x)$ has l^* such zeros. These zeros are distributed somehow in the two open intervals, $-\infty < x < c$ and $c < x < \infty$.

By the theorem of Rolle, in each of these intervals at most one zero can be lost in the passage from $G(x)$ to $G'(x)$, so that

(2.15) $$l^* \geqq l-2\,;$$

this information is correct, but insufficient for our purpose. We shall obtain, however, additional information by using the following remark (cf. [9, vol. 2, p. 39, problem 14]).

No zero can be lost in the passage from $G(x)$ *to* $G'(x)$ *in the interval* $(-\infty, c)$ *if*

(2.16) $$\operatorname{sgn} G(-\infty) = \operatorname{sgn} G'(-\infty)$$

and no zero can be lost in this passage in the interval (c, ∞) *if*

(2.17) $$\operatorname{sgn} G(\infty) = -\operatorname{sgn} G'(\infty)\,.$$

The signs mentioned in (2.16) and (2.17) refer to a certain neighborhood of $-\infty$ or ∞ and, as $G(x)$ has only a finite number of zeros, they are certainly different from 0.

(f) We know, cf. (b), that the polynomial $P(x)$ does not vanish identically. We set

(2.18) $$P(x) = b_0 x^m + b_1 x^{m-1} + \cdots + b_m$$

and distinguish two cases.

Case I. If $b_0 = 0$, there is an s such that $b_0 = b_1 = \cdots = b_{s-1} = 0$, $b_s \neq 0$ and so we easily find the initial terms in the expansions around ∞:

$$G(x) = \frac{b_s}{x^s} + \cdots, \qquad \frac{G'(x)}{G(x)} = -\frac{s}{x} + \cdots.$$

In this case, both conditions (2.16) and (2.17) are satisfied, and, by the final remark under (e), we can improve (2.15) to

(2.19) $$l^* \geqq l\,.$$

Case II. Now

(2.20) $$b_0 = \int_{-\infty}^{\infty} A(\xi)\, d\xi \neq 0\,,$$

and the expansions around ∞ begin

(2.21) $$G(x) = b_0 + \frac{mcb_0 + b_1}{x} + \cdots$$

(2.22) $$G'(x) = -\frac{mcb_0 + b_1}{x^2} + \cdots$$

where

$$mcb_0+b_1=m\int_{-\infty}^{\infty}A(\xi)(c-\xi)d\xi \ . \tag{2.23}$$

We again distinguish two cases.

Subcase II, 1. If $v=1$, c is the only point of change of sign of $A(\xi)$, the integrand in (2.23) is of constant sign, and so the integral is different from 0.

Subcase II, 2. If $v\geqq 2$, the integral (2.23) could vanish. Yet in this case $A(\xi)$ has at least another point of change of sign, c_1, and we say that (2.23) and

$$m\int_{-\infty}^{\infty}A(\xi)(c_1-\xi)d\xi$$

cannot vanish simultaneously: in fact, their difference is

$$m(c_1-c)\int_{-\infty}^{\infty}A(\xi)d\xi=m(c_1-c)b_0\neq 0$$

by our present assumption (2.20). Therefore, assuming that the point of change was properly selected from the start (which boils down to a proper choice of notation) we may assume that (2.23) is different from 0, also in the present subcase.

Finally, in both subcases, we conclude from (2.21) and (2.22)

$$\lim_{x\to\pm\infty}\frac{x^2G'(x)}{G(x)}=-\frac{mcb_0+b_1}{b_0}\neq 0$$

and we see that just one of the two conditions (2.16) and (2.17) is fulfilled. Therefore, by the final remark under (e), we can improve (2.15) to

$$l^*\geqq l-1 \ . \tag{2.24}$$

Thus, even in the less favorable of the two cases I and II, we have (2.24). Combining this with (2.13) and (2.14), we obtain

$$N^*\geqq N-1$$

and hence and from (2.10) and (2.11) we obtain

$$V-1=V^*\geqq N^*\geqq N-1$$

or $V\geqq N$, which is the desired conclusion of Lemma 4.

The foregoing somewhat involved proof becomes more understandable if it is compared with the proof for Lemma 1 given in [7] or in [9, vol. 2, p. 50, problem 79].

PART II. SOME APPLICATIONS OF THE VARIATION DIMINISHING PROPERTY OF V-MEANS

3. A theorem of Ch. Sturm and A. Hurwitz. Let $f(t)$ be a real-valued, integrable, periodic function of period 2π. Let

$$f(t)=\frac{1}{2}a_0+\sum_{\nu=1}^{\infty}(a_\nu \cos \nu t+b_\nu \sin \nu t) \tag{3.1}$$

be its Fourier expansion. Suppose that the partial sum

$$S_n(t)=\frac{1}{2}a_0+\sum_{\nu=1}^{n}(a_\nu \cos \nu t+b_\nu \sin \nu t) \tag{3.2}$$

is known. What can we say about the number $v_c(f)$ of changes of sign of $f(t)$ in a period? An answer is immediate: Knowing (3.2), we can compute (11), the nth V-mean of $f(t)$, and we must have

$$v_c(f)\geqq Z_c(V_n) \tag{3.3}$$

by Theorem 1.

The information provided by this inequality is strongest when the right hand side attains its largest value $2n$. There is a simple sufficient condition for this eventuality which we record as follows.

COROLLARY 1. *If*

$$\begin{aligned}(a_n^2+b_n^2)^{1/2}>\binom{2n}{1}(a_{n-1}^2+b_{n-1}^2)^{1/2}+\binom{2n}{2}(a_{n-2}^2+b_{n-2}^2)^{1/2}+\cdots\\ +\binom{2n}{n-1}(a_1^2+b_1^2)^{1/2}+\frac{1}{2}\binom{2n}{n}|a_0|\,,\end{aligned} \tag{3.4}$$

then every function $f(t)$ having (3.2) *as the* nth *partial sum of its Fourier series, must change sign within a period at least $2n$ times.*

Indeed, it is clear by (3.4) that the last term of the expression (11) for $V_n(t)$ so predominates that $V_n(t)$ has $2n$ simple zeros, hence $Z_c(V_n)=2n$. The statement now follows from (3.3).

We obtain a classical result [2, pp. 572-574] as a very special case.

COROLLARY 2. *If* $a_0=a_1=b_1=\cdots=a_{n-1}=b_{n-1}=0$, $a_n^2+b_n^2>0$, *then* $v_c(f)\geqq 2n$.

The following is an equivalent formulation. *If* $a_n^2+b_n^2>0$ *then*

$$v_c(f(t)-S_{n-1}(t))\geqq 2n\,. \tag{3.5}$$

This second formulation is especially interesting and intuitive because it shows that the graph of the partial sum $S_{n-1}(t)$ must cross the graph of $f(t)$ at least $2n$ times. Hurwitz's proof of Corollary 2 is direct and elementary. However, his classical argument is no longer available to establish other special cases such as the following.

If

$$f(t)=\frac{1}{2}+\sum_{\nu=1}^{n}\cos\nu t+\sum_{\nu=n+1}^{\infty}(a_\nu\cos\nu t+b_\nu\sin\nu t)$$

then

$$v_c(f)\geqq 2n\ .$$

For in this case $V_n(t)=\frac{1}{2}\omega_n(t)$, hence $Z_c(V_n)=2n$ so that (3.3) implies the result. Such particular examples are easily constructed and we see no other way of proving them except by the fundamental inequality (3.3).

4. The simplest Pólya-Wiener result concerning high order derivatives of periodic functions. Let $f(t)$ be a real function of period 2π which is infinitely often differentiable. Let us consider its zeros and also the zeros of its successive derivatives. Counting multiplicities as usual we set

$$N^{(k)}=Z_c(f^{(k)})\ ,\qquad k=0,\ 1,\ \cdots,$$

and assume all these numbers to be finite. A familiar application of Rolle's theorem shows that

$$N^{(0)}\leqq N^{(1)}\leqq\cdots\leqq N^{(k)}\leqq N^{(k+1)}\leqq\cdots\ . \tag{4.1}$$

Can this sequence remain bounded? This is surely the case if $f(t)$ is a trigonometric polynomial. The truth of the converse is stated by the following proposition due to Pólya and Wiener [8.]

COROLLARY 3. *If the sequence* (4.1) *is bounded and*

$$\lim N^{(k)}=2m\ , \tag{4.2}$$

then $f(t)$ is a trigonometric polynomial of exact order m.

Indeed, let (3.1) be the Fourier series of $f(t)$. It is known to converge under our assumptions and the expansion of $f^{(k)}(t)$ is obtained by formal differentiations of the expansion of $f(t)$. Let us assume that for a certain n

$$a_n^2+b_n^2>0\ . \tag{4.3}$$

It is clear then from the form of the Fourier series for $f^{(k)}(t)$ that this series will satisfy the inequality (3.4) of Corollary 1, provided only that

k is sufficiently large, $k>K$ say. But then by Corollary 1

$$N^{(k)} \geqq v_c(f^{(k)}) \geqq 2n, \text{ (if } k>K) .$$

Thus (4.2) and (4.3) imply that $n \leqq m$, and $f(t)$ must reduce to a trigonometric polynomial of order $q \leqq m$. On the other hand, if $f(t)$ is such a polynomial, $N^{(k)} \leqq 2q$ which implies $2m \leqq 2q$ or $m \leqq q$, hence $q=m$ and the theorem is established.

5. The graphic behavior of V-means. We now wish to discuss the shape-preserving properties of the V-means which are implicitly contained in the fundamental inequality

$$Z_c(V_n) \leqq v_c(f) . \tag{5.1}$$

It shows that $V_n(t)$ can't oscillate about zero more frequently than $f(t)$ does. But there is nothing peculiar about the level zero. Indeed, if γ is any real, then $f(t)=\gamma$ implies $V_n(t)=\gamma$. Thus we may replace in (5.1) f and V_n by $f-\gamma$ and $V_n-\gamma$, respectively, obtaining the inequality

$$Z_c(V_n-\gamma) \leqq v_c(f-\gamma) . \tag{5.2}$$

A second remark is based on the obvious known fact (see [5, p. 191]) that if $f(t)$ is absolutely continuous then $V_n'(t)$ is the V-mean of $f'(t)$. But then (5.1) immediately gives

$$Z_c(V_n') \leqq v_c(f') . \tag{5.3}$$

This operation may naturally be repeated giving

$$Z_c(V_n^{(k)}) \leqq v_c(f^{(k)}) , \tag{5.4}$$

which is valid depending on how many derivatives $f(t)$ possesses. For instance, if

$$f(t) \in C''$$

then the graph of $V_n(t)$ can't have more maxima, minima or points of inflexion than the corresponding numbers for the graph of $f(t)$.

It is desirable, however, to discuss this phenomenon for functions of a lower degree of smoothness and the following developments aim to do that. We consider the class D_0 of real periodic functions $f(t)$, of bounded variation, normalized by $2f(t)=f(t+0)+f(t-0)$. A subclass of D_0 is the class D_1 of functions satisfying the classical *Dirichlet conditions*. By $f(t) \in D_1$ we mean that the circle can be dissected into a finite number of consecutive open arcs in each of which $f(t)$ is monotone in the wide sense.

With each $f(t) \in D_1$ we associate an even non-negative integer $S(f)$, called the number of *sense-reversals* of $f(t)$ and defined as follows. Consider, for a given natural number k, the periodic sequence of ordinates

(5.5) $$f_\nu = f\left(\frac{2\pi\nu}{k}\right) = f(\nu h)\ , \qquad h = 2\pi/k\ ,$$

of period k, and the likewise periodic sequence of differences

$$\Delta f_\nu = f\left(\frac{2\pi(\nu+1)}{k}\right) - f\left(\frac{2\pi\nu}{k}\right),$$

$\nu = 0, 1, 2, \cdots, n-1$. We now define $S(f)$ by

(5.6) $$S(f) = \lim_{k\to\infty} v_c(\Delta f\nu) = \max_k v_c(\Delta f\nu)\ .$$

The reader is urged to supply proof for the statements implied in this definition; it depends on an analysis of the finitely many points which have no neighborhood in which $f(t)$ is monotone. If in addition to $f(t) \in D_1$ we assume that $f(t) \in C'$ then evidently $S(f) = v_c(f')$.

Our substitute for (5.3) for the class D_1 is given by the following.

THEOREM 3. *If* $f(t) \in D_1$ *then*

(5.7) $$v_c(V_n') = S(V_n) \leqq S(f)\ .$$

The proof is very simple. Besides the V-mean

$$V_n(t) = \frac{1}{2\pi}\int_0^{2\pi} \omega_n(t-\tau) f(\tau) d\tau$$

we consider the approximating sums

$$V_{n,k}(t) = \frac{1}{k}\sum_\nu \omega_n(t-\nu h) f_\nu\ , \qquad h = 2\pi/k.$$

Replacing t by $t+h$ we obtain

$$V_{n,k}(t+h) = \frac{1}{k}\sum_\nu \omega_n(t-\nu h) f_{\nu+1}$$

and therefore

(5.8) $$\Delta V_{n,k}/h = (V_{n,k}(t+h) - V_{n,k}(t))/h = \frac{1}{2\pi}\sum_\nu \omega_n(t-\nu h)\Delta f_\nu\ .$$

By Lemma 3, in view of (5.6), we obtain $v_c(\Delta V_{n,k}/h) \leqq v_c(\Delta f) \leqq S(f)$ or

$$v_c(\Delta V_{n,k}/h) \leqq S(f)\ .$$

Because the difference quotient (5.8) converges to $V_n'(t)$, as $k\to\infty$, for all t, the last inequality implies (5.7).

There is a similar significant substitute for (5.4) if $k=2$. In order to formulate it we define a class of functions $f(t)$ which we denote by D_2: By $f(t)\in D_2$ we mean that the circle can be dissected into a finite number of consecutive open arcs in each of which $f(t)$ is continuous and convex, or concave, or linear. It is clear that $D_2\subset D_1$.

With each $f(t)\in D_2$ we associate an even non-negative integer $T(f)$, called the number of *turn-reversals* of $f(t)$ and defined as follows: Besides the $\Delta f\nu$ we consider the periodic sequence of second differences

$$\delta^2 f_\nu = f_{\nu+1} - 2f_\nu + f_{\nu-1}$$

and define $T(f)$ by

$$T(f)=\lim_{k\to\infty} v_c(\delta^2 f_\nu)=\sup_k v_c(\delta^2 f_\nu)\ . \tag{5.9}$$

Again a proof of the equality of the last two expressions requires the consideration of the points (finite in number) which have no neighborhood in which $f(t)$ is convex, or concave. If in addition to $f(t)\in D_2$ we assume that $f(t)\in C''$ then evidently

$$T(f)=v_c(f'')\ .$$

A substitute of (5.4) for $k=2$ is given by

THEOREM 4. *If $f(t)\in D_2$ then*

$$v_c(V_n'')=T(V_n)\leqq T(f)\ . \tag{5.10}$$

The proof is so very similar to the proof of Theorem 3 that it suffices to indicate the main points. In place of (5.8) we now start from the second order difference quotient

$$\begin{aligned}\delta^2 V_{n,k}/h^2 &= (V_{n,k}(t+h)-2V_{n,k}(t)+V_{n,k}(t-h))/h^2\\ &=\frac{1}{2\pi h}\sum_\nu \omega_n(t-\nu h)\delta^2 f_\nu\ ,\end{aligned}$$

and observe that on the one hand it converges to $V_n''(t)$, on the other hand by Lemma 3 and (5.9)

$$v_c(\delta^2 V_{n,k}/h^2)\leqq T(f)\ .$$

This last inequality implies (5.10) on letting $k\to\infty$.

The following remarks concerning the simplest elements of D_1 and D_2 are called for: 1. If $f(t)=\text{const.}$ then clearly $S(f)=0$ and $T(f)=0$. Conversely, either of these relations is easily seen to imply that $f(t)=$ const. 2. The first non-trivial case is

$$S(f)=2\,. \tag{5.11}$$

Functions $f(t)$ satisfying (5.11) are in a way the simplest non-constant periodic functions and may aptly be called *periodically monotone.* Likewise functions with

$$T(f)=2 \tag{5.12}$$

may be called *periodically convex.*

It is easily shown that (5.12) implies (5.11). That these new terms are appropriate is also shown by the following two statements.

1. If the periodic function $f(t)$ is monotone (non-constant) in $-\pi<t<\pi$ then $S(f)=2$, that is, $f(t)$ is *periodically monotone.*

2. If the periodic function $f(t)$ is convex or concave (non-constant) in $-\pi<t<\pi$ then $T(f)=2$, that is $f(t)$ is *periodically convex.*

Observe that the distinction between "increasing" and "decreasing" as well as between "convex" and "concave", drops out for periodic functions.

We conclude our short excursus into "descriptive function theory" with a few examples:

$$S(\sin\ t)=T(\sin\ t)=2\,.$$

$$S(|\sin\ t|)=T(|\sin\ t|)=4\,.$$

If $f(t)=\sin\ t+1$ in $(-\pi,\ 0)$ and $f(t)=\sin\ t$ in $(0,\ \pi)$ then

$$S(f)=T(f)=6\,.$$

If $f(t)=\sin\ t+t$ in $0<t<2\pi$, then

$$S(f)=2,\ T(f)=4\,.$$

From these examples we see that

$$S(f)\leqq T(f) \tag{5.13}$$

and this inequality is generally true. We see this if we observe that for a periodic sequence (5.5) we always have

$$v_c(f_\nu)\leqq v_c(\Delta f_\nu)\leqq v_c(\delta^2 f_\nu)\,.$$

In view of (5.6), (5.9) and the corresponding relations

$$v_c(f)=\lim_{k\to\infty}\ v_c(f_\nu)=\max_k\ v_c(f_\nu)\,,$$

we conclude that

$$v_c(f)\leqq S(f)\leqq T(f)\,. \tag{5.14}$$

It is of some interest to show that the remarkable properties of the third Cesàro means established by L. Fejér in his Theorems 1, 2 and 3 [1, p. 82 and p. 86] are also enjoyed by the de la Vallée Poussin means $V_n(t)$. Thus Fejér's work suggests the following

THEOREM 5. *If $f(t)$ is an odd periodic function which is positive and concave in the range $0<t<\pi$, then*

$$0< V_n(t) \leqq f(t) \text{ if } 0<t<\pi \quad (n\geqq 1) . \tag{5.15}$$

Moreover, the function $V_n(t)$ is also concave in $0<t<\pi$.

The last statement and the first inequality (5.15) are easily proved. Indeed, it is clear that

$$v_c(f)=S(f)=T(f)=2 . \tag{5.16}$$

Observe also that $V_n(t)\not\equiv 0$ if $n\geqq 1$, for $V_n(t)\equiv 0$ would imply $S_n(t)\equiv 0$, hence also $v_c(f)\geqq 2n+2\geqq 4$ (by Corollary 2) which contradicts (5.16). By Theorem 1 and (5.16) surely

$$Z_c(V_n)=2 . \tag{5.17}$$

Since $V_n(t)$ is a sine polynomial it vanishes at 0 and π. By (5.17) these zeros are simple and the only zeros of $V_n(t)$. Also by (5.16) and Theorems 3 and 4 we conclude that

$$v_c(V_n')=v_c(V_n'')=2 .$$

These remarks show that $V_n(t)$ or perhaps $-V_n(t)$ enjoy the properties to be established. That $V_n(t)$, rather than $-V_n(t)$, has these properties is shown by observing that

$$V_n'(0)=c\int_{-\pi}^{\pi}\left(\cos\frac{\tau}{2}\right)^{2n-1}\sin\frac{\tau}{2}f(\tau)d\tau , \qquad c>0,$$

(obtained from (7) by differentiation) has a positive integrand and is therefore positive.

To establish the second inequality (5.15) or

$$V_n(t)\leqq f(t) \qquad 0<t<\pi, \tag{5.18}$$

is a little more troublesome and we resort to Fejér's own method. We consider the "roof-function"

$$\hat{f}(t)=\begin{cases} \dfrac{b}{a}t & \text{if } 0\leqq t\leqq a \\ b\dfrac{\pi-t}{\pi-a} & \text{if } a\leqq t\leqq \pi \end{cases} \qquad 0<a<\pi,\ b>0, \tag{5.19}$$

and denote again by $\hat{f}(t)$ its odd periodic extension. We now observe that indeed

$$V_n(t) < \hat{f}(t)\ , \qquad 0 < t < \pi, \tag{5.20}$$

for these special functions. Since we already know from our previous discussion that $\hat{V}_n(t)$ is positive and concave in $(0, \pi)$, the inequality (5.20) is perfectly clear as soon as we can prove that

$$\hat{V}'_n(0) < \hat{f}'(0)\ , \quad \hat{V}'(\pi) > \hat{f}'(\pi)\ , \tag{5.21}$$

These inequalities, however, follow immediately from previous remarks. Since $\hat{f}(t)$ is continuous, $\hat{V}'_n(t)$ is the V-mean of $\hat{f}'(t)$. Since $\hat{f}'(0) = \sup \hat{f}'(t)$, $\hat{f}'(\pi) = \inf \hat{f}'(t)$, we conclude, for instance from (5.2), that

$$\hat{f}'(\pi) < \hat{V}'_n(t) < \hat{f}'(0) \text{ for all } t\ .$$

The proof of the general inequality (5.18) now follows from the observation that the function $f(t)$ of Theorem 5 may be approximated by appropriate linear combinations of roof-functions with *positive* coefficients.

6. Convex, and star-shaped, conformal maps of the circle. The following introductory remark (previously made by one of us; see [10, pp. 226–227]) applies to any variation diminishing kernel $\Omega(t)$ as defined by the relations (3), (4) and (5) of our Introduction.

Let

$$f(t) = f_1(t) + i f_2(t) \qquad (f_1,\ f_2 \text{ real-valued}) \tag{6.1}$$

be a complex-valued continuous function of period 2π and let

$$g(t) = \frac{1}{2\pi}\int_0^{2\pi} \Omega(t-\tau) f(\tau)\, d\tau \tag{6.2}$$

be its transform; $g(t)$ is evidently also complex-valued periodic and we may write

$$g(t) = g_1(t) + i g_2(t)\ , \qquad (g_1,\ g_2 \text{ real-valued}). \tag{6.3}$$

Since Ω is real and (3) holds it follows that the transforms of $f_1(t)$, $f_2(t)$ and 1 are $g_1(t)$, $g_2(t)$ and 1, respectively. If A, B, C are arbitrary real constants it follows that $Ag_1(t) + Bg_2(t) + C$ is the transform of $Af_1(t) + Bf_2(t) + C$. Since $\Omega(t)$ is assumed to be a variation diminishing kernel, we conclude by (5) that the inequality

$$v_c(Ag_1(t) + Bg_2(t) + C) \leqq v_c(Af_1(t) + Bf_2(t) + C) \tag{6.4}$$

always holds.

The inequality (6.4) admits a remarkable geometric interpretation. Indeed, let us denote by $\{f\}$ the closed curve traced out by $f(t)$ in the complex plane of the variable $z=x+iy$ as t varies in the range $[0, 2\pi]$, and let $\{g\}$ be the corresponding curve described by $g(t)$. Let the following statement, too simple to be called a theorem, be referred to as a

PRINCIPLE. *The curve $\{g\}$ never crosses a straight line more often than the curve $\{f\}$ does.*

For if $Ax+By+C=0$ is the equation of a line L then the two members of the inequality (6.4) are identical with the total numbers of crossings of L by $\{g\}$ and $\{f\}$, respectively. In particular we have the

COROLLARY 4. *If the curve $\{f\}$ is convex then $\{g\}$ is interior to $\{f\}$ and $\{g\}$ is also convex.*

Indeed, $\{f\}$ being convex, it crosses any L at most twice, hence also $\{g\}$ crosses any L at most twice and is therefore convex. That $\{g\}$ has no points outside of $\{f\}$ follows already from the properties

$$\Omega(t)\geqq 0, \frac{1}{2\pi}\int_0^{2\pi}\Omega(t)dt=1 , \tag{6.5}$$

and in no way requires the sophisticated condition that $\Omega(t)$ be variation diminishing. On the other hand the conditions (6.5) are by themselves insufficient to enforce the convexity of $\{g\}$. It is also true, however, that the variation diminishing property of $\Omega(t)$ is sufficient but far from necessary for $\{g\}$ to be convex. As an example we mention the periodic kernel

$$\Omega(t)=\begin{cases} \pi/h & \text{if } -h\leqq t\leqq h \\ 0 & \text{if } -\pi\leqq t\leqq h \text{ or } h\leqq t\leqq\pi \end{cases} \quad (0<h<\pi) , \tag{6.6}$$

which is readily shown to have the "convexity preserving" property of Corollary 4. However, (6.6) is not variation diminishing because it is not periodic totally positive (see [4]).

We now turn to an application of these remarks to conformal maps of the circle, in particular to a proof of Theorem 2 of the Introduction.

Let

$$F(z)=z+c_2z^2+c_3z^3+ \cdots \tag{6.7}$$

be regular in the unit circle. For a fixed value of r we consider the complex-valued periodic function

$$f(t\,;\,r)=F(re^{it})=re^{it}+c_2r^2e^{2it}+ \cdots , \qquad 0\leqq r\leqq 1. \tag{6.8}$$

and finally from relations analogous to (11.8), express γ_n in terms of the c/c_1 and so in terms of the α and β. This leads to

$$
\begin{aligned}
&\gamma_1=\alpha_1\beta_1\\
&3\gamma_2=\alpha_2\beta_2+2\alpha_2\beta_1^2\\
(11.11)\qquad &\qquad +2\alpha_1^2\beta_2-2\alpha_1^2\beta_1^2\\
&6\gamma_3=\alpha_3\beta_3+3\alpha_3\beta_1\beta_2+2\alpha_3\beta_1^3\\
&\qquad +3\alpha_1\alpha_2\beta_3+3\alpha_1\alpha_2\beta_1\beta_2-6\alpha_1\alpha_2\beta_1^3\\
&\qquad +2\alpha_1^3\beta_3-6\alpha_1^3\beta_1\beta_2+4\alpha_1^3\beta_1^3\\
&\cdots
\end{aligned}
$$

Not all details of the general formula for γ_n are obvious ; a few features will be discussed under (e). The determinant $\boldsymbol{C}_n$ (expressed in terms of the γ as $\boldsymbol{A}_n$ is in terms of the α, cf. (11.4)) becomes by virtue of (11.11) a polynomial in the $\alpha, \bar{\alpha}, \beta$ and $\bar{\beta}$. By the theory of Carathéodory and Toeplitz, Conjecture III is equivalent to the following.

CONJECTURE IV. *The* $2n$ *inequalities*

$$\boldsymbol{A}_1>0,\ \boldsymbol{A}_2>0,\ \cdots,\ \boldsymbol{A}_n>0\ ,$$

$$\boldsymbol{B}_1>0,\ \boldsymbol{B}_2>0,\ \cdots,\ \boldsymbol{B}_n>0\ ,$$

imply the n *inequalities*

$$\boldsymbol{C}_1>0,\ \boldsymbol{C}_2>0,\ \cdots,\ \boldsymbol{C}_n>0$$

and this holds for $n=1, 2, 3, \cdots$.

This formulation excludes the case of equality in all the $3n$ inequalities considered. This is due to the fact that, without loss of generality, we may suppose $\Sigma a_n z^n$ and $\Sigma b_n z^n$ regular in $|z|\leqq 1$.

(d) The case $n=1$ of Conjecture IV is trivial. In fact, if we assume that the series are normalized, see (11.1), and introduce the coefficients of the mapping functions, see (11.8), the statement that we have to prove reduces to this :

The inequalities $\qquad |a_2|<2,\ |b_2|<2$

imply $\qquad \dfrac{|a_2b_2|}{2}<2$

which is obvious.

(e) The case $n=2$ of Conjecture IV was first established by Dr. G. A. Hummel and can be proved as follows.

We take the series as normalized, see (11.1), and set

$$a_2=a,\ a_3=A,\ b_2=b,\ b_3=B\ ;$$

that all the inclusions

$$D_1 \subset D_2 \subset \cdots \subset D_n \subset D_{n+1} \subset \cdots \tag{6.12}$$

are valid, but we are unable to prove or disprove this.

REMARK 3. Since numerous elements of the class K are explicitly known, Theorem 2 is a ready source of polynomials belonging to K. Thus

$$F(z) = \frac{z}{1-z} = z + z^2 + \cdots \tag{6.13}$$

is in K because it maps the unit circle onto the half-plane $\Re z > -\frac{1}{2}$.

The corresponding V-means

$$V_n(z) = \frac{1}{\binom{2n}{n}} \left\{ \binom{2n}{n+1} z + \binom{2n}{n+2} z^2 + \cdots + z^n \right\} \tag{6.14}$$

are a remarkable sequence of polynomials some extremal properties of which might be discussed on another occasion. Of course (6.11) holds. Here the convex boundary of D_n touches the line $\Re z = -\frac{1}{2}$ to an order of contact which increases with n. Also the inclusions (6.12) can be verified in this special case.

REMARK 4. Observe that the image D_1 of the unit circle by $V_1(z) = \frac{1}{2} z$ is the circle

$$D_1 : \ |z| < \frac{1}{2}\ . \tag{6.15}$$

By (6.11) we have $D_1 \subset D$ for every $F(z) \in K$. This proves the following proposition: *The circle* (6.15) *is covered by every convex map D and* (6.15) *is the largest circle with this property.* That D_1 is the largest circle is shown by the special function (6.13). This theorem is due to Study, [11, p. 116], and our proof is really identical with Nehari's proof in [6, pp. 223–224].

REMARK 5. A comparison of Theorem 2 with Fejér's Theorem IV [1, p. 87] again shows the extent to which the de la Vallée Poussin means of a power series are superior to its third Cesàro means as far as shape-preserving properties are concerned.

REMARK 6. In § 3 we have seen that from a knowledge of the section (3.2) of the Fourier series (3.1) of $f(t)$ we can infer the information (3.3) concerning the zeros of $f(t)$. Is there a similar result for power series? Specifically, let

$$F(z)=\sum_0^\infty c_\nu z^\nu \quad (c_0=1)$$

converge for $|z|<1$ and let, for a certain value of n,

$$V_n(z)=\frac{1}{\binom{2n}{n}}\sum_{\nu=0}^{n}\binom{2n}{n+\nu}c_\nu z$$

be given and known to have a certain number of zeros within the unit circle. Can we then draw any positive conclusion concerning the existence of zeros of $f(z)$ in the unit circle?

That the answer is negative is very simply shown as follows. With the given $c_0=1, c_1, \cdots, c_n$ derive the expansion

$$\log(1+c_1z+\cdots+c_nz^n)=b_1z+\cdots+b_nz^n+\cdots.$$

But then

$$F(z)=e^{b_1z+\cdots+b_nz^n}=1+c_1z+\cdots+c_nz^n+\cdots$$

is a *zero-free* entire function whose nth V-mean is precisely the given $V_n(z)$.

In concluding this section we wish to point out similar applications concerning the class Σ of power series $\sum_1^\infty b_\nu z^\nu$ which map the unit circle onto a univalent domain which is *star-shaped* with respect to the origin. It is well known that the two classes K and Σ are related as follows:

LEMMA 5. *$\sum_1^\infty a_\nu z^\nu \in \Sigma$ if and only if*

$$\sum_1^\infty \frac{a_\nu}{\nu} z^\nu \in K.$$

But then Theorem 2 easily implies the following.

COROLLARY 5. *For $F(z)\in\Sigma$ it is necessary and sufficient that $V_n(z)\in\Sigma$ for $n=1, 2, \cdots$.*

APPENDIX I. THE BERNSTEIN POLYNOMIALS

7. The Bernstein construction is variation diminishing. The purpose of the present appendix is to furnish for functions $f(x)$ defined in a finite interval a theory analogous to that given in Parts I and II for periodic functions. It is remarkable that such a theory is provided by the classical Bernstein polynomials. Indeed, let $f(x)$ be defined in [0, 1] and let

$$B_n(x)=\sum_0^n f\left(\frac{\nu}{n}\right)\binom{n}{\nu}x^\nu(1-x)^{n-\nu}, \qquad n\geqq 1, \tag{7.1}$$

be the corresponding Bernstein polynomial (see [5]). Let $Z(B_n)$ denote the number of zeros of $B_n(x)$ in the open range (0, 1). We now state the following

THEOREM 6. *Denoting by $v(f)$ the number of changes of sign of $f(x)$ in [0, 1] we have the inequalities*

$$v(B_n)\leqq Z(B_n)\leqq v(f). \tag{7.2}$$

This result, an analogue of Theorem 1, can be derived as a special case from a general theorem of S. Karlin [3]. It admits, however, a very simple direct proof. Indeed, with $z=x/(1-x)$ for $0<x<1$, we have

$$\frac{B_n(x)}{(1-x)^n}=\sum_{\nu=0}^n f\left(\frac{\nu}{n}\right)\binom{n}{\nu}z^\nu,$$

hence by Descartes' rule of signs

$$Z(B_n)=Z\left(\frac{B_n(x)}{(1-x)^n}\right)=\underset{(0,\infty)}{Z}\left(\sum_0^n f\left(\frac{\nu}{n}\right)\binom{n}{\nu}z^\nu\right)\leqq v\left(f\left(\frac{\nu}{n}\right)\right)\leqq v(f).$$

8. The graphic behavior of the Bernstein polynomials. If we write $B_n(x)=B_n(x;f)$ to indicate the dependence on $f(x)$, it is known that

$$B_n(x;\ Ax+B)=Ax+B. \tag{8.1}$$

But then (7.1) implies that $B_n(x)-Ax-B$ is the Bernstein polynomial of $f(x)-Ax-B$. Now (7.2) implies the

COROLLARY 6. *If $Ax+B$ is an arbitrary linear function then*

$$Z(B_n(x)-Ax-B)\leqq v(f(x)-Ax-B). \tag{8.2}$$

Intersecting the graphs of $f(x)$ and $B_n(x)$ by appropriate straight lines $y=Ax+B$, the inequality (8.2) furnishes a good deal of information concerning the shape of the graph of $B_n(x)$. Notice in particular the following

COROLLARY 7. *If $f(x)$ is convex in* [0, 1], *possibly discontinuous at the endpoints, but not linear in* [0, 1], *then*

1. *$B_n(x)$ is convex,*
2. $B_n(x) > f(x)$ if $0 < x < 1$,
3. $B_n(0) = f(0)$, $B_n(1) = f(1)$.

We may omit the simple proof based an Corollary 6.

Observe that the relation $B'_n(x\,;f) = B_n(x\,;\,f')$ is *not* valid. However a simple calculation shows that (7.1) implies

$$B'_n(x) = n\sum_0^{n-1}\binom{n-1}{\nu}\Delta f_\nu\, x^\nu(1-x)^{n-1-\nu}$$

and

$$B''_n(x) = n(n-1)\sum_0^{n-2}\binom{n-2}{\nu}\Delta^2 f_\nu\, x^\nu(1-x)^{n-2-\nu}$$

where we have set

$$f_\nu = f\left(\frac{\nu}{n}\right), \qquad (\nu = 0, \cdots, n),$$

$$\Delta f_\nu = f_{\nu+1} - f_\nu,\quad \Delta^2 f_\nu = f_{\nu+2} - 2f_{\nu+1} + f_\nu .$$

(See Natanson [5], p. 179, fifth line from the bottom). The Theorems 3, 4 and 5 have precise analogues as will now be shown with a minimum of details. The function classes D_1 and D_2 have analogues in the present situation and the numbers of sense-reversals and turn-reversals may again be defined by the relations

$$S(f) = \lim_{n\to\infty} v(\Delta f_\nu) = \sup_n v(\Delta f_\nu) \qquad f \in D_1,$$

$$T(f) = \lim_{n\to\infty} v(\Delta^2 f_\nu) = \sup_n v(\Delta^2 f_\nu) \qquad f \in D_2,$$

respectively.

As in the periodic case we obtain the following.

THEOREM 7. *If $f(x) \in D_1$ then*

$$S(B_n) = v(B'_n) \leqq S(f) .$$

If $f(x) \in D_2$ then

$$T(B_n) = v(B''_n) \leqq T(f) .$$

If $f(x)$ is odd about the point $x = \frac{1}{2}$, then $B_n(x)$ is found to share this property. As an analogue of Theorem 5 we have the following

THEOREM 8. *If* $f(x)$ *is odd about* $x=\frac{1}{2}$, *concave and non-negative in* $\frac{1}{2}\leqq x\leqq 1$, *positive in* $\frac{1}{2}<x<1$, *then also* $B_n(x)$ *is concave in* $\left[\frac{1}{2}, 1\right]$ *and*

$$(8.3) \qquad 0<B_n(x)<f(x) \ \textit{if} \ \frac{1}{2}<x<1 .$$

Indeed, let us first observe the following. Because of the invariance of linear functions expressed by (8.1), we may subtract from $f(x)$ the linear function whose graph is the chord joining the extreme points $(0, f(0))$ and $(1, f(1))$, without altering the assumptions on $f(x)$. Thus without loss of generality we may assume that $f(0)=f(1)=0$. From this point the proof is entirely similar to the proof of Theorem 5 in all details, including the use of the roof-functions. Finally notice that the equality is excluded in the second inequality (8.3). This is so because of the inequality (8.2) of Corollary 6; in the periodic case we only had the weaker analogue (5.2).

APPENDIX II. A CONJECTURE ON POWER SERIES MAPPING A CIRCLE ONTO A CONVEX DOMAIN

9. Sources and forms of the conjecture. As stated in the Introduction, a power series

$$(9.1) \qquad a_1z+a_2z^2+a_3z^3+\cdots+a_nz^n+\cdots=f(z)$$

is said to belong to the class K, if it converges in the circle $|z|<1$ and maps this circle onto a convex domain. We say that the infinite sequence of complex numbers $\lambda_1, \lambda_2, \lambda_3, \cdots \lambda_n, \cdots$ is a *convexity-preserving factor sequence* if the series $\lambda_1a_1z+\lambda_2a_2z^2+\lambda_3a_3z^3+\cdots$ necessarily belongs to K whenever (9.1) belongs to K. Let us apply such a factor sequence to the simplest power series belonging to K, to the geometric series

$$(9.2) \qquad z+z^2+z^3+\cdots=\frac{z}{1-z} .$$

We obtain

$$(9.3) \qquad \lambda_1z+\lambda_2z^2+\lambda_3z^3+\cdots+\lambda_nz^n+\cdots ;$$

if $\lambda_1, \lambda_2, \lambda_3, \cdots$ is a convexity-preserving factor sequence, the power series (9.3) must necessarily belong to K. We state the conjecture that this obvious necessary condition is also *sufficient*; that is, we formulate

CONJECTURE I. *If both power series*

$$a_1z+a_2z^2+a_3z^3+\cdots$$

$$b_1z+b_2z^2+b_3z^3+\cdots$$

belong to K, *also*

$$a_1b_1z+a_2b_2z^2+a_3b_3z^3+\cdots$$

belongs to K.[2]

In view of Lemma 5, the conjecture can be restated in other forms, equivalent to the first.

CONJECTURE II. *If the power series*

$$a_1z+a_2z^2+a_3z^3+\cdots$$

belongs to K *and*

$$b_1z+b_2z^2+b_3z^3+\cdots$$

belongs to Σ, *then*

$$a_1b_1z+a_2b_2z^2+a_3b_3z^3+\cdots$$

belongs to Σ.

CONJECTURE III. *If both power series*

$$a_1z+a_2z^2+a_3z^3+\cdots$$

$$b_1z+b_2z^2+b_3z^3+\cdots$$

belong to Σ, *also*

$$\frac{a_1b_1}{1}z+\frac{a_2b_2}{2}z^2+\frac{a_3b_3}{3}z^3+\cdots$$

belongs to Σ.

These three Conjectures I, II and III are completely equivalent, they stand and fall together. The third form brings out most clearly the relation to a conjecture that has been found, years ago and independently of each other, by two of our friends, Professor S. Mandelbrojt and Professor M. Schiffer, and which is published here with their permission:

[2] One of the "intuitive sources" of the conjecture is the feeling that (9.2) plays a "leading role" in K, that it "sets the fashion." Which one of the two authors of this paper is the author of the conjecture will be disclosed if and when the conjecture is proved.

CONJECTURE M. S. *If both power series*

$$a_1z+a_2z^2+a_3z^3+\cdots$$

$$b_1z+b_2z^2+b_3z^3+\cdots$$

are "schlicht" in the unit circle, also

$$\frac{a_1b_1}{1}z+\frac{a_2b_2}{2}z^2+\frac{a_3b_3}{3}z^3+\cdots$$

is "schlicht" in the unit circle.

Whereas III is equivalent to I or II, it appears logically independent of MS. As far as obvious conclusions from the statements go, III could be true but MS false, or MS true yet III false, or both could be true or both false. Still, the conjectures are obviously related and their joint consideration may lead to various suggestions.

The Conjectures I, II and III are more "elementary" than MS and they are certainly more accessible; we succeeded in treating several of their particular cases and consequences.

10. Verification of the conjecture in some particular cases. We shall exhibit several particular series $\sum b_nz^n$ belonging to K which, convoluted with an arbitrary series (9.1) belonging to K, generate a series $\sum a_nb_nz^n$ belonging to K.

(a) The polynomial (6.14) belongs to K. That its convolution with an arbitrary series belonging to K necessarily belongs to K is precisely what Theorem 2 asserts.

(b) If the series (9.1) belongs to K, it belongs, a fortiori, to Σ. Therefore, by Lemma 5, the series

$$\frac{a_1z}{1}+\frac{a_2z^2}{2}+\frac{a_3z^3}{3}+\cdots$$

belongs to K. This is another special case of Conjecture I; that the series

$$\frac{z}{1}+\frac{z^2}{2}+\frac{z^3}{3}+\cdots=\log\frac{1}{1-z}$$

maps the unit circle onto a convex domain follows from its relation to (9.2) and from Lemma 5 but this fact can also be established directly (see [9, vol. 1, p. 106, problem 114]).

(c) The result mentioned under (a) (Theorem 2) is due to the fact that the V-means are variation diminishing; cf. §6. Any variation diminishing transformation on the circle leads to an analogous result, and so we obtain especially the following (cf. [4]). *Let $g(z)$ be the product of $e^{-\gamma z^2}$, where $\gamma \geqq 0$, with an entire function of genus* 1, *all coeffiicients and all zeros of which are reat ; then*

$$\sum_1^\infty \frac{z^n}{g(in)}$$

belongs to K, and, provided that (9.1) *belongs to K, also*

$$\sum_1^\infty \frac{a_n z^n}{g(in)}$$

belongs to K. The term "entire function of genus 1" is used here in the comprehensive sense, that is, it is supposed to include also entire functions of genus 0 and polynomials (but, obviously, not the identically vanishing polynomial) ; the case in which $g(z)$ reduces to z was mentioned under (b).

(d) Let p and q denote two different given points on the unit circle ($|p|=|q|=1$, $p \neq q$). Assume that (9.1) belongs to K and let z describe a circle concentric with, and interior to, the unit circle. Then $f(z)$ describes a convex curve of which $f(pz)-f(qz)$ represents a moving chord ; as it is easy to see geometrically this chord turns all the time in the same sense. The argument of the complex number $f(pz)-f(qz)$ increases steadily. That is, the power series

$$\frac{f(pz)-f(qz)}{p-q} = \sum_1^\infty a_n \frac{p^n - q^n}{p-q} z^n$$

belongs to Σ (maps the unit circle onto a star-shaped domain) and so, by Lemma 5, the power series

$$\sum_{n=1}^\infty \frac{a_n}{n} \frac{p^n - q^n}{p-q} z^n \tag{10.1}$$

belongs to K (cf. [6]). The series (10.1) is the convolution of (9.1) and of that particular case of (10.1) in which $a_n = 1$; this particular series maps $|z|<1$ onto an infinite strip bounded by two parallels.

11. Verification of some consequences. In the foregoing, we have dealt mainly with form I of the conjecture, but now we shall consider its form III. We assume, therefore, that the function (9.1) belongs to the class Σ, that is, it maps the circle $|z|<1$ onto a star-shaped domain. We shall say that (9.1) is normalized if

$$a_1 = 1 . \tag{11.1}$$

(a) We are given an integer n, $n \geqq 2$. Let us consider the normalized functions of the class Σ and let us seek one for which $|a_n|$ is a maximum. We leave aside the (easy) discussion of the existence and assume that (9.1) is such a function with maximum $|a_n|$. Now we apply Conjecture III with $b_m = a_m$ for $m=1, 2, 3, \cdots$; the resulting series is again normalized and so its nth coefficient cannot have an absolute value exceeding the maximum; that is, $\frac{|a_n|^2}{n} \leqq |a_n|$, from which it follows that

$$|a_n| \leqq n .$$

For series of the class Σ this inequality is well known and easily established independently of the Conjecture III. And so our previous reasoning served only to enhance somewhat the plausibility of Conjecture III. Yet the same reasoning is also applicable to the Conjecture *MS* and reveals one of the essential sources of this Conjecture.

(b) The function $f(z)$ belongs to the class Σ if, and only if,

$$\frac{zf'(z)}{f(z)} = 1 + 2\alpha_1 z + 2\alpha_2 z^2 + 2\alpha_3 z^3 + \cdots \tag{11.2}$$

is regular in the circle $|z| < 1$ and has there a positive real part. This will be the case if, and only if, the Hermitian form of the variables $z_0, z_1, \cdots, z_n$

$$\sum_{k=0}^{n} \sum_{l=0}^{n} \alpha_{k-l} z_k \bar{z}_l \tag{11.3}$$

($\alpha_{-\nu} = \bar{\alpha}_\nu$, by definition) is positive (definite or semidefinite) for $n=1, 2, 3, \cdots$. This well known important necessary and sufficient condition is due to Carathéodory and Toeplitz. It can also be expressed in terms of the determinants

$$A_n = \begin{vmatrix} 1 & \alpha_1 & \alpha_2 & \cdots & \alpha_n \\ \alpha_{-1} & 1 & \alpha_1 & \cdots & \alpha_{n-1} \\ \alpha_{-2} & \alpha_{-1} & 1 & \cdots & \alpha_{n-2} \\ \cdot & \cdot & \cdot & \cdot & \cdot \\ \alpha_{-n} & \alpha_{-n+1} & \alpha_{-n+2} & \cdots & 1 \end{vmatrix} . \tag{11.4}$$

Now (see (9.1)) the relation (11.2) can be written in the form

$$a_1 z + a_2 z^2 + a_3 z^3 + \cdots = a_1 z \exp\left(\frac{2\alpha_1 z}{1} + \frac{2\alpha_2 z^2}{2} + \frac{2\alpha_3 z^3}{3} + \cdots \right) \tag{11.5}$$

or in the form

$$\frac{2\alpha_1 z}{1}+\frac{2\alpha_2 z^2}{2}+\frac{2\alpha_3 z^3}{3}+\cdots=\log\left(1+\frac{a_2 z}{a_1}+\frac{a_3 z^2}{a_1}+\cdots\right) \tag{11.6}$$

and so we can express both a_n/a_1 as a polynomial in the α and α_n as a polynomial in the a/a_1:

$$\begin{aligned}\frac{a_2}{a_1}&=2\alpha_1\\ \frac{a_3}{a_1}&=\frac{(2\alpha_1)^2+2\alpha_2}{2\,!}\\ \frac{a_4}{a_1}&=\frac{(2\alpha_1)^3+2(2\alpha_3)+3(2\alpha_1)(2\alpha_2)}{3\,!}\\ &\cdots\end{aligned} \tag{11.7}$$

$$\begin{aligned}2\alpha_1&=\frac{a_2}{a_1}\\ 2\alpha_2&=-\frac{a_2^2-2a_1a_3}{a_1^2}\\ 2\alpha_3&=\frac{a_2^3-3a_1a_2a_3+3a_1^2a_4}{a_1^3}\\ &\cdots\end{aligned} \tag{11.8}$$

It would be easy to write down (11.7) or (11.8) for general n, but we shall not enter into details. Using (11.8) we could express the Hermitian form (11.3) and the determinant (11.4) in terms of the coefficients of the series (9.1) and doing so we wonld render more explicit the necessary and sufficient condition for the class Σ.. Yet we postpone this consideration.

(c) Now consider, besides (9.1), two other power series with coefficients b_n and c_n respectively, and let β_n and $\boldsymbol{B}_n$ be so linked to the b, and γ_n and $\boldsymbol{C}_n$ so linked to the c, as α_n and $\boldsymbol{A}_n$ are to the a. Thus we have besides (11.5) (in all summations $n=1, 2, 3, \cdots$)

$$\sum b_n z^n=b_1 z\exp\left(2\sum\frac{\beta_n z^n}{n}\right),\quad \sum c_n z^n=c_1 z\exp\left(2\sum\frac{\gamma_n z^n}{n}\right), \tag{11.9}$$

Set

$$\frac{a_n b_n}{n}=c_n\,. \tag{11.10}$$

Now express a_n/a_1 in terms of the α from (11.7), and express analogously b_n/b_1 in terms of the β, then c_n/c_1 in terms of the α and β from (11.10)

By (6.8) and (9) its V-means are

$$V_n(t;r)=\frac{1}{2\pi}\int_0^{2\pi}\omega_n(t-\tau)F(re^{i\tau})d\tau$$
$$=\frac{1}{\binom{2n}{n}}\left\{\binom{2n}{n+1}re^{it}+\binom{2n}{n+2}c_2r^2e^{2it}+\cdots+c_nr^ne^{nit}\right\}$$

or

$$V_n(t;\ r)=V_n(re^{it})\ , \tag{6.9}$$

where $V_n(z)$ are the de la Vallée Poussin means of the power series as defined by (15), with $c_1=1$. We also record the more explicit expression

$$V_n(z)=\frac{n}{n+1}z+\frac{n(n-1)}{(n+1)(n+2)}c_2z^2+\cdots+\frac{n(n-1)\cdots 1}{(n+1)(n+2)\cdots(2n)}c_nz^n\ . \tag{6.10}$$

Our Theorem 2 seems now almost self-evident. Indeed, if $F(z)\in K$ then the curve $\{V_n(re^{it})\}$ is convex by (6.8), (6.9) and Corollary 4. This being true for every $r<1$, we conclude that $V_n(z)\in K$. Conversely, if $V_n(z)\in K$ for every n, then $\{V_n(re^{it})\}$ is a convex curve for all n and all $r<1$. From the relation

$$\lim_{n\to\infty} V_n(re^{it})=F(re^{it})$$

we conclude that also $\{F(re^{it})\}$ is convex. Hence $F(z)\in K$.

REMARK 1. In order to conclude from (17) that $F(z)\in K$ it is not necessary to assume that the power series (6.7) converges in the unit circle or that it converges at all. Rather the converse part of Theorem 2 holds for a *formal* power series (6.7). For it is known (see e.g. [9, vol. II, p. 29]) that the assumptions (17) imply that all coefficients of the polynomial (6.10) are bounded in absolute value by $n/(n+1)$. Letting $n\to\infty$ we obtain $|c_\nu|\leq 1$ $(\nu=1, 2, \cdots)$ which clearly imply the convergence of (6.7) within the unit circle.

REMARK 2. Let $F(z)\in K$ and hence $V_n(z)\in K$. Let D and D_n denote the convex domains into which the unit circle is mapped by $F(z)$ and $V_n(z)$, respectively. We know by Corollary 4 that

$$D_n\subset \mathrm{D}\ . \tag{6.11}$$

At this point it is natural to suspect that more is true, namely

we suppose, without loss of generality, that $a \geqq 0, b \geqq 0$. We have to show:

The two inequalities

$$\left| A - \frac{3a^2}{4} \right| < 1 - \frac{a^2}{4}, \quad \left| B - \frac{3b^2}{4} \right| < 1 - \frac{b^2}{4} \tag{11.12}$$

imply

$$\left| \frac{AB}{3} - \frac{3a^2b^2}{16} \right| < 1 - \frac{a^2b^2}{16}. \tag{11.13}$$

(The first inequality (11.12) results from the condition $A_2 > 0$, see (11.4), by virtue of (11.8); it *implies* $a < 2$, and so the condition $A_1 > 0$.)

Let

$$A = \frac{3a^2}{4} + u, \quad B = \frac{3b^2}{4} + v. \tag{11.14}$$

By the hypothesis (11.12) of the theorem that we are about to prove

$$|u| < 1 - \frac{a^2}{4}, \quad |v| < 1 - \frac{b^2}{4}. \tag{11.15}$$

We derive from (11.14) and (11.15)

$$AB = \frac{9a^2b^2}{16} + \frac{3a^2}{4}v + \frac{3b^2}{4}u + uv$$

$$\left| AB - \frac{9a^2b^2}{16} \right| < \frac{3a^2}{4}\left(1 - \frac{b^2}{4}\right) + \frac{3b^2}{4}\left(1 - \frac{a^2}{4}\right) + \left(1 - \frac{a^2}{4}\right)\left(1 - \frac{b^2}{4}\right). \tag{11.16}$$

We assert that

$$1 + \frac{a^2}{2} + \frac{b^2}{2} - \frac{5a^2b^2}{16} < 3 - \frac{3a^2b^2}{16}; \tag{11.17}$$

in fact, this follows from $a < 2, b < 2$, since it is equivalent to

$$\left(1 - \frac{a^2}{4}\right)\left(1 - \frac{b^2}{4}\right) > 0.$$

The right hand side of (11.16) is equal to the left hand side of (11.17), and so the combination of these two inequalities immediately yields the desired conclusion (11.13).

(e) We consider now the expression of γ_n in terms of the α and

β for general n; for the cases $n=1, 2, 3$, see (11.11). The procedure that led us to (11.11) shows that γ_n is a polynomial in $\alpha_1, \alpha_2, \cdots, \alpha_n$, $\beta_1, \beta_2, \cdots, \beta_n$ with rational coefficients. Obviously, by virtue of (11.10), γ_n is symmetric in the α and β. If we substitute ρz for z in (11.5) or, which is the same, we change α_n into $\rho^n\alpha_n$ and a_n into $\rho^n a_n$, there results a change, see again (11.10), of c_n into $\rho^n c_n$ and of γ_n into $\rho^n\gamma_n$; therefore, γ_n must be an isobaric polynomial in the α of weight n. Finally, γ_n must be of the form

$$\gamma_n=\sum_{k=1}^{p}\sum_{l=1}^{p} j_{kl}^{(n)} A_k B_l \,, \tag{11.18}$$

where

$p=p(n)$ is the number of partitions of the integer n,

$A_1, A_2, \cdots, A_p$ are the products of powers of weight n of $\alpha_1, \alpha_2, \cdots, \alpha_n$, ordered lexicographically so that

$$A_1=\alpha_n,\ A_2=\alpha_{n-1}\alpha_1,\ \cdots\ A_p=\alpha_1^n \,. \tag{11.19}$$

Generally A_k is of the form

$$A_k=\alpha_1^{k_1}\alpha_2^{k_2}\cdots\alpha_n^{k_n}\,; \tag{11.20}$$

its weight $1k_1+2k_2+3k_3+\cdots+nk_n=n$.

$B_1, B_2, \cdots, B_p$ are analogously expressed in terms of $\beta_1, \beta_2, \cdots, \beta_n$, and $j_{ik}^{(n)}$ are rational numbers, $j_{ik}^{(n)}=j_{ki}^{(n)}$.

For example $p(4)=5$ and, for $n=4$

$$A_1=\alpha_4,\ A_2=\alpha_3\alpha_1,\ A_3=\alpha_2^2,\ A_4=\alpha_2\alpha_1^2,\ A_5=\alpha_1^4 \,,$$

the B are analogously defined and the matrix of the $j_{ik}^{(4)}$ results from

9	24	9	36	12
24	24	24	−24	−48
9	24	−1	−4	−28
36	−24	−4	−136	128
12	−48	−28	128	−64

if each of the 25 numbers displayed is divided by 90.

We cannot exhibit the law of the dependence of $j_{ik}^{(n)}$ on n in some obviously useful manner, but we note here one property. If $\beta_n=1$ it is easily seen from (11.9) that $b_n/b_1=n$ and, therefore, by (11.10) $c_n/c_1=a_n/a_1$ and so finally

$$\gamma_n=\alpha_n$$

for any choice of the α_n; this must be compatible with (11.18) and so, since $B_1=B_2=\cdots=B_p=1$, by our choice of the β,

$$\sum_{l=1}^{p} j_{kl}^{(n)} = \begin{cases} 1 \text{ for } k=1 \\ 0 \text{ for } k=2, 3, \cdots, p . \end{cases} \tag{11.21}$$

(f) The system of n complex numbers $(\alpha_1, \alpha_2, \cdots, \alpha_n)$, for which we shall also use the more concise notation (α), determines a point in $2n$-dimensional Euclidean space. A point (α) belongs to the *coefficient-domain* if, and only if, it corresponds by virtue of (11.5) to the initial terms of a power series of the class Σ. The most remarkable boundary point of the coefficient domain is the "Koebe-point" which corresponds to the function

$$z+2z^2+3z^3+ \cdots = z(1-z)^{-2} .$$

Our aim is to show that, for any given n, Conjecture IV is true for two interior points of the coefficient domain which are sufficiently close to the Koebe-point.

Let us choose two arbitrary points (u) and (v) in the interior or the coefficient domain. That is, (cf. under (*b*)) both Hermitian forms

$$\sum\sum u_{k-l} z_k \bar{z}_l , \quad \sum\sum v_{k-l} z_k \bar{z}_l \tag{11.22}$$

are positive definite. Let α, β and ε denote positive numbers ; α and β are arbitrary and ε so small that $\alpha\varepsilon<1$, $\beta\varepsilon<1$. The coefficient domain is convex. Therefore, if we set

$$\alpha_\nu=(1-\varepsilon\alpha)+\varepsilon\alpha u_\nu , \quad \beta_\nu=(1-\varepsilon\beta)+\varepsilon\beta v_\nu \tag{11.23}$$

for $\nu=0, \pm 1, \pm 2, \cdots, \pm n$, the points (α) and (β) are in the interior of the coefficient domain. If A_k is given by (11.20)

$$A_k=1+\varepsilon\alpha\tilde{u}_k+O(\varepsilon^2)$$

where

$$\tilde{u}_k=k_1(u_1-1)+k_2(u_2-1)+ \cdots +k_n(u_n-1)$$

and $O(\varepsilon^2)$ denotes a quantity of order not exceeding ε^2 when ε tends to 0. There is a similar expression for B_l and finally, by (11.21),

$$\begin{aligned} \gamma_n &= \sum_{k=1}^{p}\sum_{l=1}^{p} j_{kl}^{(n)} A_k B_l \\ &= \sum_{k=1}^{p}\sum_{l=1}^{p} j_{kl}^{(n)}[1+\varepsilon\alpha\tilde{u}_k+\varepsilon\beta\tilde{v}_l]+O(\varepsilon^2) \\ &= 1+\varepsilon\alpha\tilde{u}_1+\varepsilon\beta\tilde{v}_1+O(\varepsilon^2) \\ &= 1+\varepsilon\alpha(u_n-1)+\varepsilon\beta(v_n-1)+O(\varepsilon^2) . \end{aligned} \tag{11.24}$$

By virtue of (11.24)

$$\begin{aligned}\sum\sum \gamma_{k-l} z_k \bar{z}_l &= (1-\varepsilon\alpha-\varepsilon\beta)|z_0+z_1+\cdots+z_n|^2 \\ &\quad +\varepsilon\alpha\sum\sum u_{k-l} z_k \bar{z}_l \\ &\quad +\varepsilon\beta\sum\sum v_{k-l} z_k \bar{z}_l + O(\varepsilon^2)\end{aligned}$$

and this Hermitian form is definite positive for sufficiently small ε, since the forms (11.22) are definite positive. With this, we have proved another infinitesimal part of Conjecture IV.

REFERENCES

1. L. Fejér, *Gestaltliches ueber die Partialsummen und ihre Mittelwerte bei der Fourierreihe und der Potenzreihe*, Z. f. angew. Math. und Mech., **13** (1933), 80–88.
2. A. Hurwitz, Mathematische Werke, Vol. I.
3. S. Karlin, *Pólya-type distributions* II. Ann. Math. Stat., **28** (1957), 281–308.
4. J. C. Mairhuber, I. J. Schoenberg and R. Williamson, *Variation diminishing transformations on the circle*, to be published.
5. I. P. Natanson, *Konstruktive Funktionentheorie* (German translation from Russian), Berlin, 1955.
6. Z. Nehari, *Conformal Mapping*, New York, 1952.
7. G. Pólya, *Ueber einige Verallgemeinerungen der Descarteschen Zeichenregel*, Archiv Math. Phys. III. Reihe, **23** (1914), 22–32.
8. G. Pólya and N. Wiener, *On the oscillation of the derivatives of a periodic function*, Trans. Amer. Math. Soc., **52** (1942), 249–256.
9. G. Pólya and G. Szegö, *Aufgaben und Lehrsaetze aus der Analysis*, Berlin, 1925.
10. I. J. Schoenberg, *Smoothing operations and their generating functions*, Bull. Amer. Math. Soc., **59** (1953), 199–230.
11. E. Study, *Vorlesungen ueber ausgewaehlte Gegenstaende der Geometrie*, Second Part, Leipzig and Berlin, 1913.
12. J. J. Sylvester, *Collected mathematical papers*, Vol. II, Cambridge, 1908.
13. G. Szegö, *Zur Theorie der schlichten Abbildungen*, Math. Ann. **100** (1929), 88–211.
14. Ch. J. de la Vallée Poussin, *Sur l'approximation des fonctions d'une variable réelle et de leurs dérivées par des polynomes et des suites limitées de Fourier*, Bull. Acad. de Belgique (classe des sciences), **3** (1908), 193–254.

STANFORD UNIVERSITY
THE UNIVERSITY OF PENNSYLVANIA

Reprinted from
Pacific J. Math. **8** (1958), pp. 295–334

ON A CONJECTURE CONCERNING SCHLICHT FUNCTIONS[1]

BERNARD EPSTEIN AND I. J. SCHOENBERG

Communicated March 25, 1959

Many years ago and independently of each other S. Mandelbrojt and M. Schiffer were led to the following conjecture, which has appeared in print only recently [2, p. 326]:

CONJECTURE M. S. *If two power series* $\sum_1^\infty a_\nu z^\nu$, $\sum_1^\infty b_\nu z^\nu$ *are schlicht in the unit circle, then also the power series*

$$\sum_1^\infty \frac{a_\nu b_\nu}{\nu} z^\nu$$

is schlicht in the unit circle.

This will be disproved in the following lines. Let D be the image of the unit circle by $w = \sum_1^\infty a_\nu z^\nu$. We denote by the symbols S, Σ and K the classes of such power series for which D is schlicht, schlicht and star-shaped, schlicht and convex, respectively. Evidently $K \subset \Sigma \subset S$.

Observe now that $\sum_1^\infty z^\nu \in K$. By a recent result concerning de la Vallée Poussin means [2, p. 298] we conclude that

$$\sum_1^n \binom{2n}{n+\nu} z^\nu \in K, \qquad (n = 1, 2, \cdots),$$

and therefore [2, Lemma 5, p. 321] that

$$\sum_1^n \nu \binom{2n}{n+\nu} z^\nu \in \Sigma \subset S.$$

Applying the Conjecture M. S. to this special polynomial and an arbitrary power series we obtain the following

COROLLARY OF THE CONJECTURE M.S. *If* $f(z) = \sum_1^\infty a_\nu z^\nu \in S$ *then also*

$$\sum_1^n \binom{2n}{n+\nu} a_\nu z^\nu \in S.$$

[1] This paper was prepared partly under the sponsorship of the United States Air Force, Office of Scientific Research, ARDC, under a contract with the University of Pennsylvania.

In other words: The de la Vallée Poussin means of schlicht functions are also schlicht. However, this corollary is now easily disproved as follows:

We appeal to a result of C. Loewner [1, pp. 117, 118, and 120]: To every given function $\kappa(\tau)$ which is continuous for $\tau \geqq 0$ and such that $|\kappa(\tau)| = 1$, there corresponds a power series $f(z) = z + a_2 z^2 + a_3 z^3 + \cdots$ which is in S and is such that

$$a_2 = -2\int_0^\infty \kappa(\tau)e^{-\tau}d\tau,$$

$$a_3 = 4\left(\int_0^\infty \kappa(\tau)e^{-\tau}d\tau\right)^2 - 2\int_0^\infty (\kappa(\tau))^2 e^{-2\tau}d\tau.$$

We now select

$$\kappa(\tau) = e^{-i\gamma\tau}, \qquad (\gamma \text{ real constant} \neq 0).$$

The integrals are easily evaluated and we find

$$f(z) = z - \frac{2}{1+i\gamma}z^2 + \frac{3-i\gamma}{(1+i\gamma)^2}z^3 + \cdots \in S.$$

Applying the Corollary of the Conjecture M.S. for $n = 3$ we conclude that the cubic polynomial

$$P(z) = 15z + 6a_2z^2 + a_3z^3 \in S.$$

But then *the quadratic polynomial*

$$\frac{1}{3}(1+i\gamma)^2 P'(z) = (3-i\gamma)z^2 - 8(1+i\gamma)z + 5(1+i\gamma)^2$$

can not have any zeros in the interior of the unit circle.

On the other hand we find that

$$\zeta = \frac{1+i\gamma}{3-i\gamma}(4 - (1+5i\gamma)^{1/2})$$

is one of the two zeros of this quadratic; ζ is regular for all real γ and we find its Taylor expansion at the origin to be

$$\zeta = 1 + \frac{i}{2}\gamma - \frac{9}{24}\gamma^2 + \cdots, \qquad (|\gamma| < 1/5).$$

Now

$$|\zeta|^2 = \left(1 - \frac{9}{24}\gamma^2 + \cdots\right)^2 + \left(\frac{\gamma}{2} + \cdots\right)^2 = 1 - \frac{1}{2}\gamma^2 + \cdots,$$

showing that $|\zeta| < 1$ provided that γ is sufficiently small. This contradicts our last italicized statement and completes our proof.

For the discussion of a conjecture of Polya and Schoenberg obtained from the Conjecture M.S. by replacing in its statement the term "schlicht" by "star-shaped," we refer to [2, pp. 324–334].

References

1. C. Loewner, *Untersuchungen über schlichte konforme Abbildungen des Einheitskreises*, Math. Ann. vol. 89 (1923) pp. 103–121.

2. G. Polya and I. J. Schoenberg, *Remarks on de la Vallée Poussin means and convex conformal maps of the circle*, Pacific J. Math. vol. 8 (1958) pp. 295–334.

The University of Pennsylvania

Reprinted from
Bull. Amer. Math. Soc. **65** (1959), pp. 273–275

NORM INEQUALITIES FOR A CERTAIN CLASS OF C^∞ FUNCTIONS

BY

I. J. SCHOENBERG

For Paul Montel on his 95th birthday

ABSTRACT

In 1936 the author showed that the function $\sin(\pi(x+1)/4)$ is the entire function of least exponential type ($=\pi/4$) among all entire functions $f(z)$ with the property that $f^{(n)}(z)$ vanishes somewhere in the real interval $[-1, 1]$ ($n=0, 1,2,\dots$). Now more precise results of this kind are obtained by working within the class $C^\infty[-1, 1]$.

Introduction and statement of results

In [3] the author proved long ago the following

THEOREM 1. *If $f(x)$ is an entire function of exponential type less than $\pi/4$ and is such that every one of its derivatives $f(x), f'(x), f''(x),\cdots$ vanishes somewhere in the interval $I=[-1,1]$, then $f(x)=0$ identically.*

Since the function $f(x)=\sin(\pi(x+1)/4)$ satisfies the conditions of Theorem 1 and is precisely of type $=\pi/4$, it follows that the constant $\pi/4$ can not be improved.

In 1936 Theorem 1 was a peripheral contribution towards the problem of determining the so-called Whittaker constant W (see [4, 45]). For the important more recent results of Evgrafov and Buckholtz concerning the Whittaker constant see [2], also for references. I do not pursue here this function-theoretic problem, but wish to point out that the contents of Theorem 1 can be sharpened by working within the context of functions $f(x)$ that are infinitely differentiable in the interval $I=[-1,1]$. Accordingly, the appropriate setting of our discussion is the class $\mathscr{F}$ of real or complex-valued functions defined by

(1) $$\mathscr{F}=\{f(x); f\in C^\infty(I),\ f^{(\nu)}(\eta_\nu)=0\ (\nu=0,1,\cdots),\ -1\leqq\eta_\nu\leqq 1\}.$$

Received June 10, 1971

In terms of the L_{∞} (or supremum) norm in I we may state

THEOREM 2. *If $f(x) \in \mathscr{F}$ and $\|f\| > 0$, then*

$$\|f^{(n)}\| > \omega_n \|f\|, \qquad (n = 1, 2, \cdots) \tag{2}$$

where

$$\omega_n = \begin{cases} \dfrac{n!}{2^n} \dfrac{1}{|E_n|} & \text{if } n \text{ is even,} \\[2ex] \dfrac{(n+1)!}{2^{2n+1}(2^{n+1}-1)} \dfrac{1}{|B_{n+1}|} & \text{if } n \text{ is odd.} \end{cases} \tag{3}$$

Here E_n and B_{n+1} are the Euler and Bernoulli numbers, respectively. The constants ω_n in (2) are sharp for every n, i.e. none can be improved.

We find that

$$\omega_1 = \frac{1}{2}, \omega_2 = \frac{1}{2}, \omega_3 = \frac{3}{8}, \omega_4 = \frac{3}{10}, \omega_5 = \frac{15}{64}, \omega_6 = \frac{45}{244}.$$

In §4 we show that Theorem 1 is an easy consequence of Theorem 2. In §2 we derive Theorem 2 from a result to be now described.

For every natural number n we denote by $\mathscr{F}_n$ the class of functions $f(x)$ satisfying the following two conditions:

$$f(x) \in C^{n-1}(I), \text{ and } f^{(n-1)}(x) \text{ satisfies a Lipschitz condition,} \tag{4}$$

$$f^{(\nu)}(x) \; (\nu = 0, 1, \cdots, n-1) \text{ vanishes at some point of } I. \tag{5}$$

A particular element of $\mathscr{F}_n$ is the polynomial

$$s_n(x) = \int_{-1}^{x} dx_1 \int_{1}^{x_1} dx_2 \cdots \int_{(-1)^n}^{x_{n-1}} dx_n, \tag{6}$$

in view of the relations

$$s_n(-1) = 0, \; s_n'(1) = 0, \cdots, s_n^{(n-1)}((-1)^n) = 0. \tag{7}$$

In terms of the class $\mathscr{F}_n$ and the polynomial (6) we may state the following theorem.

THEOREM 3. *If $n \geqq 1$ and*

$$f(x) \in \mathscr{F}_n \tag{8}$$

then

$$\|f^{(n)}\| \geqq \omega_n \|f\|,$$

with the equality sign if and only if

(10) $$f(x) = Cs_n(\varepsilon x), \text{ where } C \text{ is a constant and } \varepsilon = \pm 1.$$

The constant ω_n *is given by* (3) *and may also be defined as*

(11) $$\omega_n = |s_n(1)|^{-1}.$$

In §3 we show that Theorem 3 can be stated (Corollary 1) as an apparently new characteristic property of the Euler polynomial $E_n(x)$, in the interval $I^* = [0, \frac{1}{2}]$.

1. Proof of Theorem 3. We require a lemma concerning the function

(1.1) $$\Phi_n(x; \eta_0, \eta_1, \cdots, \eta_{n-1}) = \left| \int_{\eta_0}^{x} dx_1 \left| \int_{\eta_1}^{x_1} dx_2 \cdots \left| \int_{\eta_{n-1}}^{x_{n-1}} \right| dx_n \cdots \right| \right|.$$

This expression being somewhat cumbersome, we write it out completely for the case when $n = 3$:

$$\Phi_3(x; \eta_0, \eta_1, \eta_2) = \left| \int_{\eta_0}^{x} dx_1 \left| \int_{\eta_1}^{x_1} dx_2 \left| \int_{\eta_2}^{x_2} dx_3 \right| \right| \right|.$$

In [3, 14–18] the author established a lemma concerning the function (1.1) which implies readily the following

LEMMA 1. *If*

(1.2) $$-1 \leqq x \leqq 1, \ -1 \leqq \eta_\nu \leqq 1 \qquad (\nu = 0, 1, \cdots, n-1),$$

then

(1.3) $$\Phi_n(x; \eta_0, \cdots, \eta_{n-1}) \leqq \Phi_n(1; -1, 1, \cdots, (-1)^n),$$

with the equality sign if and only if

(1.4) $$x = 1, \eta_\nu = (-1)^{\nu+1}, \text{ or } x = -1, \eta_\nu = (-1)^\nu, \qquad (\nu = 0, \cdots, n-1).$$

In other words: Within the $(n+1)$ dim. cube (1.2) the function $\Phi(x; \eta)$ assumes its absolute maximum only in two opposite vertices of the cube that are described by (1.4).

Let now (8) hold and let us establish the inequality (9), where we define ω_n by (11), hence

(1.5) $$\omega_n^{-1} = |s_n(1)| = \Phi_n(1; -1, 1, \cdots, (-1)^n).$$

From the assumption (8) we know that

(1.6) $$f^{(\nu)}(\eta_\nu) = 0, \qquad (\nu = 0, \cdots, n-1),$$

for appropriate points η_ν in I. It follows that we may express $f(x)$ in terms of $f^{(n)}(x)$ by the repeated integral

$$f(x) = \int_{\eta_0}^{x} dx_1 \int_{\eta_1}^{x_1} dx_2 \cdots \int_{\eta_{n-1}}^{x_{n-1}} f^{(n)}(x_n)\,dx_n. \tag{1.7}$$

Let ξ be such that

$$|f(\xi)| = \|f\|. \tag{1.8}$$

Using Lemma 1 and (1.8) we may write the following string of equations and inequalities

$$\begin{aligned}
\|f\| &= |f(\xi)| = \left|\int_{\eta_0}^{\xi} dx_1 \int_{\eta_1}^{x_1} dx_2 \cdots \int_{\eta_{n-1}}^{x_{n-1}} f^{(n)}(x_n)dx_n\right| \\
&\leqq \left|\int_{\eta_0}^{\xi} dx_1 \left|\int_{\eta_1}^{x} dx_2 \cdots \left|\int_{\eta_{n-1}}^{x_{n-1}} |f^{(n)}(x_n)|dx_n \right| \cdots \right|\right| \\
&\leqq \|f^{(n)}\| \cdot \left|\int_{\eta_0}^{\xi} dx_1 \left| \cdots \left|\int_{\eta_{n-1}}^{x_{n-1}} dx_n\right| \cdots \right|\right| \\
&= \|f^{(n)}\|\Phi_n(\xi; \eta_0, \cdots, \eta_{n-1}) \leqq \|f^{(n)}\|\Phi_n(1; -1, 1, \cdots, (-1)^n).
\end{aligned} \tag{1.9}$$

But then (1.5) implies that

$$\|f\| \leqq \|f^{(n)}\| \cdot \omega_n^{-1}, \tag{1.10}$$

and the inequality (9) is established.

Let us now assume that (9), or (1.10), holds with the equality sign. It follows that the extreme terms of (1.9) are equal. Let us now analyze the consequences of this fact. In particular we obtain that

$$\Phi_n(\xi; \eta_0, \cdots, \eta_{n-1}) = \Phi_n(1; -1, 1, \cdots, (-1)^n),$$

and by Lemma 1 we conclude that one of the two alternatives

$$\xi = 1, \eta_\nu = (-1)^{\nu+1}, \text{ or } \xi = -1, \eta_\nu = (-1)^\nu \tag{1.11}$$

must hold. Let us assume that the first one holds. We consider now the relation

$$(-1)^{[n/2]} \int_{-1}^{1} dx_1 \int_{1}^{x_1} dx_2 \cdots \int_{(-1)^n}^{x_{n-1}} F(x_n)\,dx_n = \int_{-1}^{1} K(x)F(x)\,dx \tag{1.12}$$

which transforms the repeated integral on the left into the single integral on the right hand side. This relation is to be valid for an arbitrary continuous function $F(x)$, while $[n/2]$ has its usual arithmetic meaning. A moment's reflexion will show that $K(x) \in C(I)$, in fact $K(x) \in \pi_{n-1}$ (the polynomials of degree $n-1$), and that

$$K(x) > 0 \text{ if } -1 < x < 1. \tag{1.13}$$

In terms of $K(x)$ the equality of all members of (1.9) furnishes the relation

$$\left| \int_{-1}^{1} K(x) f^{(n)}(x) dx \right| = \| f^{(n)} \| \int_{-1}^{1} K(x)\, dx,$$

which in turn implies the relation

$$\int_{-1}^{1} K(x) \{ \| f^{(n)} \| - \varepsilon f^{(n)}(x) \}\, dx = 0, \qquad (\varepsilon = 1 \text{ or } -1). \tag{1.14}$$

The integrand of (1.14) being non-negative almost everywhere we conclude that $f^{(n)}(x) = \| f^{(n)} \|$, or perhaps $f^{(n)}(x) = - \| f^{(n)} \|$, almost everywhere in I. Moreover, the first relations (1.11) and (1.6) show that

$$f(-1) = 0,\ f'(1) = 0, \cdots, f^{(n-2)}((-1)^n) = 0,$$

which imply that

$$f(x) = \pm \| f^{(n)} \| s_n(x).$$

Similarly the second alternative (1.11) will imply that $f(x) = \pm \| f^{(n)} \| s_n(-x)$. This concludes a proof of Theorem 3.

2. Proof of Theorem 2. Let $f(x) \in \mathscr{F}$ and $\| f \| > 0$. Evidently $f(x) \in \mathscr{F}_n$, for every n, and Theorem 3 shows that $\| f^{(n)} \| \geqq \omega_n \| f \|$. Since (10) can not hold because then $f^{(n)}(x)$ would not vanish anywhere in I, we conclude that the strict inequality (2) must hold, and the inequalities (2) are thereby established.

We are still to show that the constant ω_n is best possible. This requires to show the following: *If $\delta > 0$, then there exists a function $f(x) \in \mathscr{F}$ such that*

$$\| f^{(n)} \| < (\omega_n + \delta) \| f \|. \tag{2.1}$$

This we do as follows. We already know that the polynomial

$$s_n(x) = \int_{-1}^{x} dx_1 \int_{1}^{x_1} dx_2 \cdots \int_{(-1)^n}^{x_{n-1}} dx_n \tag{2.2}$$

has the properties

$$\| s_n^{(n)} \| = 1, \ \| s_n \| = \omega_n^{-1}\ , \tag{2.3}$$

whence

$$\| s_n^{(n)} \| = \omega_n \| s_n \|. \tag{2.4}$$

We shall now perturb the function $s_n(x)$ slightly so that it will become an element of $\mathscr{F}$, while the relation (2.4) will be disturbed only slightly.

To do this we use the auxiliary function

$$h(x) = \int_0^x e^{-t^{-1}(1-t)^{-1}}\, dt \Big/ \int_0^1 e^{-t^{-1}(1-t)^{-1}}\, dt, \qquad (0 \leqq x \leqq 1), \tag{2.5}$$

having the properties

$$h^{(\nu)}(0) = 0 \qquad (\nu = 0, 1, 2, \cdots) \tag{2.6}$$

$$h(1) = 1, \qquad h^{(\nu)}(1) = 0 \qquad (\nu = 1, 2, \cdots), \tag{2.7}$$

$$h(x) \text{ increases from 0 to 1 in } [0,1]. \tag{2.8}$$

Let α be a small positive number and let us define the even function

$$g_\alpha(x) = \begin{cases} 1 & \text{if } \alpha \leqq x \leqq 1, \\ h(x/\alpha) & \text{if } 0 \leqq x \leqq \alpha, \\ g_\alpha(-x) & \text{if } -1 \leqq x \leqq 0. \end{cases} \tag{2.9}$$

Evidently

$$g_\alpha(x) \in \mathscr{F}.$$

It follows that an n-fold integral $f_\alpha(x)$ of $g_\alpha(x)$ will also be in $\mathscr{F}$, provided that each of the functions

$$f_\alpha(x), f_\alpha'(x), \cdots, f_\alpha^{(n-1)}(x),$$

vanishes somewhere in I. An n-fold integral satisfying these conditions is evidently

$$f_\alpha(x) = \int_{-1}^{x} dx_1 \int_{1}^{x_1} dx_2 \cdots \int_{(-1)^n}^{x_{n-1}} g_\alpha(x_n)\, dx_n. \tag{2.10}$$

From $f_\alpha^{(n)}(x) = g_\alpha(x)$, (2.8) and (2.9), it is clear that

$$\| f_\alpha^{(n)} \| = 1. \tag{2.11}$$

From (2.9) we see that if $x \neq 0$ then $g_\alpha(x) \to 1$ as $\alpha \to +0$. From (2.2) and (2.10) it follows that

$$\| f_\alpha \| \to \| s_n \| = \omega_n^{-1} \text{ as } \alpha \to +0,$$

and therefore

(2.12) $$\|f_\alpha\| > \frac{1}{\omega_n + \delta} \text{ provided that } \alpha \text{ is sufficiently small.}$$

For such small α we therefore have that

$$\|f_\alpha^{(n)}\| = 1 < (\omega_n + \delta)\|f_\alpha\|.$$

Since $f_\alpha(x) \in \mathscr{F}$, the inequality (2.1) is established.

3. Determination of the constants ω_n and an extremum property of the Euler polinomials. The constants ω_n were defined by (11) and (6). Let us sketch the derivation of the explicit expressions (3). In (6) we reverse the order of the lower limits of integration and define the new polynomials.

(3.1) $$r_n(x) = \int_{(-1)^n}^{x} dx_1 \int_{(-1)^{n-1}}^{x_1} dx_2 \cdots \int_{-1}^{x_{n-1}} dx_n, \qquad (r_0(x) = 1).$$

These form an Appell sequence of polynomials having the generating function

(3.2) $$\frac{2e^t}{e^{2t} + e^{-2t}} e^{xt} = \sum_0^\infty r_n(x) t^n$$

(See [3, 19]). They are related to the $s_n(x)$ by

(3.3) $$s_n(x) = \pm\, r_n((-1)^{n+1} x).$$

Now (11) and (3.3) show that

(3.4) $$(\omega_n)^{-1} = |s_n(1)| = \begin{cases} |r_n(-1)| & \text{if } n \text{ is even,} \\ |r_n(1)| & \text{if } n \text{ is odd.} \end{cases}$$

From these results and the generating function

(3.5) $$\frac{2}{e^t + e^{-t}} = \sum_0^\infty \frac{E_n}{n!} t^n$$

of the Euler numbers E_n, we easily obtain the expressions (3) for even values of n. Details may be omitted.

Similarly, the generating function

(3.6) $$\frac{2e^{xt}}{1 + e^t} = \sum_0^\infty \frac{E_n(x)}{n!} t^n$$

of the Euler polynomials $E_n(x)$, and the known relations

(3.7) $$E_n(0) = -\frac{2}{n+1}(2^{n+1} - 1)B_{n+1}$$

easily furnish the relations (3) for odd values of n.

We also mention that (3.2) and (3.6) show that

$$r_n(x) = (-1)^n \frac{4^n}{n!} E_n\left(\frac{1-x}{4}\right). \tag{3.8}$$

These relations show that our Theorem 3 may be interpreted as describing a new characteristic extremum property of the Euler polynomial $E_n(x)$ in the interval $[0,\frac{1}{2}]$. We state it as

COROLLARY 1. *Let $\mathscr{F}_n^*$ denote the class of functions $F(x)$ defined in $I^* = [0,\frac{1}{2}]$ and satisfying the three conditions*

(i) *$F(x) \in C^{n-1}(I^*)$, and $F^{(n-1)}(x)$ satisfies a Lipschitz condition,*

(ii) *$F^{(\nu)}(x)$ $(\nu = 0, \cdots, n-1)$ vanishes at some point of I^*,*

(iii) $\| F^{(n)} \| = n!$.

If $F(x) \in \mathscr{F}_n^*$, then the inequality

$$\| F \| \leqq \| E_n(x) \|$$

holds, with the equality sign if and only if

$$F(x) = \pm E_n(x) \text{ or } F(x) = \pm E_n(\tfrac{1}{2} - x).$$

Finally, we shall need below the following

LEMMA 2. *The constants ω_n satisfy the relation*

$$\lim_{n\to\infty} (\omega_n)^{1/n} = \frac{\pi}{4}. \tag{3.9}$$

A proof follows immediately from the estimate of $r_n(x)$ given in [3, formula (29)].

4. Proof of Theorem 1. Let $f(x)$ be an entire function of exponential type $=\gamma$. It is known (see [1, 11]] that

$$\overline{\lim_{n\to\infty}} |f^{(n)}(0)|^{1/n} = \gamma. \tag{4.1}$$

We first establish

LEMMA 3. *If*

$$\|f^{(n)}\| = \max|f^{(n)}(x)| \text{ in the interval } I = [-1,1], \tag{4.2}$$

then

$$\overline{\lim_{n\to\infty}} \|f^{(n)}\|^{1/n} = \gamma. \tag{4.3}$$

PROOF. If $\gamma_1 > \gamma$, then (4.1) shows that

$$|f^{(n)}(0)| < \gamma_1^n \text{ if } n > N.$$

The expansion of $f^{(n)}(x)$ in powers of x shows that, if $n > N$, then

$$\|f^{(n)}\|^{1/n} \leqq (|f^{(n)}(0)| + \frac{1}{1!}|f^{(n+1)}(0)| + \frac{1}{2!}|f^{(n+2)}(0)| + \cdots)^{1/n}$$

$$\leqq \left(\gamma_1^n + \frac{1}{1!}\gamma_1^{n+1} + \cdots\right)^{1/n} = \gamma_1 (1 + \frac{1}{1!}\gamma_1 + \cdots)^{1/n} = \gamma_1 e^{\gamma_1/n},$$

whence

$$\overline{\lim} \|f^{(n)}\|^{1/n} \leqq \gamma_1.$$

Letting $\gamma_1 \to \gamma$ we obtain (4.3).

A proof of Theorem 1 now follows immediately. Indeed, let $f(x)$ be of exponential type γ, and let $f, f', \cdots$ have the required zeros in $I = [-1, 1]$. By Theorem 2

$$\|f^{(n)}\| > \omega_n \|f\|$$

whence

$$\|f^{(n)}\|^{1/n} > (\omega_n)^{1/n} \|f\|^{1/n}.$$

Lemma 2 and 3 show that if $n \to \infty$ we obtain that

$$\gamma \geqq \pi/4.$$

REFERENCES

1. R. P. Boas, *Entire functions*, Academic Press Inc., New York, 1954.

2. J. D. Buckholtz, *The Whittaker constant and successive derivatives of entire functions*, J. Approximation Theory, **3** (1970), 194–212.

3. I. J. Schoenberg, *On the zeros of successive derivatives of integral functions*, Trans. Amer. Math. Soc., **40** (1936), 12–23.

4. J. M. Whittaker, *Interpolatory function theory*, Cambridge University Press, 1935.

MATHEMATICS RESEARCH CENTER
UNIVERSITY OF WISCONSIN
MADISON, WISCONSIN, 53706

Reprinted from *Israel J. Math.* **10** (1971), pp. 364–372

The Pólya–Schoenberg Conjecture

Conjectures make the mathematical world go round. In his "Brief account of my life and work" Schoenberg, concerning his joint paper [**55***] with Pólya, said that "the main result in that paper was a conjecture". Although this is not quite fair in view of the beautiful theorems given there, putting into the light the amazing properties of the de la Vallée-Poussin means, it pinpoints a fact in mathematical research: to invent a good conjecture can be as important for scientific progress as a bunch of theorems.

The Pólya-Schoenberg conjecture, stating that the class of convex univalent functions in the unit disc is closed under the Hadamard product, turned out to be of this type. The research towards its solution and generalization created a number of interesting other results, for instance, T. Suffridge's theorems on the convolution invariance of classes of univalent polynomials, [$\mathbf{Su_2}$], and the theory of 'dual sets of analytic functions', which permits a unified approach to many other problems—old and new—in geometric function theory. In this connection it became also very clear that the theorem proposed in the conjecture is not just an exotic flower in a desert but shares its roots with other well-known results which seem to be a fair distance from the conjecture and from each other: for instance, the celebrated theorem of Grace for apolar polynomials or the extreme point theory for certain subclasses of univalent functions. Hence, as it happens with good conjectures, it is not that the proposer just cannot prove a theorem, but much more: the conjectured result opens a door to a new field of research.

The conjecture itself was based on, I guess, three facts: 1) A classical theorem of M. S. Robertson, namely that a similar statement holds with 'convex univalent' replaced by 'typically real'. 2) An earlier conjecture of Mandelbrojt and Schiffer which stated that in Robertson's theorem one can replace 'typically real' by 'univalent'. 3) A large number of examples as evidence for its truth, growing out of the theory of variation-diminishing transformations, a theory to which Schoenberg has largely contributed.

Although, in their paper, Pólya and Schoenberg expressed the feeling that their own conjecture is "more elementary and certainly more accessible" than the Mandelbrojt-Schiffer conjecture, it turned out that the latter one was decided upon much earlier: it is false, false in an almost trivial way as extremely simple counterexamples show. This fact created doubts in many specialists in the field about the chances of the Pólya-Schoenberg conjecture holding. The preservation of a delicate geometric property of the image domains of analytic functions under a rather simple-minded algebraic operation on their MacLaurin coefficients seemed to be too unlikely to be true.

The first serious progress towards a solution was made by T. Suffridge [**Su**] in 1966, who proved that the operation in question produces, at least, a univalent function. I became interested in the conjecture in connection with a possible application for injective solutions of certain elliptic PDE's, and I wrote a little note [$\mathbf{R_1}$] in which some further special cases were verified and some structural properties towards the general problem were developed. These turned out to be the right basis for the joint research with T. Sheil-Small, who was visiting Bonn University in early 1973. He brought along a nice new analytic-geometric property of starlike functions and our work resulted in finding the missing link between our two approaches: the pieces fit together and the conjecture was settled [**RS**]. Only one year later, T. Suffridge published another independent proof [$\mathbf{Su_2}$] and in 1978, J. Lewis [**L**] gave a third one. All of these as well as the papers [$\mathbf{R_{2-3}}$,**Sh**] contain far-reaching extensions of the original statement. A survey of how far the theory had grown by 1982 is given in [$\mathbf{R_4}$].

Once T. Sheil-Small and I had found our proof we hurried to send the good news to Prof. Schoenberg and he was very happy to hear this. At this time also a secret connected with the conjecture was disclosed. In their paper containing the conjecture Pólya and Schoenberg wrote in a footnote: "Which one of the two authors of this paper is the author of the conjecture will be disclosed if and when the conjecture is proved." Well, it was, in fact, a "Schoenberg-Conjecture".

REFERENCES

[L] J. Lewis, Convolutions of starlike functions, *Indiana Univ. Math. J.* **27** (1978), 671–688.

[R_1] St. Ruscheweyh, Über die Faltung schlichter Funktionen, *Math. Z.* **128** (1972), 85–92.

[R_2] ——, Linear operators between classes of prestarlike functions, *Comment. Math. Helv.* **52** (1977), 497–509.

[R_3] ——, Some convexity and convolution theorems for analytic functions, *Math. Ann.* **238** (1978), 217–228.

[R_4] ——, Convolutions in geometric function theory, *Sém. Math. Sup.* **83** (1982), 166.

[RS] —— and T. Sheil-Small, Hadamard products of schlicht functions and the Pólya-Schoenberg conjecture, *Comment. Math. Helv.* **48** (1973), 119–135.

[Sh] T. Sheil-Small, The Hadamard product and linear transformations of classes of analytic functions, *J. Analyse Math.* **34** (1978), 204–239.

[Su_1] T. Suffridge, Convolutions of convex functions, *J. Math. Mech.* **15** (1966), 795–804.

[Su_2] ——, Starlike functions as limits of polynomials, *Lecture Notes in Math.* **505** (1974), 164–202.

Stephan Ruscheweyh
Universität Würzburg

EXTREMUM PROBLEMS

Proceedings of the International Conference on Constructive Function Theory, Varna, May 19-25, 1970

SOLUTION OF LANDAU'S PROBLEM CONCERNING HIGHER DERIVATIVES ON THE HALFLINE

I. J. Schoenberg and A. Cavaretta

For J. L. Walsh on his 75th birthday

Summary. Let $f(x)$ be defined in the interval $I = [0, \infty)$ and let $M_\nu = \sup_{x \in I} |f^{(\nu)}(x)|$, $\nu = 1, 2, \ldots, n$; $n \geq 2$. It is assumed that M_0 and M_n are finite, more precisely that $f^{(n-1)}(x)$ is absolutely continuous in I and that M_n is the essential supremum of $f^{(n)}(x)$. Then also M_ν, $0 < \nu < n$, is finite and it is known that inequalities of the form $M_\nu \leq C_{n,\nu} M_n^{(\nu/n)} M_0^{1-(\nu/n)}$ hold. In the present paper we solve the so-called Landau problem of determining the best (least) constants $C_{n,\nu}$ in the inequality.

1. Formulation of the problem. We are here concerned with real-valued functions $f(x)$ defined on a closed interval I, finite or infinite. We denote by $\|f\| = \|f\|_\infty$ the supremum, or essential supremum, of $f(x)$ on I. If the nth derivative $f^{(n)}(x)$ is mentioned, then we assume that $f^{(n-1)}(x)$ is absolutely continuous on I and that $\|f^{(n)}\|_\infty$ is finite.

In 1913 Landau [3] formulated and solved some extremum problems concerning twice differentiable functions. From among his results we single out the following.

Theorem I (Landau). *If*

$$\|f\| = 1, \ \|f''\| = 4 \ in \ I = [0, \infty), \tag{1}$$

then

$$\|f'\| \leq 4 \ in \ [0, \infty). \tag{2}$$

The constant 4 *in* (2) *is best possible.*

Landau derives Theorem I from the following

Theorem II (Landau). *If*

$$\|f\| \leq 1, \ \|f''\| \leq 4 \ in \ I_1 = [0, 1], \tag{3}$$

then

$$|f'(0)| \leq 4. \tag{4}$$

Then interval I_1 can not be replaced by any smaller interval $[0,\theta]$, nor can the constant 4 in (4) be replaced by any smaller number.

Theorem II implies Theorem I, for if we apply Theorem II to the function $f(x+x_0)(x_0 \geq 0)$, we obtain that $|f'(x_0)| \leq 4$, hence (2) holds.

In 1955 Matorin [4] generalized Landau's Theorem I as follows.

Theorem III (Matorin). *Let $T_n(x)$ denote the Chebyshev polynomial. If*

$$\|f\| = 1, \ \|f^{(n)}\| = 2^{n-1}n! \text{ in } I = [0,\infty), \tag{5}$$

then

$$\|f^{(\nu)}\| \leq T_n^{(\nu)}(1) \text{ in } [0,\infty), \quad (\nu = 1,2,\ldots,n-1). \tag{6}$$

Since $T_n^{(n)}(x) = 2^{n-1}n!$ and $T_n'(1) = n^2$, we see that Theorem III reduces to Theorem I if $n=2$. Matorin does not state a similar generalization of Landau's Theorem II. This we do now and state the result as

Theorem 1. *Let $\alpha = \cos(\pi/n)$ be the largest zero of $T_n'(x)$. If*

$$\|f\| \leq 1, \ \|f^{(n)}\| \leq 2^{n-1}n! \text{ in } [0,1+\alpha], \tag{7}$$

then

$$|f^{(\nu)}(0)| \leq T_n^{(\nu)}(1), \quad (\nu = 1,\ldots,n-1). \tag{8}$$

The interval $[0,1+\alpha]$ can not be replaced by any smaller interval $[0,1+\alpha']$ $(\alpha' < \alpha)$, nor can the constant on the right side of (8), for any ν, be replaced by any smaller constant. Moreover, if we have the equality sign in (8) for some ν, then we have equality for all ν, and this happens if and only if $f(x) = \pm T_n(x-1)$ in $[0,1+\alpha]$.

For $n=2$ our Theorem 1 reduces to Landau's Theorem II with some additional information. Clearly, Theorem 1 implies Matorin's Theorem III, just as Theorem II implies Theorem I.

Matorin observes that his Theorem III gives the best constants in (6) if $n=2$ and also if $n=3$. The main concern of the present paper is to find the best constants on the right side of (6) if $n \geq 4$. We reformulate this question as follows.

Problem 1 (Generalized Landau Problem). *For the class of functions*

$$\mathcal{F}_n = \{f(x);\ \|f\| = 1,\ \|f^{(n)}\| = 2^{n-1}n! \text{ in } I = [0,\infty)\} \tag{9}$$

we are to determine the Landau constants

$$L_n^{(\nu)} = \sup_{f \in \mathcal{F}_n} \|f^{(\nu)}\|, \quad (\nu = 1,2,\ldots,n-1) \tag{10}$$

and we are to find the extremizing functions $f(x)$, if such exist, for which

$$\|f^{(\nu)}\| = L_n^{(\nu)}. \tag{11}$$

A more familiar (but only apparently more general) formulation of this problem is as follows. If we write

$$\|f^{(\nu)}\| = M_\nu \quad (\nu = 0,1,\ldots,n) \text{ in } I = [0,\infty), \tag{12}$$

and if we assume M_0 and M_n to be finite, then inequalities of the form

$$M_\nu \leq C_{n,\nu} M_0^{1-\frac{\nu}{n}} M_n^{\frac{\nu}{n}} \quad (\nu = 1, \ldots, n-1) \tag{13}$$

hold, with appropriate numerical constants $C_{n,\nu}$ as first shown by A. Gorny (for references see [7]). The generalized Landau problem is to determine in (13) the best constants $C_{n,\nu}$ and also the functions $f(x)$, if any, for which we have equality in (13).

The problem obtained if in either of the two equivalent formulations we replace the interval $I = [0, \infty)$ by $\mathbf{R} = (-\infty, \infty)$, was solved in 1939 by Kolmogorov [2]. He showed that the best constants $C_{n,\nu} = C_{n,\nu}(\mathbf{R})$ are obtained with equality in (13), and this for all ν simultaneously, for certain functions $\mathcal{E}_n(x)$ which we propose to call *Euler splines* of degree n. This term seems justified because $\mathcal{E}_n(x)$ is a spline function whose polynomial components are essentially the Euler polynomials $E_n(x)$.

The question of unicity of the extremizing functions in Kolmogorov's case of $\mathbf{R}$ has to our knowledge not been investigated before. Using spline function theory, the first named author will publish elsewhere a new proof of Kolmogorov's result that will also show that in a certain appropriate stricter sense the functions

$$f(x) = a\mathcal{E}_n(bx + c)$$

are the only extremizing functions if $n \geq 3$.

2. The Chebyshev-Euler splines and their properties. We return to Landau's Problem 1. For its solution we need the following new spline generalization of the Chebyshev polynomials.

Let n and k be natural numbers. Moreover, let

$$-1 = \xi_0 < \xi_1 < \ldots < \xi_k < \xi_{k+1} = +1 \tag{14}$$

and let

$$S(x) = S_{n,k}(x) = S_{n,k}(x; \xi_i) \tag{15}$$

be a spline function in $[-1, 1]$, of degree n, and having the simple knots $\xi_1, \ldots, \xi_k$. This means

1° $S(x) \in C^{n-1}[-1, 1]$

2° The restriction of $S(x)$ to the interval $[\xi_i, \xi_{i+1}]$ $(i = 0, \ldots, k)$ is a polynomial of degree at most n.

We shall also assume that $S(x)$ is a *perfect spline*, using a term introduced by G. Glaeser [1]. A spline function $S(x)$ is said to be *perfect*, provided that the terms of degree n, in each polynomial component, are of the form $\pm cx^n$ (with the same constant $c > 0$), the sign of the coefficient changing from one interval (ξ_i, ξ_{i+1}) to the next. The Euler spline $\mathcal{E}_n(x)$ is a perfect example of a perfect spline having infinitely many knots. Here we are discussing its finite analogues in $[-1, 1]$.

Specifically, we assume concerning the function (15) that

$$S^{(n)}(x) = (-1)^i 2^{n-1} n! \text{ in } \xi_i < x < \xi_{i+1} \quad (i = 0, \ldots, k). \tag{16}$$

If we define

$$u_+ = \begin{cases} u & \text{if } u \geq 0, \\ 0 & \text{if } u < 0, \end{cases}$$

we can incorporate all the requirements so far mentioned by writing

$$S(x) = 2^{n-1}x^n + \sum_{i=1}^{k}(-1)^i 2^n (x-\xi_i)_+^n + \sum_{0}^{n-1} a_\nu x^\nu \text{ in } -1 \le x \le 1. \tag{17}$$

This function depends on a total of $n+k$ parameters, linearly on the a_ν, nonlinearly on the ξ_i.

Definition 1. *We define the function $T_{n,k}(x)$ as the function of the form (17) having least L_∞-norm in $[-1,1]$, all $n+k$ parameters being free to vary. Thus*

$$\|T_{n,k}\| \le \|S\| \tag{18}$$

for any S of the form (17). $T_{n,k}(\nu)$ being a perfect spline of least deviation from zero, we propose to call it the Chebyshev-Euler spline of degree n with k knots.

Observe that if k should be zero, then the first sum in (17) drops out and we obtain the Chebyshev polynomial. We may therefore write

$$T_{n,0}(x) = T_n(x). \tag{19}$$

Existence, unicity and a few descriptive properties of $T_{n,k}(x)$ are as follows.

Theorem 2. *There is a unique $T_{n,k}(x)$ and it is a perfect spline of degree n with k simple knots. $T_{n,k}(x)$ has precisely $n+k+1$ points of equioscillation. This means that if*

$$\rho_{n,k} = \|T_{n,k}\|, \tag{20}$$

then there are $n+k+1$ points

$$-1 = x_1 < x_2 < \ldots < x_{n+k} < x_{n+k+1} = +1 \tag{21}$$

such that

$$T_{n,k}(x_i) = (-1)^{n+i-1}\rho_{n,k} \quad (i = 1,\ldots,n+k+1), \tag{22}$$

and $|T_{n,k}(x)| < \rho_{n,k}$ if $x \ne x_i$ for all i.

Moreover, the equioscillation property (20), (21), (22) characterizes $T_{n,k}(x)$ within the class of perfect splines of the form (17).

Theorem 3. *$T_{n,k}(x)$ is an even function if $n+k$ is even, and it is an odd function if $n+k$ is odd.*

Let $\mu = [n/2]$ have its usual arithmetic meaning. Between the extreme points (21) of $T_{n,k}(x)$ and its knots $\xi_1,\ldots,\xi_k$, the following inequalities hold

$$\begin{aligned} x_{\mu+1} &< \xi_1 < x_{n+2-\mu} \\ x_{\mu+2} &< \xi_2 < x_{n+3-\mu} \\ &\vdots \\ x_{\mu+k} &< \xi_k < x_{n+k+1-\mu}. \end{aligned} \tag{23}$$

The deviations (20) *satisfy the inequalities*

(24) $$1 = \rho_{n,0} > \rho_{n,1} > \ldots > \rho_{n,k} > \rho_{n,k+1} > \ldots > 0$$

and

(25) $$\lim_{k\to\infty} \rho_{n,k} = 0.$$

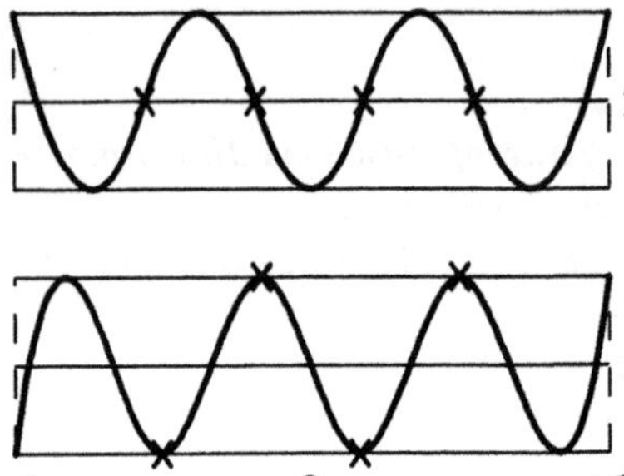

Figure 1

Just as the Chebyshev monosplines of of degree n and k knots in $[-1,1]$ were easily constructed for $n = 2, 3$, and 4, and all values of k (see [6, §1]), so the Chebyshev-Euler splines are readily found for $n = 2$ and $n = 3$ and all values of k. Without further details we refer the reader to Figure 1 for sketches of the graphs of $T_{2,4}(x)$ and $T_{3,4}(x)$. In their central portions they are identical, except for changes of scale and origin, to corresponding sections of the graphs of the Euler splines $\mathcal{E}_2(x)$ and $\mathcal{E}_3(x)$, respectively. The knots are marked by crosses. However, the construction of $T_{4,k}(x)$ is already difficult (see §6 below).

3. Solution of the generalized Landau problem. We shall now apply the Chebyshev-Euler spline $T_{n,k}(x)$ to Landau's Problem 1. This will be done by first generalizing our Theorem 1 from $T_n(x) = T_{n,0}(x)$ to $T_{n,k}(x)$. For this purpose it is convenient to perform a preliminary change of scale and origin. We consider

$$f(x) = aT_{n,k}(bx), \quad (a > 0,\ b > 0)$$

and determine a and b such that

$$\|f\| = 1 \text{ and } \|f^{(n)}\| = 2^{n-1}n! \text{ in } [-b^{-1}, b^{-1}].$$

This is evidently achieved by choosing

(26) $$a = a_{n,k} = (\rho_{n,k})^{-1}, \quad b = b_{n,k} = (\rho_{n,k})^{\frac{1}{n}}.$$

We finally translate $f(x)$ by b^{-1} and define

(27) $$S_{n,k}(x) = aT_{n,k}(bx - 1) \text{ in the interval } 0 \le x \le 2b^{-1}.$$

This is a perfect spline of degree n and k knots in $[0, 2b^{-1}]$. Using the same symbols x_i as in Theorem 2 but with a different meaning, we denote by

(28) $$0 = x_1 < x_2 < \ldots < x_{n+k} < x_{n+k+1} = 2(b_{n,k})^{-1}$$

the $n+k+1$ points of equioscillation of $S_{n,k}(x)$. By construction we therefore have

(29) $$S_{n,k}(x_i) = (-1)^{n+i-1} \quad (i = 1, \ldots, n+k+1)$$

and $|S(x)| < 1$ in all other points of $[0, 2b^{-1}]$.

Our generalization of Theorem 1 is as follows.

Theorem 4. *If*

(30) $$\|f\| \leq 1, \|f^{(n)}\| \leq 2^{n-1} n! \textit{ in the interval } 0 \leq x \leq x_{n+k}, \textit{ then}$$

(31) $$|f^{(\nu)}(0)| \leq |S_{n,k}^{(\nu)}(0)|, \quad (\nu = 1, \ldots, n-1).$$

Here the interval $[0, x_{n+k}]$ *can not be replaced by any shorter interval* $[0, \theta x_{n+k}]$ $(0 < \theta < 1)$, *nor can any of the constants on the right side of* (31) *be replaced by smaller constants.*

What follows applies only if $n \geq 3$*: If in one of the inequalities* (31) *we have the equality sign, then we have equality in all of them, and this happens if and only if*

(32) $$f(x) = \pm S_{n,k}(x) \textit{ in } [0, x_{n+k}].$$

Theorem 4 is a key result: Just as Theorem 1 implied Matorin's Theorem III, so Theorem 4 now yields

Theorem 5. *If*

(33) $$\|f\| = 1, \|f^{(n)}\| = 2^{n-1} n! \textit{ in } [0, \infty),$$

which means that $f(x) \in \mathcal{F}_n$, *then*

(34) $$\|f^{(\nu)}\| \leq |S_{n,k}^{(\nu)}(0)|, \quad (\nu = 1, \ldots, n-1).$$

The bounds in (34) are easily computed from (26) and (27) and we find that

(35) $$|S_{n,k}^{(\nu)}(0)| = ab^{\nu} |T_{n,k}^{(\nu)}(1)| = \rho_{n,k}^{-1+(\nu/n)} \cdot |T_{n,k}^{(\nu)}(1)|,$$

and in particular, for $k = 0$ when $\rho_{n,0} = 1$, we find Matorin's bounds

(36) $$|S_{n,0}^{(\nu)}(0)| = |T_{n,0}^{(\nu)}(1)| = T_n^{(\nu)}(1).$$

However, now we are in a position to improve on Matorin's bounds indefinitely, by letting k increase, and to obtain the Landau constants in the limit.

We dispose first of the simple cases when $n = 2$ or $n = 3$. In each of these cases the $S_{n,k}(x) (k = 0, 1, 2, \ldots)$ are successively extensions of each other. More precisely

(37) $$S_{n,k+1}(x) = s_{n,k}(x) \text{ in } [0, x_{n+k}^{(k)}] \text{ if } n = 2 \text{ or } 3,$$

where $x_{n+k}^{(k)}$ is the $(n+k)^{th}$ extreme point of $S_{n,k}(x)$. This can be nicely seen in Figure 1 if we interpret the two curves as the graphs of $S_{2,4}(x)$ and $S_{3,4}(x)$, respectively: In passing from $S_{2,4}$ to $S_{2,5}$, the last zero of $S_{2,4}$ simply becomes the last knot of $S_{2,5}$. In passing from $S_{3,4}$ to $S_{3,5}$, the last maximum of $S_{3,4}$ becomes the last knot of $S_{3,5}$.

These results show that

$$(38)\qquad \begin{aligned} &L_2' = |S_{2,k}'(0)|, \quad (k \geq 0)\\ &L_3' = |S_{3,k}'(0)|, \quad L_3'' = |S_{3,k}''(0)|, \ (k \geq 0), \end{aligned}$$

are the values of the corresponding Landau constants. We also see, for $k = 0$, why Matorin's bounds gave their exact values.

Matters are different if $n \geq 4$. Here we have

Theorem 6. *If*

$$(39)\qquad n \geq 4, \quad 0 < \nu < n,$$

then

(40) *the sequence* $|S_{n,k}^{(\nu)}(0)|$ $(k = 0, 1, \ldots)$ *is strictly decreasing,*

and the value of the Landau constant $L_n^{(\nu)}$ *of Problem 1 is given by*

$$(41)\qquad L_n^{(\nu)} = \lim_{k\to\infty} |S_{n,k}^{(\nu)}(0)|.$$

As one would expect, the functions $S_{n,k}(x)$ converge, as $k \to \infty$, to the extremizing function.

Theorem 7. *The limit relation*

$$(42)\qquad \lim_{k\to\infty} S_{n,k}(x) = \mathcal{E}_n^+(x)$$

holds uniformly in every finite interval $[0, A]$. *The limit function* $\mathcal{E}_n^+(x)$ *is a perfect spline of degree* n *having infinitely many knots* η_i *such that*

$$(43)\qquad 0 < \eta_1 < \eta_2 < \ldots, \ \eta_i \to \infty.$$

Moreover

$$(44)\qquad \mathcal{E}_n^+(x) \in \mathcal{F}_n$$

and we may express the Landau constants (41) *by*

$$(45)\qquad L_n^{(\nu)} = |\mathcal{E}_n^{+(\nu)}(0)|, \quad (0 < \nu < n).$$

Theorem 8. *We now assume that*

$$(46)\qquad n \geq 3.$$

If $f(x) \in \mathcal{F}_n$ *and if the equality*

$$(47)\qquad \|f^{(\nu)}\| = L_n^{(\nu)}$$

holds for one value of $\nu(0 < \nu < n)$, *then it holds for all such* ν, *and this happens if and only if*

$$(48) \qquad f(x) = \pm \mathcal{E}_n^+(x) \text{ in } [0, \infty).$$

We propose to call the functions $\mathcal{E}_n^+(x)$ *the one-sided Euler splines.*

Our expressions (45) for the Landau constants being far from explicit, Stečkin's asymptotic estimates retain their full importance (see [7]). Possibly these estimates can now be improved on the basis of some of our results.

4. A kinematic interpretation. In Theorem 8 we may evidently relax our condition $f(x) \in \mathcal{F}_n$, hence (33), and replace it by the inequalities

$$(49) \qquad \|f\| \leq 1, \ \|f^{(n)}\| \leq 2^{n-1} n! \text{ in } [0, \infty).$$

Following Newton, we think of the derivatives as "fluxions" by interpreting the variable x as *time* in the interval $[0, \infty)$. We may then think of the function $f(x)$, satisfying (49), as representing the motion of a point $f = f(x)$. The point f is to move in the interval $-1 \leq f \leq 1$ such as to have continuous ν-velocities $f^{(\nu)}(x)(0 < \nu < n)$, while its n-velocity $f^{(n)}(x)$ should never exceed $2^{n-1}n!$ in absolute value. The problem then asks for the largest ν-velocity $\|f^{(\nu)}\|$ consistent with these restrictions. Our results (Theorems 7 and 8) show that this largest ν-velocity is the *entry-ν-velocity* $L_n^{(\nu)}$. Moreover, that this maximal entry-ν-velocity $L_n^{(\nu)}$ can be realized only by the two motions (48): The point then moves at all times with the maximal n-velocity $\pm 2^{n-1}n!$ that switches sign at the times

$$x = \eta_1, \eta_2, \ldots .$$

We owe to Walter Rudin the following interesting remark made orally during a seminar lecture on the subject by one of the authors. Rudin observed that all our results, and their proofs, very likely also hold in the complex field. As a matter of fact they do, and without any additional complications of any kind. This means that $f(x)$ is now assumed to be complex-valued throughout our discussion. Then all results remain valid with the difference that now the extremizing functions $f(x)$ in Theorem 8 are not only the two functions (48), but the entire family

$$(50) \qquad f(x) = \alpha \mathcal{E}_n^+(x), \alpha \text{ constant}, \ |\alpha| = 1.$$

Returning to our kinematic interpretation, we may now interpret the complex-valued $f(x)$ as the motion of a point in the unit circle

$$(51) \qquad |f| \leq 1,$$

during the time interval $[0, \infty)$ during which its vectorial n-velocity should never exceed $2^{n-1}n!$ in magnitude. Our results show that the motion of largest entry-ν-velocity $L_n^{(\nu)}$ corresponds to the motion (50) on a diameter $\alpha u(-1 \leq u \leq 1)$ of the circle (51). It seems surprising that the moving point can make no use of its enlarged domain of variability to further increase its entry-ν-velocity.

Very likely the same result holds if we should replace the circle (51) by any plane closed convex domain having a diameter $AB = 2$, the term "diameter" being taken in the distance-theoretic sense. This is evidently so if the closed circle U, of diameter AB, should contain the given domain.

Mentioning the above results concerning the unit circle to W. Walter, he asked if they could be generalized to the unit sphere of the higher dimensional Euclidean space E_m. In fact our Theorems 4 and 5 immediately generalize to vector functions $f(x)$ from $[0,\infty)$ into E_m. The only difference is that the extremizing functions (32) are obtained by replacing the factor ± 1 by $\alpha \in E_m$ such that its magnitude $|\alpha| = 1$. The answer to Walter's question is therefore affirmative.

There is, however, a very simple geometric argument for extending our results from E_1 to E_m. For simplicity let $m = 2$. Let U be again the unit circle in the plane and let us assume that we have a motion in U such that

$$|f| \leq 1, \ |f^{(n)}| \leq 2^{n-1} n!. \tag{52}$$

Furthermore let us assume that the entry-velocity v_0 is such that

$$|v_0| \geq L'_n = |\mathcal{E}'_n(0)|. \tag{53}$$

We claim that: 1. The vector v_0 must be orthogonal to the circle; 2. That in (53) we must have equality.

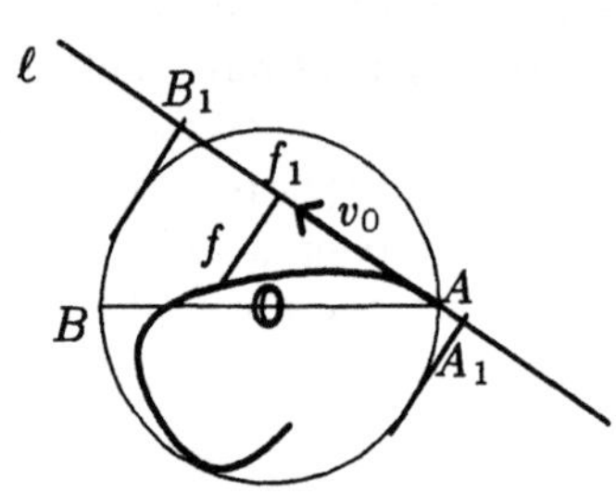

Figure 2

Indeed, let us assume that v_0 does not point towards the center 0 and let ℓ be the line carrying v_0. Let $f_1(x)$ be the orthogonal projection of $f(x)$ on ℓ. Now $f_1(x)$ is a motion on the segment A_1B_1 of length 2 whose velocities satisfy all our conditions. However, since v_0 is not the entry-1-velocity at A_1 we have a contradiction, because A is in the interior of A_1B_1. This proves both statements and v_0 lies on the diameter AB.

Let

$$f(x) = f_1(x) + ih(x), \quad (0 \leq x < \infty) \tag{54}$$

represent our motion. But then we must have

$$f_1(x) = \mathcal{E}_n^+(x) \ [0,\infty). \tag{55}$$

However, we know that $|\mathcal{E}_n^{+(n)}(x)| = 2^{n-1}n!$ for all $x \neq \eta_i$. This and the second inequality (52) imply that $h^{(n)}(x) = 0$ almost everywhere, hence $h(x)$ can only be a polynomial of degree $\leq n-1$. On the other hand, from Figure 2 it is clear that $h(x) = 0$ for such x where (55) reaches its extreme values ± 1. These are infinitely many distinct times and therefore $h(x) = 0$ for all x. This proof applies as well for the other ν-velocities and all dimensions m.

5. Proofs of a few key results are sketched. Complete proofs of our theorems will appear elsewhere. Here we sketch the main ideas concerning a few of the theorems.

We study perfect splines of the form (17) and observe that they can have at most $n + k - 1$ distinct relative extrema. We say that $S(x)$ is an *oscillatory perfect spline* (O.P.S.) provided that it has exactly this maximal number of $n + k - 1$ relative extrema. Let

$$e_1, e_2, \ldots, e_{n+k-1}$$

be the critical values of $S(x)$, i.e. the values of $S(x)$ at its extreme points in their natural order of increasing x. It is shown that the O.P.S. $S(x)$ is uniquely defined, up to horizontal translations, if these critical values e_i are preassigned at will. Of course, we must have

$$e_1 < e_2, e_2 > e_3, e_3 < e_4, \ldots \text{ if } n \text{ is even,}$$

and the opposite inequalities if n is odd.

The structure of O.P.S. of degree $n = 2$ is especially simple and proofs go by induction starting with $n = 2$ when the result concerning preassigned extreme values is perfectly trivial.

Let us assume that we already know that for a certain value of n the extreme values can be preassigned. A lemma then shows that these extreme values e_i can be so chosen *with alternating signs*, that the *areas* $A_1, A_2, \ldots, A_{n+k-1}$ of the arches of $S(x)$ (Figure 3) have *preassigned positive values.* This lemma is established by a topological argument equivalent to a version of the fixed-point theorem. By one integration we then obtain the theorem concerning preassigned extreme values for the next degree $n + 1$.

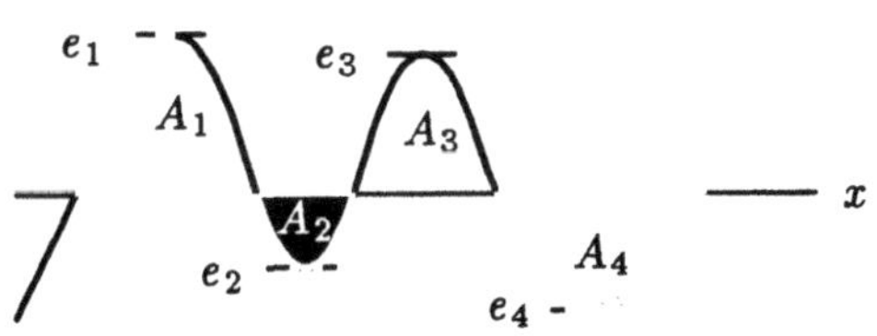

Figure 3

Another key result is Theorem 4 (of which Theorem 1 is a special case corresponding to $k = 0$). Let us sketch the proof of the inequality (31) for $\nu = 1$. We also assume that $n \geq 4$, as the case $n = 3$ requires a special discussion. We summarize the relevant properties of the function $S(x) = S_{n,k}(x)$ defined by (27): Its extreme points are given by (28) and let $\xi_1, \ldots, \xi_k$ denote its knots. Then

$$S(x_i) = (-1)^{n+i-1} \quad (i = 1, \ldots, n + k), \tag{56}$$

$$S^{(n)}(x) = (-1)^j 2^{n-1} n! \text{ in } \xi_j < x < \xi_{j+1} \quad (j = 0, \ldots, k) \tag{57}$$

where we write $\xi_0 = 0 = x_1$, and $\xi_{k+1} = x_{n+k}$.

We now consider the family

$$\mathcal{S} = \{s(x)\}$$

of spline functions $s(x)$ of degree $n - 1$ having as simple knots the k knots $\xi_1, \ldots, \xi_k$ of $S_{n,k}(x)$. This family depends on $n + k$ parameters and one would expect that there exists a unique $s(x) \in \mathcal{S}$ which is a solution of the interpolation problem

$$s(x_i) = y_i \quad (i = 1, \ldots, n + k). \tag{58}$$

This is indeed the case: In the inequalities (23) we have $\mu = [n/2] \geq 2$ and (23) imply a fortiori the inequalities

$$x_1 < \xi_1 < x_{n+1},$$
$$x_2 < \xi_2 < x_{n+2},$$
$$\vdots$$
$$x_k < \xi_k < x_{n+k}. \tag{59}$$

These inequalities (59) are the necessary and sufficient conditions for the unique solution $s(x) \in S$ of the problem (58), the y_i being preassigned (see [5, Theorem 2, 258]).

Let

$$f(x) = \sum_{1}^{n+k} f(x_\nu)L_\nu(x) + Rf \tag{60}$$

be the corresponding spline interpolation formula, the $L_\nu(x)$ being its fundamental functions. By construction we see that the interpolation formula (60) is exact if $f(x) \in S$. This means that

$$Rf = 0 \text{ if } f(x) \in S \tag{61}$$

and in particular

$$Rf = 0 \text{ if } f(x) \in \pi_{n-1}. \tag{62}$$

From (60) we obtain the approximate differentiation formula

$$f'(0) = \sum_{1}^{n+k} f(x_\nu)L'_\nu(0) + Rf \tag{63}$$

whose remainder Rf also enjoys the properties expressed by (61) and (62). Setting $A_\nu = L'_\nu(0)$ we may apply Peano's theorem and write (63) in the exact form

$$f'(0) = \sum_{1}^{n+k} A_\nu f(x_\nu) + \int_0^{x_{n+k}} K(x)f^{(n)}(x)dx. \tag{64}$$

Here $K(x)$ is a certain spline function of degree $n-1$ having as knots the nodes $x_1, x_2, \ldots, x_{n+k}$ of our formula.

At this point we must start omitting proofs and simply state the following properties of the coefficients and kernel of the formula (64):

$$(-1)^\nu A_\nu > 0 \quad (\nu = 1, \ldots, n+k); \tag{65}$$

$$(-1)^{n+j-1}K(x) > 0 \text{ in } \xi_j < x < \xi_{j+1} \quad (j = 0, \ldots, k) \tag{66}$$

where, as in (57), $\xi_0 = 0, \xi_{k+1} = x_{n+k}$. That $K(\xi_j) = 0$ follows very simply from (61). However, the precise sign properties (66) require the variation diminishing properties of spline functions if represented in terms of the so-called B-splines.

If we substitute $S(x) = S_{n,k}(x)$ into (64) and use the properties (56), (57), (65), and (66) we obtain the relation

$$|S'(0)| = \sum_{1}^{n+k} |A_\nu| + 2^{n-1}n! \int_0^{x_{n+k}} |K(x)|dx. \tag{67}$$

For this function $f(x) = S(x)$ the formula (64) is "taut". If $f(x)$ is any function satisfying the conditions (30) then (64) shows that

$$|f'(0)| \leq \sum_{1}^{n+k} |A_\nu| + 2^{n-1}n! \int_0^{x_{n+k}} |K(x)|dx = |S'(0)| \tag{68}$$

which proves (31) for $\nu = 1$. Moreover, from (30) we see that we can have equality in (68) only if $f(x)$, or perhaps $-f(x)$, satisfies the conditions

$$f(x_i) = (-1)^{n+i-1} \quad (i = 1, \dots, n+k), \tag{69}$$

$$f^{(n)}(x) = (-1)^j 2^{n-1}n! \text{ in } \xi_j < x < \xi_{j+1} \quad (j = 0, \dots, k). \tag{70}$$

Now (70) shows that $f(x)$ is not only a *spline* function of degree n, but even a *perfect* spline with the same knots and highest degree terms as $S(x)$. Therefore $f(x)$ could differ from $S(x)$ only by a polynomial of degree $n-1$ which, however, must vanish in view of the relations (69) and (56). This establishes the relation (32). Similarly we establish (31), for any ν, by working with an approximate spline differentiation formula for $f^{(\nu)}(0)$.

6. Numerical upper bounds for the Landau constants for $n = 4, 5$, and 6. We conclude the paper with numerical upper bounds for the constants $L_4^{(\nu)}, L_5^{(\nu)}$, and $L_6^{(\nu)}$. Matorin's Theorem III shows that

$$\begin{aligned}
&L_4' \leq 16,\ L_4'' \leq 80,\ L_4''' \leq 192\\
&L_5' \leq 25,\ L_5'' \leq 200,\ L_5''' \leq 840,\ L_5^{(4)} \leq 1920\\
&L_6' \leq 36,\ L_6'' \leq 420,\ L_6''' \leq 2688,\ L_6^{(4)} \leq 10368,\ L_6^{(5)} \leq 23040.
\end{aligned} \tag{71}$$

Let us improve them by using our Theorem 5 for $n = 4, 5$, and 6.

Dennis Kuba, of the Mathematics Research Center Computing Staff, using linear programming methods combined with Newton's method for finding the knots, has determined approximations to the Chebyshev-Euler splines

$$T_{n,k}(x) \text{ for } n = 4, 5, 6 \text{ and } k = 1, 2, 3, \dots, 10. \tag{72}$$

The accuracy of these approximations was a little better than 5 significant figures. These were used to compute the values of

$$|S_{n,k}^{(\nu)}(0)| \tag{73}$$

by our relation (35). By Theorem 6, the Landau constants $L_n^{(\nu)}$ are the limits of (73) as $k \to \infty$.

The results obtained are summarized in the following tables.

k	$\lvert S'_{4,k}(0)\rvert$	$\lvert S''_{4,k}(0)\rvert$	$\lvert S'''_{4,k}(0)\rvert$
0	16	80	192
1	15.9653	79.6937	191.291
2	15.9632	79.6784	191.258

k	$\lvert S'_{5,k}(0)\rvert$	$\lvert S''_{5,k}(0)\rvert$	$\lvert S'''_{5,k}(0)\rvert$	$\lvert S^{(4)}_{5,k}(0)\rvert$
0	25	200	840	1920
1	24.8796	198.212	830.909	1903.52
2	24.8680	198.043	830.057	1901.99
3	24.8673	198.028	829.978	1901.84
4	24.8666	198.024	829.963	1901.82

k	$\lvert S'_{6,k}(0)\rvert$	$\lvert S''_{6,k}(0)\rvert$	$\lvert S'''_{6,k}(0)\rvert$	$\lvert S^{(4)}_{6,k}(0)\rvert$	$\lvert S^{(5)}_{6,k}(0)\rvert$
0	36	420	2688	10368	23040
1	35.7493	414.443	2641.33	10184.1	22749.7
2	35.7118	413.628	2634.57	10157.7	22708.7
3	35.7069	413.507	2633.54	10153.6	22702.3
4	35.7063	413.485	2633.34	10152.9	22701.1

We recall that the convergence in the limit relation (41) was "instantaneous" for $n = 2$ and $n = 3$, since all terms of the sequence had the same value by (38). As a matter of fact, the rate of convergence is still very fast for $n = 4$. This is why, for $n = 4$, the available accuracy allowed only the use of the first two approximations (72). In any case, the last entries in our tables describe the corresponding Landau constants to five significant figures. At least this is the naive computer's point of view. Actually, in the absence of equally accurate *lower* bounds for the Landau constants, we can only conclude with certainty that the last entries in our tables are *upper bounds* for there constants.

REFERENCES

1. G. Glaeser, *Prolongement extrémal de fonctions différentiables*, Publications de la section de Mathématiques de la Faculte des Sciences de Rennes, 1967.
2. A. N. Kolmogorov, On inequalities between the upper bounds of the successive derivatives of an arbitrary function on an infinite interval, *Amer. Math. Soc. Transl., Series 1*, **2** (1962), 233-243.
3. E. Landau, Einige Ungleichungen für zweimal differenzierbare Funktionen, *Proc. London Math. Soc.* **13** (1913), 43–49.
4. A. P. Matorin, On inequalities between the maxima of the absolute values of a function and its derivatives on a half-line, *Amer. Math. Soc. Transl., Series 2*, **8** (1958), 13–17.
5. I. J. Schoenberg and Anne Whitney, On Pólya frequency functions III. The positivity of translation determinants with an application to the interpolation problem by spline curves, *Trans. Amer. Math. Soc.* **74** (1953), 246–259.
6. I. J. Schoenberg and Zvi Ziegler, On cardinal monosplines of least L_∞-norm on the real axis, *J. Analyse Math.* **23** (1970), 409–436.
7. S. B. Stečkin, On inequalities between the upper bounds of derivatives of functions on the halfline. (In Russian), *Mat. Zametki* **1** (1967), 665–674.

Reprinted from
Israel Journal of Mathematics
Vol. 10, No. 3, 1971

THE PERFECT *B*-SPLINES AND A TIME-OPTIMAL CONTROL PROBLEM

BY
I. J. SCHOENBERG

ABSTRACT

Let $x_\nu = \cos(\pi\nu/n)$ $(\nu = 0, 1, \dots, n)$. It is shown that the *B*-spline $M(x) = M(x; x_0, x_1, \dots, x_n)$ is such that $M_n^{(n)}(x)$ has a constant absolute value $(= 2^{n-2}(n-1)!)$ in $[-1, 1]$. Its integral $f_0(x) = \int_{-1}^{x} M(t)dt$ is shown to have an optimal property that allows to solve *explicitly* a certain time-optimal control problem.

Introduction

In [3] G. Glaeser obtains most interesting results based on the following definitions. A real-valued function $f(x)$ defined in the interval $[a, b]$ is said to be a *spline function* of degree n, provided that for an appropriate division $a = x_0 < x_1 < \cdots < x_k < x_{k+1} = b$ the following conditions are satisfied

1. $f(x) \in \pi_n$ in (x_i, x_{i+1}) $(i = 0, \cdots, k)$,
2. $f(x) \in C^{n-1}[a, b]$.

Moreover, $f(x)$ is said to be a *perfect spline*, provided that

3. $|f^{(n)}(x)| = \text{constant}$ if $a \leqq x \leqq b$, $x \neq x_1, x_2, \cdots, x_k$.

Equivalently, we can say that the $k-1$ polynomial components of $f(x)$ are of the form

$$f(x) = (-1)^i C x^n + \text{lower degree terms, in } (x_i, x_{i+1}), \quad (i = 0, \cdots, k),$$

where the constant C does not depend on i.

THEOREM I. (G. Glaeser). *If the* $2n$ *reals*

$$(1) \qquad y_0^{(\nu)}, y_1^{(\nu)}, \qquad (\nu = 0, 1, \cdots, n-1)$$

are preassigned, then the 2-point Hermite interpolation problem

Received February 23, 1971

(2) $$f^{(\nu)}(a) = y_0^{(\nu)},\ f^{(\nu)}(b) = y_1^{(\nu)}, \qquad (\nu = 0, \cdots, n-1)$$

has a unique solution $f_0(x)$ that is a perfect spline of degree n having in (a,b) k knots $x_1, \cdots, x_k$ such that

(3) $$k \leqq n - 1.$$

THEOREM II. (G. Glaeser). *The perfect spline interpolant $f_0(x)$ of Theorem I is the unique function that minimizes the sup-norm $\|f^{(n)}\|_\infty$ among all functions $f(x)$ such that $f^{(\nu)}(x)$ $(\nu = 0, \cdots, n-1)$ are absolutely continuous in $[a,b]$ and at the same time solve the interpolation problem* (2).

These results appeared first in Volume I of [4]. In Volume II of [4] there is a paper by R. Louboutin entitled *Sur une "bonne" partition de l'unité* [4, vol. II, pp. D1-D10], and concerned with the special case of the problem (2) when $y_0^{(\nu)} = 0$ for all ν, and $y_1^{(0)} = 1$, $y_1^{(\nu)} = 0$ $(\nu = 1, \cdots, n-1)$. Since we may choose $[a,b] = [-1,1]$ without loss of generality, Louboutin's results refer to the special interpolation problem

(4) $$\begin{aligned} &f^{(\nu)}(-1) = 0, &&(\nu = 0, \cdots, n-1) \\ &f(1) = 1, f^{(\nu)}(1) = 0, &&(\nu = 1, \cdots, n-1). \end{aligned}$$

For this problem Louboutin obtains an *explicit expression* for the optimal solution $f_0(x)$ by establishing the following theorem.

THEOREM 3. (R. Louboutin). *Let*

(5) $$x_\nu = -\cos\frac{\pi\nu}{n} \qquad (\nu = 1, \cdots, n-1).$$

The optimal perfect spline interpolant $f_0(x)$ of problem (4) *has the $n-1$ knots* (5) *and*

$$|f_0^{(n)}(x)| = \text{constant} = \|f_0^{(n)}\| = 2^{n-2}(n-1)!.$$

Moreover, $f_0(x)$ may be expressed by the formula

$$f_0(x) = (-1)^{n-1}2^{n-2}\int_{-1}^{x}(x-t)^{n-1}\operatorname{sgn} T_n'(t)\,dt, \qquad (-1 \leqq x \leqq 1).$$

where $T_n(x)$ is the Chebyshev polynomial.

Since the mimeographed edition of [4] was only privately circulated, the main purpose of the present paper is to make Louboutin's results better known. We derive them below independently of Glaeser's theorem. Theorems 1 and 2 and Corollaries 1 and 2 below are Louboutin's results in somewhat different notations.

In §3 we apply them to a time-optimal control problem whose solution is expressed in terms of $f_0(x)$ (Theorem 3). In §4 we describe the behavior of $f_0(x)$ as $n \to \infty$ (Theorem 4).

1. The perfect B-splines

For convenience we let $[a,b] = [-1,1]$ and write the knots x_i in decreasing order thus

$$-1 = x_n < x_{n-1} < \cdots < x_1 < x_0 = +1. \tag{1.1}$$

The corresponding divided difference of order n is

$$f(x_0, x_1 \cdots, x_n) = \sum_0^n \frac{f(x_\nu)}{\omega'(x_\nu)} \tag{1.2}$$

where $\omega(x) = (x - x_0)\ (x - x_1) \cdots (x - x_n)$. It is well known that for special divisions (1.1) the divided difference (1.2) assumes simple forms. For instance, if the division (1.1) is equidistant then we know that

$$f(x_0, \cdots, x_n) = \frac{n^n}{2^n n!} \cdot \sum_0^n\ (-1)^\nu \binom{n}{\nu} f(x_\nu).$$

It seems less known, if at all, that an even simpler form is assumed if $x_0, x_1, \cdots, x_n$ are the abscissae where the Chebyshev polynomial $T_n(x)$ assumes its extreme values ± 1. This we state as

LEMMA 1. *If*

$$x_\nu = \cos \frac{\nu\pi}{n} \qquad (\nu = 0, 1, \cdots, n) \tag{1.3}$$

then

$$\begin{aligned} f(x_0, x_1, \cdots, x_n) = \frac{2^{n-2}}{n} (f(x_0) - 2f(x_1) + 2f(x_2) - \\ \cdots + (-1)^{n-1} 2f(x_{n-1}) + (-1)^n f(x_n)). \end{aligned} \tag{1.4}$$

PROOF. The points (1.3) are ± 1 and the $n - 1$ *interior* extreme points of $T_n(x)$, hence the zeros of $T_n'(x)$. Observing that $T_n'(x) = 2^{n-1} n x^{n-1} +$ lower degree terms, we see that

$$\omega(x) = \frac{1}{2^{n-1} n} (x^2 - 1) T_n'(x). \tag{1.5}$$

The values of $\omega'(x_\nu)$ needed in (1.2), may now be easily obtained. Remembering that $T_n(x)$ satisfies the differential equation

$$(1 - x^2)\,T_n''(x) - xT_n'(x) + n^2T_n(x) = 0 \tag{1.6}$$

we obtain from (1.5) and (1.6)

$$\begin{aligned} -\,2^{n-1}n\omega'(x) &= \frac{d}{dx}((1-x^2)T_n'(x)) = (1-x^2)\,T_n''(x) \\ &\quad -\,2x\,T_n'(x) = -\,xT_n'(x) - n^2T_n(x). \end{aligned}$$

If $x = x_\nu$ $(\nu = 1,\cdots,n-1)$ then $T_n'(x_\nu) = 0$ and therefore

$$\omega'(x_\nu) = \frac{n}{2^{n-1}}T_n(x_\nu) = (-1)^\nu\frac{n}{2^{n-1}} \qquad (\nu = 1,\cdots,n-1). \tag{1.7}$$

If $x = 1$, then $T_n = 1$ and $T_n' = n^2$, and therefore

$$\omega'(1) = \frac{1}{2^{n-1}n}\cdot 2n^2 = \frac{n}{2^{n-2}}, \text{ and similarly } \omega'(-1) = \frac{(-1)^n n}{2^{n-2}}. \tag{1.8}$$

Substituting the values (1.7), (1.8), into (1.2) we obtain (1.4).

It is well known that for the division (1.1) we may express the divided difference of a function $f \in C^n$ in the form

$$f(x_0, x_1, \cdots, x_n) = \frac{1}{n!}\int_{-1}^{1} M(x)f^{(n)}(x)\,dx \tag{1.9}$$

where $M(x)$ is a certain special spline function called a B-spline. This fundamental relation has been known and was used for many years by a number of analysts in particular by D. V. Ionescu (see [5, Chap. III] and [6]). The linear functional $Rf = f(x_0,\cdots,x_n)$ has the property that $Rf = 0$ if $f \in \pi_{n-1}$. We may therefore apply Peano's theorem (see [2, §3.7]) and conclude that we can write

$$\frac{1}{n!}M(x) = R_t\,\frac{(t-x)_+^{n-1}}{(n-1)!}. \tag{1.10}$$

Here we use the function $u_+ = u$ if $u \geqq 0$ and $u_+ = 0$ if $u < 0$, while the subscript "t" means that the operation of differencing is to be performed on the variable t, leaving x as a parameter. By (1.2) we obtain that

$$M(x) = \sum_0^n \frac{n(x_\nu - x)_+^{n-1}}{\omega'(x_\nu)}. \tag{1.11}$$

If we write

$$M(x;t) = n(t-x)_+^{n-1} \tag{1.12}$$

we may also express (1.11) symbolically in Steffensen's notation in the form

$$M(x) = M(x; x_0, x_1, \cdots, x_n). \tag{1.13}$$

Observe that (1.11) defines $M(x)$ for all real x and that it has the following properties.

1° $M(x)$ is piecewise polynomial in the intervals $(-\infty, x_n), (x_n, x_{n-1}), \cdots, (x_1, x_0), (x_0, \infty)$, all these polynomials being of degree $\leqq n-1$.

2° $M(x) \in C^{n-2}$

3° $M(x) = 0$ if $x < x_n$ or if $x > x_0$.

From the properties 2° and 3° it follows that

$$M^{(\nu)}(\pm 1) = 0, \qquad (\nu = 0, 1, \cdots, n-2). \tag{1.14}$$

If we substitute $f(x) = x^n$ into (1.9) we see that

$$\int_{-\infty}^{\infty} M(x)\,dx = 1. \tag{1.15}$$

A function $M(x)$ with the properties 1° and 2° is called a *spline function* of degree $n-1$ having the *knots* $x_0, x_1, \cdots, x_n$. The further property 3° together with the fact that we have precisely $n+1$ knots identifies $M(x)$ as a so-called *B-spline* (see [1, §4]). Finally (1.15) expresses the usual normalization. Concerning the descriptive properties of B-splines see [1, §1]. B-splines were the original source of the so-called Polya frequency functions as shown in [1, Part II].

In view of Lemma 1 our attention is naturally drawn to the very particular B-spline

$$M(x) = M(x; x_0, \cdots, x_n), \text{ where } x_\nu = \cos\frac{\nu\pi}{n} \qquad (\nu = 0, 1, \cdots, n). \tag{1.16}$$

Let us study it a little closer.

The relation (1.11) may alternatively be also written in the form

$$M(x) = (-1)^n \sum_0^n \frac{n(x - x_\nu)_+^{n-1}}{\omega'(x_\nu)}, \tag{1.17}$$

in view of the identity $(x_\nu - x)_+^{n-1} = (x_\nu - x)^{n-1} + (-1)^n (x - x_\nu)_+^{n-1}$ and the fact that

$$\sum_0^n \frac{n(x_\nu - x)^{n-1}}{\omega'(x_\nu)} = 0 \qquad \text{for all real } x,$$

the left side being a divided difference of order n of an element of π_{n-1}. Differentiating (1.17) we obtain

$$M^{(n-1)}(x) = (-1)^n n! \sum_0^n \frac{(x - x_\nu)_+^0}{\omega'(x_\nu)}$$

and using (1.7) and (1.8) we get

$$M^{(n-1)}(x) = 2^{n-2}(n-1))!\{(x-x_n)_+^0 - 2(x-x_{n-1})_+^0 + \cdots \tag{1.18}$$
$$+ (-1)^{n-1}2(x-x_1)_+^0 + (-1)^n(x-x_0)_+^0\}.$$

Therefore

$$M^{(n-1)}(x) = \begin{cases} 2^{n-2}(n-1)! & \text{if } x_n < x < x_{n-1} \\ -2^{n-2}(n-1)! & \text{if } x_{n-1} < x < x_{n-2} \\ \vdots & \\ (-1)^{n-1}2^{n-2}(n-1)! & \text{if } x_1 < x < x_0. \end{cases} \tag{1.19}$$

We have therefore established the following

THEOREM 1. *The B-spline*

$$M(x) = M(x; x_0, x_1, \cdots, x_n) \tag{1.20}$$

of degree $n-1$ *based on the knots*

$$x_\nu = \cos\frac{\pi\nu}{n}, \qquad (\nu = 0, 1, \cdots, n) \tag{1.21}$$

is a perfect B-spline *such that*

$$|M^{(n-1)}(x)| = 2^{n-2}(n-1)! \text{ if } -1 < x < 1, \ x \neq x_1, \cdots, x_{n-1}. \tag{1.22}$$

From (1.14) and (1.15) we immediately obtain the following

COROLLARY 1. *If we define*

$$f_0(x) = \int_{-1}^{x} M(t)\,dt \quad \text{in } -1 \leqq x \leqq 1, \tag{1.23}$$

Then $f_0(x)$ *is a perfect spline function in* $[-1,1]$ *with knots* $x_1, x_2, \cdots, x_{n-1}$, *such that*

$$|f_0^{(n)}(x)| = 2^{n-2}(n-1)! \quad \text{in } (-1,1), \quad x \neq x_i. \tag{1.25}$$

and satisfying the boundary conditions

$$f_0^{(\nu)}(-1) = 0, \qquad (\nu = 0, 1, \cdots, n-1),$$
$$f_0(1) = 1, \quad f^{(\nu)}(1) = 0, \qquad (\nu = 1, \cdots, n-1),$$

From the properties of its derivative $M(x)$ we conclude that $f_0(x)$ is *strictly increasing* in $[-1,1]$ and that it is *convex* in $[-1,0]$ and *concave* in $[0,1]$.

2. An optimal property of perfect *B*-splines

So far we have established only the existence part of Glaeser's Theorem I for the special interpolation problem (4), by exhibiting its solution (1.23) explicitly (that $a = -1$ and $b = 1$ are evidently insignificant restrictions). The *unicity* of the solution will be the subject of Corollary 2 below.

THEOREM 2. *The perfect spline function $f_0(x)$ defined by (1.23), has the following optimal property: Within the class F_n of functions $f(x)$ defined in $[-1,1]$ and satisfying the conditions*

$$f^{(\nu)}(x)\ (\nu = 0,\cdots,n-1) \quad \text{are absolutely continuous,} \tag{2.1}$$

and

$$\begin{aligned} &f^{(\nu)}(-1) = 0, && (\nu = 0,1,\cdots,n-1),\\ &f(1) = 1, \quad f^{(\nu)}(1) = 0, && (\nu = 1,\cdots,n-1), \end{aligned} \tag{2.2}$$

the function $f_0(x)$ is the unique function that minimizes the norm $\|f^{(n)}\|_\infty$ giving it its minimal value

$$\|f_0^{(n)}\|_\infty = 2^{n-2}(n-1)! \tag{2.3}$$

REMARK. For $n = 1$ Theorem 2 reduces to the statement: If $f(x)$ is absolutely continuous in $[-1,1]$, while $f(-1) = 0$, $f(1) = 1$, then $\|f'\|_\infty \geqq \frac{1}{2}$ with equality iff $f(x) = (x+1)/2$. This is very easily shown directly. However, already for $n = 2$ Theorem 2 is far from trivial.

PROOF OF THEOREM 2. Let $f(x) \in F_n$ and assuming that

$$\|f^{(n)}\| \leqq \|f_0^{(n)}\| = 2^{n-2}(n-1)!. \tag{2.4}$$

let us show that

$$f(x) = f_0(x) \quad \text{in } [-1,1]. \tag{2.5}$$

From Theorem 1 we know that

$$f_0^{(n)}(x) = M^{(n-1)}(x) \tag{2.6}$$

is a perfect stepfunction in $[-1,1]$ which is discontinuous at the points

$$x_\nu = \cos\frac{\pi\nu}{n}, \qquad (\nu = 1,\cdots,n-1). \tag{2.7}$$

We also observe that the polynomial $T_n'(x)$ has precisely the zeros (2.7). By integrations by parts, and using (2.2), we obtain

$$\begin{aligned} \int_{-1}^{1} T'f^{(n)}dx &= \int_{-1}^{1} T'df^{(n-1)} = -\int_{-1}^{1} T''f^{(n-1)}dx = \cdots \\ &= (-1)^{n-1}\int_{-1}^{1} T_n^{(n)}f'dx \\ &= (-1)^{n-1}2^{n-1}n!\int_{-1}^{1} f'dx = (-1)^{n-1}2^{n-1}n! \end{aligned}$$

whence

$$\int_{-1}^{1}(-1)^{n-1}T_n'(x)f^{(n)}(x)\,dx = 2^{n-1}n!. \tag{2.8}$$

Since $f_0(x) \in F_n$, we may apply this to $f_0(x)$ and remembering (2.6) we obtain

$$\int_{-1}^{1}(-1)^{n-1}T_n'(x)M^{(n-1)}(x)\,dx = 2^{n-1}n! \tag{2.9}$$

Here the integrand is > 0 in $[-1,1]$ if $x \neq x_1,\cdots,x_{n-1}$. In fact using (1.19) we may write (2.9) as

$$\int_{-1}^{1}|T_n'(x)|2^{n-2}(n-1)!\,dx = 2^{n-1}n!. \tag{2.10}$$

From (2.4), (2.8) and (2.10) we now obtain the string of relations

$$\begin{aligned} 2^{n-1}n! &= \int_{-1}^{1}(-1)^{n-1}T_n'(x)f^{(n)}(x)\,dx \\ &\leqq \int_{-1}^{1}|T_n'(x)|\cdot 2^{n-2}(n-1)!\,dx = 2^{n-1}n!. \end{aligned} \tag{2.11}$$

The equality of the extreme terms shows that the two integrals are equal, a fact that we may write as

$$\int_{-1}^{1}\{|T_n'(x)|2^{n-2}(n-1)! - (-1)^{n-1}T_n'(x)f^{(n)}(x)\}\,dx = 0. \tag{2.12}$$

However, our assumption (2.4) shows that the integrand in (2.12) is $\geqq 0$ almost everywhere (a.e.). Therefore the vanishing of the integral (2.12) implies that the integrand must vanish a.e., whence

$$f^{(n)}(x) = (-1)^{n-1}2^{n-2}(n-1)!\,\mathrm{sgn}\,T_n'(x) \quad \text{a.e.} \tag{2.13}$$

However, (2.6) and (1.19) show that also

$$f_0^{(n)}(x) = (-1)^{n-1}2^{n-2}(n-1)!\,\mathrm{sgn}\,T_n'(x) \quad \text{a.e.}$$

Therefore $f^{(n)}(x) - f_0^{(n)}(x) = 0$ almost everywhere, so that

$$f(x) = f_0(x) + P(x) \text{ in } [-1,1], \text{ where } P(x) \in \pi_{n-1}. \tag{2.14}$$

Now the boundary properties (2.2) and (1.25) show that $P^{(\nu)}(-1) = 0$ for $\nu = 0,1,\cdots,n-1$, and therefore $P(x) = 0$ for all x. Finally (2.14) establishes the desired relation (2.5). This completes a proof of Theorem 2.

COROLLARY 2. *The perfect spline $f_0(x)$ of Corollary 1 is the unique perfect spline of degree n in $[-1,1]$ having at most $n-1$ knots in $(-1,1)$ and such as to satisfy the boundary conditions* (1.25).

PROOF. Let $\tilde{f}(x)$ be a perfect spline of degree n in $[-1,1]$ having k knots, $-1 < \tilde{x}_k < \tilde{x}_{k-1} < \cdots < \tilde{x}_1 < 1$, where

$$k \leqq n-1. \tag{2.15}$$

and also satisfying the boundary conditions

$$\begin{aligned} &\tilde{f}^{(\nu)}(-1) = 0, && (\nu = 0, \cdots, n-1), \\ &\tilde{f}(1) = 1, \quad \tilde{f}^{(\nu)}(1) = 0, && (\nu = 1, \cdots, n-1). \end{aligned} \tag{2.16}$$

We observe first that

$$\tilde{M}(x) = \tilde{f}'(x) \tag{2.17}$$

is a spline function of degree $n-1$. If we extend the function $\tilde{M}(x)$ for all real x by defining $\tilde{M}(x) = 0$ if $x < -1$, or if $x > 1$, then we see, by (2.16), that $\tilde{M}(x)$ is a spline function of degree $n-1$, of class C^{n-2} for all real x, having the knots

$$-1, \tilde{x}_k, \cdots, \tilde{x}_1, 1 \tag{2.18}$$

and vanishing in the complement of $(-1,1)$. It is known that such spline functions must have *at least* $n+1$ knots (See [1, Theorem 4]). By (2.18) we must therefore have $k+2 \geqq n+1$ or $k \geqq n-1$. Now our assumption (2.15) shows that $k = n-1$. Again Theorem 4 of [1] shows that $\tilde{M}(x)$ is a B-spline that (in our previous notation) we may write as

$$\tilde{M}(x) = M(x; 1, \tilde{x}_1, \cdots, \tilde{x}_{n-1}, -1). \tag{2.19}$$

Moreover, let

$$|\tilde{M}^{(n-1)}(x)| = |\tilde{f}^{(n)}(x)| = C \qquad \text{for all } x \neq \tilde{x}_\nu. \tag{2.20}$$

We can now establish for $\tilde{f}(x)$ an optimal property by repeating the proof of Theorem 2. This property will show that

$$C = \inf_{f \in F_n} \|f^{(n)}\|,$$

while $\tilde{f}(x)$ is the unique extremizing function. In this proof, which we omit, the role of $T_n'(x)$ would be played by the polynomial

$$\tilde{T}'(x) = (x - \tilde{x}_1) \cdots (x - \tilde{x}_{n-1}).$$

This optimal property of $\tilde{f}(x)$ having been established, we conclude by Theorem 2 that $C = 2^{n-2}(n-1)!$ and that $\tilde{f}(x) = f_0(x)$ in $[-1, 1]$.

3. A time-optimal control problem

The problem we have in mind is as follows.

PROBLEM 1. *A particle F moves on the y-axis such that $y = F(t)$ and that the velocities of different orders*

$$F^{(\nu)}(t), \qquad (\nu = 0, 1, \cdots, n-1)$$

are all absolutely continuous, while the nth *velocity $F^{(n)}(t)$ satisfies at all times the inequality*

$$(3.1) \qquad |F^{(n)}(t)| \leqq A \qquad (A \text{ is preassigned}).$$

We assume that the particle F starts from rest at $y = 0$ at the time $t = 0$, i.e.

$$(3.2) \qquad F^{(\nu)}(0) = 0 \qquad (\nu = 0, 1, \cdots, n-1),$$

and that F reaches the point $y = l$ also at rest at the time $t = T$ (> 0), i.e.

$$(3.3) \qquad F(T) = l, \ F^{(\nu)}(T) = 0 \quad (\nu = 1, \cdots, n-1).$$

We are to find the shortest time T_0 during which this motion can be performed and are to describe the nature of this optimal motion.

SOLUTION. If we recall that the perfect spline $f_0(x)$, defined by (1.23), satisfies the boundary conditions (1.25), it seems clear that

$$(3.4) \qquad F(t) = l f_0\left(\frac{2t}{T} - 1\right), \qquad 0 \leqq t \leqq T,$$

is a function satisfying the conditions (3.2) and (3.3). It is likewise clear that $F(t)$ is a perfect spline in $[0, T]$. Also, by (2.3), that

$$(3.5) \qquad \| F^{(n)} \| = l\left(\frac{2}{T}\right)^n 2^{n-2}(n-1)!\,.$$

If we also wish to meet the condition (3.1), as we must, we obtain the inequality

$$l\left(\frac{2}{T}\right)^n 2^{n-2}(n-1)! \leqq A.$$

Selecting here the least such T, producing the *equality* here, we obtain

$$(3.6) \qquad T_0 = 2(2^{n-2}(n-1)!\, l/A)^{1/n}.$$

THEOREM 3. *The motion corresponding to*

$$F_0(t) = lf_0\left(\frac{2t}{T_0} - 1\right), \qquad 0 \leqq t \leqq T_0, \tag{3.7}$$

is the optimal one and T_0, *given by* (3.6), *is the least time in which the motion of Problem* 1 *can be performed.*

We prepare a proof of Theorem 3 by first generalizing Theorem 2 in an obvious way that may be described as the introduction of superfluous parameters.

COROLLARY 3. *Among all functions* $G(t)$ *defined in* $[0, T_0]$ *and satisfying the conditions of absolute continuity of the derivatives and the interpolatory conditions*

$$\begin{aligned} &G^{(\nu)}(0) = 0, && (\nu = 0, \cdots, n-1), \\ &G(T_0) = l, \quad G^{(\nu)}(T_0) = 0, && (\nu = 1, \cdots, n-1), \end{aligned} \tag{3.8}$$

the function $F_0(t)$, *defined by* (3.7), *is the unique function minimizing the norm* $\| G^{(n)} \|$ *and giving it its minimal value*

$$\min \| G^{(n)} \| = \| F_0^{(n)} \| = l\left(\frac{2}{T_0}\right)^n 2^{n-2}(n-1)! = A. \tag{3.9}$$

PROOF OF COROLLARY 3. Let $G(t)$ satisfy (3.8). Then evidently the function

$$f(x) = l^{-1}G\left(\frac{T_0}{2}(x+1)\right), \qquad (-1 \leqq x \leqq 1),$$

satisfies the assumptions (2.1) and (2.2) of Theorem 2 while

$$\| f^{(n)} \| = l^{-1}\left(\frac{T_0}{2}\right)^n \| G^{(n)} \|.$$

By Theorem 2 we conclude that this is $\geqq 2^{n-2}(n-1)!$ and therefore

$$\| G^{(n)} \| \geqq l\left(\frac{2}{T_0}\right)^n 2^{n-2}(n-1)! = A,$$

while $\| F_0^{(n)} \| = A$. The unicity of $F_0(t)$ also follows from Theorem 2.

PROOF OF THEOREM 3. Let $F(t)$ be a motion in the time interval $[0, T]$ satisfying the conditions (3.1), (3.2) and (3.3) of Problem 1 and let us show that

$$T \geqq T_0, \tag{3.10}$$

with equality iff $F(t)$ is the motion (3.7). To show this we consider the motion

$$G(t) = F(tT/T_0), \qquad 0 \leqq t \leqq T_0,$$

which evidently satisfies the conditions (3.8) of Corollary 3. However

$$G^{(n)}(t) = \left(\frac{T}{T_0}\right)^n F^{(n)}(tT/T_0)$$

and therefore by Corollary 3 and (3.1)

(3.11) $$A \leqq \| G^{(n)} \| = \left(\frac{T}{T_0}\right)^n \| F^{(n)} \| \leqq \left(\frac{T}{T_0}\right)^n A.$$

Hence $A \leqq (T/T_0)^n A$ and (3.10) is established. Moreover, again by Corollary 3, if $T = T_0$ then (3.11) shows that $\| G^{(n)} \| = A$ and therefore $F(t) = F_0(t)$.

REMARK. The solution of the optimal control Problem 1 for the special case when $n = 2$ is well known. For its discussion from the point of view of general optimal control theory we refer to [7, 23–27] and [8, 233–236].

4. The behavior of the perfect spline $f_0(x)$ as $n \to \infty$

The function $f_0(x)$ defined by (1.23) is a distribution function whose frequency function is defined by (1.20) and (1.21). In order to indicate their dependence on the integer n we write

(4.1) $$f_0(x) = f_{0,n}(x) \text{ and } M(x) = M_n(x).$$

Their behavior as $n \to \infty$ is described as follows.

THEOREM 4. *The distributions described by the functions* (4.1) *converge, as* $n \to \infty$, *to the unit point-mass placed at the origin. This means that for arbitrarily small positive* δ *we have*

(4.2) $$\lim_{n\to\infty} M_n(x) = 0 \textit{ uniformly in } x \textit{ if } |x| \geqq \delta,$$

and therefore

(4.3) $$\lim_{n\to\infty} f_{0,n}(x) = 0 \textit{ uniformly in } -1 \leqq x \leqq -\delta .$$

(4.4) $$\lim_{n\to\infty} (1 - f_{0,n}(x)) = 0 \textit{ uniformly in } \delta \leqq x \leqq 1.$$

It was pointed out in [1, Footnote on page 74] that the mean and variance of the general B-spline

(4.5) $$M(x) = M(x; x_0, x_1, \cdots, x_n)$$

are given by

(4.6) $$\mu_1 = \int_{-\infty}^{\infty} x M(x)dx = \frac{1}{n+1} \sum_0^n x_\nu$$

and

(4.7) $$\sigma^2 = \int_{-\infty}^{\infty} (x - \mu_1)^2 M(x)\, dx = \frac{1}{(n+1)^2(n+2)} \sum_{i<j} (x_i - x_j)^2.$$

From these formulas we derive the following general

LEMMA 2. *Let*

(4.8) $$\tilde{M}_n(x) = M(x; x_{0,n}, x_{1,n}, \cdots, x_{n,n}) \qquad (n = 1, 2, \cdots)$$

be a sequence of B-splines subject only to the restriction that its knots are all in $[-1, 1]$, *i.e.*

(4.9) $$-1 \leqq x_{n,n} \leqq x_{n-1,n} \leqq \cdots \leqq x_{0,n} \leqq 1, \qquad x_{n,n} < x_{0,n}.$$

Let $\tilde{\sigma}_n$ *denote the standard deviation of* $\tilde{M}_n(x)$. *Then*

(4.10) $$\tilde{\sigma}_n < \frac{1}{\sqrt{n}} \text{ for all } n.$$

PROOF OF LEMMA 2. It is clear that the right side of (4.7), if viewed as a quadratic function of the variables x_ν, all contained in $[-1, 1]$, will be maximized if, as nearly as possible, one half of the points x_ν are at -1 and the other half at $+1$. The details depend on the parity of $n + 1$.

1. $n + 1 = 2k$ *is even.* We can then place k points at each of the two endpoints to obtain

$$\max \sum_{i<j} (x_k - x_j)^2 = 2^2 k \cdot k.$$

By (4.7) we have

$$\max \sigma^2 = \frac{1}{2k+1} = \frac{1}{n+2} < \frac{1}{n}.$$

2. $n + 1 = 2k + 1$ *is odd.* We must place k points at one endpoint and $k + 1$ points at the other to obtain

$$\max \sum_{i<j} (x_i - x_j)^2 = 2^2 k(k+1),$$

and (4.7) shows that

$$\max \sigma^2 = \frac{2k}{(2k+1)^2} = \frac{n}{(n+1)^2} < \frac{1}{n}.$$

The inequality (4.10) evidently follows.

PROOF OF THEOREM 4. We return to our sequence of perfect *B*-splines $M_n(x)$ having variances $\sigma_n^2 < 1/n$ by Lemma 2. By (1.21) and (4.6) their means are $= 0$. Moreover, $M_n(x)$ is an even function that is decreasing in the interval $[0, 1]$. Let

$$\delta = 2\varepsilon.$$

By comparison of areas and Chebyshev's inequality we see that

$$\varepsilon M_n(2\varepsilon) < \int_\varepsilon^1 M_n(x)\,dx = \frac{1}{2}\int_{|x|\geqq\varepsilon} M_n(x)\,dx \leqq \frac{1}{2}\frac{\sigma_n^2}{\varepsilon^2} < \frac{2}{\delta^2 n}$$

whence

$$M_n(\delta) < \frac{4}{\delta^3 n} \to 0 \text{ as } n \to \infty,$$

and Theorem 4 is thereby established.

References

1. H. B. Curry and I. J. Schoenberg, *On Pólya frequency functions IV. The fundamental spline functions and their limits*, J. Analyse Math. **17** (1966), 71–107.
2. P. J. Davis, *Interpolation and Approximation*, New York, 1963.
3. G. Glaeser, *Prolongement extrémal de fonctions différentiables d'une variable*, to appear in J. Approximation Theory.
4. G. Glaeser (ed.), *Le Prolongateur de Whitney*, vol. I (1966), vol. II (1967), Université de Rennes.
5. D. V. Ionescu, *Cuadraturi Numerice*, Bucharest, 1957.
6. D. V. Ionescu, *Introduction à la théorie des "fonctions spline"*, Acta Math. Acad. Sci. Hungar. **21** (1970), 21–26.
7. L. S. Pontryagin, V. G. Boltyanskii, R. V. Gamkrelidze and E. F. Mishchenko, *The Mathematical Theory of Optimal Processes*, New York, 1962.
8. L. C. Young, *Lectures on the Calculus of Variations and Optimal Control Theory*, Philadelphia, 1969.

Mathematics Research Center
The University of Wisconsin

I. J. SCHOENBERG

The Landau problem I.
The case of motions on sets

Dedicated to N. G. de Bruijn on the occasion of his 60th birthday

Communicated by Prof J. H. van Lint at the meeting of December 17, 1977

Department of Mathematics, United States Military Academy, West Point, New York 10996

1. STATEMENT OF THE PROBLEM

Let $f(t)$ be real-valued, $-\infty<t<\infty$. As we think of t as time, we denote its derivatives by $\dot f(t)$ and $\ddot f(t)$. We assume that $f\in C^1(\mathbb{R})$, hence $\dot f\in C(\mathbb{R})$, and that $\ddot f(t)$ is piecewise continuous with dicontinuities only of the first kind. Using the supremum norm on $\mathbb{R}$, we have the well known inequality of Landau

(1) $\quad \|\dot f\|\leqq\sqrt{2\cdot\|f\|\cdot\|\ddot f\|}$, (See [1], [2]).

If $a>0$ and $f(t)$ satisfies the inequalities

(2) $\quad -a\leqq f(t)\leqq a$ for all t,

then (1) shows that

(3) $\quad \|\dot f\|\leqq\sqrt{2a}\sqrt{\|\ddot f\|}.$

Here $\sqrt{2a}$ is the best constant, as shown by the special motion $\bar f(t)=4a(t-t^2)$ if $0\leqq t\leqq 1$, which is extended to all real t by the functional equation $\bar f(t+1)=-\bar f(t)$, for all real t, with the result that $\|\bar f\|=a$, $\|\dot{\bar f}\|=4a$, $\|\ddot{\bar f}\|=8a$. This motion produces the equality sign in (3).

We may interpret (2) by saying that $f(t)$ describes a motion in the interval

(4) $\quad I_a=\{-a\leqq x\leqq a\}.$

Reprinted from Proceedings of the Koninklijke Nederlandse Akademie van Wetenschappen, Amsterdam, series A, volume 81 (2), June 9, 1978

In terms of the functional

$$(5) \qquad F(f) = \frac{\|\dot{f}\|}{\sqrt{\|\ddot{f}\|}}$$

we may restate the result (3) by writing that

$$(6) \qquad \sup_f F(f) = \sqrt{2a}$$

for all motions f within the interval I_a.

It seems rather natural to replace the interval (4) by a given *closed and connected set* S of the complex plane $\mathbb{C}$, or belonging to a finite-dimensional euclidean space. Accordingly, we consider the class of motions

$$(7) \qquad (S) = \{f(t);\ f(t) \in S \text{ for all } t,\ f(t) \not\equiv \text{constant}\}.$$

Notice that we have excluded the trivial case when $f(t) = \text{constant}$ for all t, when $F(f) = 0/0$ is undefined.

We can now formulate

LANDAU'S PROBLEM FOR THE SET S. *To determine the quantity*

$$(8) \qquad L(S) = \sup_{f \in (S)} F(f).$$

We call $L(S)$ the Landau constant of the set S.

2. A FEW SPECIAL SETS S WHEN $L(S)$ IS KNOWN

In terms of (1.4) and the definition (1.8), we may restate the result (1.6) by writing that

$$(1) \qquad L(I_a) = \sqrt{2a}.$$

Here are a few trivial further examples. If $S = \{0 \leqq x < \infty\}$, then $f(t) = t^2 \in (S)$, and $\|\dot{f}\| = \infty$, $\|\ddot{f}\| = 2$. Therefore $L(S) = \infty/2 = \infty$. If $S = \mathbb{R}$, then $f(t) = t \in (S)$, and $L(S) = 1/0 = \infty$.

An obvious and fundamental property of the Landau constant is the following

MONOTONICITY PROPERTY. If S_1 *and* S_2 *are two sets, then*

$$(2) \qquad S_1 \subset S_2 \text{ implies that } L(S_1) \leqq L(S_2).$$

A few further non-trivial examples when $L(S)$ is known are the following. If S is the *circular disc*

$$(3) \qquad D_a = \{|z| \leqq a\},$$

then $I_a \subset D_a$ implies, by (2), that $L(I_a) \leqq L(D_a)$. As a matter of fact we have here equality, so that

$$(4) \qquad L(D_a) = \sqrt{2a}.$$

For the *circle*

$$(5) \qquad C_a = \{|z| = a\}$$

we will show that

$$(6) \qquad L(C_a) = \sqrt{a}.$$

It seems worthwhile to look at the two results (1) and (6) and compare their respective extremizing motions. Let $\mathscr{E}_2(t)$ be the so-called *quadratic Euler spline* (See [2, Section 2]) defined by

$$(7) \qquad \mathscr{E}_2(t) = 1 - 4t^2 \text{ if } -\tfrac{1}{2} \leqq t \leqq \tfrac{1}{2}, \text{ and } \mathscr{E}_2(t+1) = -\mathscr{E}_2(t) \text{ for all } t.$$

We found above that $\bar{f}(t) = a\mathscr{E}_2(t) \in (I_a)$ and $F(\bar{f}) = \sqrt{2a}$. For $S = C_a$ the extremizing motion is the *uniform rotation*

$$\bar{f}(t) = ae^{it}.$$

Indeed, here $\|\dot{\bar{f}}\| = a$, $\|\ddot{\bar{f}}\| = a$, hence $F(\bar{f}) = \sqrt{a}$, and this is the extremizing motion in view of (6).

Let now S be the *circular ring*

$$(8) \qquad R_{a,b} = \{b \leqq |z| \leqq a\}, \quad (0 < b < a).$$

Here *we shall find that* *

$$(9) \qquad L(R_{a,b}) = \sqrt{a + \sqrt{a^2 - b^2}}.$$

We would expect to obtain from this the result (6) when $b = a$, and (4) when $b = 0$.

Our last special example is the *solid shell*

$$(10) \qquad SoSh_{a,b} = \{b \leqq \sqrt{x^2 + y^2 + z^2} \leqq a\}, \quad (0 < b < a).$$

As this will yield most of the previous results, we formulate it as

THEOREM 1. *For the solid shell* (10) *we have that*

$$(11) \qquad L(SoSh_{a,b}) = \sqrt{a + \sqrt{a^2 - b^2}}.$$

Actually Theorem 1, with the same value for the Landau constant, holds for the solid spherical shell in n dimensions, $n \geqq 2$. Observe again, that $R_{a,b} \subset SoSh_{a,b}$, while their Landau constants are equal by (9) and (11).

Of an entirely different nature is

THEOREM 2. *If the set S is closed, bounded and convex, then*

$$(12) \qquad L(S) = \sqrt{\text{diameter of } S}.$$

---

* This was first shown in [3]. In Section 5 below we first derive the stronger relation (11) which is somewhat easier to prove than (9). Theorem 2 below was stated without proof in [3].

Clearly Theorem 2 implies as special cases the results (1) and (4). Again Theorem 2, and its proof below, remains valid, as it stands, in any finite-dimensional euclidean space.

3. A LEMMA

The lemma to be now discussed, will show that an extremizing function $\bar{f}(t)$, i.e. producing the equality $F(\bar{f})=L(S)$, may also be characterized as a solution of a problem of Optimal Control.

LEMMA 1. *If the motion*

(1) $\qquad \bar{f}(t)\in(S)$

has the

PROPERTY A. *If*

(2) $\qquad f(t)\in(S)$ *and* $\|\ddot{f}\|\leqq\|\ddot{\bar{f}}\|$

then

(3) $\qquad \|\dot{f}\|\leqq\|\dot{\bar{f}}\|,$

then $\bar{f}$ *also has the*

PROPERTY B.

(4) $\qquad F(\bar{f})=L(S),$

and conversely, if $\bar{f}$ *enjoys the Property B, then* $\bar{f}$ *also has the Property A.*

PROOF: 1. *Property A implies Property B.* We assume (1) and Property A and we are to establish (4). If $f\in(S)$, then for the motion $g(t)=f(t)\sqrt{\|\ddot{\bar{f}}\|/\|\ddot{f}\|)}$ we find that $g(t)\in(S)$ and also that

$$\|\ddot{g}\|=\|\ddot{f}\|(\|\ddot{\bar{f}}\|/\|\ddot{f}\|),\text{ hence }\|\ddot{g}\|=\|\ddot{\bar{f}}\|.$$

Hence $g(t)$ satisfies the assumptions (2), and by Property A we conclude that $\|\dot{g}\|\leqq\|\dot{\bar{f}}\|$, and by the definition of $g(t)$, that

$$\|\dot{f}\|\sqrt{\|\ddot{\bar{f}}\|/\|\ddot{f}\|}\leqq\|\dot{\bar{f}}\|,\text{ and so }F(f)\leqq F(\bar{f}).$$

But then

$$L(S)=\sup_{f\in(S)}F(t)\leqq F(\bar{f}).$$

From $F(\bar{f})\leqq L(S)$ we now conclude that (4) holds.

2. *Property B implies Property A.* Now we assume (1) and (4), and that f satisfies (2), and we are to show that (3) holds. From (4) we obtain

that $F(f) \leqq F(\bar{f})$, or $\|\dot{f}\|/\sqrt{\|\ddot{f}\|} \leqq \|\dot{\bar{f}}\|/\sqrt{\|\ddot{\bar{f}}\|}$, whence

$$\|\dot{f}\| \leqq \|\dot{\bar{f}}\|\sqrt{\|\ddot{f}\|/\|\ddot{\bar{f}}\|} \leqq \|\dot{\bar{f}}\|$$

in view of the second inequality (2), and (3) is established. Lemma 1 was stated without proof in [3].

4. A PROOF OF THEOREM 2

Let A and B be points of S such that

(1) $\quad |B-A| = d = \text{diameter of } S, \ (A \in S, \ B \in S).$

From the relations (2.7) we find that $-1 \leqq \mathscr{E}_2(t) \leqq 1$ for all t, and that $\mathscr{E}_2(t)$ represents a to-and-fro motion on $[-1, 1]$. We now define on the segment $[A, B]$ the motion

(2) $$\begin{aligned} \bar{f}(t) &= \tfrac{1}{2}A(1-\mathscr{E}_2(t)) + \tfrac{1}{2}B(1+\mathscr{E}_2(t)) \\ &= \tfrac{1}{2}(B+A) + \tfrac{1}{2}(B-A)\mathscr{E}_2(t). \end{aligned}$$

From $\|\dot{\mathscr{E}}_2\| = 4$, $\|\ddot{\mathscr{E}}_2\| = 8$, and (1), we find that

(3) $\quad \|\dot{\bar{f}}\| = 2d, \ \|\ddot{\bar{f}}\| = 4d,$

and therefore

(4) $\quad F(\bar{f}) = \sqrt{d}.$

We wish to prove that $F(\bar{f}) = L(S)$, which is the relation (4) of Property B of Lemma 1. By Lemma 1 it suffices to prove the following:

If $f \in (S)$ *such that*

(5) $\quad \|\ddot{f}\| \leqq \|\ddot{\bar{f}}\| = 4d,$

then we also have

(6) $\quad \|\dot{f}\| \leqq \|\dot{\bar{f}}\| = 2d.$

This we do as follows: We consider an arbitrary but fixed t_0 such that $\dot{f}(t_0) \neq 0$, and let L be a fixed straight line such that

(7) $\quad$ *L is parallel to the vector $\dot{f}(t_0)$.**

Let the segment $[P, Q] \subset L$ be the orthogonal projection of S onto the line L. Clearly

(8) $\quad |Q-P| \leqq |A-B| = d.$

Let $f_p(t)$ denote the orthogonal projection of the motion $f(t)$ onto L. By (5) we have

$$\|\ddot{f}_p\| \leqq \|\ddot{f}\| \leqq \|\ddot{\bar{f}}\| = 4d$$

* The reader is asked to sketch a diagram.

280

and by (8) and Landau's inequality (1.1) we find that

$$\|\dot{f}_p\| \leqq \sqrt{|Q-P|}\cdot\sqrt{\|\ddot{f}_p\|} \leqq \sqrt{d}\cdot\sqrt{4d} = 2d.$$

Now (7) implies that

$$|\dot{f}(t_0)| = |\dot{f}_p(t_0)| \leqq \|\dot{f}_p\| \leqq 2d.$$

and therefore $\|\dot{f}\| \leqq 2d$, since t_0 was arbitrary. This proves (6) and our proof is complete.

REMARK. Our proof shows that for Theorem 2 to be valid, the set S need not be convex. All that is needed is that (1) should hold, and that $[A, B] \subset S$. These assumptions already imply that $L(S) = L([A, B])$. It does seem curious that we can not use the greater freedom of motions in S to increase the value $\sqrt{d}$ of the Landau constant.

5. A PROOF OF THEOREM 1

We now turn to the solid shell

(1) $$S = SoSh_{a,b} = \{b \leqq \sqrt{x^2+y^2+z^2} \leqq a\},\ (0<b<a).$$

Just as in the case of the convex set A we identified the motion (4.2) as the motion that maximizes $F(f)$ among all motions within S, so we will now construct a motion again to be denoted by $\tilde{f}(t)$ $(t \in \mathbf{R})$, that will maximize $F(f)$ for the solid shell (1).

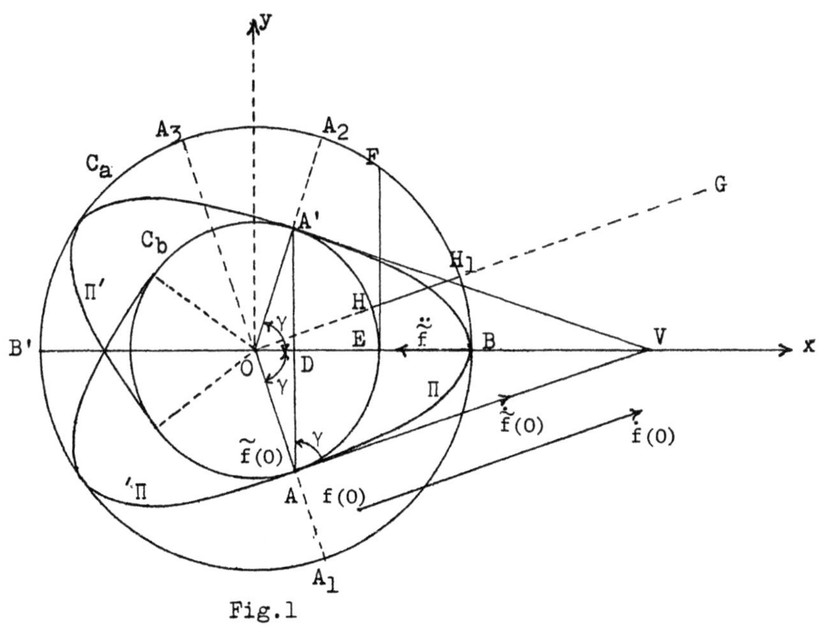

Fig.1

Fig. 1 shows the ring $R_{a,b}$ obtained by intersecting the solid shell with the plane $z=0$. We consider the parabolic arc $\Pi=\widehat{ABA'}$ having its vertex B on the circle C_a, and tangent to C_b at the points A, A'. It will aid our computations if we first show how Π can be constructed. Erect EF perpendicular to OB and intersecting C_a at F. Mark the two points D and V such that

(2) $\qquad DB=BV=EF,$

and erect the perpendicular to OB at D, intersecting C_b at A and A'. Join V to A and to A'. We claim that

(3) $\qquad \angle OAV=\angle OA'V=90°.$

PROOF: From the triangle OEF we find that $(EF)^2=a^2-b^2$. However, by (2)

$$OD=a-DB=a-EF,\ OV=a+BV=a+EF,$$

and so $OD\cdot OV=a^2-(EF)^2=a^2-(a^2-b^2)=b^2=(OA')^2$, or $OD\cdot OV=(OA')^2$. This is a characteristic relation for a right-angled triangle and establishes (3).

Incidentally, we have shown that in fig. 1 we have

(4) $\qquad OV=a+\sqrt{a^2-b^2}.$

From $DB=BV$ it also follows that the parabolic arc Π with vertex at B and passing through A and A', is also tangent to C_b at A and A'.

Next we consider on the arc $\Pi=\widehat{ABA'}$ a motion $f(t)$ $(\alpha\leq t\leq\beta)$, which we call *Galilean*; by this we mean that the point $f(t)$ describes the arc Π, $f(\alpha)=A$, $f(\beta)=A'$, and that the point $f(t)$ has a *constant acceleration* $\ddot{f}$ which must, of course, be real and negative. For this motion the modulus $|\dot{f}(t)|$ of the velocity vector assumes its maximal value at the endpoints A and A'. Let $v=|\dot{f}(\alpha)|=|\dot{f}(\beta)|$ denote this maximal value. Along Π we therefore have

(5) $\qquad F(f)=v/\sqrt{|\ddot{f}|}.$

This functional being invariant if we change the scale in t, without loss of generality we may assume $\alpha=0$, $\beta=1$, and now

(6) $\qquad f(t),\ (0\leq t\leq 1)$

represents the Galilean motion along the arc Π.

The quantity $F(f)$ is of dimension $LT^{-1}(LT^{-2})^{-1/2}=L^{1/2}$. Let us show that *in terms of fig.* 1 *we have*

(7) $\qquad F(f)=\sqrt{OV}.$

282

PROOF: It is well known that a material particle projected vertically upward with the initial velocity v_0, will reach a maximal height h given by the relation

(8) $v_0^2 = 2gh.$

Using (5), the relation (7), to be established, amounts to

(9) $v^2 = |\ddot{f}| \cdot OV.$

The motion $\bar{f}(t)$ may be regarded as a motion in the field of constant acceleration $\ddot{\bar{f}} = -|\ddot{f}|$, while its initial "vertical" velocity is $v_0 = v \sin \gamma$, where $\gamma = \angle DAV$. By (8) we have

$$(v \sin \gamma)^2 = 2|\ddot{f}| \cdot DB = \ddot{f} \cdot DV,$$

in view of (2). It follows that

$$v^2 = |\ddot{f}| \frac{DV}{(\sin \gamma)^2} = |\ddot{f}| \cdot \frac{AV}{\sin \gamma} = |\ddot{f}| \cdot OV$$

proving (9).

We now *extend the motion* (6) *to all real* t *by the functional equation*

(10) $\bar{f}(t+1) = e^{2i\gamma} \bar{f}(t)$ *for all real* t.

This amounts to successive rigid rotations of the motion on Π through the angle $2\gamma = \angle AOA'$, and shows that the curve Γ described by the extended motion $\bar{f}(t)$ $(t \in \mathbf{R})$, is an infinite sequence of parabolic arcs

$$\ldots \Pi'' \Pi' \Pi \Pi' \Pi'' \ldots,$$

all congruent to Π. Therefore (10) defines a motion $\bar{f}(t)$ having the properties

(11) $\bar{f}(t) \in (R_{a,b}),$

(12) $\bar{f}(t) \in C^1(\mathbf{R}),$

(13) $F(\bar{f}) = \sqrt{OV} = \sqrt{a + \sqrt{a^2 - b^2}},$

the last by (7) and (4).

Notice that the acceleration vector $\ddot{\bar{f}}(t)$ is piecewise constant, with discontinuities for all integer values of t, while for all other values of t, its modulus is constant, in fact

(14) $|\ddot{\bar{f}}(t)| = |\ddot{f}|.$

After these extensive preliminaries, we can pass to *a proof of Theorem* 1. *Let* $f(t)$ *be a motion within the solid shell* (1) *subject to the condition*

(15) $\|\ddot{f}\| \leq \|\ddot{\bar{f}}\| = -\ddot{\bar{f}},$

and we are to show that this implies that

(16) $\quad \|\dot{f}\| \leqq \|\dot{\bar{f}}\| = |\dot{\bar{f}}(0)| = v.$

By lemma 1 this would imply that $L(S) = F(\bar{f})$, and prove Theorem 1 that

(17) $\quad L(SoSh_{a,b}) = F(\bar{f}) = \sqrt{a + \sqrt{a^2 - b^2}}.$

To prove (16) we assume that for some t_0 we have $|\dot{f}(t_0)| > v$ and get a contradiction. We may as well assume that $t_0 = 0$, hence

(18) $\quad |\dot{f}(0)| > v = |\dot{\bar{f}}(0)|.$

We begin by rotating the motion $f(t)$ rigidly about 0, hence within the shell S, so that the vector

(19) $\quad \dot{f}(0)$ *is parallel to the vector* $\dot{\bar{f}}(0)$.

If $f(0)$ is in the half-space containing the arc Π and bounded by the vertical plane π that intersects the horizontal plane xOy in the line A_1OA_3, we do nothing further. However, if $f(O)$ is not in the half-space, then we reflect the entire motion $f(t)$ with respect to π, so as to have now $f(O)$ in that half-space; notice that (19) still holds after this reflection. Let $OG \perp A_1A_3$, hence $OG \perp \pi$. Now we rotate the motion $f(t)$ rigidly about the axis of rotation OG so that $f(O)$ falls with the quadrant A_1OG. This places $f(O)$ within the (partly) curvilinear quadrilateral

(20) $\quad Q = AA_1H_1H.$

Assuming that

(21) $\quad 2\gamma \geqq \dfrac{\pi}{2},$

which is the case of fig. 1, we find that

(22) $\quad f(O) \in Q' = AA_1A_2A',$

and it follows that

(23) $\quad \operatorname{Re} f(O) \geqq OD.$

Notice that (19) still holds. We may even assume that the vector

(24) $\quad \dot{f}(O)$ is parallel and *points in the same direction as* $\dot{\bar{f}}(O)$.

For if not we replace $f(t)$ by $f(-t)$, thereby reversing the direction of $\dot{f}(O)$.

We now consider the motions $\bar{f}_p(t)$ and $f_p(t)$ obtained by projecting orthogonally on Ox the motions $\bar{f}(t)$ and $f(t)$, respectively. Clearly

(25) $\quad \bar{f}_p(t) = \operatorname{Re} \bar{f}(t), \; f_p(t) = \operatorname{Re} f(t).$

284

Let us now compare these rectilinear motions during the time interval from $t=0$ to $t=\frac{1}{2}$. The motion $\bar{f}_p(t)$ starts from D, at $t=0$, with initial velocity $\dot{\bar{f}}_p(O)=v\sin\gamma$, and, while uniformly decelerated with constant deceleration $|\ddot{\bar{f}}|$, just manages to arrive at B, at $t=\frac{1}{2}$, with *zero velocity*. The motion $f_p(t)$ starts at $\operatorname{Re} f(O)$, *ahead of* $\bar{f}_p(O)$, because of (23), with a *larger initial velocity*, because of (24) and (18), and moves during $0 \leqq t \leqq \frac{1}{2}$ with an acceleration $\ddot{f}_p(t)$ that nowhere exceeds in modulus the value $|\ddot{\bar{f}}|$, because of (15). It follows that *the point $f_p(t)$ must have reached the point B, for some $t<\frac{1}{2}$, with a positive velocity*. This being cinematically impossible, we have obtained the desired contradiction.

There is a slight additional difficulty if (21) does not hold, but rather

$$(26) \qquad 2\gamma<\frac{\pi}{2}.$$

Indeed now we would have in fig. 1 that

$$Q=AA_1H_1H \supset Q'=AA_1A_2A'.$$

It follows that while $f(O)$ is in Q, $f(O)$ need not be in Q'. In this case, let δ denote the *positive* angle $\delta=\angle A'Of(O)$. For the last time we rotate the motion $f(t)$ rigidly *clockwise* about Oz by the angle δ. As a result the new position of $f(O)$ lies on the segment $[A', A_2]$, and our previous arguments will apply. Indeed observe that (23) will now hold, and that the value of $\dot{f}_p(O)$ has only increased by the last rotation.

REMARK. In fig. 1 we let $b \to 0+$. As a result we find that

$$\lim_{b\to 0+} \bar{f}(t)=a\cdot\mathscr{E}_2(t),$$

uniformly in every finite interval of the t-axis: The extremizing motion for the solid shell converges to the extremizing motion of $I_a=[-a, a]$.

6. COROLLARIES OF THEOREMS 1 AND 2

It was already mentioned that the relations (2.1) and (2.4) are special examples of Theorem 2.

PROOF OF (2.9). From $R_{a,b} \subset SoSh_{a,b}$, the monotonicity property (2.2) and Theorem 1, we obtain that

$$L(R_{a,b}) \leqq \sqrt{a+\sqrt{a^2-b^2}}.$$

On the other hand we know that $\bar{f}(t) \in (R_{a,b})$, and therefore

$$F(\bar{f})=\sqrt{a+\sqrt{a^2-b^2}} \leqq L(R_{a,b}).$$

PROOF OF (2.6). From $C_a \subset R_{a,b}$ and (2.9) we obtain that

$$L(C_a) \leqq \sqrt{a+\sqrt{a^2-b^2}}.$$

Letting $b \to a$ we obtain that

$$L(C_a) \leqq \sqrt{a}.$$

However, in section 2 we found that for $\bar{f}(t) = ae^{it}$ we have $F(\bar{f}) = \sqrt{a}$.

REFERENCES

1. Landau, E. – *Einige Ungleichungen für zweimal differentiierbare Functionen*, Proc. London Math. Soc. (2), 13, 43–49 (1913).
2. Schoenberg, I. J. – *The elementary cases of Landau's problem of inequalities between derivatives*, Amer. Math. Monthly, 80, 121–158 (1973).
3. Schoenberg, I. J. – *The Landau problem for motions in a ring and in bounded continua*, Amer. Math. Monthly, 84, 1–12 (1977).

MATHEMATICS Proceedings A 84 (3), September 28, 1981

The Landau problem. II

The case of motions on curves.

by I.J. Schoenberg

Math. Research Center, 610 Walnut street, Madison, Wisconsin 53706, U.S.A.

Communicated by Prof. J.H. van Lint at the meeting of September 27, 1980

1. INTRODUCTION

The present article is a continuation of [2] to which we refer as L.P.I. We facilitate its reading by repeating a few key definitions.

Let f: $R \to R^n$, where $f = f(t) \in C^1(R)$, with the second derivative $\ddot{f}(t)$ piecewise continuous and having only discontinuities of the first kind. We also assume that $\dot{f}(t)$ has only finitely many zeros in every finite t-interval. *We define the class $\mathcal{M}$ of motions $f(t)$ in R^n as the totality of functions $f(t)$ satisfying the above conditions.*

Let S be a closed and connected set of R^n. We are concerned with the subclass of motions

(1.1) $(S) = \{f(t); f(t) \in S \text{ for all } t, f(t) \in \mathcal{M}\}.$

With the supremum norm on R we consider the functional

(1.2) $F(f) = \|\dot{f}\| / \sqrt{\|\ddot{f}\|}.$

The problem is to determine the quantity

(1.3) $L(S) = \sup_{f \in (S)} F(f),$

Sponsored by the United States Army under Contract Nos. DAAG29-75-C-0024 and DAAG29-80-C-0041.

called *the Landau constant of the set S*. A few examples of sets S, when $L(S)$ can be explicitly determined, were given in L.P.I §§ 1 and 2.

Evident from the definition (1.3) is the *monotonicity property*:

(1.4) *If* $S_1 \subset S_2$, *then* $L(S_1) \leqq L(S_2)$.

A result from L.P.I. (Theorem 2) worth repeating here is the following

THEOREM FOR CONVEX SETS. *If the set S is convex and bounded, then*

(1.5) $L(S) = \sqrt{d}$, *where* $d =$ *the diameter of S.*

An immediate consequence, not mentioned in L.P.I, is the

COROLLARY 1. *If S is a closed and bounded set, then*

(1.6) $L(S) \leqq \sqrt{d}$, *where* $d =$ *the diameter of S.*

For if K is the convex extension of S, then S and K have the same diameter d, while (1.4) and (1.5) show that $L(S) \leqq L(K) = \sqrt{d}$. In the next paper [3] we determine the Landau constant $L(C_{R,l})$ of a circular arc $C_{R,l}$ of radius R and length l and then show the following:

(1.7) *There are sets S of diameter d having arbitrarily small* $L(S)$.

The results of L.P.I, II, and III, were given a few years ago in [1], except the theorems of L.P.III on time-optimal control problems.

2. THE MAIN RESULTS FOR MOTIONS ON CURVES

For the remainder of our discussion we assume that the set S reduces to a curve Γ. As expected, we get in this case simpler and more definite results. To start with, we assume that Γ is in the complex plane, that it is rectifiable with a continuously turning tangent, and that it is endowed with a piecewise continuous radius of curvature $R(s)$ as function of its arc-length s. The function $R = R(s)$ should also have only discontinuities of the first kind and should satisfy the Dirichlet condition; on some portions of Γ we may well have $R(s) = \infty$, meaning that Γ has straight segments.

We classify arcs Γ into four cases.

CASE 1. Γ is a *bi-infinite arc*, a parabola being a good example; measuring s from a point 0 on Γ we have $-\infty < s < \infty$.

CASE 2. Γ is a *half-infinite arc*. If 0 is its endpoint, then we think of Γ as a bi-infinite arc (Case 1) doubled-up on itself at 0. Thus again $-\infty < s < \infty$, with $\pm s$ corresponding to the same point of Γ.

CASE 3. Γ is a *closed curve* of total length $2l$. A circle, or an ellipse are good examples. It is convenient to think of the real $s -$ axis as being wrapped around Γ, so that s and $s + 2lk$ represent the same point on Γ. Thus again $-\infty < s < \infty$.

CASE 4. Γ is a *finite arc* of length l. It is convenient to think of Γ as a closed curve (Case 3) with its two halved $0 \leqq s \leqq l$ and $0 \geqq s \geqq -l$ identified. Now the three arc-lengths s, $s+2l$, and $-s$, correspond to the same point of Γ. Again, along Γ, $-\infty < s < \infty$.

As in the general case of sets S we select a fixed constant

(2.1) $\quad A > 0,$

and consider the subclass of motions

(2.2) $\quad (\Gamma)_A = \{f(t);\, f(t) \in \Gamma,\ |\ddot{f}(t)| \leqq A \text{ for all } t\}.$

Our main result is

THEOREM 1. *There is in* $(\Gamma)_A$ *a unique motion*

(2.3) $\quad \bar{f}(t) = \bar{f}_A(t),$

called the Landau motion on Γ, corresponding to the constant A, *and having the following four properties:*

(i)

(2.4) $\quad |\ddot{\bar{f}}(t)| = A$ *for all real* t.

(ii) *In all four cases of arcs* Γ *the arc-length* $\bar{s} = \bar{s}(t)$, *corresponding to the point* $\bar{f}(t)$, *increases steadily with* t *from* $-\infty$ *to* $+\infty$. *We assume that*

(2.5) $\quad \bar{f}(0) = 0.$

(iii) *In Case* 2 *we have*

(2.6) $\quad \bar{f}(-t) = \bar{f}(t),$

so that $f(t)$ *is an even function. Observe that the continuity of* $\dot{\bar{f}}(t)$ *implies that at the endpoint of* Γ *we have*

(2.7) $\quad \dot{\bar{f}}(0) = 0.$

In Case 3 *the motion* $\bar{f}(t)$ *along the closed curve* Γ *is periodic. If* $2T$ *is its smallest positive period, we have*

(2.8) $\quad \bar{f}(t+2T) = \bar{f}(t).$

In Case 4 *we have again the periodicity* (2.8), *and* $f(t)$ *is also* even, *so that*

(2.9) $\quad \bar{f}(-t) = \bar{f}(t).$

Again the continuity of $\dot{\bar{f}}(t)$ *implies in this case that*

(2.10) $\quad \dot{\bar{f}}(0) = 0,\ \dot{\bar{f}}(T) = 0.$

It follows from (2.6), (2.8), *and* (2.9), *that if for some* $t = t_0$ *we have* $\bar{f}(t_0) = P \in \Gamma$, *then* $|\dot{\bar{f}}(t_0)|$ *assumes at* P *the same value, independent of* t_0, *and depending only on* P. *We denote this value by*

(2.11) $\quad |\dot{\bar{f}}(t_0)| = |\dot{\bar{f}}_P|,$

to indicate its dependence on P *only.*

So far we have merely endowed the Landau motion $\tilde{f}(t)$ with a qualitative property which may be described by the term "steady", *in the sense that it assumes at every point P, of Γ, always the same speed $|\dot{\tilde{f}}_P|$, whether the point $\tilde{f}(t)$ goes around Γ* (Case 3), *or to and fro* (Case 4).

(iv) This is the quantitative characteristic property of $\tilde{f}$: *If*

$$(2.12) \quad f(t) \in (\Gamma)_A,$$

and $t = t_1$ is such that

$$(2.13) \quad f(t_1) = P,$$

then

$$(2.14) \quad |\dot{f}(t_1)| \leqq |\dot{\tilde{f}}_P|.$$

In words: *The speed $|\dot{\tilde{f}}_P|$ of the Landau motion at the point $P \in \Gamma$ is never exceeded by the speed at P of any motion $f(t)$ on Γ, such that $|\ddot{f}(t)| \leqq A$ for all t. Thus $\tilde{f}$ maximizes the speed at every point P of Γ, within the class of motions $(\Gamma)_A$.*

There is an immediate consequence worth recording:

COROLLARY 2. *If $\tilde{f}(t)$ is the Landau motion on Γ corresponding to A, then*

$$(2.15) \quad L(\Gamma) = F(\tilde{f}) = \|\dot{\tilde{f}}\| / \sqrt{A}.$$

PROOF: Here $F(f)$ is the functional (1.2), which is the subject of Lemma 1 of L.P.I § 3. Lemma 1 states that if a motion $\tilde{f}(t)$ satisfies its Property A, then it also has its Property B to the effect that $L(\Gamma) = F(\tilde{f})$. Since the Property A of $\tilde{f}(t)$ is guaranteed by our Theorem 1, the relation (2.15) follows.

3. THE CONDITION $|\ddot{f}(t)| \leqq A$ AMOUNTS TO A DIFFERENTIAL INEQUALITY

Let $f(t) \in (\Gamma)$ and let $s = s(t)$ denote the arc-length corresponding to the point $f(t)$ of Γ. Also the speed

$$(3.1) \quad v = \frac{ds}{dt} = \pm |\dot{f}(t)|$$

is a function of t. Let us assume that

$$(3.2) \quad \ddot{f}(t) \text{ and } R(s) \text{ are continuous if } \alpha < t < \beta.$$

At the point $f = f(t) = (x, y)$ we draw a half-line tangent to Γ in the direction of increasing s, and denote by φ its angle with the x-axis. Then by (3.1) the following classical relations hold

$$(3.3) \quad \frac{d\varphi}{dt} = \frac{d\varphi}{ds} \cdot \frac{ds}{dt} = \frac{1}{R} v,$$

$$(3.4) \quad \dot{x} = \frac{dx}{dt} = \frac{dx}{ds}\frac{ds}{dt} = v \cos \varphi$$

$$(3.5) \quad \dot{y} = v \sin \varphi.$$

Differentiating again the last two relations, we obtain by (3.3) that

$$(3.6)\qquad \begin{cases} \ddot{x} = \dot{v}\cos\varphi - \dfrac{v^2}{R}\sin\varphi, \\ \\ \ddot{y} = \dot{v}\sin\varphi + \dfrac{v^2}{R}\cos\varphi. \end{cases}$$

These give the well-known components of the acceleration $\ddot{f}$ in the direction of the tangent and the normal to Γ at the point $f=(x,y)$. Since

$$|\ddot{f}(t)|^2 = \ddot{x}^2 + \ddot{y}^2 = \dot{v}^2 + \frac{v^4}{R^2},$$

we immediately conclude the following: *The inequality*

$$(3.7)\qquad |\ddot{f}(t)|^2 \leqq A^2$$

holds, if and only if

$$(3.8)\qquad \dot{v}^2 \leqq A^2 - \frac{v^4}{R^2}$$

is satisfied.

Since $\dot{v} = dv/dt = (dv/ds)(ds/dt) = v(dv/ds)$ we may also write (3.8) as

$$(3.9)\qquad v^2(dv/ds)^2 \leqq A^2 - v^4R^{-2}.$$

This we multiply by 16 v^4 to obtain

$$(3.10)\qquad \left(4v^3\frac{dv}{ds}\right)^2 \leqq 16v^4\left(A^2 - \frac{v^4}{R^2}\right).$$

A glance at this shows that we should introduce the new variable

$$(3.11)\qquad u = v^4$$

to obtain the inequality

$$(3.12)\qquad \left(\frac{du}{ds}\right)^2 \leqq 16u\left(A^2 - \frac{u}{R^2}\right).$$

Clearly, as shown, (3.9) implies (3.12), but the converse is true only if $v \neq 0$. However, for motions $f(t)$ satisfying

$$(3.13)\qquad f(t) \in (S) \cap \mathscr{M},$$

we have the important.

LEMMA 1. *We assume $R(s)$ $(\neq 0)$ and $\ddot{f}(t)$ to be continuous at the time t. In terms of the variable $u = v^4$ the inequality*

$$(3.14)\qquad |\ddot{f}(t)| \leqq A$$

is equivalent with the differential inequality

$$(3.15) \quad \left(\frac{du}{ds}\right)^2 \leqq 16u\left(A^2 - \frac{u}{R^2(s)}\right).$$

Moreover, the equation

$$(3.16) \quad |\ddot{f}(t)| = A$$

is equivalent with the differential equation

$$(3.17) \quad \left(\frac{du}{ds}\right)^2 = 16u\left(A^2 - \frac{u}{R^2(s)}\right).$$

4. THE NATURE OF THE SOLUTIONS OF THE D.E. (3.17)

The main features of our problem will arise clearly if we consider the Case 3 of a simple closed curve Γ of length $2l$. We assume moreover $R(s)$ to have only isolated and finitely many maxima and minima. As we are interested in motions satisfying (3.14), the inequality (3.15) shows that we are to consider in the (s, u)-plane the open domain

$$(4.1) \quad \Omega: 0 < u < A^2R^2(s), \qquad (-\infty < s < \infty),$$

a period $(0, 2l)$ of which is shown in fig. 1.

Writing

$$(4.2) \quad U(u, s) = 4\sqrt{u(A^2 - u/R^2(s))},$$

we observe the following:

1°. That the D.E. (3.17) reduces to the two D.E.

$$(4.3) \quad \frac{du}{ds} = U(u, s)$$

and

$$(4.4) \quad \frac{du}{ds} = -U(u, s).$$

2°. The upper boundary of Ω is the curve

$$(4.5) \quad \gamma: u = A^2(R(s))^2, \qquad (-\infty < s < \infty).$$

At every point (s, u) of Ω the D.E. (4.3) and (4.4) define two line elements of slopes $\pm U(u, s)$. For a fixed s the slope $U(u, s)$ is the square root of a quadratic polynomial in u, and therefore assumes its maximal value at the midpoint $u = \frac{1}{2}A^2R^2(s)$ of the ordinate of γ, and decreases to zero as u approaches one of its endpoints $u = 0$, or $u = A^2R^2(s)$. (See fig. 1).

3°. That each of the D.E. (4.3) and (4.4) satisfies a Lipschitz condition $|U(u_1, s) - U(u_2, s)| < K|u_1 - u_2|$ in every compact subset of Ω, with the constant K depending on the subset.

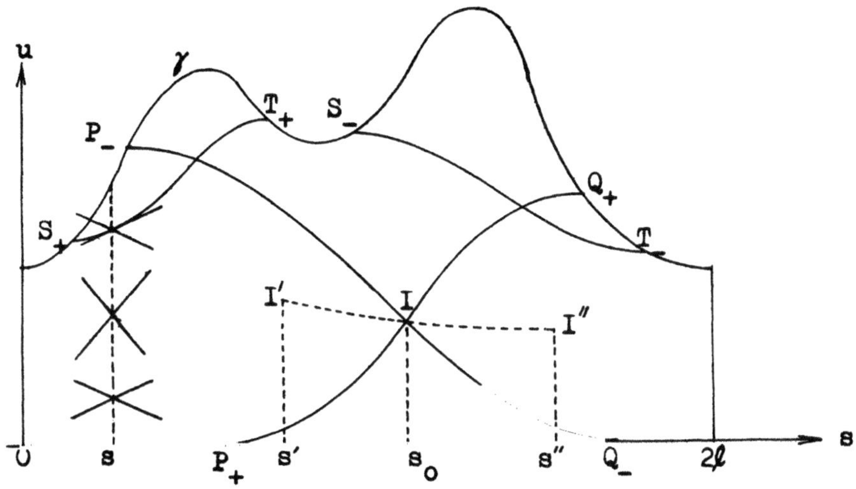

Fig. 1

4°. It follows that Ω is covered simply by a family of solutions of the D.E. (4.3) which are also the *increasing solutions* of the D.E. (3.17). Let

(4.6) $\quad V_+$ *denote the class of solutions of* $\dfrac{du}{ds} = U(u, s).$

Likewise, let

(4.7) $\quad V_-$ *denote the class of solutions of* $\dfrac{du}{ds} = -U(u, s).$

These are the *decreasing solutions* of (3.17).

5°. Through every point (s_0, u_0) of Ω there passes a unique increasing solution $u = u_+(s)$, and a unique decreasing solution $u = u_-(s)$. Also that $u = u_+(s)$ has a *left endpoint* on either the line $u = 0$, or on the curve γ; *its slopes at these endpoints must vanish*. The same holds for the *right endpoint* of the graph of $u = u_-(s)$.

6°. The last remark shows that there is a partition

(4.8) $\quad V_+ = V_+^I \cup V_+^{II},$

where V_+^I denotes the subclass of increasing solutions having their *left* endpoint on $u = 0$, while V_+^{II} is the subclass with *left* endpoint on γ. In fig. 1 we show the graphs

$$P_+Q_+ \in V_+^I \text{ and } S_+T_+ \in V_+^{II}.$$

Similarly, we have the dichotomy

(4.9) $\quad V_- = V_-^I \cup V_-^{II},$

where the criterion is the location of the *right endpoint* of the graph. In fig. 1 we have

$$P_-Q_- \in V^I_- \text{ and } S_-T_- \in V^{II}_-.$$

7°. Let $f(t)$ be a motion on Γ satisfying the conditions

(4.10) $|\ddot{f}(t)| \leqq A$ and $|\dot{f}(t)| > 0$, if $\alpha < t < \beta$.

From $|v| = |\dot{f}(t)| > 0$ we conclude that its arc-length $s = s(t)$ is *strictly monotone*. Replacing $f(t)$ by $f(-t)$, if necessary, we may even assume that $v = ds/dt > 0$. Let therefore $f(t)$ move on the arc

(4.11) $I'I'': s' = s(\alpha) < s < s'' = s(\beta)$,

and let

(4.12) $u = (v(t))^4 = u(s), \qquad (s' < s < s'')$

be the corresponding curve in the (s, u)-plane.

If $I = (s_0, u(s_0))$, $(s' < s_0 < s'')$, is a point of the arc (4.11), and

(4.13) $P_+IQ_+ \in V_+, \qquad P_-IQ_- \in V_-,$

are the elements of V_+ and V_- passing through I, then

(4.14) $\left\{\begin{array}{l}\textit{the arc } I'I'' \textit{ can not cross either of the arcs } P_+Q_+ \textit{ and } P_-Q_-,\\ \textit{except, of course, at the point } I.\end{array}\right.$

This important remark follows from Lemma 1, in particular the inequality (3.15), which implies that *along the arc $I'I''$ we must have the inequalities*

(4.15) $-U(u, s) \leqq \dfrac{du}{ds} \leqq U(u, s).$

This follows from well-known properties of differential inequalities.

5. THE CRITICAL MOTIONS ON Γ

We begin by defining the so-called *critical motions* (abbreviated to C.M., plural C.Ms.)

(5.1) $u = u_M(s)$,

one such C.M. being associated with every minimum point M of the curve

(5.2) $\gamma: u = A^2R^2(s), \qquad (-\infty < s < \infty).$

Let all its minimum points within the period $0 \leqq s \leqq 2l$ be

(5.3) $M_i = (\sigma_i, A^2R^2(\sigma_i)), \qquad (i = 0, \ldots, k; \sigma_0 = 0, \sigma_k = 2l),$

and let M denote one of them (fig. 2).

By integrations of the equations (4.3) and (4.4), respectively, we determine the two arcs

(5.4) $M^-M \in V^{II}_-$ and $MM^+ \in V^{II}_+$,

332

terminating at the points M^- and M^+ of γ. Let

(5.5) $\quad u = \tilde{u}_M(s), \qquad (\sigma^- \leqq s \leqq \sigma^+),$

denote the function having as graph the arc M^-MM^+. It also defines a motion

(5.6) $\quad \tilde{f}_M(t)$ on the arc $\sigma^- \leqq s \leqq \sigma^+$ of Γ.

The connection with the variable t is established by the equations $ds/dt = v = (\tilde{u}_M(s))^{\frac{1}{2}}$ which show that

$$(5.7) \qquad t = \int_{\sigma}^{s} \frac{ds}{(\tilde{u}_M(s))^{\frac{1}{2}}} + t_0, \qquad (\sigma^- \leqq s \leqq \sigma^+),$$

when $t = t_0$ at the point M. The motion (5.6) is defined only on the arc $\sigma^- \leqq s \leqq \sigma^+$ and has the property that $|\ddot{\tilde{f}}(t)| = A$ in all its points.

What happens before $s = \sigma^-$, *or after* $s = \sigma^+$? The answer is that the motion $\tilde{f}_M(t)$ is not defined there. It is important to realize that *no matter how we extent the definition of* (5.5) *for* $s < \sigma^-$, *or* $s > \sigma^+$, *continuously, we get a motion such that*

(5.8) $\quad |\ddot{f}(t)| > A$ *just before* $s = \sigma^-$, *and just after* $s = \sigma^+$.

This follows from Lemma 1, because the inequality (3.15) must be violated immediately before $s = \sigma^-$, and immediately after $s = \sigma^+$.

We already know from Lemma 1 that the (s, u)-graph of a motion $f(t) \in (\Gamma)_A$ can not possibly have a point above the curve γ defined by (5.2). However, much more is true:

LEMMA 2. *It is impossible that the* (s, u)*-graph*

(5.9) $\quad u = u(s)$ *of a motion* $f(t) \in (\Gamma)_A$

should contain a point J which lies between the graph M^-MM^+ *of the critical motion* $\tilde{u}_M(s)$ *and the curve* γ.

PROOF: Assuming J to be above M^-M, we consider through J the arc

(5.10) $\quad N^-JN \in V^{II}_{-}$,

hence a solution of the D.E. (4.4) which must terminate, by (4.14), at a point N of γ with vanishing slope. Indeed by (4.14) the graph of (5.9) must cross γ and thereby lead to the forbidden inequality

(5.11) $\quad |\ddot{f}(t)| > A.$

The following is a

COROLLARY 3. *Let*

(5.12) $\quad u = \tilde{u}_M(s), \qquad (\sigma^- < s < \sigma^+)$

be the critical motion associated with the minimum point M. Let

(5.13) $u = u(s)$

be in the (s, u)*-plane the function for the motion* $f(t) \in (\Gamma)_A$. *If at the times* t_1 *and* t_2 *the points* $f(t_1)$ *and* $f_M(t_2)$ *coincide at a point of the arc* $\sigma^- < s < \sigma^+$, *then we have the inequality*

(5.14) $|\dot{f}(t_1)| \leqq |\dot{f}_M(t_2)|$.

In words: *The critical motion* $f_M(t)$, *of* (5.6), *is the fastest motion satisfying* $|\ddot{f}(t)| \leqq A$, *and at the same time capable of rounding* the "corner" M.

6. CONSTRUCTION OF THE LANDAU MOTION $\tilde{f}(t)$

We consider the $k+1$ minimum points of γ within the period $[0, 2l]$, and the corresponding critical motions

(6.1) $\tilde{u}_{M_i}(s), \ \sigma_i^- < s < \sigma_i^+, \qquad (i = 0, 1, \ldots, k).$

Now *we define the Landau motion by*

(6.2) $\tilde{u}(s) = \min_i \tilde{u}_{M_i}(s), \qquad (0 \leqq s \leqq 2l).$

This is by definition the (s, u)-function of the Landau motion $\tilde{f}(t)$ of Theorem 1, and we are to show that it enjoys the properties claimed for it by Theorem 1. Referring again to fig. 2, the graph of the function (6.2) is the composite curve $M_0C_1M_2C_2MC_3M_4$, assuming that $k = 4$.

The first thing to observe is that (6.2) defines $u(s)$ throughout the period $[0, 2l]$. This requires showing that

(6.3) $\{0 \leqq s \leqq 2l\} \subset \bigcup_{i=0}^{k} \{\sigma_i^- < s < \sigma_i^+\}.$

This becomes apparent as soon as we realize in fig. 2 that the graphs of the consecutive critical motions intersect each other so that their domains of

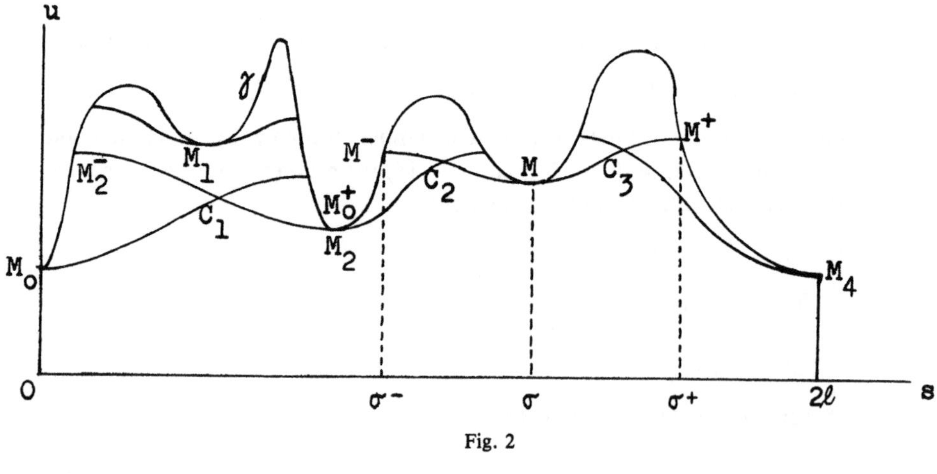

Fig. 2

334

definition cover the entire period $[0,2l]$. To be sure, it is well possible that some of the critical motions will play no role in the formation of the minimum (6.2). Such is the case for the minimum point M_1 of fig. 2.

The intersection property of the C.Ms. is seen as follows: We start from M_0 and obtain the arc $M_0M_0^+ \in V_+^{II}$. The arc $M_0M_0^+$ has zero slope at M_0^+ and therefore M_0^+ is on a *descending branch* of the curve γ. We proceed from M_0^+ along γ to the next strict minimum point M_2. This means that γ decreases strictly along $M_0^+M_2$. But then the arc $M_2^-M_2 \in V_-^{II}$ is sure to intersect $M_0M_0^+$ at some point C_1, because it can not possibly terminate on $M_0^+M_2$ with a vanishing slope, as it should.

From $v = ds/dt = (\tilde{u}(s))^4$ we obtain the dependence on the time variable t by setting

$$t = \int_0^s \frac{ds}{(\tilde{u}(s))^{\frac{1}{4}}},$$

which holds for all t if we first extend $\tilde{u}(s)$ to all s with the period $2l$. That the motion $\tilde{f}(t)$ just defined has all the properties required by Theorem 1 for Case 3 is readily seen, including the essential extremum property (iv). All these follow from appropriate applications of Corollary 3.

The case when $R(s)$ has jump-discontinuities presents no special difficulty: The vertical segments brigging the discontinuities of γ should be treated as integral parts of the curve γ.

In Cases 2 and 4 we must assign to $R(s)$ the value zero at the endpoints of Γ and treat them as minimum points of γ. In the next paper L.P.III we discuss some special cases which will help to clarify these matters.

REFERENCES

1. Schoenberg, I.J. – The Landau problem for motions on curves, MRC T.S. Report No. 1809, November 1977, Madison, Wisconsin.
2. Schoenberg, I.J. – The Landau problem I. The case of motions on sets, Indagationes Math. 276–286 (1978).
3. Schoenberg, I.J. – The Landau problem III. Motions on special curves and time-optimal control problems, to appear in the Proceedings of the Koninklijke Nederlandse Akademie van Wetenschappen.

Proceedings A 84 (3), September 28, 1981

The Landau problem. III

Motions on special curves and time-optimal control problems

by I.J. Schoenberg

Math. Research Center, 610 Walnut street,
Madison, Wisconsin 53706, U.S.A.

Communicated by Prof. J.H. van Lint at the meeting of September 27, 1980

1. INTRODUCTION

This is the third and last of my papers on the Landau problem (L.P.) (See [2] and [3]) and is a direct continuation of L.P.II. In Part I of this paper we apply the general approach of L.P.II to the Landau motions on a few simple curves Γ. In Part II we collect as consequences of Theorem 1 of L.P.II several theorems on time-optimal control problems. Part III deals with skew motions, a conjecture , and a question from Murray Klamkin.

I. MOTIONS ON SPECIAL CURVES

2. Γ IS A CIRCLE C_R, OR AN ARC OF A CIRCLE

Circular uniform motions have been around since the dawn of Astronomy. The following theorem seems to describe a new extremum property of these simple motions.

THEOREM 2. *If $\Gamma = C_R$ is a circle of radius R, then the Landau motion for the control constant A is the uniform circular rotation*

$$\tilde{f}(t) = Re^{it\sqrt{A/R}}. \tag{2.1}$$

Sponsored by the United States Army under Contract Nos. DAAG29-75-C-0024 and DAAG29-80-C-0041.

PROOF: The statement that the uniform rotation (2.1) is the Landau motion (or L-motion) on C_R for the control constant A, means the following: *That the motion* (2.1) *is the fastest motion on C_R, at every point of C_R, within the class of motions*

(2.2) $\quad (C_R)_A = \{f(t);\ f(t) \in C_R \text{ and } |\ddot{f}(t)| \leqq A \text{ for all } t\}.$

From (2.1) we obtain $\dot{f}(t) = \sqrt{AR}\ i \exp(it\sqrt{A/R})$ and therefore that $|\dot{f}(t)| = \sqrt{AR}$. To establish Theorem 2 we are therefore to show that

(2.3) $\quad f(t) \in (C_R)_A$ *implies that* $|\dot{f}(t)| \leqq \sqrt{AR}$ *for all* t.

Following our general approach of L.P.II we prove (2.3) by passing from $v = ds/dt$ to the new variable $u = v^4$. Then the inequality (2.3) amounts to $u^{\frac{1}{4}} \leqq \sqrt{AR}$, or

(2.4) $\quad u \leqq A^2R^2.$

That (2.4) holds for all motions of the class (2.2) follows from Lemma 1 of L.P.II § 3, in particular from its differential inequality (3.15). For the L-motion we have $ds/dt = \sqrt{AR}$, whence

$$t = \int_0^s \frac{ds}{\sqrt{AR}} = \frac{s}{\sqrt{AR}} \text{ for all } t.$$

Therefore $s = t\sqrt{AR}$, and for the angular variable $\theta = s/R = t\sqrt{A/R}$. This establishes (2.1).

From $\dot{f}(t) = \sqrt{AR}\ i \exp(it\sqrt{A/R})$ we find that $\|\dot{f}\| = \sqrt{AR}$, and the Corollary 2 of L.P.II § 2 shows that

(2.5) $\quad L(C_R) = \sqrt{AR}/\sqrt{A} = \sqrt{R},$

in agreement with the equation (6) of L.P.I § 2.

We pass now to the case when

(2.6) $\quad \Gamma = C_{R,l}$ *is an arc of the circle of length* l.

This problem gives rise to the simplest possible fig. 2 of L.P.II § 6. Because $R = R(s)$ is constant, the upper boundary γ of the domain Ω is the horizontal line $u = A^2R^2$, except that we have the discontinuities

$$R(s) = 0, \text{ if } s = 0, \pm l, \pm 2l, \ldots.$$

Being in Case 4 we may even consider only one half of the period, which is the interval $[0, l]$. Thus we deal with fig. 2 where

(2.7) $\quad \gamma\colon u = \begin{cases} A^2R^2 & \text{if } 0 < s < l, \\ 0 & \text{if } s = 0, \text{ or } s = l. \end{cases}$

338

Within this rectangular domain Ω we are to solve our basic D.E.

(2.8) $$\left(\frac{du}{ds}\right)^2 = 16u\left(A^2 - \frac{u}{R^2}\right).$$

Writing $U(u,s) = +4\sqrt{u(A^2 - uR^{-2})}$, this splits into the two D. equations

(2.9) $$\frac{du}{ds} = U(u,s) \text{ and } \frac{du}{ds} = -U(u,s).$$

Since variables separate, we easily find that the solutions of the first D.E. (2.9) are given by the translates of the *increasing branch* of the function

(2.10) $$\begin{cases} u = g(s) = \frac{1}{2}A^2R^2\left(1 - \cos\frac{4s}{R}\right) = A^2R^2\sin^2\frac{2s}{R}, \\ (-R\pi/4 \leqq s \leqq R\pi/4), \end{cases}$$

while the solutions of the second equation (2.9) are the translates of the *decreasing branch* of (2.10). (The reader is asked to draw a rough diagram).

In order to carry out the construction of the function $\tilde{u}(s)$ for the L-motion as given by equation (6.2) of L.P.II, we must realize that we have in $[0,l]$ precisely *three* critical motions:

1) $\tilde{u}_0(s) = g(s)$ for $0 \leqq s \leqq R\pi/4$,
2) $\tilde{u}_{l/2}(s) = A^2R^2$ for $0 \leqq s \leqq l$,
3) $\tilde{u}_l(s) = g(s-l)$ for $l - R\pi/4 \leqq s \leqq l$.

The minimum (6.2) of L.P.II evidently depends on the size of l. There are two cases: If

(2.11) $l \geqq R\pi/2$,

i.e. the arc $C_{R,l}$ is a quarter circle or larger, then in the half-period $[0,l]$ of $\tilde{u}(s)$ we have

(2.12) $$\tilde{u}(s) = \begin{cases} g(s) & \text{if } 0 \leqq s \leqq R\pi/4, \\ A^2R^2 & \text{if } R\pi/4 \leqq s \leqq l - R\pi/4, \\ g(s-l) & \text{if } l - R\pi/4 \leqq s \leqq l. \end{cases}$$

However, if

(2.13) $l \leqq R\pi/2$,

then

(2.14) $$\tilde{u}(s) = \begin{cases} g(s) & \text{if } 0 \leqq s \leqq l/2, \\ g(s-l) & \text{if } l/2 \leqq s \leqq l. \end{cases}$$

If (2.11) holds, then $\|\dot{f}\| = (\max \tilde{u}(s))^{\frac{1}{4}} = \sqrt{AR}$, while if (2.13) holds, then $\|\dot{f}\|^4 = g(l/2) = A^2R^2\sin^2(l/R)$, whence $\|\dot{f}\| = \sqrt{AR\sin(l/R)}$. Now Corollary 2 of L.P.II § 2 establishes our

THEOREM 3. *The Landau constant of the arc* $C_{R,l}$ *is given by*

(2.15) $$L(C_{R,l}) = \begin{cases} \sqrt{R}\sin(l/R) & \text{if } l \leqq R\pi/2, \\ \sqrt{R} & \text{if } l \geqq R\pi/2. \end{cases}$$

If we let $R \to \infty$ then $C_{R,l}$ becomes a straight segment $I_{l/2}$ of length l; letting $R \to \infty$ in (2.15) we obtain $L(I_{l/2}) = \sqrt{l}$, which is Landau's original result (1) of L.P.I § 2.

As an application of (2.15) let OD be a segment of length d; we divide OD into n equal parts, and on each part as diameter we construct a half-circle alternatingly above and below OD. These n semi-circles of radii $R = d/(2n)$ form a (corrugated) smooth arc Γ_n of diameter $= d$. By Theorem 3 we find that $L(\Gamma_n) = = \sqrt{R} = \sqrt{d/(2n)}$. For large n this establishes our statement (1.7) of L.P.II.

3. THE ARC Γ IS A PARABOLA

The parabolic motion of a material particle in a constant field of forces is also an old motion due to Galileo. We state an apparently new extremum property of this motion as

THEOREM 4. *On the parabola*

(3.1) $$\Pi\colon y = \frac{1}{2p} x^2$$

the Landau motion $\tilde{f}(t)$ *for the control constant* A, *is identical with the Galilean motion*

(3.2) $$f_G(t) = t\sqrt{Ap} + i \cdot \tfrac{1}{2} A t^2$$

having the constant acceleration $\ddot{f}_G(t) = i \cdot A$.

PROOF: We select the vertex V of Π as origin of its arc-length s. The radius of curvature $R(s)$ of Π is an elementary but complicated function whose explicit expression we do not need. We only use the readily verified fact that in the (s, u)-plane the graph of the curve

(3.3) $$\gamma\colon u = A^2 R^2(s), \quad (-\infty < s < \infty),$$

is a convex curve symmetric in the u-axis (fig. 1). From $R(0) = 2p$, the curve γ has a unique minimum at the point $M = (0, 4p^2A^2)$. There being therefore only a single critical motion $\tilde{u}_M(s)$, we conclude that the L-motion $\tilde{f}(t)$ is defined by

(3.4) $$\tilde{u}(s) = \tilde{u}_M(s) \text{ for all real } s.$$

According to our general discussion of L.P.II § 5, the graph $L'ML$ of $\tilde{u}_M(s)$ passes through the point M of γ and is composed of two arcs

(3.5) $$L'M \in V_-^{II}, \text{ and } ML \in V_+^{II},$$

340

which are decreasing and increasing solutions, respectively, of the D.E.

(3.6) $$\left(\frac{du}{ds}\right)^2 = 16u(A^2 - uR^{-2}(s)).$$

The function $\tilde{u}(s)$, defined by (3.4) has the following properties:

(3.7) $\tilde{u}(s) > 0$ *for all s,*

(3.8) $u'(s)$ *is everywhere continuous,*

(3.9) $\tilde{u}(s)$ *satisfies the D.E.* (3.6) *for all real s.*

Also the Galilean function

(3.10) $u_G(s) = |\dot{f}_G(t)|^4,$

arising from (3.2), *enjoys the three properties* (3.7), (3.8), (3.9). Let us point out that there are many functions $u(s)$, made up of a succession of arcs from V_+ and V_-, which satisfy the two conditions (3.7) and (3.9). An example is the function $u(s)$ defined by the curve *BCDEFGH* of fig. 1. However, *the function $\tilde{u}(s)$ is among them the only one whoses graph is free* of corners. The reason is that the point M is the only smooth junction point of an arc from V_+ (V_-) with an arc from V_- (V_+). It follows that we must have that

$$u_G(s) = \tilde{u}(s) \text{ for all } s,$$

proving Theorem 4.

Fig. 1

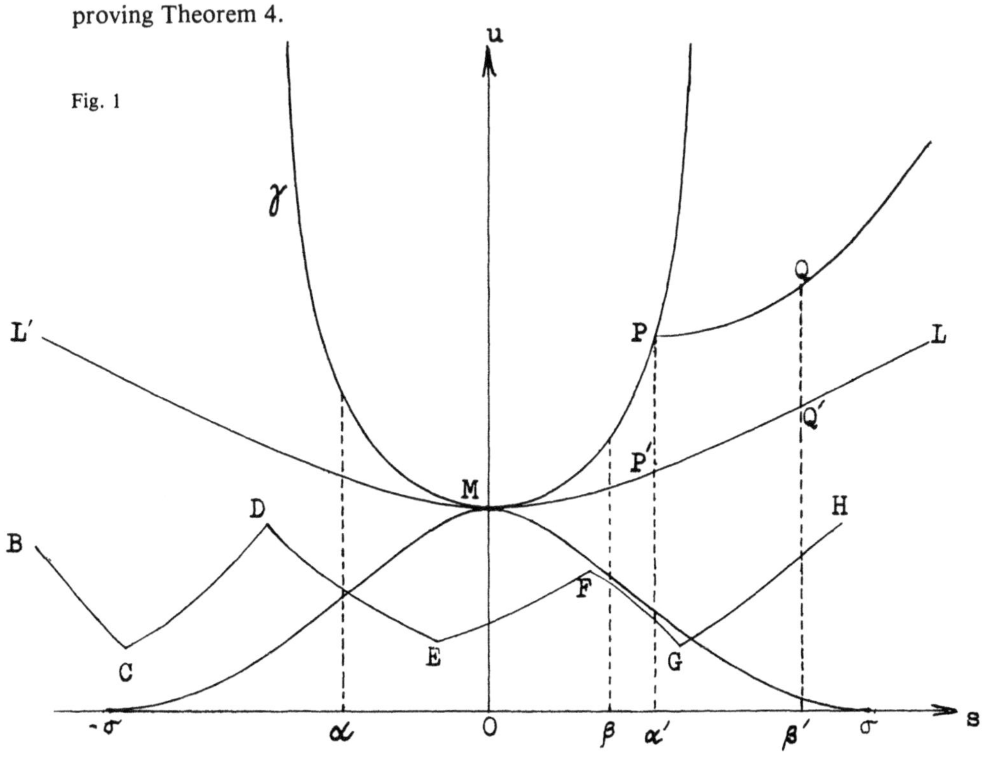

341

4. THE CURVE Γ IS AN ELLIPSE

Let it be

$$(4.1) \qquad E = E_{a,b}: \frac{x^2}{a^2} + \frac{y^2}{b^2} = 1, \quad (0 < b < a).$$

This is an example of Case 3. A period $[0, 2l]$ of the curve

$$(4.2) \qquad \gamma: u = A^2R^2(s), \quad (-\infty < s < \infty),$$

is sketched in fig. 2. Taking on E its vertex V as origin for the arc-length, we have $u(0) = A^2R^2(0) = A^2b^4/a^2$ and $u(l/2) = A^2R^2(l/2) = A^2a^4/b^2$. Fig. 2 also shows a sketch of the periodic graph $M_0C_1M_1C_2M_2$ of the Landau motion according to § 6 of L.P.II.

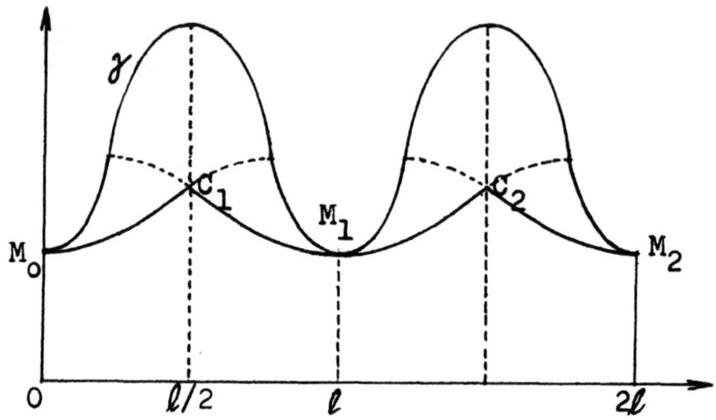

Fig. 2

Because $R(s)$ is an elementary but complicated function of s, we use the representation

$$(4.3) \qquad E: x = a \cos \theta, \; y = b \sin \theta$$

in terms of which we find

$$(4.4) \qquad R = (a^2 \sin^2 \theta + b^2 \cos^2 \theta)^3 a^{-2} b^{-2}.$$

Passing to the independent variable θ we find the D.E. for the increasing solutions of the class V_+ to be

$$(4.5) \qquad \frac{d\tilde{u}}{d\theta} = 4\sqrt{\tilde{u}\left(A^2(a^2 \sin^2 \theta + b^2 \cos^2 \theta) - \frac{a^2b^2\tilde{u}}{(a^2 \sin^2 \theta + b^2 \cos^2 \theta)^2}\right)}$$

with the initial condition for the arc M_0C_1 being

$$(4.6) \qquad \tilde{u}|_{\theta=0} = A^2b^4/a^2.$$

By Corollary 2 of L.P.II § 2 we find that

$$(4.7) \qquad L(E_{a,b}) = (\tilde{u}(l/2))^{\frac{1}{4}} A^{-\frac{1}{2}},$$

and it suffices to determine the value of $\tilde{u}(l/2)$.

Here is a simplification: Observe that by similitude

$$f(t) \in (E_{a,b}) \text{ iff } f_1(t) = \frac{1}{a} f(t) \in (E_{1,b/a}).$$

In terms of the functional $F(f) = \|\dot{f}\| / \sqrt{\|\ddot{f}\|}$ we have that

$$F(f) = \sqrt{a} F(f_1).$$

By taking suprema on both sides we find that

$$(4.8) \qquad L(E_{a,b}) = \sqrt{a} L(E_{1,r}), \text{ where } r = b/a.$$

To determine the value of $u(l/2)$ we have chosen in (4.5)

$$A = 1, \ a = 1, \text{ and } b = r = .1, .2, \ldots, .9$$

and integrated numerically the D.E. (4.5) from $\theta = 0$ to $\theta = \pi/2$. This was beautifully done by C. Vargas, of the MRC Computing Staff, by a Runge-Kutta method. A source of difficulty is that (4.5) does *not* satisfy a Lipschitz condition in a neighborhood of the starting point $\theta = 0$, $\tilde{u} = A^2 b^4 / a^2$. For this reason the solution was started by an appropriate power series expansion. The results are shown in table 1 for the values of the Landau constant.

Table 1

r	$L(E_{1,r})$	r	$L(E_{1,r})$
.0	$1.41421 = \sqrt{2}$	.5	1.30011
.1	1.40978	.6	1.24811
.2	1.39643	.7	1.18609
.3	1.37394	.8	1.11782
.4	1.34199	.9	1.05409
		1.0	1.00000

A glance at fig. 2 shows the Landau motion $\tilde{f}(t)$ to be accelerated on the arcs VW and $V'W'$ of fig. 3, and descellerated on WV' and $W'V$, the maximal speeds occuring at W and W'. At these points are the only discontinuities of $\ddot{\tilde{f}}$, as indicated in fig. 3.

We mention the following three limiting cases in fig. 3:

1. If b decreases and tends to zero, then the angle $\alpha = \angle(\ddot{\tilde{f}}_+, \ddot{\tilde{f}}_-)$ increases to 180° and $\tilde{f}$ becomes the to and fro Landau motion on the segment VV'.

2. If b increases and tends to the value a, then the angle α decreases to zero, while $\tilde{f}$ becomes the uniform rotation on the circle $E_{a,a} = C_a$.

3. If we keep fixed the focus F of E, and also its vertex V, while we let its second focus F' tend to $-\infty$, then E approaches a parabola Π. Also the acceleration pattern $\ddot{\tilde{f}}$ on the arc $W'VW$ is fanning out and approaches in the limit the horizontal acceleration vectors of the Galilean motion on the parabola Π.

The last example of a closed curve Γ which we mention here, leaving out the simple details, is the following. Let $RT_{a,b}$ denote a *racetrack* composed of a rectangle $(2a) \times (2b)$, whose sides of length $2b$ have been replaced by two semi-

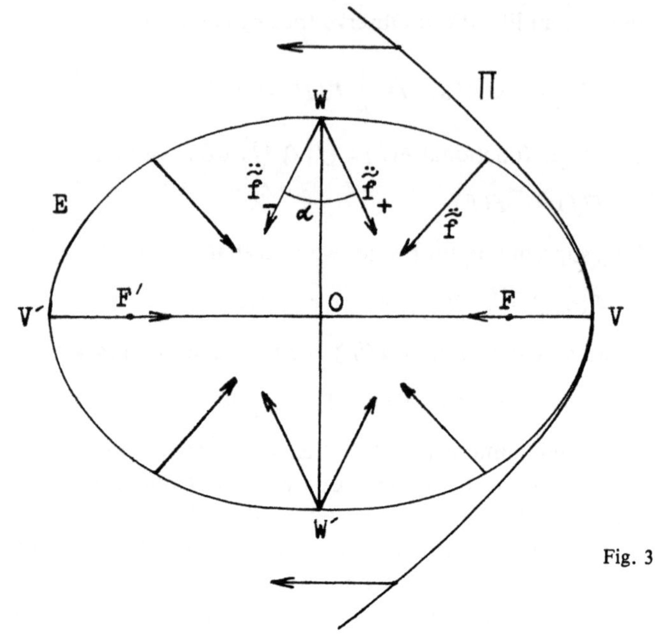

Fig. 3

circles of radius b. Using our approach and the results of § 2, it is easy to show that

$$L(RT_{a,b}) = \sqrt{2a+b}.$$

For details see § 9_4 of Reference [1] of L.P.II.

II. TIME - OPTIMAL CONTROL PROBLEMS ON CURVES

5. THE CASE WHEN Γ IS A SEGMENT AND ITS GENERALIZATIONS

Let $\Gamma = \{0 \leqq s \leqq l\}$ be a straight segment of length l. Let

(5.1) $\quad s = f(t), \quad (0 \leqq t \leqq T),$

be a motion on Γ restricted by the requirement that

(5.2) $\quad |\ddot{f}(t)| \leqq A$, where A is prescribed,

and satisfying the boundary conditions

(5.3) $\quad f(0) = 0, \dot{f}(0) = 0$, and $f(T) = l, \dot{f}(T) = 0,$

the problem being *to determine the shortest time T in which such a motion is possible.*

This is a well-known problem (See e.g. [6, pp. 233–236]). Its connection with the Landau problem is shown by the general

THEOREM 5. *Let $\Gamma = 0L$ be an arc of length l* (Case 4), *and let $\bar{f}(t)$ be its Landau motion for the control constant A. We know $\bar{f}(t)$ to be an even function*

344

of period $2\tilde{T}$, say. *In particular, its restriction*

(5.4) $\tilde{f}(t)$, $(0 \leqq t \leqq \tilde{T})$.

is known to satisfy the boundary conditions

(5.5) $\tilde{f}(0) =$ *the endpoint* 0 *of* Γ, *with* $\dot{\tilde{f}}(0) = 0$,

and

(5.6) $\tilde{f}(\tilde{T}) =$ *the endpoint* L *of* Γ, *with* $\dot{\tilde{f}}(\tilde{T}) = 0$.

Furthermore, let

(5.7) $f(t)$, $(0 \leqq t \leqq T)$,

be any motion along Γ *such that* $f(t) \in (\Gamma) \cap \mathscr{M}$ *and*

(5.8) $|\ddot{f}(t)| \leqq A$.

We also assume that

(5.9) $f(0) =$ *the endpoint* 0 *of* Γ, *with* $\dot{f}(0) = 0$,

and

(5.10) $f(T) =$ *the endpoint* L *of* Γ, *with* $\dot{f}(T) = 0$.

Also that $f(t)$ *moves throughout* $0 \leqq t \leqq T$ *in the direction of increasing arc-length* s, *so that*

(5.11) $$v = \frac{ds}{dt} \geqq 0.$$

Then

(5.12) $\tilde{T} \leqq T$,

with equality only if the motions $\tilde{f}$ *and* f *are identical.*

This shows that the Landau motion is the solution of this time-optimal control problem.

PROOF: Following L.P.II § 3 we introduce the variable $u = v^4$, where $v = ds/dt$. Let $\tilde{u}(s)$ and $u(s)$ be the functions corresponding to $\tilde{f}$ and f, respectively. Moreover, let the two motions be at the same point of arc-length s at the times $\tilde{t}$ and t, respectively, so that $\tilde{f}(\tilde{t}) = f(t)$. This implies the relations

(5.13) $$\tilde{v} = \frac{ds}{d\tilde{t}} = (\tilde{u}(s))^{\frac{1}{4}} = |\dot{\tilde{f}}(\tilde{t})|, \quad v = \frac{ds}{dt} = +(u(s))^{\frac{1}{4}} = |\dot{f}(t)|.$$

Notice that in the second relation we have used the assumption (5.11). From (5.13) we derive for the corresponding times the values

(5.14) $$\tilde{T} = \int_0^l \frac{ds}{(\tilde{u}(s))^{\frac{1}{4}}}, \quad T = \int_0^l \frac{ds}{(u(s))^{\frac{1}{4}}}.$$

From Theorem 1 of L.P.II § 2 we know that $|\dot{\tilde{f}}(\tilde{t})| \geqq |\dot{f}(t)|$, and therefore that $\tilde{u}(s) \geqq u(s)$, for all s. Now (5.14) evidently implies (5.12), including the condition for the equality sign to hold.

6. THE CURVE Γ IS A CLOSED CURVE, OR AN ARC

Let Γ be a closed curve (Case 3). We propose the problem of determining the shortest time $2T$ in which a motion along Γ restricted by

(6.1) $|\ddot{f}(t)| \leqq A$ for all t,

can perform one complete revolution around Γ. The solution is given by

THEOREM 6. *Let $\tilde{f}(t)$ be the Landau motion of period $2\tilde{T}$ along Γ for the constant A. Then*

(6.2) $\quad 2\tilde{T} \leqq 2T,$

with equality only if the motions $\tilde{f}(t)$ and $f(t)$ coincide, except for a shift in the time-origin.

A proof which is much like the proof of Theorem 5, may be omitted.

THE CASE WHEN Γ IS AN ARC. Somewhat different from the two problems so far discussed is the following: Let

(6.3) $\quad \Gamma = XY$ be an arc of length l, $\alpha \leqq s \leqq \beta = \alpha + l$, say.

What is the shortest time T for a motion $f(t)$, $(0 \leqq t \leqq T)$ to move from $s=\alpha$ to $s=\beta$, satisfying (6.1) *and the boundary conditions $f(0)=X$, $f(T)=Y$?*

Notice that our previous boundary conditions $\dot{f}(0)=0$, $\dot{f}(T)=0$, of § 5, have been omitted.

The solution is given by a slight modification of the construction of the L-motion as given in L.P.II §§ 5 and 6. Let

(6.4) $\quad \gamma\colon u = A^2R^2(s), \quad (\alpha \leqq s \leqq \beta),$

be the curve (5.2) of L.P.II § 5, but restricted to our arc $\Gamma = XY$. The modification is this: *If the point $s=\alpha$ happens to be a relative minimum point of γ in the interval $[\alpha, \beta]$, we include in the construction* (6.2) (L.P.II) *of $\tilde{u}(s)$, the critical motion attached to the minimum at $s=\alpha$. Similarly for the endpoint $s=\beta$.*

We discuss this problem when $\Gamma = XY$ is an arc of the parabola Π defined by (3.1). Referring to its (s,u)-diagram of fig. 1, there are two cases:

1. If the arc XY contains the vertex V of the parabola, hence $\alpha \leqq 0 \leqq \beta$, then the least time T is furnished by the Landau motion. By Theorem 4 this is identical with the Galilean motion (3.2).

2. If the vertex V is outside the arc XY, $0 < \alpha' < \beta'$ say, then we must construct the arc $PQ \in V_{+}^{II}$, defined as the graph of

(6.5) $\quad$ *the solution $u = u(s)$ of the D.E. $du/ds = +U(u,s)$*

such that $u(\alpha')=A^2R^2(\alpha')$. This graph defines the fastest motion, and the shortest time is given by

$$(6.6) \qquad T=\int_{\alpha'}^{\beta'} \frac{ds}{(u(s))^{\frac{1}{2}}}.$$

Observe that the time T, given by (6.6), is *below the corresponding traversal time for the Galilean motion given by the arc $P'Q'$ of* fig. 1. The reason is that now the motion need not worry about negotiating safely the perilous vertex V of the parabola.

III. MOTIONS ON SKEW CURVES, A CONJECTURE, AND AN ANSWER TO A QUESTION FROM MURRAY KLAMKIN

Here we collect three disconnected subjects.

7. MOTIONS ON SKEW CURVES

Let Γ be a curve in the space R^3 belonging to any of the four cases of L.P.II § 2. If we apply our analysis used for plane curves, and use for this purpose the Frenet formulae (See [5, Chap. III, § 7]) we find that the acceleration $\ddot{f}(t)$ depends on the curvature $\varrho=1/R(s)$ of Γ, but disregards entirely its torsion τ. It follows that the Landau motion on Γ depends on the arc-length s in the same way as the L-motion on a plane curve Γ^* having, for all relevant values of s, the same curvature as Γ. For example, the L-motion on a circular helix of curvature $1/R$, represented by $s=\tilde{s}(t)$, is the same as for a circle C_R of same curvature.

Let us determine the Landau constant $L(S_a)$ of the spherical shell

$$(7.1) \qquad S_a\colon x^2+y^2+z^2=a^2.$$

If Γ is any smooth curve on S_a, it is known [5, Chap. IV, § 12] that for its radius of curvature we have $R(s)\leqq a$, for all s. It follows that the curve γ defined by (5.2) of L.P.II, will never rise above the horizontal line $u=A^2a^2$. But then for the L-motion on Γ we must have $\tilde{u}(s)\leqq A^2a^2$, for all s. By Corollary 2 of L.P.II § 2 we conclude that

$$L(S_a)=\sqrt{a}.$$

This also follows from Theorem 1 of L.P.I concerning the L-constant of a solid spherical shell on letting $b=a$.

8. A CONJECTURED GENERAL EXTREMUM PROPERTY OF UNIFORM ROTATION

The content of Theorem 2, for $R=1$, is evidently the following: For the uniform rotation $\tilde{f}(t)=e^{it}$ we have $\|\dot{\tilde{f}}\|=1$, $\|\ddot{\tilde{f}}\|=1$. *If $f(t)$ is any smooth motion on the circle C_1 such that*

$$\|\ddot{f}\|\leqq\|\ddot{\tilde{f}}\|=1, \textit{ then } \|\dot{f}\|\leqq\|\dot{\tilde{f}}\|=1.$$

Let me state the rather bold.

CONJECTURE 1. *Let n be an integer* $\geqq 2$. *If* $f(t)$ *is any smooth motion on the circle* C_1, such that

$$(8.1)\qquad \|f^{(n)}\| \leqq \|\bar{f}^{(n)}\| = 1,$$

then

$$(8.2)\qquad \|f^{(k)}\| \leqq \|\bar{f}^{(k)}\| = 1, \textit{ for } k = 1, 2, \ldots, n-1.$$

We may rephrase this in connection with the famous result of Kolmogorov: *If* $f\colon R \to \mathbb{C}$ *is a smooth map, and we write* $M_k = \|f^{(k)}\|$, *then*

$$(8.3)\qquad M_k \leqq C_{n,k} M_0^{1-(k/n)} M_n^{k/n}, \textit{ for } k = 1, \ldots, n-1.$$

Here we have equality for the Euler spline $\mathscr{E}_n(t)$ (For references see [4]). The Conjecture 1 is equivalent with the statement that for *circular motions* $f(t)$ we may sharpen (8.3) to

$$M_k \leqq M_0^{1-(k/n)} M_n^{k/n}, \textit{ for } k = 1, \ldots, n-1.$$

9. AN AFFIRMATIVE ANSWER TO A QUESTION FROM MURRAY KLAMKIN

On the occasion of a lecture of mine at Edmonton, Alberta, Murray Klamkin raised the following interesting question:

The acceleration $\ddot{\bar{f}}(t)$ of the Landau motion on a circle is *continuous*, while it is *discontinuous* for a non-circular ellipse. *Are there closed convex curves* Γ, *besides circles, having a continuous* $\ddot{\bar{f}}$?

We give here an affirmative answer by constructing a family of closed convex Γ having this property. Let α be a parameter such that

$$(9.1)\qquad 0 \leqq \alpha \leqq \tfrac{1}{2},$$

and let us prescribe $\ddot{f}(t)$ by

$$(9.2)\qquad \ddot{f}(t) = e^{i(\pi + t - \alpha \sin 2t)}.$$

A function $f(t)$ is readily recovered by using Bessel functions. From the identity (See [1, page 361])

$$(9.3)\qquad \exp\{\tfrac{1}{2}\alpha(z - z^{-1})\} = \sum_{k=-\infty}^{\infty} J_k(\alpha) z^k, \quad (z \neq 0),$$

we find that

$$(9.4)\qquad \ddot{f}(t) = \sum_{-\infty}^{\infty} (-1)^{k+1} J_k(\alpha) e^{(2k+1)ti},$$

whence by two integrations

$$(9.5)\qquad f(t) = \sum_{-\infty}^{\infty} (-1)^k \frac{J_k(\alpha)}{(2k+1)^2} e^{(2k+1)ti}.$$

If $\alpha = .4$, say, this series converges very fast as seen from the table

k	$J_k(.4)$
0	.96040
1	.19603
2	.01973
3	.00132
4	.00007

The curve

$$(9.6) \qquad \Gamma\colon z = f(t), \quad (0 \leqq t \leqq 2\pi), \quad (\alpha = .4)$$

is an oval symmetric in both coordinate axes, because the function $\operatorname{Re} f(t)$ is even and $\operatorname{Im} f(t)$ is odd. Moreover, the function $\psi(t) = \pi + t - \alpha \sin 2t$ is *strictly increasing*, and this implies by (9.2), that the curves described by the points $\dot{f}(t)$, and $f(t)$, are both *strictly convex*. We have plotted accurately the transcendental curve (9.6) and found it to differ little from an ellipse, which it can not be. Why? The reason is that we will show in (9.7) that the motion $f(t)$ is its own Landau motion, and then the *continuity* of $\ddot{\tilde{f}}(t)$ shows that Γ must be different from an ellipse, whose L-motion has a *discontinuous* acceleration. Concerning Γ we found

$$f(0) = -f(\pi) = 1.13330, \quad f\left(\frac{\pi}{2}\right) = -f\left(\frac{3\pi}{2}\right) = (.78470)i.$$

I claim: *The motion*

$$(9.7) \qquad f(t) = \tilde{f}(t)$$

is also the Landau motion of the curve (9.6).

PROOF: Here $f(t)$ is the motion on Γ defined by (9.5), while $\tilde{f}(t)$ denotes the L-motion on Γ for $A = 1$.

As mentioned before, the vector $\ddot{f}(t)$, of (9.2), turns steadily by 360° as t varies from 0 to 2π. This being the hodograph of the motion of the point $\dot{f}(t)$, it follows that also $z = \dot{f}(t)$ describes a closed and convex curve with center of symmetry at 0. The symmetry follows from $\dot{f}(t+\pi) = -\dot{f}(t)$ and implies that

$$(9.8) \qquad \dot{f}(t) \neq 0, \text{ if } 0 \leqq t \leqq 2\pi.$$

This already implies the identity (9.7) for the following reasons: From $|\ddot{f}(t)| = 1$ for all t, we infer that the graph $u = u(s)$ for the motion $f(t)$ is composed of a succession of arcs from the classes V_+ and V_-. A junction point M_1 of two such arcs C_1M_1 and M_1C_2 say, must be a stationary point of the curve

$$(9.9) \qquad \gamma\colon u = A^2R^2(s), \quad (A = 1),$$

because there is no corner at M_1.

Suppose this junction point M_1 to be a minimum point of γ. It is then impossible that $C_1M_1 \in V_+$ and $M_1C_2 \in V_-$, because this would forcibly lead to a point where $u=0$, in contradiction to (9.8). We must therefore have the situation shown by fig. 4, where M_0, M_1, M_2 are the minimum points of γ. This already shows that $u(s)=\tilde{u}(s)$ gives rise to the L-motion, and the claim (9.7) is established.

Notice that we have placed the junction points C_1 and C_2 at the maximum points of γ. This follows from the continuity of $u'(s)$ implied by the continuity of $\ddot{f}(t)$. Indeed $M_0C_1 \in V_+$ and $C_1M_1 \in V_-$ can only join smoothly at a maximum point of γ, because of the web-like structure of the solutions of the D.E.

$$\left(\frac{du}{ds}\right)^2 = 16u(A^2 - uR^{-2}(s)).$$

CONCLUDING REMARKS. Let us say that a Landau motion f is *a Landau-Klamkin motion*, or *L-K* motion, *provided that the acceleration $\ddot{f}$ is continuous for all t.* Now that Murray Klamkin mentioned them, we can see many *L-K* motions, not necessarily on closed curves. A nice example is a symmetric parabolic arc Γ which is part of the parabola Π of (3.1). We refer to fig. 1 and let the interval $(-\sigma, \sigma)$ be the support of the bell-shaped curve formed by the arc of V_+ with endpoint at M, and by the arc of V_- starting at M. I remind the reader that these arcs are, respectively, the solutions of the differential equations (2.9) for the parabola Π, both passing through M. Let

(9.10) $\quad u=\tilde{u}(s), \quad (-\sigma \leqq s \leqq \sigma),$

denote the function having the bell-shaped graph of fig. 1. Writing

(9.11) $\quad t=t(s)=\int_0^s \frac{ds}{(\tilde{u}(s))^{\frac{1}{4}}}, \quad (-\sigma \leqq s \leqq \sigma),$

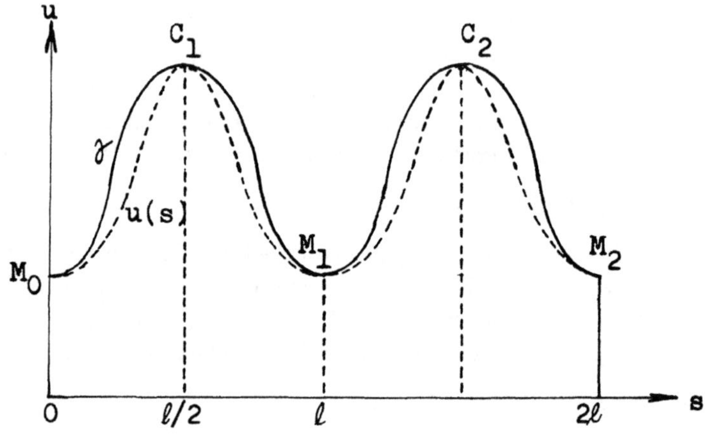

Fig. 4

350

and solving for s to get

(9.12) $s=s(t), \quad (-T/2 \leqq t \leqq T/2),$

we obtain precisely a half-period of the Landau motion $\vec{f}(t)$ on the arc $\Gamma = \Pi[-\sigma, \sigma] = XY$ of Π (Case 4). This is a to and fro motion $\vec{f}$ on Γ with

(9.13) $\dot{\vec{f}}(-T/2) = \dot{\vec{f}}(T/2) = 0.$

Of course, we do know that $|\ddot{\vec{f}}(t)| = A$ for all t. However

(9.14) *$\ddot{\vec{f}}(t)$ is continuous for all times t.*

PROOF: This is evident for all t except if $t = (T/2) + kT$. However, Taylor's series shows that we must have

(9.15) $$\ddot{\vec{f}}\left(\frac{T}{2}+0\right) = \ddot{\vec{f}}\left(\frac{T}{2}-0\right),$$

the common value being a vector of length A which is tangent to Π at the point Y and pointing into the arc. Here we have used the equations (9.13).

It is interesting to compare the L-motion for $\Gamma = \Pi[-\sigma, \sigma]$ with Landau's original motion on a straight segment I_σ of the same length 2σ. *In passing from I_σ to Γ the discontinuity of $\ddot{\vec{f}}$ has disappeared.* As a matter of fact we achieved the same effect in § 2, without pointing it out, that the discontinuity for I_σ disappears if we bend the segment into a circular arc of radius $R \leqq 4\sigma/\pi$.

Let me close with the following problem: *Can the above L-K motion on the parabolic arc $\Pi[-\sigma, \sigma]$ be explicitly determined in terms of known functions?*

REFERENCES

1. Abramowitz, M. and I.A. Stegun – Editors, Handbook of Mathematical functions, National Bureau of Standards, Washington, 1965.
2. Schoenberg, I.J. – The Landau problem I. The case of motions on sets, Indagationes Math., 276–286 (1978).
3. Schoenberg, I.J. – The Landau problem II. The case of motions on curves, Indagationes Mathematicae, **43**, 325–335 (1981).
4. Schoenberg, I.J. – The elementary cases of Landau's problem of inequalities between derivatives, American Math. Monthly, **80**, 121–158 (1973).
5. Stoker, J.J. – Differential Geometry, Wiley-Interscience, New York, 1969.
6. Young, L.C. – Lectures on the Calculus of Variations and Optimal Control Theory, W.B. Saunders Co., Philadelphia, 1969.

HISTORY

Proceedings of the Royal Society of Edinburgh, **90A**, 195 207, *1981*

On polynomial interpolation at the points of a geometric progression

I. J. Schoenberg
Mathematics Research Center, University of Wisconsin, Madison, Wisconsin 53706, U.S.A.

(MS received 29 January 1981. Revised MS received 28 April 1981)

Dedicated to Professor D. V. Ionescu on his 80th birthday on 14 May 1981

Synopsis

This note pursues two aims: the first is historical and the second is factual.

1. We present James Stirling's discovery (1730) that Newton's general interpolation series with divided differences simplifies if the points of interpolation form a *geometric progression*. For its most important case of extrapolation at the origin Karl Schellbach (1864) develops his algorithm of q-differences that also leads naturally to theta-functions. Carl Runge (1891) solves the same extrapolation at the origin, without referring to the Stirling–Schellbach algorithm. Instead, Runge uses "Richardson's deferred approach to the limit" 20 years before Richardson.

2. Recently, the author found a close connection to Romberg's quadrature formula in terms of "binary" trapezoidal sums. It is shown that the problems of Stirling, Schellbach, and Runge, are elegantly solved by Romberg's algorithm. Numerical examples are given briefly. Fuller numerical details can be found in the author's MRC T.S. Report #2173, December 1980, Madison, Wisconsin. Thanks are due to the referee for suggesting the present stream-lined version.

1. Introduction

It is well known that Newton's interpolation series

$$f(x)=f(x_0)+(x-x_0)f(x_0,x_1)+(x-x_0)(x-x_1)f(x_0,x_1,x_2)+\dots \tag{1.1}$$

simplifies considerably in the case where the points $x_n=a+nh$ form an arithmetic progression, for the divided differences reduce to ordinary differences. It seems much less well known that a similar simplification occurs in the case when the points of interpolation form a geometric progression; at least this is not mentioned in the standard treatises on the subject.

The main contribution of the present paper is to call attention to the references [**6**, **5**, **3**], to the works of Stirling, Schellbach, and Runge, which seem now to be practically forgotten; these have been known to the author since 1943. Stirling expands the function $F(z)$ in the form

$$F(z)=a_0+a_1r^z+a_2r^{2z}+\dots,\qquad (r>1),$$

determining the coefficients by interpolation at $z=0,1,2,\dots$, and then extrapolates at $z=-\infty$. Schellbach retains Stirling's approach casting the method in an elegant algorithmic form. The obvious change of variable $x=r^z$ transforms the problem into our polynomial interpolation in the points of a geometric progression (Theorem 1 below).

Quite recently I noticed the close connection with the elegant algorithm of Romberg (Theorem 2), for which we refer to the important paper [**1**] of Bauer, Rutishauser and Stiefel. Also recently I noticed that Runge [**3**] solves the same interpolation problem without stating this fact explicitly (Theorem 3). Instead, he applies the idea of the "Richardson deferred approach to the limit" 20 years before Richardson, working out error estimates. We illustrate these connections with numerical examples.

2. The Stirling–Schellbach algorithm

Let

$$x_k = aq^k, \quad (k=0,1,\ldots), \quad a \neq 0, \quad 0<|q|<1, \tag{2.1}$$

be a geometric progression. Let $f(x)$ be analytic and regular at the origin $x=0$, the problem being to describe explicitly Newton's series (1.1). It is of particular interest for applications to obtain approximations to the value of $f(0)$ by polynomial extrapolation at $x=0$ from the $n+1$ data

$$u_k = f(aq^k), \qquad (k=0,1,\ldots,n). \tag{2.2}$$

Following Schellbach [**5**, §157, p. 280] we define the q-differences $\mathscr{D}^m u_k$ recursively by

$$\mathscr{D}^m u_k = \mathscr{D}^{m-1} u_{k+1} - q^{m-1}\mathscr{D}^{m-1} u_k, \qquad (m=1,2,\ldots), \quad \mathscr{D}^0 u_k = u_k, \tag{2.3}$$

and arrange these in a triangular array

$$\begin{array}{cccc}
u_0 & & & \\
 & \mathscr{D}u_0 & & \\
u_1 & & \mathscr{D}^2 u_0 & \\
 & \mathscr{D}u_1 & & \mathscr{D}^3 u_0 \\
u_2 & & \mathscr{D}^2 u_1 & \\
 & \mathscr{D}u_2 & & \\
u_3 & & & \\
\vdots & & &
\end{array} \tag{2.4}$$

In terms of these differences we may state

LEMMA 1. *For the interpolation points* (2.1) *the divided differences are*

$$f(a, aq, \ldots, aq^n) = \frac{\mathscr{D}^n u_0}{a^n(q^n-1)(q^n-q)\ldots(q^n-q^{n-1})}. \tag{2.5}$$

We omit an easy proof by induction in n.

Using Newton's expansion (1.1) we obtain

THEOREM 1. (Stirling–Schellbach) *For the interpolation points* (2.1) *Newton's series* (1.1) *becomes*

$$f(x)=\sum_{n=0}^{\infty} P_n(x)\mathscr{D}^n u_0, \tag{2.6}$$

where

$$P_n(x)=\frac{(x-a)(x-aq)\dots(x-aq^{n-1})}{a^n(q^n-1)(q^n-q)\dots(q^n-q^{n-1})}, \qquad P_0(x)=1. \tag{2.7}$$

In particular, for $x=0$ *we obtain the expansion*

$$f(0)=u_0+\frac{\mathscr{D}u_0}{1-q}+\frac{\mathscr{D}^2u_0}{(1-q)(1-q^2)}+\dots+\frac{\mathscr{D}^n u_0}{(1-q)\dots(1-q^n)}+\dots \tag{2.8}$$

Consider the sequence to sequence transformation

$$(u_n)\to(v_n) \tag{2.9}$$

which transforms the sequence (u_n) into the sequence (v_n) of partial sums of the series (2.8), hence

$$v_n=u_0+\frac{\mathscr{D}u_0}{1-q}+\dots+\frac{\mathscr{D}^n u_0}{(1-q)\dots(1-q^n)}, \qquad (n=0,1,\dots). \tag{2.10}$$

We refer to this transformation as *the Stirling–Schellbach algorithm.*

It is not difficult to show that (2.9) is limit preserving in the sense that if $u_n\to l$ then also $v_n\to l$. Moreover, the converse is also true: if $v_n\to l$ then also $u_n\to l$. The reason is that the inverse transformation $(v_n)\to(u_n)$ is easily shown to be also limit preserving. But then it is clear than (2.9) cannot be used as a limiting method that changes some divergent sequences (u_n) into convergent ones. Instead, the importance of (2.9) lies in another direction: *it speeds up the convergence of some slowly convergent sequences.*

A dramatic example of acceleration of convergence was given by Stirling himself.

3. Stirling's computation of π

We interpolate the entire function

$$f(x)=\sin\frac{\pi\sqrt{x}}{2}\Big/\frac{\sqrt{x}}{2}=\pi-\frac{1}{24}\pi^3x+\dots \tag{3.1}$$

at the points

$$x_k = q^k, \quad (k=0,1,\ldots,n), \quad \text{where} \quad q=\tfrac{1}{4}. \tag{3.2}$$

Observing that the area $A(m)$ of a regular polygon of m sides inscribed in the unit circle is given by $A(m)=m\sin(\pi/m)\cos(\pi/m)$, we find that

$$u_k = f(4^{-k}) = 2^{k+1}\sin(\pi/2^{k+1}) = 2^{k+2}\sin(\pi/2^{k+2})\cos(\pi/2^{k+2})$$

and therefore

$$u_k = f(4^{-k}) = A(2^{k+2}). \tag{3.3}$$

Stirling computes the areas $A(4)$, $A(8)$, $A(16)$, $A(32)$, $A(64)$, each requiring a square root extraction to derive it from the previous one. Working with 8 decimals we find that (2.4) becomes

$$\begin{array}{lllll}
u_0=A(4)\ =2 & & & & \\
 & 0.82842712 & & & \\
u_1=A(8)\ =2.82842712 & & 0.02593356 & & \\
 & 0.23304034 & & 0.00009675 & \\
u_2=A(16)=3.06146746 & & 0.00171760 & & 0.00000006 \\
 & 0.05997769 & & 0.00000157 & \\
u_3=A(32)=3.12144515 & & 0.00010892 & & \\
 & 0.01510334 & & & \\
u_4=A(64)=3.13654849 & & & &
\end{array} \tag{3.4}$$

We find the relevant q-differences $\mathscr{D}^k u_0$, $(k=0, 1, 2, 3, 4)$ in the top diagonal of (3.4). Forming the partial sums (2.10) we find the approximations

$$\begin{aligned}
v_0 &= 2 \\
v_1 &= 3.10456949 \\
v_2 &= 3.14145277 \\
v_3 &= 3.14159256 \\
v_4 &= 3.14159265.
\end{aligned} \tag{3.5}$$

Notice the rapid convergence of the v_n to π, the value of v_4 having all its decimals correct.

Remarks. 1. Stirling [**6**, p. 133], and Schellbach who reproduces Stirling's computations [**5**, p. 286], also uses the next area $u_5 = A(128)$. For some reason they use in their algorithm these values in their reverse order $A(128), \ldots, A(4)$. This reversal requires the replacement of $q=\tfrac{1}{4}$ by $q=4$ (extrapolation at $x=$

$+\infty$!). How does Schellbach justify the choice of $q=4$? He argues as follows: "Since each difference is nearly 4 times as large as the previous difference, the choice of $q=4$ will result in small values of the higher differences and lead to rapid convergence." They compute with 14 decimals and obtain π with 14 correct decimals. Arranging the data in the natural order of (3.4), the standard error estimate for polynomial interpolation shows the error to satisfy $|\pi - v_5| < (1.068)10^{-16}$. This shows that they should have computed with 17 or 18 places to obtain π with 16 correct decimal places.

Schellbach devotes an entire chapter [5, pp. 275–294] to Stirling's interpolation series and its applications and concludes the chapter by saying "... this series, which seems to have escaped so far the attention of mathematicians, appears to be of exceptionally high practical importance". This statement is still true today (1981).

2. We want to estimate the error $f(0)-v_n$ of polynomial interpolation of $f(x)$ at $x=0$. From (1.1) we know the error at x to be

$$(x-x_0)\dots(x-x_n)f(x_0,\dots,x_n,x)=(x-x_0)\dots(x-x_n)\frac{1}{2\pi i}\int_\Gamma \frac{f(z)\,dz}{(z-x_0)\dots(z-x_n)(z-x)}$$

where Γ is a simple closed curve containing x and the interpolation points. In particular, for $x=0$ we obtain the estimate

$$|f(0)-v_n| < C\cdot K^{n+1}\cdot q^{n(n+1)/2}, \tag{3.6}$$

valid for all n; here C depends on Γ and $f(x)$, while K depends only on Γ. (3.6) clearly shows that we have so-called *quadratic convergence.*

4. The Romberg algorithm

We use with slight modifications the notation of [1]. Let r be a constant such that

$$|r|>1. \tag{4.1}$$

Starting from the column of values $R_0^{(m)}$ we form the Romberg triangular array

$$\begin{array}{cccc}
R_0^{(0)} & & & \\
 & R_1^{(0)} & & \\
R_0^{(1)} & & R_2^{(0)} & \\
 & R_1^{(1)} & & R_3^{(0)} \\
R_0^{(2)} & & R_2^{(1)} & \\
 & R_1^{(2)} & & \\
R_0^{(3)} & & & \\
\vdots & & &
\end{array} \tag{4.2}$$

the general recursive definition being

$$R_m^{(k)} = \frac{r^m R_{m-1}^{(k+1)} - R_{m-1}^{(k)}}{r^m - 1}. \tag{4.3}$$

Our main result is

THEOREM 2. *The Stirling–Schelbach algorithm* (2.10) *is equivalent to the Romberg algorithm* (4.2), (4.3), *with*

$$r = 1/q. \tag{4.4}$$

This means the following: if we identify the first columns of the arrays (2.4) *and* (4.2), *so that*

$$u_m = R_0^{(m)}, \qquad (m = 0, 1, \ldots), \tag{4.5}$$

then the elements of the leading diagonal of (4.2) *are respectively equal to the partial sums of the series* (2.8), *i.e.*

$$v_m = R_m^{(0)}, \qquad (m = 0, 1, \ldots). \tag{4.6}$$

Proof. My colleage C. de Boor pointed out to me the following fact: if we apply the Neville algorithm for a geometric progression $x_m = aq^m$ and for interpolation at $x = 0$, then Neville's fractions simplify and become identical with the elements of Romberg's algorithm. This connection was mentioned in [**4**, pp. 301–302].

We omit a second direct proof that operates by induction in m.

Remark. Even though the two algorithms (2.10) and (4.2), with $u_m = R_0^{(m)}$ and $r = 1/q$, solve the same interpolation problem, it is clear that the elegant Romberg algorithm is much to be preferred. As a convincing example we can recompute the Stirling–Schellbach approximations (3.5) to π: if we apply the Romberg algorithm, with $r = 1/q = 4$, to the first column of the triangular array (3.4) we obtain directly, by (4.6), the approximations (3.5) to π.

5. Runge's solution of the interpolation problem of Stirling and Schellbach

Runge [**3**] also solves the problem of finding approximations to $f(0)$ in terms of the $n+1$ values

$$u_k = f(aq^k), \qquad (k = 0, \ldots, n), \tag{5.1}$$

where

$$0 < |q| < 1, \quad 0 < |a| < \rho, \tag{5.2}$$

while $f(x)$ admits a power series expansion

$$f(x)=a_0+a_1x+\ldots+a_kx^k+\ldots \qquad (|x|<\rho). \tag{5.3}$$

Of course, Runge wishes to approximate $a_0=f(0)$, but nowhere does he mention Stirling, or Schellbach, nor does he speak of polynomial interpolation. Instead, he solves the problem directly as follows.

Setting

$$r=1/q, \tag{5.4}$$

Runge considers the $n+1$ equations

$$f(aq^{n-k})=a_0+a_1aq^nr^k+a_2a^2q^{2n}r^{2k}+\ldots$$
$$+a_na^nq^{n^2}r^{nk}+a_{n+1}a^{n+1}q^{(n+1)n}r^{(n+1)k}+\ldots \qquad (k=0,\ldots,n). \tag{5.5}$$

He multiplies this equation by C_k and sums over all k, the objective being to so choose the C_k as to annihilate the n terms whose leading coefficients are $a_1, a_2, \ldots, a_n$. This is seen to be achieved, provided the C_k are the coefficients of the polynomial

$$\phi_n(x)=C_0+C_1x+\ldots+C_nx^n=\frac{(r-x)(r^2-x)\ldots(r^n-x)}{(r-1)(r^2-1)\ldots(r^n-1)} \tag{5.6}$$

defined by the conditions

$$\phi_n(r)=\phi_n(r^2)=\ldots=\phi_n(r^n)=0, \qquad \phi_n(1)=1. \tag{5.7}$$

Thus Runge obtains from (5.5) the equation

$$\mathscr{R}_n=\sum_0^n C_kf(aq^{n-k})=a_0+a_{n+1}a^{n+1}\,q^{(n+1)n}\phi_n(r^{n+1})+\ldots. \tag{5.8}$$

Since $\phi_n(r^{n+1})=(-1)^nr^{1+2+\ldots+n}=(-1)^nq^{-n(n+1)/2}$, we conclude from (5.8) that

$$\mathscr{R}_n=a_0+\mathcal{O}(q^{n(n+1)/2}).$$

Thus Runge's approximation $\mathscr{R}_n$ to a_0 gives the same order of approximation (3.6) as the v_n of Stirling and Schellbach. The following result should therefore come as no great surprise.

THEOREM 3. *Runge's approximation $\mathscr{R}_n$ to $f(0)=a_0$, defined by (5.8), is identical with the result of polynomial extrapolation and hence*

$$\mathscr{R}_n=v_n=R_0^{(n)}. \tag{5.9}$$

B

A proof of this is fairly easy if we define the extrapolated value v_n by Lagrange's interpolation formula, and also use the Gaussian identity

$$(x-r)(x-r^2)\dots(x-r^n)=\sum_0^n(-1)^i\begin{bmatrix}n\\i\end{bmatrix}r^{i(i+1)/2}x^{n-i}$$

where

$$\begin{bmatrix}n\\i\end{bmatrix}=\frac{(r^n-1)(r^{n-1}-1)\dots(r^{n-i+1}-1)}{(r^i-1)(r^{i-1}-1)\dots(r-1)}.$$

We omit the details.

Remarks. Besides the Stirling–Schellbach computation of π of §3, further attractive examples are provided by the following functions.

1. Our best example is the function

$$f(x)=(1+x)^{1/x}=e+a_1x+\dots. \tag{5.10}$$

Of course $\lim(1+1/n)^n=e$, but now watch how Romberg's algorithm (4.2) speeds up the convergence. We choose $a=1/16=2^{-4}$, and $q=1/2$, so that $r=1/q=2$. Therefore $x_k=aq^k=2^{-4-k}$, and the computation of $u_k=f(aq^k)=(1+2^{-4-k})2^{4+k}$ only requires successive squaring. We work to the capacity of a hand-held calculator. Using (4.5), and (4.6) for $m=4$, we find that $e\approx R_4^{(0)}=2.718281828$. This is also the exact 10-figure value of e.

2. The entire function

$$f(x)=\frac{1}{x}(2^x-1)=\log 2+a_1x+\dots \tag{5.11}$$

will give approximations to $\log 2=.69314718055$. We apply Romberg's algorithm choosing $a=1/4$, $q=1/2$, hence $r=1/q=2$, and so the computation of $u_k=f(aq^k)$ $=2^{2+k}(2^{2^{-2-k}}-1)$, $(k=0,1,2,3,4)$ only requires successive square root extractions. We find that $v_4=R_4^{(0)}=0.6931471805$, with nearly 10 correct decimals. In (5.11) the integer 2 may be replaced by any integer N, the result being $\log N$.

3. *The case when* $f(x)$ *is an even function of* x. If the function (5.3) is even, so that

$$f(x)=a_0+a_2x^2+a_4x^4+\dots, \tag{5.12}$$

we define the new function

$$g(x)=f(\sqrt{x})=a_0+a_2x+a_4x^2+\dots \tag{5.13}$$

and observe that

$$g(a^2q^{2k})=f(aq^k)=u_k, \qquad (k=0,1,\dots).$$

Since $g(x)$ is also regular at $x=0$, we obtain

THEOREM 4. *If the function* (5.3) *is an even function, then we may apply to the data* (5.1) *the Romberg algorithm with r replaced by* r^2.

Replacing q by q^2 will clearly accelerate the convergence of the v_n to $f(0)=a_0$. This is seen from the estimate (3.6). This important device has already been used in §3: instead of interpolating the even function

$$(2/x)\sin(\pi x/2) \quad \text{at} \quad x=1, 1/2, 1/2^2, \ldots$$

we interpolated the entire function (3.1) at $x=1, 1/4, 1/4^2, \ldots$. In the statement (7.12) we shall again use this device to good advantage.

6. Runge's computation of an inverse function

Let

$$y=f(x)=a_0+a_1x+a_2x^2+\ldots, \qquad (|x|<\rho), \tag{6.1}$$

be a function whose coefficients a_i are all known, with $a_1 \neq 0$. Runge's problem is to compute x if the value of y is prescribed.

He reduces this to our old problem of interpolation at the points of a geometric progression, provided that $f(x)$ has the property that the value of $f(x/2)$ can be expressed in terms of the value of $f(x)$. On iteration it follows that

$$f(x/2^k) \quad \textit{can be expressed in terms of} \quad f(x) \quad \textit{for} \quad k=1,2,\ldots. \tag{6.2}$$

A simple example is the function $f(x)=\sin x$, as shown by the identity

$$\sin^2(x/2)=(1-\cos x)/2=(1-(1-\sin^2 x)^{\frac{1}{2}})/2. \tag{6.3}$$

In terms of the expansion

$$f(xt)=a_0+a_1xt+a_2x^2t^2+a_3x^3t^3+\ldots$$

we define the new function

$$g(t)=\frac{f(xt)-a_0}{a_1t}=x+\frac{a_2}{a_1}x^2t+\frac{a_3}{a_1}x^3t^2+\ldots \tag{6.4}$$

which depends also on the fixed value of x. Setting $t=2^{-k}$ we obtain that

$$u_k=g(2^{-k})=2^k\cdot\frac{f(x2^{-k})-a_0}{a_1}=x+\frac{a_2}{a_1}x^2 2^{-k}+\frac{a_3}{a_1}x^3 2^{-2k}+\ldots, \qquad (k=0,1,\ldots). \tag{6.5}$$

Notice that the values of the u_k can be evaluated recursively, due to our assumption (6.2). Notice also from (6.4) that

$$g(0)=x. \tag{6.6}$$

We may therefore state the following

Solution of Runge's problem. We set

$$y_k=f(x2^{-k}), \qquad (k=0,1,\ldots,n) \tag{6.7}$$

and compute these by virtue of (6.2). *We obtain approximations to* $g(0)=x$ *by applying Romberg's algorithm with* $r=2$ *to the data*

$$u_k=g(2^{-k})=2^k\cdot\frac{y_k-a_0}{a_1}, \qquad (k=0,1,\ldots,n). \tag{6.8}$$

Then

$$x\approx R_n^{(0)}. \tag{6.9}$$

Runge applies this method to invert the trigonometric and the hyperbolic functions. However, his more interesting application is to the evaluation of the incomplete elliptic integral. See [**3**, pp. 222–223] and also Schellbach [**5**, pp. 88–90].

7. The evaluation of the incomplete elliptic integral of the first kind

Let

$$x=\int_0^y\frac{d\theta}{\sqrt{(1-c^2\sin^2\theta)}}, \qquad 0<c<1, \quad 0<y\leqq\pi/2, \tag{7.1}$$

where y is prescribed and we are to compute x. In terms of the inverse function

$$y=f(x)=\sin am x=x+a_3x^3+a_5x^5+\ldots, \tag{7.2}$$

we have precisely Runge's problem, observing that it is an odd function of x. That the assumption (6.2) is satisfied follows from the following

THEOREM OF LEGENDRE. *If we determine acute angles* γ *and* y_1 *from the equations*

$$\sin\gamma=c\sin y, \quad \sin y_1=\sin\frac{y}{2}\Big/\cos\frac{\gamma}{2}, \tag{7.3}$$

then

$$\frac{x}{2}=\int_{0}^{y_1}\frac{d\theta}{\sqrt{(1-c^2\sin^2\theta)}}. \tag{7.4}$$

(See [**2**, vol. 1, §21, pp. 25–26].)

We may iterate this result. Determining successively acute angles γ_{n-1} and y_n by the equations

$$\sin\gamma_{n-1}=c\sin y_{n-1},\quad \sin y_n=\sin\frac{y_{n-1}}{2}\Big/\cos\frac{\gamma_{n-1}}{2}$$

$$(\mathrm{n}=1,2,\ldots;\quad \gamma_0=\gamma,\quad y_0=y), \tag{7.5}$$

we find that

$$\frac{x}{2^k}=\int_{0}^{y_k}\frac{d\theta}{\sqrt{(1-c^2\sin^2\theta)}},$$

or equivalently

$$y_k=f(x/2^k). \tag{7.6}$$

By (7.2) we have that $a_0=0$ and $a_1=1$, so that (6.8) reduces to

$$u_k=g(2^{-k})=2^k y_k. \tag{7.7}$$

A numerical example. Let us evaluate the integral

$$x=\int_{0}^{\pi/4}\frac{d\theta}{\sqrt{(1-\frac{1}{4}\sin^2\theta)}},\quad \text{where}\quad c=\tfrac{1}{2}\quad\text{and}\quad y=\tfrac{1}{4}\pi. \tag{7.8}$$

From Legendre's Table VIII [**2**, vol. 3, p. 339] we obtain the 12-place value

$$x=.80436\,61012\,32. \tag{7.9}$$

The recurrence relations (7.5) were solved, for $n=1,2,3,4$, in double precision (20 decimal places) by Fred W. Sauer, of the MRC Computing Staff. We reproduce here only the values of the quantities (7.7) to 15 places:

$$\begin{aligned} u_0&=\ \ y\ \ =.78539\,81633\,97448\\ u_1&=\ 2y_1=.79912\,68518\,62251\\ u_2&=\ 4y_2=.80302\,23930\,93970\\ u_3&=\ 8y_3=.80402\,80071\,93103\\ u_4&=16y_4=.80428\,14415\,06865 \end{aligned} \tag{7.10}$$

where the last value u_4 already gives the crude 4-place approximation $x\approx.8043$.

We can now improve this approximation in two different ways.

1. The Romberg algorithm applied to the data (7.10) with

$$r=2 \text{ gives the approximation } R_4^{(0)}=.80436\,56250 \tag{7.11}$$

which is seen to be correct to 6 decimal places. However, we can do better by applying Theorem 4. To do this we observe that the function $g(t)$, which is being interpolated, admits, by (6.4) and (7.2), the expansion

$$g(t)=x+a_3x^3t^2+a_5x^5t^4+\ldots,$$

showing that it is an *even* function of t. By Theorem 4 we can therefore choose $q=\frac{1}{4}$, so that $r=4$. Sauer's computations show that applying the Romberg algorithm to the data (7.10) with

$$r=4 \text{ gives the approximation } \bar{R}_4^{(0)}=.80436\,61012\,29163 \tag{7.12}$$

which is seen to be correct to nearly 12 decimals.

In conclusion it seems worthwhile to recall that Romberg invented his algorithm, also with $r=4$, for the evaluation of a definite integral like (7.8), in terms of its binary trapezoidal sums (see [**1**]). Dividing the interval of integration $[0,\frac{1}{4}\pi]$ into $2^4=16$ equal parts, we can compute the 5 equidistant trapezoidal sums

$$T_1=\frac{\pi}{4}\left(\tfrac{1}{2}+\sqrt{2/7}\right), T_2, T_4, T_8, T_{16}. \tag{7.13}$$

If we apply to the column of the sums (7.13) the Romberg algorithm with $r=4$ we find the approximation

$$\tilde{R}_4^{(0)}=.80436\,61012\,31069. \tag{7.14}$$

A comparison with Legendre's value (7.9) shows that (7.14) is even slightly more accurate than (7.12).

It does seem remarkable that the same Romberg algorithm applied to the two sets of data (7.10) and (7.13), derived for the same integral (7.8), but otherwise seemingly unrelated, should both furnish equally good approximations to the integral.

References

1 F. L. Bauer, H. Rutishauser and E. Stiefel. New aspects in numerical quadrature. *Proc. Sympos. Appl. Math.* **15** (1963), 199–218.
2 A. M. Legendre. *Exercises de Calcul Intégral*, 3 volumes (Paris, 1816).
3 C. Runge. Über eine numerische Berechnung der Argumente der cyclischen, hyperbolischen und logarithmischen Funktionen. *Acta Math.* **15** (1891), 221–247.
4 R. Sauer and I. Szabó (Eds). *Mathematische Hilfsmittel des Ingenieurs* (Berlin: Springer 1968).
5 K. H. Schellbach. *Die Lehre von den elliptischen Integralen und den Theta-Funktionen* (Berlin: Georg Reimer, 1864).

6 James Stirling. *The differential method or a Treatise concerning summation and interpolation of infinite series*, London, 1749 (translation by F. Holliday of the original Latin edition of 1730).

(*Issued* 25 *January* 1982)

Permissions

Birkhäuser Boston thanks the original publishers of the papers of I.J. Schoenberg for granting permission to reprint specific papers in this collection

Number Theory

[1] Reprinted from *Math. Zeit.* **28,** © 1928 by Springer-Verlag

[18] Reprinted from *Trans. AMS* **39,** © 1936 by the American Mathematical Society

[22] Reprinted from *J. London Math. Soc.* **12,** © 1937 by the London Mathematical Society

Linear Inequalities

[7] Reprinted from *Bull. AMS* **38,** © 1932 by the American Mathematical Society

[50] Reprinted from *Canadian J. Math.* **6,** © 1954 by the University of Toronto Press

Metric Geometry

[16] Reprinted from *Annals of Math.* **36,** 1935

[25] Reprinted from *Trans. AMS* **44,** © 1938 by the American Mathematical Society

[26] Reprinted from *Annals of Math.* **39,** 1938

[28] Reprinted from *Trans. AMS* **50,** © 1941 by the American Mathemtical Society

[29] Reprinted from *Duke Math. J.* **9,** 1942

[49] Reprinted from *Amer. Math Monthly* **60,** © 1953 by the Mathematical Association of America

Real Analysis

[24] Reprinted from *Bull. AMS* **44,** © 1930 by the American Mathematical Society

[51] Reprinted from *Acta Math.* **91,** © 1954 by Akademiai Kiado

[70] Reprinted from *Studies in Mathematical Analysis and Related Topics:* Stanford U. Press, © 1962 by the Board of Trustees of the Leland Stanford Junior University

Complex Analysis

[44] Reprinted from *Proc. AMS* **3,** © 1952 by the Mathematical Society

[53] Reprinted from *Annals of Math.* **62,** © 1955 by Princeton U. Press

[55] Reprinted from *Pacific J. Math.* **8,** © 1958 by Pacific Journal of Mathematics

[60] Reprinted from *Bull AMS* **65,** © 1959 by the American Mathematical Society

[106] Reprinted from *Israel J. Math.* **10,** © 1971 by the Weizmann Science Press

Extremum Problems

[100] Reprinted from *Proc. Int. Conf. on Constructive Function Theory*, Varna, May 1972

[105] Reprinted from *Israel J. Math.* **10,** © 1971 by the Weizmann Science Press

[141] Reprinted from *Indagationes Math.* **20,** © 1978 by Springer-Verlag

[149–150] Reprinted from *Indagationes Math.* **43,** © 1981 by Springer-Verlag

History

[157] Reprinted from *Proc. Royal Soc. Edinburgh* **90A,** © 1981 by the Royal Society of Edinburgh.

GPSR Compliance

The European Union's (EU) General Product Safety Regulation (GPSR) is a set of rules that requires consumer products to be safe and our obligations to ensure this.

If you have any concerns about our products, you can contact us on ProductSafety@springernature.com

In case Publisher is established outside the EU, the EU authorized representative is:

Springer Nature Customer Service Center GmbH
Europaplatz 3
69115 Heidelberg, Germany

Batch number: 10578397

Printed by Printforce, the Netherlands